Technologies for the Wireless Future

Technologies for the Wireless Future

Wireless World Research Forum (WWRF)

Edited by

Rahim Tafazolli

John Wiley & Sons, Ltd

Other Wiley Editorial Offices

John Wiley & Sons Inc., 111 River Street, Hoboken, NJ 07030, USA

Jossey-Bass, 989 Market Street, San Francisco, CA 94103-1741, USA

Wiley-VCH Verlag GmbH, Boschstr. 12, D-69469 Weinheim, Germany

John Wiley & Sons Australia Ltd, 33 Park Road, Milton, Queensland 4064, Australia

John Wiley & Sons (Asia) Pte Ltd, 2 Clementi Loop #02-01, Jin Xing Distripark, Singapore 129809

John Wiley & Sons Canada Ltd, 22 Worcester Road, Etobicoke, Ontario, Canada M9W 1L1

Wiley also publishes its books in a variety of electronic formats. Some content that appears in print may not be available in electronic books.

British Library Cataloguing in Publication Data

A catalogue record for this book is available from the British Library

ISBN 0-470-01235-8

Typeset in 10/12pt Times by Laserwords Private Limited, Chennai, India
Printed and bound in Great Britain by Antony Rowe Ltd, Chippenham, Wiltshire
This book is printed on acid-free paper responsibly manufactured from sustainable forestry in which at least two trees are planted for each one used for paper production.

Contents

List of Figures

Preface

It is expected that in the future, users will use wireless communication services in every walk of their lives. This includes a wide range of applications and services satisfying broad categories of health, wealth, education and entertainment, with applications suitable for individuals and groups including family members. The success of the system is decided by cost, efficiency, reliability and simplicity of use of such services.

This mandates a multi-disciplinary approach to the design of future wireless communication systems and services, consisting of social sciences, computer science and telecom engineering. Social sciences are particularly important in the design of services and ways of offering them to the end-user in terms of simplicity of use, catering different people with different degree of technical capabilities and different requirements.

Based on the user requirements, the computer scientist and telecom engineers are to research into cost effective and efficient technical solutions for the identified enabling technologies. This same approach has also been adopted in the organisation of this book and serves to provide a comprehensive single point of reference for the future wireless world research. It is hoped to be beneficial to everyone who currently is or plans to be involved in Beyond 3G, and 4G research. This includes researchers from both academia and industry, engineers at all levels, management, strategy setters, regulators and economists.

The opening chapter offers the vision for a Wireless World adopted and pursued by the Wireless World Research Forum (WWRF). The chapter introduces membership, structure and mission of the Forum. It also provides a comprehensive overview of other international initiatives and fora and their liaison and cooperation with WWRF.

Chapter 2 underlines the importance of understanding what capabilities people will find valuable, and how to make those capabilities simply usable. It proposes new methods, processes and best practices for user-centered research and design; reference frameworks for modelling user needs within the context of wireless systems; user scenario creation and analysis; and user interaction technologies.

Chapter 3 presents the I-Centric communication paradigm which puts the user at the centre of service provisioning as opposed to the current practice where service provisioning is carried out, irrespective of the actual user's needs or situation. It argues for a new service infrastructure for communication systems where it can model each individual in his/her environment with his/her preferences, and offers the capability to adapt services according to different situations and resources in real-time.

Based on the identified user needs, a range of technical problems, essential research issues and possible solutions have been identified which are the subjects of Chapters 4, 5, 6 and 7. The solutions and approaches proposed are aimed at basic required functionalities needed in a terminal, air-interface and network protocols for support of efficient cooperation between heterogeneous networks, ease of upgrading and introduction of new services, spectrally efficient air-interface techniques and new deployment strategies to make the system cost-efficient from an operator's point of view.

Chapter 4 identifies the core research problems in facilitating seamless cooperation between different radio access technologies, through which an average user of the future could traverse in their average daily routine. This chapter identifies a number of very interesting research topics and proposes the functionalities and protocols needed to provide the cooperation between the identified radio access technologies that are heterogeneous in nature.

Chapter 5 addresses research issues and state of art techniques for software reconfigurable systems. SDR technology has the potential to be the main enabler for adaptation and reconfiguration of sub-systems such as RF/IF, baseband and network protocols to new services, applications and various operational environments and frequency bands. The chapter presents a reference architectural model, as well as a reconfiguration management architecture. It also defines some of the basic required system interfaces to facilitate adaptation and reconfiguration.

Based on expected high data rates and higher frequency bands of operations of future systems, chapter 6 argues that the conventional star-topology based cellular architectures does not always provide adequate range and capacity for a cost-efficient system solution. It proposes alternative deployment strategies based on multihop architectures that are extensions of AdHoc and infrastructure-based networking concepts. It covers different approaches of exploiting the benefits of multihop communications via relays, such as solutions for radio range extension in mobile and wireless broadband cellular networks (trading range for capacity), and solutions to combat shadowing at high radio frequencies. It is shown that multihop relaying can enhance the capacity in cellular networks, e.g. through the exploitation of spatial diversity.

Chapter 7 provides a comprehensive list and discussion on a choice of techniques for spectrally- and power-efficient air-interface architectures. The solutions and research issues encompass broadband multi-carrier solutions, mixed OFDM/Single-carrier access schemes, smart antenna and related technologies, solutions for short-range wireless networks and state of art in UWB technology and the associated research issues.

Chapter 8 outlines and analyses, in detail, possible business models and scenarios for the future wireless market. The study is based on the current status and provides predictions for evolution of the wireless market in the year 2010.

Finally, chapter 9 proposes a base line system reference model and technology roadmap based on predicted user requirements and expectations to the future wireless world. It identifies and explains the basic elements of the reference model and their interdependencies.

List of Contributors

Chapter 1 editors	Mikko Uusitalo and Sudhir Dixit (Nokia)
Authors	Mikko Uusitalo, Sudhir Dixit (Nokia)
	Ken Crisler (Motorola)
	Radu Popescu-Zeletin, Stefan Arbanowski (Fraunhofer FOKUS)
	Rahim Tafazolli, Klaus Moessner (The University of Surrey)
	Bernard Walke, Ralf Pabst (Aachen University, ComNets)
Chapter 2 editor	Ken Crisler (Motorola, US)
Authors	Ken Crisler (Motorola, US)
	Andrew Aftelak (Motorola Labs, UK)
	Satu Kalliokulju, Tea Liukkonen-Olmiala, Juha Matero, Minna Asikainen, Harri Mansikkamäki (Nokia, Finland)
	Manfred Tscheligi (CURE, Austria)
	Ken Crisler (Motorola, US)
	Angela Sasse (University College London, UK)
	Mikael Anneroth (Ericsson, Sweden)
	Petri Pulli (University of Oulu, Finland)
	Michele Visciola (Polytechnic of Milan, Italy)
	Thea Turner (Motorola, US)
	Petri Pulli (University of Oulu, Finland)
	Peter Excell (University of Bradford, UK)
	Axel Steinhage (Infineon Technologies, AG Germany)
	Martin Rantzer (Ericsson, Sweden)
	Bruno von Niman (Ericsson, Sweden)
	Lele Dainesi, Antonella Zucchella (University of Pavia, Italy)
Chapter 3 editors **Authors**	Radu Popescu-Zeletin, Stefan Arbanowski (Fraunhofer FOKUS)
Section 1	Stefan Arbanowski (Fraunhofer FOKUS, Germany)

	Stephan Steglich (Technical University Berlin, Germany)
	Radu Popescu-Zeletin (Fraunhofer FOKUS, Technical University Berlin, Germany)
Section 2	Pieter Ballon (TNO Strategy, Technology and Policy, the Netherlands)
	Stefan Arbanowski (Fraunhofer FOKUS, Germany)
Section 3	Olaf Droegehorn, Klaus David (University of Kassel, Germany)
	Herma van Kranenburg, Johan de Heer (Telematica Instituut, The Netherlands)
	Stefan Arbanowski (Fraunhofer Fokus, Germany)
	Erwin Postmann (Siemens AG, Austria)
	Axel Busboom (Ericsson Research, Germany)
	Kimmo Raatikainen (Nokia, Finland)
Section 4	H. van Kranenburg (eds.), J. de Heer (Telematica Instituut, The Netherlands)
	S. Arbanowski (Fraunhofer FOKUS, Germany)
	E. Postmann (Siemens AG, Austria)
	J. Hjelm (Ericsson, Sweden)
	F. Hohl (Sony, Germany)
	S. Gessler (NEC Europe Ltd., UK)
	H. Ailisto, A. Tarlano (VTT, Finland)
	W. Kellerer (DoCoMo Communications Laboratories Europe GmbH, Germany)
	F. Carrez (Alcatel, France)
	K. Kawano, S. Sameshima (Hitachi SDL, Japan)
Section 5	Kimmo Raatikainen (Nokia Research Center, University of Helsinki)
	Fritz Hohl (Sony International (Europe))
	Sasu Tarkoma (Helsinki Institute for Information Technology)

Chapter 4 editors **Authors**	Rahim Tafazolli, Klaus Moessner (CCSR, The University of Surrey)
Section 1	Hong-Yon Lach (CRM)
	Armardeo Sarma (NEC)
	Raimo Vuopionperä (Ericsson)
Section 2	Toshikane Oda (Ericsson)
	Christoph Lindemann (University of Dortmund)
	Christos Politis (University of Surrey)
	Dorgham Sisalem (Fhg-Fokus)
	Gerard Hoekstra, Harold Teunissen (Lucent)
	Johan Nielsen (Ericsson)
	Josef Urban (Siemens)
	Marko Palola (VTT Electronics)
	Masahiro Sawada (NTT DoCoMo)

	Osvaldo A. Gonzalez (Motorola)
	Peter Schoo (DoCoMo Eurolabs) Tasos Dagiuklas (Intracom)
	Wouter Teeuw (Telematica Instituut)
Section 3	Takashi Koshimizu (NTT DoCoMo)
	Sami Uskela (Nokia)
	Tasos Dagiuklas, Dionysios Gatzounas (Intracom)
	Johan Nielsen (Ericsson)
	Henning Sanneck (Siemens)
	Andreas Schieder, Carl-Gunnar Perntz, Raimo Vuopionperä (Ericsson)
	Djamal Zeghlache, Hossam Afifi (INT)
Section 4	Andreas Schieder (Ericsson)

Chapter 5 editors Authors	Rahim Tafazolli, Klaus Moessner (CCSR, The University of Surrey)
Section 1 to 4	Nancy Alonistioti (University of Athens)
	Didier Bourse (Motorola CRM-Paris)
	Soodesh Buljore (Motorola CRM-Paris)
	Antoine Delautre (Thales Communications)
	Markus Dillinger (Siemens AG)
	Rainer Falk (Siemens AG)
	Dieter Greifendorf (IMS Fraunhofer Gesellschaft)
	Mirsad Halimic (Panasonic PMDO)
	Tim Hentschel (Technische Universität Dresden)
	Quiting Huang (ETH ISL)
	Apostolos Kountouris (Mitsubishi Electric ITE-TCL)
	John MacLeod (University of Bristol)
	Ashok Marath (ICR)
	Linus Maurer (DICE)
	Stefano Micocci (Siemens MC)
	Klaus Moessner (CCSR, The University of Surrey)
	Nikolas Olaziregi (CTR, King's College London)
	Parbhu D. Patel (Panasonic PMDO)
	Santhosh Kumar Pilakkat
	Christian Prehofer (DoCoMo Communications Labs Europe)
	Tapio Rautio (VTT)
	Joachim Sachs (Ericsson)
	Andreas Schieder (Ericsson)
	Wolfgang Schott (IBM)
	Matthias Siebert (Comnets)
	Joerg Stammen (IMS Fraunhofer Gesellschaft)
	Shiao-Li Tsao (CCL/ITRI)
	Paul Warr (University of Bristol)

Thomas Wiebke (Panasonic European Laboratories)
Manfred Zimmermann (Infineon)

Chapter 6 editors   Bernhard Walke, Ralf Pabst (Aachen University, ComNets)
Authors

Section 1   Srdjan Krco (Ericsson Systems Expertise, Ireland)
                             Frank Fitzek (acticom, Germany)

Section 2   Bernhard H. Walke, Ralf Pabst, Daniel Schultz (Aachen
                                 University, ComNets)
                             Patrick Herhold, Gerhard P. Fettweis (Dresden University)
                             Halim Yanikomeroglu, David D. Falconer (Carleton University,
                                 Ottawa)
                             Sayandev Mukherjee, Harish Viswanathan (Lucent
                                 Technologies, US)
                             Matthias Lott, Wolfgang Zirwas (SIEMENS ICM, Munich)
                             Mischa Dohler, Hamid Aghvami (King's College London)

Section 3   Pekka Ojanen (Nokia, Finland)

Chapter 7 editors   Bernhard Walke, Ralf Pabst (Aachen University, ComNets)
Authors

Section 1   Stefan Kaiser (German Aerospace Research Center (DLR))
                             Germany Moshe Ran (Holon Academic Institute of Technology)
                             Israel Arne Svensson (Chalmers University of Technology)
                             Sweden Bernard Hunt (Philips Research, UK)

Section 2   Rui Dinis (Instituto Superior Técnico, Lisbon, Portugal)
                             David Falconer (Carleton University, Ottawa, Canada)
                             Tadashi Matsumoto (University of Oulu, Finland)
                             Moshe Ran (Holon Academic Institute of Technology, Israel)
                             Andreas Springer (University of Linz, Austria)
                             Peiying Zhu Nortel Networks (Ottawa Canada)

Section 3   Martin Haardt (Communications Research Laboratory, University
                                 of Ilmenau, Germany)
                             Angeliki Alexiou (Bell Labs, Wireless Research, Lucent
                                 Technologies, UK)

Section 4   Karine Gosse (Motorola Labs, France)

Section 5   Walter Hirt (IBM Zurich Research Laboratory, Rüschlikon,
                                 Switzerland)
                             Domenico Porcino (Philips Research, UK)

Chapter 8 editors   Klaus Moessner (CCSR, The University of Surrey, Guildford,
                                 Surrey, UK) Ross Pow (Analysys, UK)

Chapter 9 editor	Andreas Schieder (Ericsson)
Authors	Stefan Arbanowski (Fraunhofer FOKUS – Institute for Open Communication Systems)
	Michael Lipka (Siemens AG, Information and Communication Mobile, Germany)
	Klaus Moessner (The University of Surrey, Centre for Communication Systems Research, UK)
	Karl Ott (Ericsson Research, Corporate Unit, Ericsson Eurolab Deutschland, Germany)
	Ralf Pabst (Aachen University, ComNets, Germany)
	Petri Pulli (University of Oulu, Dept. of Information Processing Science, Finland)
	Andreas Schieder (Ericsson Research, Corporate Unit, Ericsson Eurolab Deutschland, Germany)
	Mikko A. Uusitalo (Nokia Research Center, Finland)

Acknowledgements

I would like to thank and acknowledge contributions of all the researchers and experts from all over the world who passionately and openly shared their ideas in WWRF meetings and influenced the contents of this book. The number is so large to be practically listed here.

The book would not have come to existence without the hard work and continuous encouragement of the WWRF Steering Board members and in particular Mikko Uusitalo (Nokia), Pierre Chevillat (IBM), Werner Mohr (Siemens) and Fiona Williams (Ericsson).

I would like to acknowledge the hard work of the past Working Groups Chairs in driving and directing the vision within their working groups and assisting me in editing the chapters of this book. These are;

Mr Ken Chrisler-Motorola- Chair of Working Group 1,

Professor Radu Popescu-Zeletin – Fraunhofer FOKUS, The chair of Working Group 2,

Professor Bernhard Walke-Aachen University, ComNets- The chair of Working Group 4

Professor Rahim Tafazolli
Past chair of WWRF-WG3
CCSR,
The University of Surrey

Foreword

Dr Joao Schwarz da Silva
Director – Communications Networks, Security, Software and Applications
European Commission

I remember back in the year 2000, when a small group of researchers and scientific staff from key European industries, Universities and research laboratories got together as a Think Tank with the purpose of setting a long term vision for mobile and wireless communications beyond 3G. Very soon, thanks to the commitment and foresight of the Think Tank members, the idea was put on the table to enlarge the participation of the group to a much wider forum, well beyond Europe.

Indeed, by their very nature, mobile communications markets are global. Different regions of the world therefore need to cooperate, not only at the research level but well beyond, into the domains of standards and frequency allocations. The debate that followed showed the firm belief that it is of paramount importance to transcend national and regional borders and achieve interoperability of services within and across distinct networks.

A further dimension, contributing to the establishment of an open level playing field for all market actors in the wireless value chain, thereby significantly enhancing the quality of life in the mobile information and communications age, was also at the root of the discussions.

The first blueprint of the long-term vision of the Wireless World was issued in 2001. It was immediately recognised by the research community as the most significant compendium of research ideas ever to be placed in the public domain. It led in particular to a much more structured and better coordinated approach to R&D within the IST pillar of the 5th and 6th EU Framework Research Programmes. Key European players proceeded to work in a much more integrated manner, paying particular attention to going beyond technological development towards meeting user service requirements.

What had started as a Forum for discussion within the industrial and academic communities of Europe very quickly evolved into an organisation with members from all world regions, committed to collaborate and lay the foundations of the wireless future.

In setting the scene for the development of converged cooperative networks that will be carrying phenomenal quantities of traffic at speeds orders of magnitude beyond today's cellular capabilities, in raising the key issue of adaptability and reconfigurability without which service provision and management will reach unprecedented levels of complexity, and notably in bringing forward the need to develop spectrum efficient mechanisms and novel radio interfaces, the Book of Visions provides a unique glimpse of tomorrow's challenges. Quite remarkably, the perspective of the user and its likely interactions with wireless devices, systems and applications, is a core element of the debate.

With this second edition of the Book of Visions, the result of the pioneering work of so many experts, WWRF provides us with a snapshot of the various technological issues at stake, together with a view of the current thinking and initiatives of key players worldwide. From a European perspective, the Book of Visions should help mobilise resources and aggregate efforts to better address the major challenges ahead.

I am convinced that the "wireless minded" reader will enjoy as much as I did, this the WWRF's very valuable contribution to the overall debate on future wireless communication systems. This work will strongly contribute to extending the phenomenal success encountered by the early generations of mobile digital systems, and pave the way towards another wave of innovation with significant impact on employment, growth, and societal issues, notably in the context of the digital divide.

Foreword

Dr Kohei SATOH
Director
Association of Radio Industries and Businesses (ARIB), Japan

Congratulations on the successful publication of *"The Book of Visions 2004"*.

In contrast to the previous version *"The Book of Visions 2001 – Vision of Wireless World"* published in December 2001 to give an overview on the planned activities of Wireless World Research Forum (WWRF), this edition, as illustrated by its subtitle *"Technologies for the Wireless World"* provides in-depth descriptions on the results of WWRF activities undertaken during the last three years. While the research fields covered by WWRF are diverse and extensive – ranging from service infrastructure to radio and network technologies, this book is far more than a mere list of deliverables, providing readers with easy-to-follow technical descriptions and a complete list of reference documents enabling them to advance these research themes building upon the results contained herein. Each section is written by active WWRF participants who are researchers, engineers or university professors at the forefront in their respective fields, hence, I believe the book will come in useful particularly for experts working on similar themes.

In Japan, Mobile IT Forum (mITF) was inaugurated in June 2001 for the purpose of facilitating the research and development and standardization of future mobile communication systems and services such as the fourth-generation mobile communications systems and mobile commerce services. As described in Chapter 1.1 of this book, WWRF and mITF (the 4G Mobile Communications Committee) concluded a Memorandum of Understanding in March 2003 with an aim to further mutual understanding and collaboration. The two parties have hitherto convened five joint meetings, and most recently, commenced a joint activity that handles topics discussed at ITU-R Working Party 8F. I sincerely hope that the two organizations will continue to work closely with each other to add momentum to the global research and development and standardization activities pertaining to future mobile communication systems.

1

Introduction

Edited by Mikko Uusitalo and Sudhir Dixit (Nokia)

Third generation mobile radio systems are currently being deployed in different regions of the world. Future systems beyond the third generation are already under discussion in international bodies and forums such as ITU, WWRF and global R&D programmes. These systems will determine the research and standardization activities in mobile and wireless communication in future years. Based on the experience of the third generation, future systems will be developed mainly from the user perspective with respect to potential services and applications including traffic demands. Therefore, the Wireless World Research Forum (WWRF) was launched in 2001 as a global and open initiative of manufacturers, network operators, SMEs, R&D centres and the academic domain. The WWRF is focused on the vision of such systems – the wireless world – and potential key technologies.

One of the main drivers for the WWRF vision is the introduction of new services and the transformation of the network usage model. I-centric services, adjustable to a vast range of user profiles and needs, along with seamless connectivity anywhere at any time, are considered crucial to the vision of the emerging wireless world. In addition, the cost/benefit ratio will make such services less expensive than any alternative or traditional solutions. Flexibility, adaptability, reusability, innovative user interfaces and attractive business models will be the keys to the success of the systems envisioned for deployment beyond year 2010. This book outlines this vision of the future considering the environmental, contextual and technical aspects.

The first version of The Book of Visions [1] was publish in 2001 (this is commonly referred to as *The Book of Visions 2001*) and was an early attempt to document the long-term vision of the wireless world that each one of us will live in. Since 2001, WWRF working groups have received a constant flow of contributions, and the membership strength has increased significantly. Meetings have offered unique opportunities for both industry and academia to harmonize their views on the various topics. Some of these topics resulted in several contributors jointly writing common white papers. This new Book of Visions is a collection of those white papers and the missing information necessary to produce a cohesive book. Hundreds of researchers around the world from every sector of

Technologies for the Wireless Future.   Edited by R. Tafazolli
© 2005 John Wiley & Sons, Ltd   ISBN: 0-470-01235-8

the information and communications industry have collaborated to envision the wireless world ahead of us, and this book summarizes their collective wisdom.

The Forum has already played the initiator role in the establishment of the Wireless World Initiative (WWI), a European research project, as an initial step towards the basis of future standardization of beyond 3G systems. WWI has collected some 156 organizations under an umbrella of research actives of the order of hundreds of millions of euros annually. Now the WWRF is turning its attention to other regions in order to establish research initiatives there as well.

The work of the WWRF continues at an ever increasing volume. It offer a unique means of harmonizing views from both industry and academia. Through harmonization, the participants can pool their resources in research, and consequently reduce risk in investments at an early stage.

1.1 Goals and Objectives

The major objectives of the WWRF are:

- to develop and maintain a consistent vision;
- to generate, identify, and promote research and trends for mobile and wireless systems;
- to identify and assess the potential of new technologies and trends;
- to contribute to the definition of research programmes;
- to ease future standardization by harmonizing and disseminating views; and
- to inform a wider audience about research activities that are focused on the wireless world.

Furthermore, the WWRF's scope of work is as follows.

- to contribute to the development of a common and comprehensive vision for the wireless world, concentrating on the definition of research relevant to the future of mobile and wireless communications, including preregulatory impact assessments;
- to invite worldwide participation and be open to all actors;
- to disseminate and communicate wireless world concepts; and
- to provide a platform for the presentation of research results.

The WWRF supports the UMTS Forum, ETSI, 3GPP, IETF, ITU, IEEE and other relevant bodies on commercialization and standardization issues derived from the research work. However, the WWRF is not a standardization body. Despite this, liaison agreements have been established with the UMTS Forum (January 2003), mITF (May 2003) and IEEE Communication Society (October 2003).

1.2 Evolution Beyond 3G

Figure 1.1 shows the development towards the third generation and beyond and related research framework programmes of the European Union (EU). European research on 3G started in 1989 within the RACE I programme and continued in 1991 in RACE II.

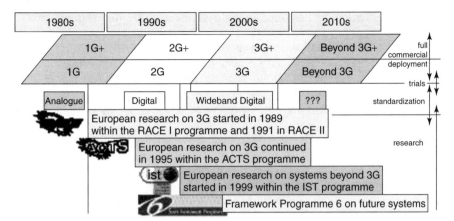

Figure 1.1 Evolution of mobile communication systems

Several years later, European research on systems beyond 3G started in 1999 within the IST programme.

The development of 3G systems started in Europe around 1988/89 in the EU RACE I framework research programme nearly two years ahead of the first deployment of GSM. In the beginning, a vision of UMTS was developed including preliminary estimations of the frequency spectrum demand for UMTS. This was the basis for WARC'92 (World Administrative Radio Conference) which identified the initial IMT-2000 bands. The first radio interface concepts and network architectures were investigated in research projects in the EU RACE II framework programme. The research projects in the EU ACTS programme towards radio interface proposals and network architectures had a significant impact on the international standardization of third generation mobile communication systems [2–5]. At that time the development of 3G systems was mainly driven by technology in terms of supported radio environments, data rates, mobile speed, etc. The development of services, applications, business models and the new value chain started around 1997, close to the decision of ETSI SMG on the selected UTRA concept (UMTS Terrestrial Radio Access) in January 1998. This work was facilitated by the activities of the UMTS Forum, which was established in 1996. It took some ten years for the next generation to come to market. The future will show whether this pattern will be repeated or whether there will be gradual evolution instead of a clear new generation.

1.3 Vision of the Wireless World

One important lesson from the development of 3G systems since the beginning of the nineties is that potential future services and applications, including the expected user behaviour, should be taken into account from the very beginning to derive technical requirements. This approach is essential for the economic success of future systems. On the other hand, the mobile community has problems developing technical requirements for new systems for an unpredictable future more than ten years hence. Therefore, future systems should enable sufficient flexibility to match the operator and user expectations from a service and economic perspective as closely as possible.

Important questions to be answered in the development of future systems are:

- What essential demand will the wireless world address?
- How can advances in technologies be combined in a consistent manner?
- How can wireless communications become universally available for both people and things?
- What business models will drive the wireless world (what are its fundamental laws)?

The WWRF established a Vision Committee to provide technical steering. The Vision Committee identified the key principles for the WWRF vision:

- Users are in control through intuitive interactions with applications, services and devices.
- Services and applications are personalized, ambient-aware, adaptive (I-centric) and ubiquitous from the point of view of the user.
- Seamless services should be provided to users, groups of users, communities and machines (autonomously communicating devices), irrespective of place and network and with agreed quality of service.
- Users, application developers, service and content providers, network operators and manufacturers can create new services and business models efficiently and flexibly, based on the component-based open architecture of the wireless world.
- There is awareness of, and access to, appropriate levels of reliability, Security and trustworthiness, in the wireless world.

The key aspects of the wireless world in the future will be as follows.

1.3.1 Fulfilling User Needs and Enhancing User Experience

Future communication systems will provide the intelligence required for modelling the communication space of each individual. The future service architecture will be I-centric. I-centric communication considers human behaviour as a starting point by which to adapt the activities of communication systems. Human beings do not want to employ technology but rather to interact with their environment. They communicate with objects in their environment in a certain context. Objects pertaining to a certain context can be active or passive at given moments in time depending on the situation of the user.

Furthermore, the end user expects a seamless, holistic experience as a member of the e-community. In addition to the universal user needs, there are culturally specific needs. To ensure that concepts and products are not designed based only on the requirements of the most important market areas, an investigation of how to obtain better usability in emerging culture areas is necessary. Moreover, mobile consumers can be segmented based on their mindsets, lifestyles, cultural aspects, etc.

1.3.2 Ultra High Bit Rates

The maximum peak rate is anticipated to be 1 Gbps for local, typically indoor, coverage under excellent radio conditions. This scales down to a maximum of 100 Mbps for high-speed wide-area coverage. These figures are not based on an analysis of demand in 2012 and beyond. Rather, they represent a bold goal for a future wireless system, a necessary step for the adoption of a new telecom standard, and compare favourably with current wireline systems (primarily Ethernet) and likely WLAN evolution. It can be argued that once such bit rates are available with good mobility, the demand will grow to match the capabilities. The request for access and transmission of high quality multimedia content at any time, anywhere is one such application that is expected to gain momentum in the emerging consumer market [6].

1.3.3 Ubiquitous Coverage Via Heterogeneous Access

Low-cost ubiquitous presence of all broadband services, with bit rates comparable to those offered by wireline systems, forms a compelling package for the end user and can truly make the mobile terminal a centrepiece of people's lives (this will also require major leaps in user interface technology). Therefore, it is essential to develop an architecture that is scalable and can cover large geographical areas and adapt to various radio environments with highly scalable bit rates, while encompassing the personal space (BAN/PAN) for virtual reality at faraway places (WAN).

1.3.4 Low Cost

As we are targeting an increase in bit rates (and the volume of traffic) by a factor of 100–1000 from 3G, the cost per bit of such a future wireless network must be reduced by approximately the same factor. One factor that may alleviate this requirement is that the actual amount of traffic may not increase as much as the air interface capacity, instead the extra air interface capacity could be used to boost QoS (such as low latency).

1.3.5 Machine-to-Machine and Sensor Networks

In addition to humans generating traffic, most of the traffic will be generated by machines, sensors and agents on behalf of humans. Machine-to-machine (M2M) and wireless sensor networks (WSNs) are characterized by a need to understand and carefully integrate features and functions, which have been, until now, treated as separate. This would help to optimize cost, performance and efficiency, especially with respect to the quality of decision making (its own or on behalf of humans, e.g., the semantic web, software agents), energy consumption and device/system management. This area is also closely related to the software configurable systems discussed later in this book. Therefore, close cooperation in research from different areas is extremely desirable.

Note that the above key aspects are merely a subset of what is around the corner. The rest of the chapter and the book provide a more comprehensive vision of the wireless world.

1.4 The International Context and 4G Activities

The WWRF is a consortium of many leading mobile telecommunications companies, research organizations and universities, and was founded in 2001 [1, 7]. The goal of this new body is to secure momentum, strategic orientation and impact of the research on wireless communication beyond 3G. The WWRF emphasizes that systems in the wireless world should integrate 2G, IMT-2000 and its enhancements, other systems such as WLAN type systems, short range connectivity, broadcast systems and new radio interface concepts on a common IP-based platform. Continuous connectivity between end users and a variety of services will play a vital role in enhancing the wireless technology. Other issues, such as short-range spontaneous networking, system coexistence and interworking will also help to bring about a revolution in wireless systems.

In addition to the WWRF, several other initiatives are currently underway with somewhat different focuses, but they all point to the same long-term wireless vision of the world. Due to the long time frame for the development and standardization of new systems, the identification and availability of new spectrum and the regulatory issues, such as the licensing process, the research work on systems beyond third generation mobile communications started in around 1999 in international projects and bodies, [7–11]. Several projects under the EU IST framework programme (5th Framework Programme of the EU) addressed different aspects of future systems. Recently, the EU launched the 6th Framework Programme with more strategically oriented projects. Similar activities are ongoing in other regions. China initiated the FuTURE Project under the umbrella of the state funded 863 programme. In Japan the Ministry of Public Affairs, Home Management, Post and Telecommunications – MPHPT–supports these research activities. The mobile IT Forum – mITF–plays a coordinating role between the different actors from industry, the operator and provider side and the academic and R&D domain. Korea has established the NGMC Forum (Next Generation Mobile Communication Forum). Under this umbrella, related research activities are coordinated. In North America, activities are ongoing in the industry following market expectations. DARPA's XG programme is the main government initiative to launch flexible spectrum allocation. Other US organizations that are active are the IEEE and the IETF. The following sections provide more detailed information on the key initiatives.

1.4.1 mITF (Mobile IT Forum) – Japan

The mITF was established in June 2001 with the aim of realizing future mobile communication systems and services, such as the fourth generation mobile communications systems and mobile commerce services, at an early date. It aimed to do this performing research and development activities, pushing for standardization, conducting coordination with related bodies, collecting information and carrying out promotional and educational activities, thereby contributing to a healthy utilization of radio spectrum.

Its near-term activities are to establish a framework for R&D and standardization of the 4G mobile communication systems in a comprehensive manner with a view to creating a new business market and spectrum needs beyond 2010.

1.4.2 4G at the International Telecommunication Union (ITU)

ITU-R WP8F (International Telecommunication Union – Radio Sector, Working Party 8F) and ITU-T SSG (ITU – Telecommunication Sector, Special Study Group) have undertaken high-level vision work to help facilitate global consensus on basic system concepts [12]. These activities are a prerequisite for the World Radio Conferences (in particular WRC '07) to discuss and to identify new spectrum for future systems. Its visions and recommendations for technical realization are built on expected user requirements for future mobile telecommunication systems. ITU-R concentrates on the technical aspect of IMT 2000 and systems beyond, and is addressing future market and services aspects, spectrum needs beyond 2010 and high-level goals for the new radio interface(s). The vision stresses seamless service provisioning across a multitude of wireless systems as being an important feature of future generation systems. In general, the views of ITU-R and the WWRF are aligned to a large extent, such as the research goals of 100 Mb/s high mobility and 1 Gb/s low mobility. The ITU-R Radio Assembly approved a related recommendation at the beginning of June 2003 on the future framework for IMT-2000 and systems beyond (the vision), which serves as an important foundation in developing global consensus and for the forthcoming process of spectrum identification at WRC '07. One of the main activities in the future will be to prepare for the WRC '07 and facilitate global studies on the future market, services and spectrum needs beyond 2010.

1.4.3 NGMC Forum – Korea

The NGMC Forum is a relatively new initiative by the Korean government and companies, and is a reincarnation of an earlier initiative by ETRI [13, 14]. Its objectives are:

- technical and social trends analysis and vision establishment;
- international cooperation;
- steering of advanced R&D strategies;
- studies on spectrum use.

1.4.4 FuTURE Project under the China Communications Standards Association (CCSA)

To adapt to reform of the telecommunications industry and the liberalization of the telecommunications market, the Ministry of Information Industry (MII) approved, in succession, from April 1999, six standards groups for carrying out standards related R&D. They were CWTS, TNS, IPSG, NSSG, NMSG and power supply. In addition, two ad hoc groups on telecommunication terminals and mobile Internet (CMIS), and a Technical Coordination Department of Telecommunication Standards were approved as well. However, in order to establish a nationally unified standards organization that could adapt to the growing market, keep pace with the global industry and was relevant to Chinese situations, it was proposed to establish a single unified communications standards organization which combined the previously established six standards groups. With the approval of the MII, the Standardization Administration of China and the Civil Affairs Ministry, the China Communications Standards Association (CCSA) was founded on December 18, 2002 [15, 16].

FuTURE (Future Technology for Universal Radio Environment) is a key project of the wireless communication branch of the communication theme of the National High Technology Research and Development Programme (863 Programme). It carries out investigations on key technologies for the air interface of beyond 3G/4G mobile communication systems, sets up demonstration systems to verify the key technologies that can support future wireless services and promotes international cooperation. The project aims to contribute to the standardization and intellectual property for beyond 3G systems and promote the rapid development of China's future wireless telecommunication industry. In the short run, the FuTURE project will focus on wireless transmission technology for B3G/4G, self-organization mobile network technology and technology in the multi-antenna wireless telecommunication environment.

1.4.5 3GPP (3rd Generation Partnership Project)

3GPP is a collaboration agreement that was established in December 1998. The collaboration agreement brings together a number of telecommunications organizational partners; the current organizational partners are ARIB, CCSA, ETSI, T1, TTA, and TTC [17, 18].

The 3GPP system is a combination of the new wideband CDMA air interface and related radio access interface (RAN) architecture, and evolved GSM and GPRS core networks. 3G network evolution (systems beyond 3G) to meet new challenges (high bit rates, lower communication latency, ubiquitous service availability and low service cost) is the vision of 3GPP. This vision will be achieved by adopting an all IP solution (e.g., IP RAN) and improving the current logical architecture for optimal delivery of IP based services [19, 20, 21].

On the other hand, 3GPP2 was established in January 1999 by a joint group of TIA in the US, ARIB and TTC in Japan, CWTS (now CCSA) in China and TTA in Korea, with a goal to set specific standards of an ANSI-41 network and CDMA2000 technologies. 3GPP2 is now working on improving wireless connection technologies, applying new services such as MMS (Multimedia Messaging Service) and developing IP systems for all networks.

1.4.6 IEEE 802

The IEEE 802 LAN/MAN Standards Committee (LMSC) develops LAN and MAN standards, mainly for the lowest two layers of the Reference Model for Open Systems Interconnection (OSI) [22]. LMSC coordinates with other national and international standards groups, with some standards now published by ISO as international standards. The standards activities of the IEEE 802 project are typically short- to medium-term.

There are several IEEE 802 standards that will play an integral part in the wireless world of the future, mainly because of their large embedded base of users and manufacturers. The standards that are potentially relevant are: 802.11 Wireless LAN (WLAN), 802.15 Wireless Personal Area Network (WPAN), 802.16 Broadband Wireless Access (BBWA) and 802.20 Mobile Wireless Access (MWA).

The IEEE 802.11 WLAN Working Group has a Wireless Next Generation Subcommittee (WNG SC), whose mission is to facilitate discussions of new ideas on how to respond to, and anticipate, market expectations of improved WLAN technology. Some topics of ongoing interest are: VoIP over WLAN, lossless data compression, software defined radio, radio regulatory, ETSI/BRAN, MMAC, requirements of WLAN ten years from now and coexistence with other technologies (e.g., 802.19). Some of the key outputs from the WNG SC are:

- IEEE 802.11j;
- IEEE 802.11n;
- IEEE 802.11k;
- WAVE SG;
- Fast roaming SG;
- Mesh networking SG;
- Wireless performance prediction SG;
- WLAN interworking with external IP networks;
- Security management;
- Management of wireless devices;
- Access point functional behaviour.

1.4.7 Internet Engineering Task Force (IETF)

The IETF is a large open international community of network designers, operators, vendors and researchers concerned with the evolution of the Internet architecture and the smooth operation of the Internet [23]. The work being done at the IETF could potentially impact the wireless world of the future. The IETF work is divided into several areas: Applications, General, Internet, Operations and Management, Routing, Security, Sub-IP and Transport. The IETF operates somewhat differently from the standards organizations, where only RFCs are produced with heavy emphasis on implementation, and it is up to the market to adopt the RFCs. If the adoption is large-scale they become de facto industry standards.

1.4.8 National Science Foundation (NSF)

The National Science Foundation, NSF, has launched a multidisciplinary research initiative throughout the United States to address fundamental research issues [24] which are critical for future generation wireless systems.

The research areas where the NSF has shown interest include:

- Network topologies for supporting integrated services.
- Adaptive data flow at the physical and higher layers.
- Joint optimization of multiple adaptive subsystems and dynamic resource allocation schemes.
- Security/privacy across the layers (physical through applications) and dynamic support for quality of service.
- Development of fully adaptive mobile wireless communications and information systems, including coordination of adaptation between information, networks, multiple access links and physical layers.
- Ubiquitous information processing, virtual collaboration and visualization, semantic routing for information discovery and protocols for content/capability adaptation.

- Application and system adaptability to network bandwidth variability (e.g., error rates in transition and handoffs, fault tolerance and recovery, query optimization, management of power limitations, transaction management and security).
- Ability to continue operation during disconnection (e.g., via caching), smart push/pull techniques based on user profiles using multicast network protocols and mixed bandwidth links, and seamless environments for remote sensing, including use of GIS and GPS.

1.4.9 Department of Defense (DoD) DARPA

Government institutions, such as the Department of Defense (DoD) DARPA, have launched research initiatives for advancing Next Generation Wireless Networks (NGWN) xG Wireless Systems [25, 26]. DARPA's XG Communications programme is developing technology to allow multiple users to share use of the spectrum through adaptive mechanisms that deconflict users in terms of time, frequency, code and other signal characteristics. DARPA's goals are to enable an increase of a factor of ten in the usage of typical spectrum. The DARPA programme is developing technologies that are applicable to not only the military, but also potentially for civil use.

1.4.10 4G Mobile Forum

The 4GMF (Fourth Generation Mobile Forum) is run by the Delson Group [27]. Its mission is to provide a technical forum to promote exchange of technological advancement, resulting from academic and industrial research and development efforts, in order to facilitate the realization of the 4G Mobile Vision. The Forum's objective is to define an open wireless platform architecture supporting the convergence of broadband wireless mobile and wireless access. Its focus is on the next generation broadband wireless mobile communications which converge wireless access, wireless mobile, wireless LAN and packet-division-multiplexed (PDM) networks. This integrated 4G mobile system provides wireless users an affordable broadband mobile access solution for the applications of secured wireless mobile Internet services with value-added quality of service (QoS) from the application layer all the way to the media access control (MAC) layer. The 4GMF has been launched on a new different platform so as to complement (and not to compete with) the WWRF, DARPA XG, mITF, Korean 4G, Japanese 4G, FuTURE 4G, IEEE 802.20, etc. Its work has been divided into 15 working groups and the Forum has held several annual summits.

1.5 Working Groups of the WWRF

1.5.1 WG1 – Human Perspective

Since the beginning of mobile communications, voice has been the single dominant application. As a result, the application of human factors principles or user centred design has been limited to the ergonomics and usability of the terminal device, and to the study of usage patterns, primarily for marketing purposes. However, the mobile communications environment is changing. Data services are beginning to gain traction worldwide as 2.5G and 3G systems offer more sophisticated data applications, higher data rates and greater capacity for data services. As the wireless world evolves, it will be critical to understand people's needs for new and enhanced wireless services. There is a need to anticipate which broad application categories (such as video, email, browsing or m-commerce, etc.) will be important, and what attributes will be most critical to optimizing the user experience. The answers to these questions are based in the science of Human Factors, which can identify underlying human needs and desires, the human value placed on a service and how to provide those needs given the limitations of the technology.

The Human Perspective Working Group (WG1) is a working group within the WWRF focused on discovering and promoting research areas that strive to understand the users' needs for future wireless systems and how users will interact with devices, systems and applications in the wireless world. An understanding of user requirements affords the opportunity to guide the research and development of applications, services and underlying technologies consistent with a primary goal of meeting user needs. The working group has been active since early 2001 and has gathered many inputs and views from industry and academia. A major objective of WG1 has been to synthesize these views and document them in the form of white papers, to influence future visions and research priorities and to share results across the forum. These white papers are included in Chapter 2 and cover the following areas:

- Methods, processes and best practice for user-centred research and design.
- Reference frameworks for modelling user needs within the context of wireless systems.
- User scenario creation and analysis.
- User interaction technologies.

1.5.2 WG2 – Service Architecture

Since September 2001, Working Group 2 has worked in the area of service architectures for future wireless systems. WG2 follows the approach of WG1 in making the user the driving force in future communication systems. Based on an IP-centric, always connected world, WG2 investigates how the user can be provided with I-centric service environments by future systems. The services addressed by WG2 are configured on top of an all-IP environment, addressing the fulfilment of any user demand.

The vision of I-centric communications has been developed. This puts the individual user ('I') in the centre of all activities a communication system has to perform. WG2 is studying communication systems, which in the future will be able to model each individual, his preferences, and adapt to different situations and resources in time. The developed reference model for I-centric communications addresses the individual user interacting in his personal communication space.

The reference model for I-centric communication follows a top-down approach starting with the introduction of individual communication spaces, related contexts and objects. It is a common understanding that I-centric services have to support at least three different features, namely: ambient-awareness; personalization; and adaptability. These are discussed in detail in the following sections.

Following this view, WG2 identifies necessary research and enabling technologies. The white papers provided by Working Group 2 explain in detail what future service architectures have to provide and how such service architectures can be established. To come up with a complete picture, WG2 is analysing future service architectures from a business point of view, breaking them down to already available technologies and necessary research activities.

1.5.3 WG3 – Cooperative and Ad Hoc Networks

The WWRF Working Group 3 has been, since the start of the Forum, concerned with interoperability issues. The two main tracks of discussions are the cooperative networking and reconfigurability issues. Two subgroups have been formed to further the technical discussions and to work towards the feasibility and implementation of the vision of a seamlessly interoperating wireless world.

The current view in the WWRF is the vision that mobile communication systems beyond the 3rd generation will offer a multitude of services over a variety of available and future narrow and wideband access technologies. To make this vision come true and to accomplish these objectives for future networks that support heterogeneity in network access, communication services, mobility, user devices, etc. is the objective of the work described in Chapter 4, the chapter is divided into four sections:

- The Cooperative Networks Vision and Roadmap;
- Research Challenges in Cooperative Networks;
- CoNet Architectural Principles;
- Network Component Technologies for Cooperative Networks.

The other part of the work of WG3, the aim of defining a set of WWRF SDR Reference Models for the different parts, from end-to-end of a system, is covered in Chapter 5. The chapter analyses the overall problem of the SDR architectures and identifies the main SDR research themes to be investigated in the next decade. It documents a top-down approach and the reference model targeted encompasses system and network (including core and access network/base stations), the hardware issues in both RF and BB sides and the data and control/management interfaces between the various building blocks of the reconfigurable environment. The chapter consists of the following four sections:

- Reconfigurable Systems and Reference Models;
- RF/IF Architectures for Software Defined Radios;
- SDR Baseband Architectures;
- Reconfiguration Management and Interfaces.

1.5.4 WG4 – New Radio Interfaces, Relay-based Systems and Smart Antennas

The contributions made to WG4 have been filtered by small editorial groups to identify substantial contents and contributions to be used as input for the development of a total of eight white papers in well-defined thematic areas. These white papers (WPs) have been included in Chapter 6 and Chapter 7 and form the basis of the section within those chapters. These are:

- Ad hoc Networking;
- Relay-based Deployment Concepts;
- Spectrum Issues;
- Broadband Multicarrier Techniques;
- Mixed OFDM plus Single Carrier Techniques;
- Smart Antennas, MIMO Systems;
- Short-range Wireless Technologies;
- Ultra Wideband Techniques.

The existence of Moore's Law for bandwidth is more and more perceptible – the end users tend to embrace services and applications utilizing ever faster data rates. The service bandwidth (offered/required) is doubling every 12 months or so. The constant increase in 'users' injects further positive feedback into the system, thus sending the bandwidth demand spiralling up.

This potential growth scenario needs to be evaluated in the light of the following:

1. The extent of good quality radio coverage is inversely proportional to the transmitted bit rate. The cost of 'continuous' and 'all time everywhere' radio coverage increases very sharply with the transmitted bit rate. Spectrum will be necessary in the low frequency range, which provides the required coverage in sparsely populated areas.
2. The higher the service bit rate, the larger the required bandwidth and the higher the frequency range where some additional spectrum might be available for offering the required capacity.
3. The present users of the already allocated frequency bands (systems and service providers) would like to make the most of their allocation. As a result, the sophistication of standardized air interfaces, e.g., by introducing space–time coding, smart antenna systems and multihop links to improve the radio coverage, appear to be the direct consequence of frequency spectrum shortage for mobile radio use.
4. A proliferation of competing wireless information services is expected in the near future.
5. The variety of networks for provision of seamless services in private to public and short-range localized coverage to wide area coverage will be limited by the necessity of cost effectiveness of the corresponding business case.
6. The requirement of a very wide consensus (much more extensive than technology standardization) shall be motivated by the globalization of markets. The international regulatory frame needs to be enhanced to handle the coexistence of wireless networks operating in distinct, adjoining and partially overlapping frequency bands and the interworking of permanently established and spontaneously created networks.

Based on the usually employed artefacts for improving the situation of spectrum congestion, the findings contained in the WPs are addressed later in this book.

1.6 References

[1] WWRF, *'The Book of Visions: Visions of the Wireless World'*, Version 1.0, December 2001.

[2] Prasad, R., Mohr, W. and Konhäuser, W., *'Third Generation Mobile Communications Systems'*, Artech House, Boston, 2000.

[3] Mohr, W., 'Internationale Standardisierungsaktivitäten zur Defintion der dritten Mobilfunkgeneration', *Frequenz*, **54**(3–4), 2000, 97–105.

[4] Chaudhury, P., Mohr, W. and Onoe, S., 'The 3GPP Proposal for IMT-2000,' *IEEE Communications Magazine*, **37**(12), 1999, 72–81, initiated by the European Commission.

[5] Wisely, D., Eardley, P. and Burness, L., *'IP for 3G: Networking Technologies for Mobile Communications'*, John Wiley & Sons, 2002.

[6] Li Zhen *et al*, 'Consideration and Research Issues for the Future Generation of Mobile Communications', *IEEE Canadian Conference on Electrical and Computer Engineering*, September 2002.

[7] WWRF – Wireless World Research Forum: http://www.wireless-world-research.org/.

[8] Mohr, W. and Konhäuser, W., 'Access Network Evolution Beyond Third Generation Mobile Communications', *IEEE Communications Magazine*, **38**(12), 2000, 122.

[9] Wisely, D., Mohr, W. and Urban, J., 'Broadband Radio Access for IP-based Networks (BRAIN) – A key enabler for mobile Internet access', *11th IEEE International Symposium on Personal, Indoor and Mobile Radio Communications (PIMRC) 2000*, London, September 18 to 21, 2000, pp. 431–436.

[10] Becher, R., Dillinger, M., Haardt, M. and Mohr, W., 'Broad-Band Wireless Access and Future Communication Networks', *Proceedings of the IEEE*, **89**(1), 2001, 58.

[11] IST WSI Project: http://www.ist-wsi.org/.

[12] ITU: http://www.itu.int/imt.

[13] Han, K.C., 'A Study on Systems beyond IMT-2000 in Korea', ETRI.

[14] http://www.ngmcforum.org

[15] Lu, W.W., '4G Mobile Research in Asia', *IEEE Communication Magazine*, March 2003.

[16] http://www.ccsa.org.cn/english/

[17] http://www.3GPP.org

[18] http://www.3GPP2.org

[19] Uskela, S., 'Key Concepts for Evolution Beyond 3G Networks', *IEEE Wireless Communications*, February 2003.

[20] Dixit, S., 'Evolving to Seamless All IP Mobile/Wireless Networks', *IEEE Communication Magazine*, **9**(12), December 2001.

[21] Dixit, S. and Prasad, R., *'Wireless IP: Building the Next Generation Mobile Internet'*, Artech House, Boston, 2002.

[22] http://www.ieee802.org

[23] http://www.ietf.org

[24] National Science Foundation, http://www.nsf.gov/pubs/1999/nsf9968/nsf9968.htm

[25] Department of Defense: www.DoD.mil

[26] http://www.darpa.mil/ato/programs/XG/index.htm

[27] http://4Gmobile.com

1.7 Credits

The sections on the WGs in this chapter have been provided by the respective chairs of the Wireless World Research Forum Working Groups. The contributors include: Ken Crisler (Motorola, US); Radu Popescu-Zeletin and Stefan Arbanowski (Fraunhofer FOKUS, Germany); Rahim Tafazolli and Klaus Moessner (The University of Surrey, UK); Bernard H. Walke and Ralf Pabst (Aachen University, ComNets, Germany).

2

A User-Centred Approach to the Wireless World

Edited by Ken Chrisler (Motorola)

Overviewing the Human Perspective

The application of science and engineering to mobile communications has been largely focused on technological issues such as developing technologies enabling radio access networks, terminal devices or engineering radio sites. Because, until recently, voice has been the single dominant application for mobile communications, there has been relatively little need to understand the user and what he/she wants to do with the mobile service. Application of human factors principles or user-centred design has been limited to the ergonomics and usability of the terminal device, and in the study of usage patterns, primarily for marketing purposes.

However, the mobile communications market is changing. Data services are beginning to gain traction worldwide, but particularly in Europe and Asia. 2.5G systems are being deployed to support more sophisticated data applications, such as web browsing and email. Third generation (3G) systems are going to be deployed soon, offering higher data rates to support more applications and greater capacity for data services. It will be critical to understand what people are going to use this capability for, and what they are prepared to pay for. We need to anticipate which broad application categories (such as video, email, browsing or m-commerce, etc.) will be important and also to project which attributes will be most critical to optimizing the user experience of a particular application. The answers to these questions are based in the science of Human Factors, which can identify underlying human needs and desires, the human value we place on any service and how to provide those needs given the limitations of the technology.

The Wireless World Research Forum (WWRF) is a forum whose stated aim is to build a vision of the future. It has recognized that the vision and technology of the future needs to be defined by understanding the user. The Human Perspective Working Group (WG1) is a working group within the WWRF focused on discovering and promoting research areas that strive to understand the users' needs for future wireless systems and

Technologies for the Wireless Future. Edited by R. Tafazolli
© 2005 John Wiley & Sons, Ltd   ISBN: 0-470-01235-8

how users will interact with devices, systems and applications in the wireless world. An understanding of user requirements affords the opportunity to guide the research and development of applications, services and underlying technologies consistent with a primary goal of meeting user needs. The working group exists to gather input and views from industry and academia, to synthesize these views to influence future visions and research priorities and to share results across the forum.

The subject matter scope of the Human Perspective Working Group includes all areas relevant to user needs for future wireless systems and the interaction of users with technologies embodied in these systems. Recently, the working group has concentrated its efforts in the following areas:

- methods, processes and best practice for user-centred research and design;
- reference frameworks for modelling user needs within the context of wireless systems;
- user scenario creation and analysis;
- user interaction technologies.

This chapter is comprised of the collected white papers from WG1, providing a thorough discussion of the issues and viewpoints in each of these four areas.

Considering the user in the vision of the wireless world will be critical to realizing the benefits of the vision for all interested parties (users, operators, manufacturers, etc.). The well-known principles of user-centred design (UCD) provide a solid foundation, although some refinements and enhancements are appropriate to drive the user perspective in the early phase research that can drive the vision. Section 2.1 addresses this topic, presenting in a straightforward and understandable fashion the basics of UCD processes that should be followed to make technology research user driven. In order to contribute to the formulation of a comprehensive reference model for systems beyond 3G, WG1 sought to develop a model from the perspective of the wireless user. Section 2.2 describes a proposed model that portrays user requirements in two planes, the Value Plane and the Capability Plane. By considering user needs in terms of both values and capabilities in a systematic way, there is the opportunity to link the user perspective on requirements to the more traditional view of system technical requirements.

Scenarios provide an intuitive and powerful means to express user expectations in an evocative manner. Thorough analysis and evaluation of such scenarios holds the key to discovering where the compelling value lies and enumerating detailed user requirements. An engineering approach for the management of scenarios is described in Section 2.3. While such approach involves three phases: scenario generation, scenario prototyping and scenario evaluation, the treatment here concentrates on analysis of existing scenarios, attempting to find patterns and draw conclusions from what has already been described.

Finally, the user interface is critical as it represents the physical point at which the user engages with the technology. Interfaces of the future must leverage new technologies, employ intuitive elements and engage users from multiple segments to drive towards new levels of ease of use. Section 2.4 describes visions for future user interfaces and some examples of existing solutions that are advancing the art. In addition, this section provides an overview of the research issues connected with the development of robust user interaction.

2.1 UCD Processes for Wireless World Research

2.1.1 Introduction

The Wireless World Research Forum (WWRF) was established to create a set of visions of the future of wireless in the next 10 to 15 years. In creating these visions, members of the forum identified research areas that would be critical for the realizations of the systems envisaged for beyond 3G wireless services. The forum identified and strongly supported the concept that future wireless systems, their capabilities and technologies should be driven by the needs and desires of users. Working Group 1 of the WWRF was established to identify the research and methods needed to make the visions truly user centred. The work of WG1 is based in the realm of User-Centred Design, where the requirements of the user are elicited first, the system concepts are prototyped and the prototypes are evaluated against the requirements through user trials. This principle of starting and finishing with the user is vital to making the research for the future wireless world user centred. It is not enough for the technology developers to simple consider how users in the abstract might use the technologies that they are creating. From a UCD perspective, the user research has to:

1. Identify the technology agnostic requirements of wireless systems of the future and translate these into technology requirements of classes of wireless systems.
2. Transfer these requirements to the individual technology projects.
3. Identify the requirements of specific technologies from a user perspective.
4. Evaluate or translate the performance of the technologies against the user requirements.
5. Evaluate the likely services and performance of the systems against user requirements.

The purpose of this section is to present the basics of UCD processes which should be followed to make technology research user driven. Its objectives are to:

1. Summarize the basic user-centred research principles, as the majority of the technologists are unaware of these concepts.
2. Suggest a process for applying UCD as the driver and evaluator of technical research as opposed to a design process.
3. Suggest a process model that applies UCD techniques to elicit user requirements of classes of wireless systems for users that are largely technology agnostic, and a basic model for translating those requirements into more technical system requirements.
4. Suggest a process model for accepting the gross user requirements into the user-centred research part of a technology specific project.
5. Define the basic approach to the evaluation of the technologies and systems.

2.1.2 Overview of the User-Centred Design Process

2.1.2.1 Principles and Benefits of User-Centred Research

The basic principle of user-centred design is to consider the user when designing a new system or product. When considering how to conduct technological research for future systems, we can use the term *user-centred research*. The basic benefit of applying user-centred research is that the technologies are not developed for their own sake, but to fulfil a user need or desire, thus maximizing the chance that technology research will be used in products and systems that people will use, and thus building value in the market place. Specifically, the reasons that we consider the user are:

1. People use the systems and products.
2. User-centred design ensures that systems and products meet the user's needs ('useful'); that they enable the users to carry out their tasks and demands and/or meet their goals; and that they are easy to use ('usable').
3. Enabling technologies are developing rapidly, so that we can conceive that almost any system or capability will be possible. Only some of these possibilities will succeed in the market, so UCD fulfils the need to understand what future mobile communications systems can offer, and how they will be used.
4. UCD will facilitate the development of innovative and usage optimized interaction modes and modalities to overcome the boundaries of mobile technology drawbacks.

User-centred research is developing a deeper understanding of people and what they do to establish applications that will be truly valuable to people in the future, establish and verify requirements for future products and identify and establish the requirements for future technologies. The basic UCD process is a three-step process, shown in Figure 2.1.

User-centred design incorporates the activities of deriving the requirements from analysing the user and the context of use, following a structured design approach including prototyping against the user requirements and evaluating the design against the requirements. Note that the whole process is highly iterative and interrelated, as are the steps between the process elements, so that analysis, design and evaluation can be modified whilst the process runs.

Analysis
The characteristics of the users, tasks and the organizational and physical environment define the context in which the system is used. It is important to understand and identify the details of this context in order to guide early design decisions, and to provide a basis for evaluation. Optimized usage of mobile services and technologies is highly dependent on the methodological consideration of context conditions based on the fact that mobile contexts are different and change rapidly (the essence of mobile usage).

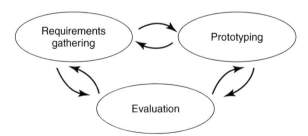

Figure 2.1 The basic UCD process

Structured analysis activities incorporate the following most important steps: user analysis, task analysis, environment analysis and comparative analysis. User analysis identifies the most important characteristics of the persons who will use the product/system in order to accommodate human diversity (e.g. physical abilities, previous experiences, cognitive capabilities, intercultural diversity, education, skills). Task analysis identifies the tasks that will be performed with the future systems from the user's perspective. Tasks have a close relationship to end user oriented services to be accomplished with mobile technology. The analysis of the environment includes the technical conditions in which a system is used, the organizational and social influence variables and the physical conditions. The comparative analysis identifies strengths and weaknesses from the user's perspective in order to learn valuable information for the development of a product.

Design
A well structured design process supports the user orientation of mobile systems development. A large amount of qualitative knowledge exists in the fields of UCD, Human–Computer Interaction, Cognitive Ergonomics and other disciplines. In particular, the mobile field has been very active in providing user interface guidelines. This guiding knowledge very much supports ease of use.

Design steps should start with conceptual design, which defines the overall logical structure of the product (e.g. general paradigms applied, navigation logic and principles, information architecture). If the conceptual design is agreed and evaluated the detailed design of the system starts (e.g. layout details, design details, icons, menu details).

Design is a highly iterative process. Design solutions should be assessed and evaluated early in the design process and improved until the requirements are met. Design iterations have to be planned in advance and the progress of the design iteration should be documented. Prototypes (there are various kinds of prototype) and simulations can make the outcome and interaction scenario more tangible. Interim design solutions can be explored and evaluated by the design team and by users. As an example, the evaluation of voice dialogues for mobile devices could be accomplished by using a 'Wizard of Oz' methodology.

Evaluation
During evaluation activities, the design solutions are assessed and feedback is given into the design. There are two possible objectives of evaluation: one is to identify usability problems; the second objective is to assess the degree to which the user requirements of the analysis phase are met.

Evaluation starts by providing an evaluation plan. There are many evaluation methods. Each method has its strengths and weaknesses. So the right mixture of methods and adaptation to the specific conditions of a development project is crucial. Generally, user-based methods (i.e. users are involved in the investigation) and expert-based methods can be distinguished. Streamlining the methods for mobile systems is under development (e.g. mobile devices are used much more in freedom to move than desktop systems). A multitude of techniques exists to accomplish the different steps in various phases, examples of which are given below.

Analysis
• contextual enquiry models;

- focus groups and scenario-based interviews;
- external data gathering;
- constraints analysis.

Design
- conceptual design;
- paper prototyping;
- wizard of Oz prototyping;
- computer-based prototypes;
- physical prototypes.

Evaluation
Early Development Stages
- 'walk through';
- iterative prototype development;
- experimental study;
- subjective evaluation;
- usability study.

In-field Evaluations
- functional prototype;
- ergonomics;
- social impact.

2.1.2.2 ISO 13407 and its Application to User-Centred Research

One widely adopted standard for UCD is the ISO standard ISO 13407. This standard provides guidance on human-centred design activities throughout the life cycle of inter-active computer-based systems. It is a tool for those managing design processes and provides guidance on sources of information and standards relevant to the human-centred approach. It describes human-centred design as a multidisciplinary activity, which incorporates human factors and ergonomics knowledge and techniques with the objective of enhancing effectiveness and efficiency, improving human working conditions, and counteracting possible adverse effects of use on human health, safety and performance. The recommended process is shown in Figure 2.2.

ISO 13407 is a particular instantiation of the three-phase UCD model introduced earlier in this section. The standard itself documents in some detail the application of the model. In applying ISO 13407 to a user-centred research programme, Figure 2.3 shows how the basic ISO 13407 model changes in order to drive technology research.

ISO 13407 is a good example of the principles of user-driven research, and provides a valuable framework.

When applying UCD techniques to drive a research programme, we suggest the following basic guidelines.

1. The process starts with a top-down approach, preferably running in advance of the technology development.

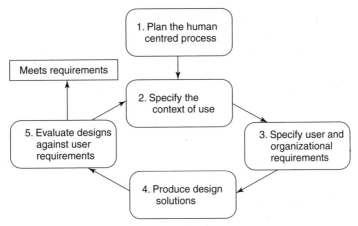

Figure 2.2   ISO 13407 user-centred design process

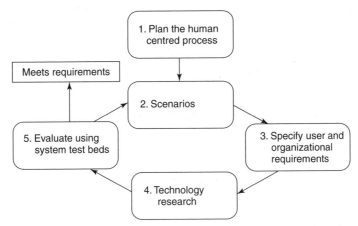

Figure 2.3   Modifying ISO 13407 to drive technology research projects

 — Elicit user needs, wants and requirements (analysis).
 — Technology agnostic – a systems view.
2. Define the requirements of broad technology areas from the user requirements. This translation of requirements from user requirements to technology requirements would typically go through the phases of translating the:
 — User requirements into system requirements.
 — System requirements into a systems architecture or reference model.
 — System architecture into requirements on the elements that make up the system.
3. A user-centred approach can then be run within a technology development area, using the broad technology requirements to:
 — Gain an understanding of technological capabilities to understand how they may or may not meet user needs.
 — Allow multidisciplinary development.

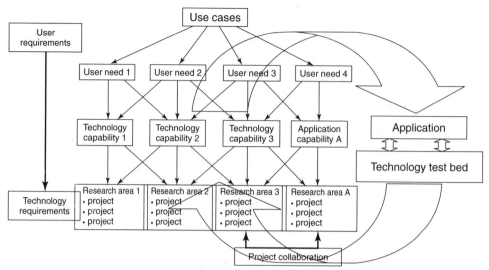

Figure 2.4   A representation of applying UCD to a research programme

4. Develop design solutions transforming technological needs and requirements to the user, based on the user requirements.
5. Use prototype design solutions as a basis for test beds and evaluation.
6. The prototype technologies should be brought together into a systems test bed either by implementation or emulation of their capabilities.
7. Evaluate design solutions and test beds and feed back into requirements, design and technology development.

Figure 2.4 shows a representation of how this might be achieved.

2.1.3 A UCD Process for Wireless World Research

2.1.3.1 Introduction

The purpose of this section is to describe the general user-centred design (UCD) process for wireless technology and conceptualizing projects, i.e. the processes that can be followed in the course of research and design of an application, a technology or a system. Processes are described in terms of purpose and outputs, i.e. why a particular process exists and what its actual outputs are. The primary reference model is based upon the KESSU UCD Processes [1], and is briefly introduced in the next section. However, changes have been made to the reference model based on experiences and the literature [2, 3].

Before using the described process, there needs to be a high-level idea about the system/technology to be developed, i.e. what is the main purpose and the benefit of the system/technology to its end user? It is also necessary to have some kind of idea about the targeted user group(s) since the process requires certain, and actually quite specific, information concerning the user group(s), their goals and usage environment(s) of the selected application/system domain.

Finally, this section presents the linkage between the user-focused and technology-focused design phases.

2.1.3.2 Overview of the Primary Reference Model – KESSU UCD Processes Model

This UCD process follows the guidelines presented in ISO 13407, ISO 18529 and their modifications [1, 2, 3]. It contains the following primary processes and outputs as illustrated in Figure 2.5:

Context of Use Process
The purpose of the context of use process is to establish and communicate the characteristics of the user groups, their tasks and the technical, organizational and physical environment in which the product/system will operate.

User Requirements Process
The purpose of this process is to define usability and UI design requirements for the product/system. The deliverables should drive decision making in the design processes.

User Task Design Process
The purpose of this process is to design how users carry out their tasks with the new product/system that is being developed.

Product–User Interaction Solutions Process
The purpose of this process is to design those parts of the product/system that users interact with. These parts of the product/system include interaction and graphical design of the user interface, user documentation, user training and user support.

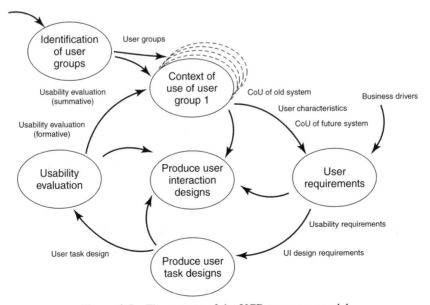

Figure 2.5   The outputs of the UCD processes model

Usability Evaluation Process
The purpose of this process is to evaluate the product/system against the requirement in terms of the context of use.

Description of the Processes
In this chapter, the processes that should be carried out in the course of a concept/system design project are described in terms of purpose and actual outputs. The primary reference model is the KESSU UCD Processes model, but other models, such as ISO 18529 and ISO 13407, as well as experience gathered, have also been utilized.

Context Of Use Process
The purpose of the Context of Use process is to describe the characteristics of the users, their tasks and the technical, organizational and physical environment in which the concept/system is intended to be used. This information will be used as a basis for early decision making and evaluation in the design process.

The outcomes of the Context of Use process are:

- *User characteristics.* The characteristics of the user group(s) are identified and documented. These may include the following information: language, skills, education, physical and mental capabilities, product and application experience, preferences, habits, values, attitude towards technology, age, etc.
- *Context of Use description of the new application(s).* The description contains the following things:
  — *Goals.* User accomplishments in the concept/system domain should be identified and documented. These accomplishments should relate to overall goals of the users, not to features or functions of the current products.
  — *Task descriptions.* Identify what actual steps users currently need to take in order to achieve their goals. Allocation of functions/activities between system and user should also be described, as well as the manual steps – documenting tasks solely in terms of product/application functions is not enough. Task descriptions could include things such as (sub-) intent(s), triggers, loops, steps, activity chunks, different strategies, etc.
  — *Task characteristics.* The characteristics that can affect the usability of the application should be identified, e.g. frequency, 'obligatoriness', duration, error criticality, etc. Also, indications pointing out that users would clearly benefit from achieving the goal(s) of the task while in mobile context (i.e. perform the task on the spot when needed, 'regardless' of the place) or the task is currently done while in mobile context should be identified and documented.
- *Environment descriptions.* The characteristics of the cultural (e.g. values, attitudes, laws, directives), physical (e.g. place, furniture) and larger technical environments (e.g. GSM 900) that affect the new application should be described. Documentation should also include the equipment, products (software, hardware) and materials that affect use. In the case of mobile usage environments (possibly numerous) the extent to which it is possible to identify the key environments and issues related to them should be described.
- *Task/feature analysis of comparable products in the market.* Identify and analyse possible competitors' or own comparable products in the current market, for example in

terms of user tasks, UI structure, UI presentation etc. This analysis can function as an input to the requirement definition in the User and Design Requirements process. In the case where there are no such products available, one may try to identify other tasks that are similar (in terms of goals, characteristics and usage environment) to the ones under development. This helps in determining 'reasonable' usability measures to the requirements in the User and Design Requirements process.

User and Design Requirements Process

The purpose of the user requirement process is to define usability and UI design requirements for the application. This process utilizes the information gathered in the Context of Use process. The outcomes of this process should drive decision making in the design processes. They also provide additional data to the concept/system definition (high-level design).

The outcomes for the User and Design Requirements process are:

- *Usability requirements specification.* The documentation describes prioritized and measurable usability requirements, which application implementation should fulfil. Primary usability requirements (central user goals) are gathered for each identified user group and they mean the required performance of the product against context of use. These requirements are typically given in terms of effectiveness, efficiency, and satisfaction in the specific context of use. So-called secondary usability goals (desired properties of the product that contribute to the usability – e.g. consistency) and the criteria for their successful implementation might also need to be determined.
- *UI design and technical requirements.* Those guidelines and specifications that affect the design should be identified and documented.
- *UI guidelines.* UI guidelines concerning the concept or system under development should be documented.
- *Specification of the customer product.* The specification of the customer product should be available and the implications of the document should be identified and documented.
- *Description of the requirements/restrictions that selected technologies (e.g. Java) set for UI design.* Enabling technologies and their implications on the concept/system should be clarified and documented.

Other Stakeholder Requirements

- *Third party requirements for the design.* If the project involves third party collaboration, their requirements concerning the design must be identified and documented.
- *Legislative requirements.* The project should consider that there might be some legislative requirements, such as EU directives or national laws, that might have an impact on the design. In such cases the implications of the legislative requirements should be identified and documented.
- *Safety/health issues.* The project should consider safety and health issues that might have an impact on the usage situation and set requirements accordingly.

User Task Design Process

The purpose of this process is to design how users are to carry out their tasks with the new application. Task design models should be developed based on Context of Use analysis and designers should take all the requirements/restrictions set in the User and Design Requirements process into account to the extent that they are relevant or applicable at this stage.

The outcomes for the User Task Design process are:

- *Task design.* The description should include things such as goal(s) and subgoals that will be accomplished by performing the redesigned task, description of the redesigned steps of the task, allocation of functions between user and system, etc.
- *Example methods and data sources for gathering required information.*
- *Output.* Methods and data sources for gathering required information.
- *Set of task-driven designs.* Storyboards, written scenarios and narratives, conceptual design.

Product–User Interaction Solutions Process

The purpose of this process is to design those elements of the product that users interact with and make them concrete by creating mock-ups, simulations etc. Making design concrete helps the communication between designers, and between designers and users. Producing alternative solutions, taking user and design requirements into account and changing design in the light of feedback from evaluation is an essential part of this process.

The outcomes of the User Interaction Solutions process may include:

- *Functional specifications.* These describe the UI functionality, but are not task-driven nor UI specific. They form the ground floor design of the system.
- *UI designs.* Possible outputs include:
  — paper layouts;
  — low-fi prototypes (rough drawings of the actual UI) which allow designers to do rapid design iterations;
  — graphical layouts;
  — simulations.

Evaluation Process

The purpose of this process is to provide feedback, which can be used to improve the design and assess whether the usability and other goals set in the User and Design Requirements process have been meet. Evaluation is split into two categories: formative evaluation, giving qualitative feedback, and summative evaluation, determining whether the set requirements have been achieved.

The outcomes for the Evaluation process are:

- *Formative evaluation report.* The aim is to get qualitative feedback on design from different perspectives. For example, this could include identification of usability problems and strengths, as well as qualitative assessment from different technology points of view.
- *Summative evaluation report.* This evaluates to what extent the design meets the set requirements in the User and Design Requirements process. In other words
  — Have the usability requirements been met?
  — Have the design and technical requirements been met?
  — Have the third party requirements been met?

— Have the legislation-based requirements been met?
— Have the healthy/safety requirements been met?

2.1.3.3 Overall UCD Process Linkage to Technology and System Design

Task design is carried out in cooperation with system analysis. Figure 2.6 illustrates the process of task design and application/functionality design linkage to system analysis and design.

Usage model work consists of producing:

1. *Usage narratives.* These provide a rough scenario-level description of the feature/system in use from the end user's point of view. They describe why and where a user uses the system. They refer to the 'Context of Use' phase in the UCD process. It is important that usage narratives are based on end user data as much as possible. Narratives are windows to the end user's world. Stories can present user goals (why to use the system) and provide some kind of description of the end user's world (age, lifestyle etc.). When narratives are ready, they are shared with the whole design team. Multidisciplinary work should be started at the latest when moving to task flow design. A core team is formed, and it is important that everyone in the team understands the results of the usage analysis. Everyone should share the same vision of the designed feature.
2. *End user task flow.* This describes user actions step by step, presenting the main success scenario. Technical use cases, end-to-end and terminal architecture analysis are created at the same time with end user use cases. 'Context of Use' continues in this phase. The main steps in using the system are shared with the team working on the system analysis side. System analysis can give input to task analysis by evaluating the 'durability' of the scenario, i.e. what might be the critical factors in implementing the main success scenario. The main success scenario should also consider the possible extensions related to the main scenario (e.g. operation is cancelled by the user before

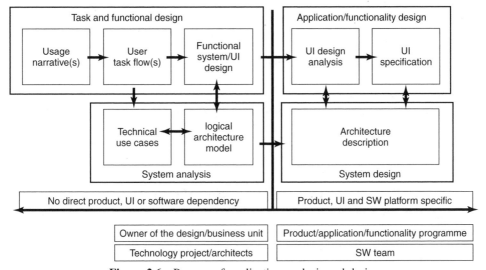

Figure 2.6   Process of application analysis and design

the main success scenario ends). Formulating the main success scenario might also produce some alternative ways of completing the task. In other words, there might be several ways to complete the task. The design team should decide if these alternative ways should be written down as their own task flows. When working with task flows, high-level requirements start to appear. These requirements can be described in more detail when the design process proceeds. One high-level requirement can be split into several subrequirements. Finally, when the task flow seems to be ready it can be tested by 'walking it through' with the core team. Output (task flows produced) is communicated outside the core team and delivered to relevant stakeholders.

3. *Functional system design.* This describes the structure of the system (application) and the functionality; also includes the functions that the system does automatically on behalf of the user or as a response to user actions.

4. *UI design analysis.* This describes the 'user interaction' with selected UIs. UI design analysis is not a complete UI specification. It starts with the main success scenario. Collected user/usability requirements should be kept in mind because they form the basis for usability testing – usability criteria are based on them. Results of UI design analysis can be tested, e.g. with paper prototypes. Feedback received is then mainly formative. Test users can be experts or represent real end users (selected target group.

An application/functionality design model describes the feature/system mapped to a certain UI. It can be used as a basis for the official UI specification process. This step consists of:

1. UI design analysis.
2. UI specifications.

2.1.4 Conclusions

The WWRF has identified that considering the user is essential to the success of technologies and systems research in preparation for the future wireless world. In order to make this happen, it is necessary to adopt a user-centred approach, which goes beyond technologists simply considering the user in the abstract when specifying and carrying out their research. We have shown that a user-centred approach involves following a process both on a systems and a broad technology level that studies the user to elicit requirements, carries out the research and prototypes systems or capabilities, and finally evaluates the capabilities against the requirements.

ISO 13407 provides a good general guideline for user-centred design and can be modified to provide similar guidelines for user-driven research. Finally, we have presented a candidate model, the KESSU UCD Processes model, as a suitable reference model for user-driven research in the wireless world.

2.2 A User-Focused Reference Model for Wireless Systems Beyond 3G

2.2.1 Introduction

In line with the general aims of the WWRF, Working Group 1 aims to define a human-centred vision of strategic future research directions in the wireless field, and to promote a human-centred design approach for future mobile and wireless technologies and services. Advancing from the work presented in the Book of Visions [4], Working Group 1 has established a reference model from the perspective of the wireless user for systems beyond 3G. The following part of this section presents that reference model and describes its various components in detail, discussing the implications for future mobile technologies and services.

This part of the chapter is structured into four main sections. Section 2.2.2 presents an overview of the proposed reference model, including descriptions of its general structure, how the model relates to the results of user scenario analysis (see Section 2.3), and high-level descriptions of the two 'planes' of the model: the Value Plane (described in Section 2.2.3) and the Capability Plane (described in Section 2.2.4). Finally, Section 2.2.5 offers a summary and conclusions.

2.2.2 Overview of the Reference Model

This section of the chapter provides a high-level view of the reference model and the structural constructs that have been adopted to describe the components of the model.

2.2.2.1 General Structure

We propose a reference model consisting of two main areas or planes (see Figure 2.7). The two planes offer the opportunity to reveal characteristics at different levels of abstraction that are relevant to a human-centred view of wireless systems. The *Value Plane* addresses the core human needs, such as safety or belonging, that products and services should address. To deliver such products and services, certain functionalities must be provided by mobile technology beyond 3G. The functionalities are the subject of the model's

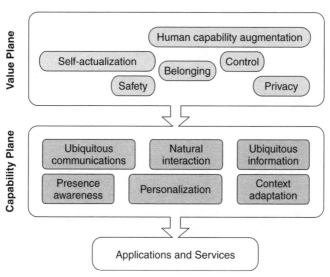

Figure 2.7   Reference model

Capability Plane. Thus, the reference model is able to link into the reference models produced by the other WWRF Working Groups, to describe how these applications and services are enabled.

For example, Working Group 2 (WG2) of the WWRF has developed a model for representing a service architecture (see Figure 3.2 in the following chapter). Later in this chapter, when we illustrate a linkage to supporting services, we generally refer to aspects of the WG2 model. An alternative model, based on the work of the Wireless Strategic Initiative (WSI) project is known as the Cyberworld Model. A Cyberworld reference model incorporating the two user-centred planes is illustrated in Figure 2.8.

The structure of our reference model provides a framework for describing the characteristics of each component in the model and its relationship to other components. To describe the components of the model, we have adapted the format and formalism of the User Environment Design (UED) model as described by Beyer and Holtzblatt [5]. The UED provides an excellent starting point as it is intended to document the organization of a system from the user's point of view, and capture the structure and function of the system without straying into the realm of implementation. This approach keeps our focus on a high-level view of what the system does.

We sought to retain the benefits of the UED model, while adapting the formalism slightly. At the highest level, the reference model shows a set of focus areas and their relationships. In the standard application of a UED, the focus areas are action-oriented (e.g. format a page, copy a text box, etc), while the components of our model are more like topic domains. Since we are attempting to describe a reference model, not a design, we adopt the following descriptive elements for each model component:

- *Summary*. A brief description of the scope of the focus area.

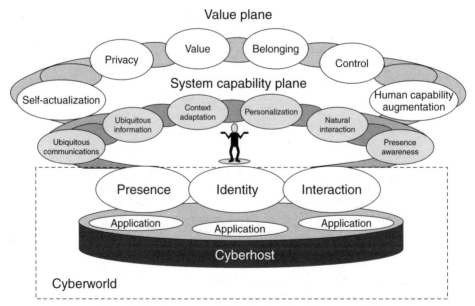

Figure 2.8   A user-centred Cyberworld reference model

- *Functions.* In the UED formalism, these are the actions that can be invoked by the user or that are done by the system automatically within the area. Here is where we document linkages to the other parts of an overall beyond-3G reference model. For example, the focus areas within the Capability Plane list the functions drawn from the other reference models (e.g. that of WG2) that are relevant to the capability. Likewise, a focus area within the Value Plane lists components from the Capability Plane here.
- *Dependencies.* These illustrate the relationships between the function areas. In a standard UED, these relationships or links indicate trajectories through the system. For example, a use case could be generated by following a path from area to area according to these links. Since our focus areas are not actions, this approach is not appropriate. Instead, we use this to represent dependencies, for example complementary areas (such as context adaptation and personalization) or potentially conflicting ones (such as belonging and privacy).
- *Constraints.* Key assumptions that influence the model.
- *Issues.* A listing of critical issues that need to be resolved through future research.

2.2.2.2 Linkage To Scenario Analysis Results

Another focus for the work of WG1 has been to collect, examine and analyse scenarios and use cases for future wireless systems (see Section 2.3). One of the outputs of this scenario analysis is a catalogue of requirements and critical capabilities for systems beyond 3G. This enumeration of capabilities, in turn, has driven the definition of the components of our reference model. Thus, our reference model is ultimately a reflection of the requirements as derived from user-focused scenarios. As our catalogue of scenarios expands over time, it will be important to update the reference model to account for any new or changing requirements implied by new scenarios.

2.2.2.3 Value Plane Overview

The Value Plane of the model describes the core human needs that wireless systems must meet (or at least not interfere with) in order to be successful. Research on technology and change has established that applications and services must address, and fit in with, human values to be successful [6]. Our initial analysis has indicated that we can group these core needs into six focus areas: safety, belonging, control, privacy, self-actualization and human capability augmentation. The details of each of these components and their interrelationships are described in Section 2.2.3.

2.2.2.4 Capability Plane Overview

Meeting the needs described in the Value Plane demands that the system provide a set of basic functionalities to the user. These are described as components within the Capability Plane. Our initial analysis led us to group the capabilities into six focus areas: ubiquitous communications, ubiquitous information, context adaptation, personalization, natural interaction and presence awareness. The details of each of these components and their interrelationships are described in Section 2.2.4.

2.2.3 The Value Plane

The Value Plane of the model describes the core human needs that wireless systems must satisfy (or at least not degrade) in order to be successful. Research on technology and change has established that successful applications and services must address and fit in with human values to be successful [6]. This is especially true for applications and services used in a personal (as opposed to work) context, and paid for directly by end users. The majority of scenarios that have been envisioned in the context of the WWRF fall within this sector. Our initial analysis has indicated that we can group these core needs into six focus areas: safety, belonging, control, privacy, self-actualization and human capability augmentation. These focus areas are based on Maslow's hierarchy [7] of human needs, and in line with recent reconceptualizations proposed by Jordan [8] and Shneiderman [9].

2.2.3.1 Safety

Summary
Safety is the most basic human need. Most people, most of the time, value their own safety, and that of other people, especially those close to them. Physical safety is the most immediate and obvious concern, but mental well-being can be affected through fear, harassment, etc. At the most essential level, this means that mobile devices, applications and services must be safe to use, both in the short and long term. Any perceived safety risks of mobile technology – whether they have a factual basis or not – will result in either nonacceptance, or give rise to worrying, which can be distracting (and thus interfere with other values). Prolonged or severe worrying is likely to affect mental health/safety.

On the other hand, mobile technology can provide or enhance safety of persons or property, so applications and services that provide these are likely to be highly valued. New technology and services should enhance protection of people and their property wherever possible, e.g. by providing warnings of approaching danger ('you are approaching an area where muggings occur frequently'), identifying safe or safer routes and options, or summoning assistance. Technology should, in particular, aim to enhance safety of vulnerable persons, such as children, the elderly and disabled. This could be achieved through continuous monitoring and detection of states and raising alerts either directly to the person monitored ('your blood pressure is dangerously high – seek medical assistance') or to others ('a person previously convicted of sex offences is following your child').

Functions
Ubiquitous communication, information access and presence awareness are pre-requisites for detecting threats to safety, alerting users, and summoning assistance. Context awareness and adaptation is likely to be required to interpret the information correctly. Furthermore, natural interaction will be required since the users will include children and elderly people.

Dependencies
- *Belonging.* Since awareness about the physical or mental state of others, even when they are not close, will increase empathy and communication. The ability to provide assistance and support when needed reinforces relationships with others. All of

this contributes to self-actualization, since many people define themselves – at least partly – through their relationships with others.

- *Privacy*. Since constant monitoring generates data about people's whereabouts, activities and states.
- *Control*. This is an important related value as it is affected by who has access to the data and how they use it.

Constraints

Users must be able to feel a very high degree of confidence in such technologies. Services must always be available when they are needed and function reliably. Systems must be designed to safety-critical standards, and users and operators must be faced with simple, transparent choices and receive appropriate feedback – they should never have to worry whether they have operated the technology correctly when in an emergency situation. The design of such applications will have to consider (and test) the impact of safety alerts, because over-zealous alerting could have the opposite effect to the reassurance most users seek.

Some applications and services in this area can be envisaged as traditional mobile services, but most are likely to require cooperation or joint ventures with other public or commercial service providers, such as emergency services, security and healthcare providers.

Issues

For most people, most of the time, safety ranks higher than any other value. It is also a highly emotive issue, especially when vulnerable persons (and especially children) are involved. Other values – such as privacy and control – tend to be sacrificed when there is a perception of serious risk to safety. More research is required on the balance of perceived risk vs. privacy and control for applications, and under which circumstances that perception changes. Also, regulation or standard policies for acceptable use need to be developed regarding the use of data generated.

2.2.3.2 Belonging

Summary

This value has been captured by John Donne's observation that 'no man is an island'. Communication and relationships with others are the most basic human need after safety. Arguably, past successes of mobile communications have been almost exclusively sustained by meeting needs springing from these two basic values. Belonging goes beyond the traditional, synchronous, person-to-person communication provided by a phone call. Most individuals want to, and do, belong to a multitude of groups and social networks, such as family, local communities, social and professional peer groups. Applications and services must therefore support the interactions and activities that define these groups. Shneiderman [9] identified four types of high-level interaction that pertain to almost all 'belonging' activities: collect, relate, create, donate. However, different communities can have highly specific needs, so specific support for different professional and recreational interactions is needed. As well as supporting communication within groups and communities, support for communication across existing geographical, language and social barriers should be provided. Opening new channels of communication between individuals and groups is likely to reduce stereotyping and build shared social capital.

Functions
Satisfying this value may require ubiquitous support for one-to-one, one-to-many, and many-to-many synchronous and asynchronous communications and interactions through a variety of media (text, speech, graphics, still and moving images, 3D, haptic, olfactory). Ubiquitous communication, ubiquitous information access and presence awareness are key prerequisites to support initiation and maintenance of relationships and communities. To be truly effective, communication between participants must be supported by natural interaction.

Dependencies
- *Safety and security.* Since engaging in a relationship always involves making oneself vulnerable (by disclosing information, emotive state, etc.). Participants in relationships and communities must be able to trust each other not to exploit these vulnerabilities.
- *Control.* Since participants must be able to control access to communities and the information generated by its members.

Constraints
Technology must support interactions in an effective, efficient, and enjoyable manner – successful human communication and interaction requires focus on the communication partner. A high degree of availability, and quality of service necessary to sustain the type of communication and media used will be essential.

Issues
Whilst virtual communities have thrived on the early Internet technologies – such as chatrooms and the web – more advanced sharing and interaction will require more sophisticated virtual spaces. Sophisticated virtual spaces are currently limited to a small number of locations. Is it possible to support participation in virtual spaces whilst on the move? More advanced input–output devices and low-delay connections would be required to support this. Haptic and olfactory interfaces are still in their infancy; yet, for applications such as remote dating, smell is essential.

Integration and trans-coding of different media and languages will be needed to facilitate community building on a global scale.

Most people belong to more than one community, and many people want more than one identity to represent themselves in different communities and types of interaction – stay anonymous or hide behind different identities. This can be very liberating and empowering in perfectly legitimate contexts – e.g. some patients are more able to open up and discuss problems with a therapist when they can remain anonymous. However, anonymity and false or multiple identities can also be used for criminal purposes, which conflicts with safety – legal and/or policy frameworks will need to be created to regulate this. Certain services may need to operate stringent controls when issuing users with identities for participating in services, which increases cost.

2.2.3.3 Control

Summary
Not 'feeling in control' is a state that causes distress and anxiety in most people. This lack of control can be real or perceived. Ubiquitous information and communication access without sufficient control by the individual could lead to constant interruptions. As Meier [10] states, this must be avoided:

> 'Observation of human interaction suggests that a prime cause of stress in human behaviour is the appearance of signals or cues calling for the initiation of a new operation before the current one is completed. A choice must be made as to which is more important. At the present time the telephone and intercom systems almost always win, and a flurry of calls leaves behind a debris of incompleted sequences of behaviour upon which effort has been expended but for which personal rewards have not yet been realised. Increasing interruptions seem to be associated with increasing stress.'

The second major threat to control is access to data/information about individuals and groups, and their activities.

Functions
Individuals must be able to 'switch off' and 'opt out' when they feel like it; communities must be able to screen participants and restrict access. Personalization and customization can be ways of allowing individuals to control interruptions and disclose data selectively (essential to maintain privacy). Devices that can detect and adapt to contexts can reduce the need for users to remember to switch, but context recognition needs to work very well to be effective. This means accurate recognition of physical and social contexts, and user goals and emotional states must be possible before this can be contemplated. For personalization to be manageable, natural interaction is required.

Dependencies
Controlling one's availability is essential in one respect, but at the same time it prevents the person who is trying to reach that individual from satisfying their goal. Not being available when others want to communicate can jeopardize belonging, because unavailability and endless 'message-chasing' causes frustration in others. Individuals who over-control their availability are less likely to be open to spontaneous and serendipitous experiences and influences, which can, in the long-term, jeopardize self-actualization.

Constraints
Extensive customization and personalization of devices or service profiles may require too much effort from users. Also, if the user has to remember to switch between different states, it increases the mental load and the possibility of error.

Issues
The desire to 'switch off' and 'opt out' may conflict with social, professional and legal obligations – legal frameworks and/or policies may be needed to regulate availability for different applications.

2.2.3.4 Privacy

Summary

Privacy is often described as a basic human right – the need for privacy only arises when there is 'a public'. Mobile technology and services create a wealth of data that can be used to – intentionally or unintentionally – invade the privacy of individuals or groups. In the design of any new application or service, the privacy implication of data generated must be considered. Most invasions of privacy occur when data is used for a purpose other than the primary one supported by the application – e.g. when data are accumulated over a period of time to establish patterns of activity of preferences. Policies specifying the use of information (who can use what data for which purpose) must be established, and implementation and operating practice must strictly follow those policies. Policies and risks must be clearly communicated to users.

Functions

To give individuals the ability to guard their privacy, they need control, which will leverage personalization capabilities as a means to make personal choices about who they give access to which of their data, and for what purpose. Ubiquitous information access would allow individuals to locate their personal data, and establish who is using it, and for what purpose.

Dependencies

- *Safety.* Information needed to support safety (location, mental or physical state of a person, presence of others) can almost always be used for other purposes, many of which would be regarded as an invasion of privacy.
- *Belonging.* In communicating and interacting with others, we reveal information about ourselves.
- *Control.* The most basic means of protecting one's privacy is by controlling what information is released to whom, and for what purpose. Similarly, being over-protective of one's privacy can lead to missed opportunities for self-actualization and human capability augmentation.

Constraints

Privacy is subject to legislation, which differs in different countries. There is a fundamental trade-off between privacy and other values (see above). Much of the published literature on privacy – and most legal frameworks – currently concentrate on protecting certain types of data, without establishing what people regard as private information. Adams and Sasse [11] have provided a user-perception based model of privacy; essentially, to design for privacy it is necessary to determine what data is generated, and implement policies that clearly state who can use the data, and for what purpose. The policies, and any risks to privacy of individuals or groups, must be made clear to users before they commence a service, to avoid emotional backlashes when invasions of privacy do occur.

Issues

The relationship between technology and privacy is complex and more research is needed to understand when and how new technologies can afford invasions of privacy. In the past, privacy was often not considered in the design of technology, and only discussed after the privacy of individuals or groups had been invaded. Then, privacy is discussed in emotional – rather than rational – terms [12]. Legal prescriptions currently differ in different countries – how will this be enforced with different providers? Privacy-enhancing technologies have been a research topic for a number of years, but what has been developed is not sufficient to provide a balance between data subjects and data users. More radical ideas need to be investigated – e.g. data that self-destruct after a period of time, or agents that help users track down and 'reclaim' their data.

2.2.3.5 Self-actualization

Summary

When basic needs are taken care of, most humans look to achieve and create. Some people find self-actualization in their relationships with others but for many people, self-actualization means achieving goals in a chosen field of education, or in a profession. Many people also have a strong need to create – poetry, paintings, music, needlework. Such activity may be a form of communication – i.e. a way of expressing themselves and sharing that expression with others. For many, however, such activity may be primarily about an internal sense of achievement, both in the final product, and in mastering the skills needed to create it.

Functions

Ubiquitous communication and information access provide the foundation for self-actualization activities, since they allow people to access the human and data resources they need for these activities. At the same time, these capabilities prevent self-actualization. Ubiquitous communication can lead to constant interruptions, which interfere with goals and give people no time for concentration and reflection. Ubiquitous information access can lead to too much or too complex data, which can jeopardize the quality of decision making and/or prevent satisficing.

   Context adaptation and personalization can make technologies smart enough to recognize and support individual goals, while natural interaction allows people to focus on their goals and activities, rather than on how to operate the technology.

Dependencies

Since self-actualization is the highest-order value, for most people, the other – more basic – values have to be fulfilled before self-actualization becomes their main concern. This means a person's self-actualization activities must not be allowed to compromise their own, or others', safety, belonging, ability to control or privacy. It is highly likely that many individuals will consider, and opt for, human capability augmentation to reach goals they could not otherwise attain.

Constraints

The freedom to express oneself is limited by legal and social conventions.

Issues

Ubiquitous communication and information access can both enable and hinder self-actualization. The possibility of jeopardizing satisficing should be of particular concern, since it will not only lead to dissatisfaction among users, but affect individual and collective performance in the long term. Research is needed to determine how these need to be implemented to support people's activities in different areas.

Context awareness and personalization could be harnessed to find the right configuration for different people and activities. Dertouzos [13] suggests that machine learning techniques can be developed to support this, and support for managing multiple and/or conflicting goals.

Another area of research is how professional responsibilities are likely to develop. The more technology supports us in our professional activities, the more we delegate responsibility. When wrong decisions are made, who is responsible – the person who made the decision, or those who developed the technology that led him to make that decision? After human error judgements in aircraft accidents in the mid-eighties placed the blame on the pilots flying highly automated aircraft, today, designers of safety-critical systems have to assume responsibility when their design makes it difficult for a user to 'do the right thing'.

2.2.3.6 Human Capability Augmentation

Summary
The notion of human capability augmentation means that technology should amplify, enhance, or develop human capabilities – physical, mental, or social. Applications and services should aim to substitute when such capabilities are below par, or missing, in an individual or group. Mobile devices should include enhancements that wholly or partially substitute capabilities missing in older or disabled individuals – e.g. allowing partially deaf people to hear. Mobile devices should also aim to enhance human capabilities that do not function perfectly in most human adults; e.g. most adults do not have perfect memory, which can impair performance and/or lead to social embarrassment. Mobile devices could amplify human memory performance by, for example, discreet, automatic capture and retrieval of images and names of people and places, and thus enhance mental performance and social interaction.

Functions
Any functions substituting, enhancing or amplifying physical, mental, or social capabilities can support this value. There are already enhancements for blind and (partially) deaf people, and a range of technologies that allow severely disabled people to communicate; essentially, these are highly personalized natural interaction capabilities.

Dependencies
- *Safety.* Any such technology must be safe in prolonged use.
- *Belonging.* Human capability augmentation will enhance the ability to communicate and interact, and make some user groups able to participate in a much wider range of relationships and activities than before.
- *Control.* Individuals may want to choose when to use augmentation, and customize augmentation to suit their own needs and preferences.

Constraints

For many capabilities to be effective, functions have to be highly customized/persona-
lized – this carries a potentially high cost implication.

Issues

Human capability augmentation as discussed/presented is often considered the most
advanced set of values to tackle. Whilst this may be the case when aiming to augment
the capabilities of the 'average' human, augmentation for those with impaired physical,
mental or social capability is part of universal access, meaning that new applications
and services must, wherever possible, be accessible to all. Technologies and services
must not create new access barriers for specific social groups or those with impairments.
Human capability augmentation will raise a number of ethical questions that need to
be researched – e.g. should people be allowed to opt for an implant that improves their
capabilities, but at the same time shortens their lifespan?

2.2.4 *The Capability Plane*

If we look at the total user experience, it is determined by how well we meet some or all
of the user's needs. In order to meet the needs presented in the Value Plane of this model,
some basic functions (capabilities) need to be present in any system. This chapter describes
these basic functions, and tries to provide both a linkage between the Value Plane and
the Capability Plane, as well as a linkage between the Capability Plane functions and the
technical reference models (see Figure 3.2). In our initial analysis of the Capability Plane,
we have identified six key focus areas: ubiquitous communications access, ubiquitous
information access, context adaptation, personalization, natural interaction and presence
awareness. The description of each component in the Capability Plane follows the structure
given in Section 2.2.2.1.

Examining capabilities in more detail raises a number of issues. In this chapter, we will
only list a few of the most evident issues, some of which have previously been identified
in the *Book of Visions 2001* [4].

Considering the constraints for the different research areas, some similarities can be
found regarding those that deal with different aspects of information (e.g. all areas except
ubiquitous communication access and natural interaction). In each of these areas, laws
and regulations governing the use, ownership and distribution of personal information
will present significant constraints on the use of information in different situations and
the deployment of information based communication systems and services.

2.2.4.1 Ubiquitous Communications Access

Summary

With the development of ever smaller, faster and cheaper communications devices, they
will become embedded in everyday artefacts: clothing, walls, furniture, white goods,
cars, and many different kinds of handheld gadget that we will carry around with us.
It will make anytime/anywhere communication possible. Communication both between

persons and between persons and machines (or machine-to-machine for that matter) will be enabled.

Functions
- *Value Plane mapping.* Communication access available whenever needed is the basis for services aimed at satisfying the values of safety and belonging for the user.
- *System reference model mapping.* This focus area will link into the IP-based communication subsystem. It puts requirements upon things like network wide handover, distributed network decision making, integrated management and seamless dynamic service creation.

Dependencies
One of the conflicting dependencies within this focus area is the matter of privacy versus availability. When and how can the user be left alone if the whole world is communication enabled? How can the end user control it?

Within the Capability Plane, solutions from the natural interaction focus area will create new means of interaction to communication appliances that surround us. The area also has interdependency with the ubiquitous information access focus area, whereas the services building on the communication access feed on information access.

Constraints
The extent of deployment of communication systems in different areas of our society will limit the degree of ubiquity achieved.

Issues
In order to realize the notion of ubiquitous communication access, further experience and understanding needs to be acquired regarding issues of privacy, security, addressing and presence awareness.

There is also a need to explore the requirements and needs of users in different communication environments. How will the possibilities and complexity of ubiquitous communication systems be managed, both from an end user perspective and from a communication provider perspective?

2.2.4.2 Ubiquitous Information Access

Summary
Ubiquitous information access describes the notion that information should be available for use by services or consumption by users independent of the application at hand or interaction device in use. The data could be anything from personalization information published by a user, presence information available to a group of users, charging preferences from an Internet Service Provider (ISP) to digital content ready for consumption via wireless services. Data and metadata should be understood by all services, regardless of underlying data formats. In the world of ubiquitous information access there is also the notion of autonomous agents processing information on behalf of users or services. Thus, automating tedious information gathering tasks or negotiations for the best deals for a particular product or service.

Functions
- *Value Plane mapping.* When looking at services satisfying the user needs of belonging and control, ubiquitous information access is a key. Being able to share data between a closer community of users, or always being able to get information of the present location are just examples of the types of service in this area.
- *System reference model mapping.* The function area will link into the Service Semantic level of the WG2 framework. Requirements will, for instance, be on meta-definitions of services, service information and content, media adaptation and knowledge representation.

Dependencies
This focus area shares the same conflicting dependency of privacy versus openness as most of the Capability Plane focus areas. There is a threshold when the information I convey about myself is starting to inflict upon my personal integrity.

Another dependency concerns the availability of the information versus the ownership of the information. Should all information be available to every entity present within the ubiquitous communication world, and what is the price associated with using information? Who is the owner of my location and other context dependent data that describes my current situation?

As potential sources of information to be shared, there is a dependency relationship between personalization, context adaptation and presence awareness with the ubiquitous information access area.

Constraints
No unique constraints identified.

Issues
Credibility and trust are two perceived qualities that are important concepts for all systems. Credibility equals believability and takes presentation of information into account, while trust is about having a positive belief in the objects and/or processes. If these factors are not considered properly, it may affect efficiency, and even usage of an application, negatively.

There is also a need to explore and define the user requirements and technologies available for handling the complexity of all data that will surround us in the future. Will there be a need for virtual space management tools like avatars and virtual personal assistants in order to cope with information?

Other issues important to this area are information security, privacy, control over information, information overload, adaptability, quality, information management, Digital Rights Management (DRM), publication, information access and active vs. passive sharing of information.

2.2.4.3 Context Adaptation

Summary
Context is information that can be used to characterize the situation of an entity – the entity being a person, location or object – that is important to the user-interaction with an application. This, of course, includes the user and applications themselves.

A system is context-aware if it uses context to provide relevant information and/or services to the user, where relevancy depends on where the user is, what he is doing, what he has done or is thinking of doing in the future. Also to detect change, the system has to remember past values. Thus, context-aware systems are about what is currently happening in the present situation, as well as maintaining a context memory about the past and a predictive model of the future.

Context adaptation is about adapting capability (of a service, device or the communication network) and information to elements of context information, including location, situation, personal role, task and environment.

Functions
- *Value Plane mapping.* Towards the Value Plane this capability is used for services towards the values of Security and Human Capability Augmentation.
- *System reference model mapping.* This area hooks into the Service Semantic as well as the generic services elements/service platform within the WG2 reference model.

Dependencies
Again we face the conflict between privacy and openness. Links can also be found to areas such as ubiquitous information access, personalization and presence awareness.

Constraints
No unique constraints identified.

Issues
Privacy issues underlie most context-aware applications: the more the application knows, the better it can perform, but the more it knows, the more invasive it is of privacy. We also need to trust the information to be credible.

Another issue concerns the consistency and predictability of data – how to infer appropriate action based on data concerning the user's present context and comparing this with historical data of the user's past context.

There are many issues related to interpretation of contextual data. There will always be possible cultural differences in interpretation of contextual data. Even within the same cultural setting, the same contextual data may be interpreted differently by different users depending on their unique point of reference. This points to the need to intelligently translate contextual data into a form with relevant semantic meaning to users and applications. As humans we have senses of sight, hearing, touch, smell and taste. Current contextual sensors can only capture a tiny part of these.

2.2.4.4 Personalization

Summary
Personalization is very closely coupled to the capability of context adaptation. One can see that mobile phone users are showing a great interest in customization of their phones with different coloured covers, ring tones, background images, etc. All of this effort is to make their mobile phone stand out and mirror their own personality. This is also starting to be reflected in the services accessible through the fixed or mobile Internet.

In an attempt to build customer loyalty, personalization is used to build a meaningful one-to-one relationship by knowledgeably addressing each individual's needs and goals in a specific context [14].

To be able to provide a service tailored to a user's specific needs, concepts including ambient information, content filtering, user profiles and adaptation to terminal capabilities play an important role. Traditionally, the implementations of personalization have been based upon the use of user and service profiles. Now, new elements can be considered to incorporate a sense of context-awareness into services. For example, concepts like the use of intermediaries [14] can be employed to create a dynamically adapted personalized mobile service.

Functions
- *Value Plane mapping.* Self-actualization and belonging are two values addressed by the use of the personalization capability.
- *System reference model mapping.* The function area will link into the Service Semantic level of the WG2 framework. Requirements will be on profile format and categories, profile-learning functionality, standards to exchange profiles and secure privacy-sensitive parts, etc.

Dependencies
Service personalization is not limited to content and information only, but needs to address service logic and communication mechanisms as well. We find close dependencies with both ubiquitous communication and information access, as well as context adaptation.

Constraints
No unique constraints identified.

Issues
Specification and modelling of generic and dynamic personalization information and representation are issues to consider. Privacy, security, cultural variations, consistency and predictability, information filtering and fusion are additional issues that need to be highlighted within this capability area. Others cover dynamic information gathering, history and behaviour coverage, knowledge-management and personalized assistance.

2.2.4.5 Natural Interaction

Summary
In a future where people are surrounded by means of communication and information access at any time, there is a need to encapsulate the underlying, sometimes complex, technology with intelligent intuitive interfaces that are embedded in all kinds of objects. Within natural interaction there is an extension of traditional user interfaces using more of our senses (speech, vision, touch, smell, taste etc), i.e. multimodal interaction. We also include issues like sensors, virtual and augmented reality. The interaction type should always be optimized and appropriate to the context of use and personal preferences.

Functions

- *Value Plane mapping.* It is obvious that natural interaction capabilities will help enable services offering human capability augmentation (e.g. information presentation tailored to the user's sight capabilities) as well as safety (e.g. interaction with mobile services while driving a car). There are also possibilities to satisfy values around the control of self-actualization.
- *System reference model mapping.* Within the WG2 framework, natural interaction will link into the Service Semantic level. Requirements will be on adaptation and personalization.

Dependencies

Incorporating natural interaction capabilities into mobile services, there is a need to interwork with capabilities within the ubiquitous information access, personalization and context adaptation areas.

Constraints

The possibilities within this capability area are mostly controlled by the evolution of the technologies at hand, some of which do not even exist today (e.g. olfactory input and output devices).

Issues

Broad research issues exist, such as quality, device technology, sensor and recognition techniques, ease of use, user group dependencies, cultural variations and mixed modality interaction. How will new interaction techniques like haptic interfaces, multimodal interaction, conversational and ambient interfaces change the way we experience the ubiquitous communication and information space in the future?

There is a need to further understand the implications of body area networks and how they will enable more natural user interaction models in the future. Also, augmented and virtual spaces are two neighbouring areas where novel interaction techniques will be a necessity.

2.2.4.6 Presence Awareness, Including Personal Safety and Security

Summary

Presence is not something you measure; it is a subjective experience that you must feel. However, information about a person's whereabouts or the occurrence of an event can help to create a feeling of presence for the user receiving the information.

Presence can be divided into two categories: social and environmental presence. Social presence has been defined as 'the extent to which other beings also exist in the world and appear to react to you'. Environmental presence has been defined as 'the extent to which the environment itself appears to know you are there and react to you'. So far, presence features for communication services have been strongly associated with the first category within different instant messaging applications. Applications include features to enable the user to see whether or not other users within a group are available for communication. This information can be actively or automatically published to all persons within a group. The second category applies more to the context adaptation focus area and feeds the system with information about the user's whereabouts. For more information refer to Lombard and Ditton [15].

Awareness during a communication session is sometimes referred to as *virtual presence* or *telepresence*. In certain communication applications, a more or less detailed representation of other persons is continuous throughout the whole session [16, 17]. Continuous virtual presence, both within and outside communication, is a form of social awareness.

Functions
- *Value Plane mapping.* Maps into the Value Plane regarding belonging, self-actualization, control and human capability augmentation.
- *System reference model mapping.* This function area will link into the generic service element and service platform subsystem of the technology reference model. It puts requirements upon things like environmental monitoring and service control.

Dependencies
There are two design trade-offs that presence-based applications face. The first one is informativeness vs. privacy and considers how much information about a person's status can be shown before it violates the privacy of that person. Increasing the privacy most of the time means decreasing the presence information. The second trade-off is overhead vs. control and is about the amount of effort that is required to maintain the accuracy of the presence information. A largely automatic presence update mechanism will require little overhead work, but will also give the user little control over his/her own presence.

Looking at presence information, the concepts of credibility and trust are highly important [18]. A user must be able to trust the presented information in, e.g., a buddy list. In order to do that, he/she must fully understand it. The user must also feel certain about how much presence information is displayed to other users. Fast updates of both your own and others' presence information are required for the information to be perceived as credible.

Natural interaction will also impact upon how well the end users will adopt presence awareness capabilities.

Constraints
Regulations defining ownership for presence information may constrain the ways in which this information can be revealed and used.

Issues
Issues around presence awareness deal with unobtrusiveness, privacy, identification, authentication, trust, ethics, collection and management of data, data fusion and information overload, to name but a few.

2.2.5 Summary and Conclusions

In this section, we have described a proposal for a user-focused reference model for systems beyond 3G. The general structure of the proposed model involves two 'planes': the Value Plane and the Capability Plane. The two planes offer the opportunity to reveal characteristics at different levels of abstraction that are relevant to a human-centred view of wireless systems. The Value Plane addresses the core human needs, such as safety or belonging, that products and systems need to support. Addressing these core needs demands that certain functionalities exist in the system. The functionalities are the subject of the system Capability Plane. The system Capability Plane places requirements on the system, e.g. applications and services that are needed to realize the needed capabilities. In this way, this reference model is envisioned to link into other models that describe how these applications and services are enabled.

The structure of our reference model provides a framework for describing the characteristics of each component in the model and its relationship to other components. The definition of our components is driven by the enumeration of system requirements and capabilities resulting from analyses of a catalogue of future wireless world scenarios. Thus, our reference model is ultimately a reflection of the requirements as derived from user-focused scenarios.

To describe the components of the model, we have adapted the format and formalism of the User Environment Design (UED) model. The UED provides an excellent starting point as it is intended to document the organization of a system from the user's point of view and capture the structure and function of the system without straying into the realm of implementation. This approach keeps our focus on a high-level view of what the system does.

We hope that this model will serve as an appropriate artefact for continuing discussions on critical issues and research needs to enable a compelling and user-centred vision of the wireless world of the future.

2.3 The Use of Scenarios for the Wireless World

2.3.1 Introduction and Objectives

Increasingly over the past several years, manufacturers, carriers, analysts and various industry pundits have offered visions for what the future of mobile communications will be like. These predictions have become more numerous as 3G systems have become real; it is quite fashionable to predict what might lie 'beyond 3G'. Rather than propose yet another set of scenarios for the wireless world, we instead pause to listen to what has already been said about the future, and attempt to make sense of it. Are the futures people are painting wildly divergent, or are there commonalities that may be identified? What are the implications for what has been proposed – can we find a way to agree on methodological grounds how to assess and validate emerging visions?

With this as prelude, the accomplishment of four tasks is attempted.

1. We survey several of the more well-thought-out and user-centred visions that have been proposed for the wireless world, and catalogue their common aspects.
2. We explore ways to manage the analysis of these scenarios in order to derive key technological requirements for their implementation.
3. We propose key research questions that will need to be answered to implement the identified common visions for the wireless world. (This task has to be completed in the next versions of the white paper).

4. We propose an implementation model to support the management of research and development projects that draw from the scenario-based design approach and, moreover, from the UCD approach as outlined from the WG1.

2.3.2 Scenario Catalogue

To prepare for this effort, a number of scenarios for the wireless world were reviewed. A large majority of the 'scenarios' or 'use cases' focused explicitly on 3G or earlier technologies, mostly to make a business case or sell equipment for 3G. A large proportion of those focused almost entirely on technological capability, not user benefits. We noticed a dearth of 'validated' scenarios; there was often no evidence that scenarios or visions were based on real user data. Much of the 3G hype was devoted to 'more is better': more connectivity, more 'bandwidth', faster connections, without any real thought given to how communications could fundamentally change with these new systems.

In contrast, we did find several scenarios that did go beyond the standard 3G fares. These scenarios were, in the opinions of the authors, the most well-developed and thought out, and also the most user-centred. We chose this subset for further analysis. The list of these scenarios and their sources is shown below.

- ISTAG scenarios for ambient intelligence in 2010 [19];
- 'Usage scenes of the future mobile communication systems', Mobile IT Forum [20];
- ETSI/TIA Project MESA scenarios [21];
- MIT Project Oxygen scenarios [22];
- NTT DoCoMo 'Vision 2010' scenarios [23];
- University of Oulu/Project Paula CyPhone 'Taxicab' scenario [24];
- Wireless World Research Initiative scenarios [25][1].

Before we review these scenarios, a word is in order about what we mean by the term 'scenario'. Unfortunately, the term is overloaded and has many meanings. In our usage, we follow the human–computer interaction literature and define scenario as a narrative describing a specific user interacting with a product, service, or system to do specific things (e.g. see [23]). This is in sharp contrast to the 'scenario' used in 'scenario planning', which is a more macroscopic view of larger scale market trends. In our usage, scenarios are stories about real people doing specific things.

2.3.2.1 Scenario Overviews

In this section, we describe, in overview, the scenarios that were analysed. The interested reader is referred to the sources for additional information. The scenarios listed are just a small cross section of scenarios available, additional scenarios, i.e. the Wireless World Research Initiative scenarios, are listed in Chapter 8.

---

[1] Recently, the WWRF developed a set of scenarios – 'Blue', 'Red' and 'Green' – about how the wireless market might develop over the years up to 2010 as part of the Wireless World Research Initiative. These scenarios are discussed separately and in more detail in Chapter 8. However, these scenarios were not available as of the time of the analysis described in the following sections and thus will not be further discussed here.

ISTAG Scenarios

The IST Advisory Group (ISTAG) sponsored a set of scenarios to be developed in order to hasten the development of information and communication technologies (ICTs), and give European development efforts heightened focus on future services. The scenarios themselves were developed in 2000 by the IPTS (part of the European Commission's Joint Research Centre), DG Information Society and with the active involvement of 35 experts from across Europe [19].

The ISTAG scenarios describe life with what they term 'ambient intelligence'. In this world, users are literally 'bathed' in connectivity and intelligence, having full access to a range of services and applications anywhere, at any time. Four scenarios were developed [19].

Mobile IT Forum Scenarios

The Mobile IT Forum is an industry consortium, sponsored by Japan's ARIB. The forum conducts studies and research on technologies and standardization. The forum's goal is to realize an early implementation of future mobile communication systems, including systems beyond IMT-2000 and mobile commerce.

The mITF scenarios are excerpted from the forum's Information and Communications Council 'Outlook for Future Mobile Communication System'. The scenarios take the rather unconventional form of cartoons. There are nine scenarios, covering aspects of daily life, work and medicine.

Project MESA Scenarios

Project MESA is a joint ETSI/TIA effort aimed at the future of public safety communications. MESA stands for Mobile Broadband for Emergency and Safety Applications. Project MESA began as an ETSI and TIA agreement to work collaboratively by providing a forum in which the key players can contribute actively to the elaboration of MESA specifications for an advanced digital mobile broadband standard much beyond the scope of currently known technologies.

The Project MESA 'scenarios' are the least user-centred of the scenarios considered here. However, they were included for breadth of coverage, as they form one of the only future-oriented treatments of public safety communications available.

MESA's scenario areas include:

- *Remote patient monitoring.* To provide effective medical assistance to injured persons allowing timely and correct treatment of patients. This is of interest to emergency management, emergency medical services (EMS), military, and other similar governmental functions that have a need for aeronautical and terrestrial, high-speed, broadband, digital, mobile wireless communications.
- *Mobile robotics.* Robots designed in both micro- and macroscale may be used to assist such applications as rescue of people from hazardous areas, automated inspection of nonaccessible areas, antiterrorist actions, urban warfare and land mine clearing. Interested parties include all criminal justice services, emergency management and medical services, fire, land, natural resource and wildlife management, military, transportation, and other similar governmental functions that have a need for aeronautical and terrestrial, high-speed, broadband, digital, mobile wireless communications.

- *MESA firefighter.* To provide services and applications that will reduce the personal risk to firefighters, particularly in tall building fires. Interested parties include all emergency management, emergency medical services (EMS), fire, land management, natural resource management, military and other similar governmental functions that have a need for aeronautical and terrestrial, high-speed, broadband, digital, mobile wireless communications.

MIT Project Oxygen Scenarios

Oxygen is a project housed in MIT's Laboratory for Computer Science (LCS). The Project's slogan is: 'Bringing abundant computation and communication, as pervasive and free as air, naturally into people's lives.' Oxygen's vision is one of pervasive, human-centred computing, similar to the ISTAG's ambient intelligence. MIT's specific vision is one of ubiquitously available networks, dubbed 'N21s'. Users can avail themselves of services using 'H21s', the handheld devices in this future. In addition, computing devices, called 'E21s' are embedded in everyday objects and communicate across the N21 networks.

MIT's scenarios (see their website) include:

- *Business conference.* Three co-workers collaborate using rich media on three continents.
- *Guardian angel.* Jane and Tom, an elderly couple, are able to stay in their home through an augmented environment.
- *Field trip.* Students take a field trip to New York City, where they visit a museum, check into a hotel, see a Broadway play, etc.

NTT DoCoMo 'Vision 2010' Scenarios

As part of a far-reaching corporate strategic plan, NTT DoCoMo have assembled an extensive video scenario set entitled 'Vision 2010'. The scenarios follow a Japanese family from the 2003 world of 3G systems to a 2010 world of the same sorts of ubiquitous computing and communication that appear in the ISTAG and MIT scenarios.

The NTT DoCoMo 2010 scenario scenes include the following, excerpted from the video:

- *Multipoint video conference with simultaneous language interpretation.* Keita is attending a conference with people from different countries. They are connected by a convenient mobile network and different languages are simultaneously translated into English. Location and language will no longer be barriers to education thanks to mobile networks supporting automatic interpretation technology.
- *Point-to-point 3D telepresence.* Mami is talking to her boyfriend Jean via the wall-mounted display that is wireless connected to the mobile terminals. The three-dimensional images of Jean are displayed to provide a 3D telepresence. Wall-mounted displays will support communications with mobile terminals and project three-dimensional images. Users will be able to talk to these images as they would to an actual person.
- *Multiparty mobile telepresence.* Today is Mami's wedding day. The whole family is gathered in Paris. Her best friend from high school, Tomoco, and her loving grandmother, were not able to be in Paris for the wedding, but they joined the party from Tokyo. By the advances in mobile communications technology, people are able to keep in touch no matter where they are.

- *Virtual golf game.* My husband is playing a virtual golf game with a friend at another location. The swing can be replayed in the virtual game. A network of wearable PC technology will allow people to participate in a communal virtual environment. You will be able to play games with people at every location.
- *Automatic driving.* We are going out to buy wedding gifts for our daughter Mami. 'I want to go to Banlamphu market, choose the widest road, please', my husband speaks to the digital assistant. 'Roger, it'll take about 25 minutes', the digital assistant answers. Navigation systems transform to communicate with both the car and the road, using personalized information to realize automatic driving.
- *Smart coffee kiosk.* Mami is using the mobile terminal on her finger to order a cup of coffee. Her personal data is then identified through the secure identification system and the coffee is then ready to collect. Contact mobile terminals will communicate with user identification systems, they will also support the exchange of customized data.

University of Oulu CyPhone Scenario
CyPhone is an interdisciplinary project located at the University of Oulu in Finland. The CyPhone concept is a hybrid wearable computer/smartphone. The CyPhone scenario includes:

- *Airport indoor service discovery.* For example, finding a printer in a public space.
- *Outdoor hotel service discovery.* Discovering hotel rooms to rent.
- *University indoor navigation.* Finding a room using conformally mapped guidance.
- *Outdoor hotel booking.* Reserving a hotel room.
- *Electronic taxi fee payment.* Electronic payment of a taxi fare.

2.3.2.2 Scenario Analysis

In this section two approaches have been taken. The first is to try to analyse the above collection of scenarios as a whole for patterns and commonalities, and the second is to propose ways to analyse individual scenarios.

Scenarios: Patterns and Commonalities
In order to assess common themes in the collection of scenarios outlined in the previous section, we compiled a list of over 80 individual discrete elements. These concepts appeared in more than one scenario and were user behaviour, environment or technology based. The final goal is to propose a general purpose taxonomy of socio-technical scenarios that will foster the collecting and processing of data, according to one or more schemata of representation. This is a typical academic approach that might feed forward basic research in the field.

The most common elements in this analysis are outlined in Table 2.1.

Not surprisingly, many of the same themes emerge from this analysis as were postulated for 3G: a heavy emphasis on multimedia and associated technologies (e.g. display), interoperability, advanced input/output user interface technologies (e.g. natural language), intelligent, autonomous agents, ubiquitous interconnected networks, etc.

In a later section, we will return to these themes and propose key research questions in these areas.

Table 2.1   The most frequently named common elements across the scenarios

Common element	Frequency
Wireless device used for communication and network access	20
Visual and display technology	18
Seamless hybrid wireless and hardwired network	13
Augmented objects	12
Reliable broadband wireless	12
Secure encryption and deletion	12
Biometry or chip implant	10
Device coordination	10
Ease of use, intuitive	9
Speech technologies	8
Wearable personal device	7
Intelligent agents, agent technology	7
Adapts to owner through observation and/or personalization	7
Intelligent infrastructure or environment	7
Natural language comprehension and translation	7
Interoperability between various wireless protocols and/or platforms	6
Personal software assistants	6
Ad hoc network management	6
Health monitoring	6
Government support for infrastructure	6
Secure ID	5
Sensors in environment	5
Multi-X (voice, pattern, object emotion, language, recognition	5

2.3.2.3 Individual Scenario Analysis

Guidelines for Scenario Analysis Management
In the following, we define an incomplete set of guidelines that can be followed to manage
the analysis of scenarios.

- A scenario analysis should enable one to single out differences between the current
  socio-technical status and that following the technological innovation.
- A scenario analysis should highlight relevant and significant changes from the user(s)
  perspective when moving from one status to the innovative one. For instance, it can be
  useful to investigate aspects concerning communication patterns, behaviours, working
  habits and cooperation modalities.
- Further, the scenario should investigate and supply with documentary evidence why the
  changes can be accepted by users and what makes the changes acceptable to the users.
- A scenario analysis should allow one to identify key actors and roles within a group
  or community that will be involved in the innovation processes. The analysis should
  emphasize different points of view when necessary.
- A scenario analysis should put into evidence the interdependencies among wireless
  and wired technologies in dynamic situations and contexts of use, so that users may
  usefully plan actions and not be faced with surprises.

- A scenario analysis should be a creative exercise and therefore it should come from a collective thinking. Roles within groups and communities should be identified and involved in the scenario(s) development process.
- A scenario analysis should be validated by more people than those involved in its development, according to representative criteria.
- A scenario analysis should foresee the most convenient way to represent data, results, and hypotheses. For instance, animation and simulation of scenarios offer the most powerful way to build credibility and persuasion.
- A scenario analysis should give place to the agreement of a set of requirements. In order to achieve this result, a proper set of tools has to be identified so that different analysts can obtain similar and comparable results at different levels of familiarity with the domain.
- A scenario analysis should incrementally specify the proper level of detail. In other words, beyond the definition of requirements, it should also trigger data collection.

Applying Static and Dynamic Modelling to Mobile Services

Static Modelling
A static mobile scenario reference model is demonstrated in Figure 2.9. The three building blocks, value, capability and service provisioning, together with the building blocks communication and access, construct the static reference model. In this model the emphasis is placed on mobile service value functions, shown at the top of the model. The user-centred mobile services can then be developed and deployed with the employment of this five building block model.

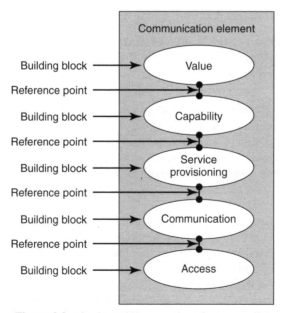

Figure 2.9   Static mobile scenario reference model

1. *Value.* The value function deals with user-centred mobile service development. It emphasizes the user by placing the user in the centre of the cyberworld. All the mobile services are driven by user values. Research in this area includes studies to show how the key characteristics of mobile users form the rationale for developing services that are wanted and easily approved by the consumers. Mobile users can be modelled relying on the analysis of cognitive cultural schemes, emphasizing transfer, social group factors and motivation. Quality function deployment (QFD) is a powerful method to collect, prioritize and balance user needs and effectively build them into the mobile services.

2. *Capability.* Mobile service application capabilities reflect the internal and external mobile service requirements and characteristics. Mobile service application capabilities consist of the following functions: ambient awareness, personalization, natural interaction and adaptation. These application capabilities represent implementation requirements and characteristics of mobile service applications. They address the values of the service system or network beyond the user needs and act as a linkage between user values and mobile service development.

3. *Service provisioning.* Mobile service provisioning provides developer automatic provisioning of, and access to, wireless value-added services for application development, testing and deployment. Mobile service provisioning can be divided into application support and application execution. The application support layer includes typical middleware products like application servers, virtual machines, web servers, databases and run-time libraries. It also covers other necessary tools and adapters. The application execution layer consists of all the basic services allowing the service developers to focus on the essential.

4. *Communication.* Communication deals with connection and data transmission between networks and mobile terminals. It involves connectivity and reachability functions, caching, proxy, gateway, and adaptation functions in network entities.

5. *Access.* Access comprises mobile terminal access and network access. Unified interfaces are needed for different access technologies in terminals. Both types of access have data transport with signalling, RT, NRT and BE classes and related control functions.

Dynamic Modelling
Dynamic modelling indicates the dynamics of the objects and subsystems and their changes in state. By exploring the behaviour of the objects over time and the flow of control and events among the objects, the dynamic model captures the essential dynamic behaviour of the system.

In mobile scenarios, dynamic modelling concludes by building enhanced mobile scenarios between the subsystems and events. In contrast to the object modelling technique (OMT), a subsystem is treated as a black box, i.e., the interactions are with the subsystem as a whole and not with its internals. However, in a mobile scenario, several environment objects or complete subsystems may interact with the subsystem being analysed. A mobile scenario may represent a complete operation, a set of operations, or only a part of an operation. It shows valid sequencing of events observable in a particular situation, which may be an error situation. A mobile scenario may show the time outs between the events, the changes in the subsystem states and also the expected suboperations carried out by the subsystem between a sequence of events. The mobile scenarios are built for all important and complex operations of the subsystem, including the error cases where the handling of errors is not trivial.

The message sequence chart (MSC) notation specified in ITU standard Z.120 (Z120-94) is used to present the mobile scenarios. The standard also provides a textual notation, but here the description is restricted only to part of the graphical notation and it is demonstrated using the following CyPhone example.

Putting Static and Dynamic Models to Use: CyPhone Example
We use a personal navigation example to find out interesting places, such as museums or shops offering discounts, in both unfamiliar and familiar cities. This kind of application can be easily implemented with GPS and orientation sensors.

Let us have in our scenario a hungry user, who wants to eat some fast food in an urban outdoor environment. The user already has his user profile in the CyPhone, this details that the user prefers some sorts of fast food. He selects a food category in his phone, which starts to calculate the shortest routes for a few closest places (the maximum searching distance is defined in the user profile). Then the phone can give one to three choices for the user. For example, one option is to go to the nearest place, and another to go a little further, where a special discount, suitable for the student profile, is offered. After selecting the destination, the shortest path to the destination will be displayed, along with other information like distance and estimated travelling time. In our scenario, a screen shot of the navigation service is given, which is the view through a see-through head-mounted display. On the top left corner there is a map showing the route with red dots. Street names, distance to destination and estimated time of arrival (ETA) are shown on the top right corner.

The vertical axis in the dynamic ladder model in Figure 2.10 represents increasing time from top to bottom, and hence interactions are understood to follow each other in time. Communicating entities (the subsystem being analysed and objects and other subsystems from the environment) are represented by vertical lines ending with small solid rectangles, whereas their names are entered into boxes attached to the top of the lines. An interaction (event, message) between the entities is represented by a normal horizontal arrow and is labelled by a name. The time out is represented by an arrow symbol and is labelled by the value of the time difference. Suboperations are shown in normal boxes attached to the entity that performs them. The box with angular sides is the condition symbol that can be used to describe a state of an entity or of a group of entities. In the latter case, the symbol is extended over the entities' symbols. If an entity is not involved with the condition, then its symbol appears through the condition symbol.

Reading the dynamic ladder model is easy. In Figure 2.10 we have five communicating entities, namely User (with HMD support), Access (CyPhone), Communication, Cyberworld (Food store) and Clock. The user first turns on the CyPhone by the message 'Power on CyPhone' and it is ready within five seconds. This is then followed by the command 'Search for shortest fast food store' to the CyPhone. The phone begins to search the user profile food data, which is preinstalled in its memory. After processing and communication with the selected fast food store, it will calculate the shortest route to the store and the travelling time. These messages are sent back to the user and displayed through the HMD. This whole process takes three seconds to complete. The CyPhone then sets ten

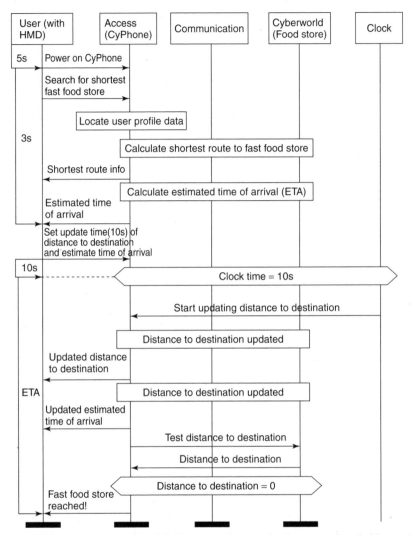

Figure 2.10 Dynamic model of CyPhone outdoor navigation scenario – ladder model

seconds as the regular updating time for distance to destination and estimated time of arrival. Every ten seconds the above calculating process will repeat, and updated information is fed back to the user. When the distance to destination reaches zero, the user gets to the fast food store and the navigation scenario is complete.

The dynamic ladder model clearly describes the operation of a communication entity across its application boundary and addresses the real-time, reactive aspects. It explains under what conditions the operations are performed and how long they are allowed to take. It captures the order of interactions between the entities and their environment necessary to perform an operation, shows what operations are possible in different states and how different operations affect each other.

2.3.2.4 Scenario Key Capabilities and Research Questions

In the previous section, we outlined several different ways to analyse individual scenarios to elicit key requirements. It is possible to identify key capability groupings that are implied in the scenarios as they stand today, based on the common elements identified earlier.

Before going further, however, it should be noted that there is an inherent methodological inconsistency with the concept of user centred design if we actually specify systems based on this type of meta-analysis. Instead, the purpose herein should be to identify common proposed sets of functionality and, in so doing, identify the key research questions that should be addressed in the process of specifying the systems.

With this in mind, let us return to our meta-analysis of the scenarios. The following is a more condensed list of the most often included capabilities:

- ubiquitous, interoperable connectivity;
- visual communication and telepresence;
- intelligent and autonomous agents, 'ambient awareness';
- device coordination/synchronization;
- ease of use.

While it is recognized that this list is somewhat high level and sparse, there are still important human-related research questions that should be addressed, even without knowledge of specific application designs and use cases.

- Ubiquitous, interoperable connectivity
  — How ubiquitous is ubiquitous? In other words, how will users react to coverage 'holes'? Is there a usage model that may dictate less than 'perfect' service coverage?
  — Privacy vs. ubiquitous access. Users will want to protect themselves against intrusions in their daily lives, yet also want to be always connected. How much connectivity is too much? What security/preferences/profiles/agents can be developed to help filter user access?
- Visual communication and telepresence
  — What quality levels (refresh rate, resolution, colour depth, compression, latency) are required to achieve usable/desirable telepresence? Are there other aspects that need to be added to the experience, above today's videoconference experience, that will finally make it like 'being there'?
  — What are the value functions in multimedia? Is there any additional value in visual communication?
  — When does a user really want to make a communication 'just like being there'? When does a user NOT want that?
  — How to negotiate calling party vs. called party desires to be seen?
  — Privacy and seeing vs. being seen.
- Intelligent and autonomous agents, 'ambient awareness'
  — How will people react to an intelligence that is always there – will it be satisfying or disturbing?
  — How will the ambient intelligence react to inevitable sensor errors, and how will users react to the environment's 'illogical' results?

- — How will the user develop trust with autonomous agents acting on their behalf?
- — What level of protection will users want for themselves in dealing with the ambient awareness?
- — What is the interaction paradigm with the awareness? Speech/haptic/text?
- Device coordination/synchronization
  - — How will people manage their digital possessions across multiple devices and locations? How will people ensure security of their sensitive information in a world of ubiquitous connectivity and device coordination?
- Ease of use
  - — How to design natural and easy to learn and use interaction styles for the wireless world.
  - — How to include all people – differently-abled (blind, deaf, physically handicapped) and the elderly in the wireless world. Designers must design for all levels of cognitive ability.
- Technology overlaps
  - — Ambient awareness. How to acquire ambient information. How to process ambient information.
  - — Context adaptation. What technologies can be used in location and situation dependence adaptation?
  - — Movement toward richer communication. What are the value functions in multimedia?
  - — Importance of latency in the application. Real-time vs. non real-time.
  - — Personal devices and public devices. Enhanced security protocols and authentication systems for public devices.

2.3.2.5 Towards an Implementation Model of Scenarios in the Design for Innovation

Scenarios are Managerial Tools

The contents of this section are connected to the contents of Section 2.1. A schema of the model is proposed in Figure 2.11.

Scenarios can be envisaged and generated both for demo purposes and for assigning values to visions of cultural innovation. In both circumstances, scenarios can be considered as tools that help to manage the technological innovation at different levels of project development. Further, they are helpful in creating cohesion around innovation perspectives, by appealing to shared cultural values and use requirements. The analysis of scenarios can play a powerful role when it is used as a way to iteratively increase awareness on current technological limitations and emerging patterns of behaviours within the

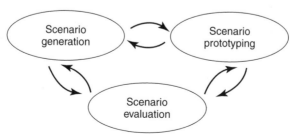

Figure 2.11 Relationships between the various phases of scenario analysis

referred population of users. Therefore, it is worth defining an engineering approach for the management of scenario analysis. In the following we briefly discuss three phases of scenario analysis management, as outlined in Figure 2.11.

Scenario Generation

The generation of persuasive scenarios for engineering and designing purposes is a creative process that can be derived from a careful assessment of a given previous situation. Without an analysis of current limitations and conditions, it is very likely that innovation and visions of change will be technology-driven. In those visions, humans have to adapt themselves to the technological changing conditions, and not vice versa. Moreover, given our bounded rationality and limited capability to keep a coherent set of preferences, we cannot rely on methods that just elicit preferences and beliefs. The generation of scenarios has the potential to overcome some of these limitations, as it allows one to compose different views and to augment them throughout the design process, by allowing lifecycle participation and vivid descriptions of end user experiences.

The number of scenarios, and of details that are necessary to depict use situations can be considered as a function of the design phase's level. For high-level design, a few scenarios and details may be enough to solicit thoughts and reactions. These scenarios could be based on the assessment of current technological limitations and highlight opportunities of change by analysing viewpoints and typical problems that innovation could help to tackle. In other words, the generation of high-level scenarios could fruitfully be descended from socio-technical analyses that help identify typical scenarios and critical scenarios of use. Typical scenarios are the most representative and frequent situations. Critical scenarios are, instead, rare situations in which surprises (especially bad ones) are caused. Furthermore, a sort of hierarchy of problems and related priority might give place to a corresponding hierarchy of scenarios.

Depending upon the design phase's level, the proper set of techniques can be selected from among the manifold methods that are available. For instance, ethnographic studies are very useful and relatively easy to conduct at early phases of scenario analysis. At more advanced levels of analysis, techniques such as 'task analysis' might be more appropriate to feed forward scenarios generation. This means that as the design gets to a more comprehensive level, more requirements are defined and, therefore, more details need to be clarified.

Scenario Evaluation

There is no better way to clarify the details during a scenario analysis than passing through an evaluation phase. At this phase, one of the main goals is to let a collective vision emerge in order to get participation and prepare the ground for future acceptance of innovation.

The evaluation of scenarios is an iterative process and is aimed at identifying and involving areas incrementally within an organization. The scope of the analysis can be extended out of an organization in the case scenarios involving communities of practice, professional groups and large corporations. During the evaluation phase, analysis of costs and return on investments, analysis of communication patterns and analysis of lifestyles might take place and be used to better focus on the benefits to be achieved by innovating situations of use.

Again, the choice of the proper set of techniques is strictly dependent upon the type and quantity of scenarios that will be evaluated and, moreover, upon the design phase's level. For instance, large-scale projects might involve more than 50 scenarios. Therefore, the elicitation of views, the identification and naming of involved scenarios, the definition of functional and user requirements need to be evaluated through the right number of subjects and a balance needs to be struck between the opportunity to have many subjects for a few scenarios or a few subjects for many scenarios.

A variety of use cases could be envisaged for the same high-level scenarios and close comparisons between the different solutions could be made, so as to get approval and further suggestions on the scenario that fits better behaviours, values and preferences. Evaluation by comparison is aimed at determining which advanced services, applications and underlying technologies are appropriate.

Scenario Prototyping

Prototypes are used for a variety of purposes depending on the managerial culture within an organization. For example, prototypes may be used for encouraging exploration of design solutions and incrementally building on them. As an alternative, they may be used to advocate current design achievements. The main goal of scenario prototyping should be the validation of the emerging innovation model and, later, the building of consensus on that model.

The prototyping of scenarios allows making one or more simulations of the basic requirements that have emerged during the generation and evaluation of the scenarios. Simulations are powerful tools at a reasonably low cost. Furthermore, simulation studies give more realism to the envisaged solutions. Simulations might foresee the use of cases, storyboards, video-clips, pictures and flowcharts. Animated simulations are at the top end of simulation studies since they stress the dynamics of scenarios and can help to put emphasis on opportunities for innovation design.

It is worth mentioning that prototypes do help to introduce stop rules in scenario analysis; i.e., an agreed number of scenarios, the level of detail that needs to be achieved, and so on. Finally, due to the fact that prototypes introduce constraints in the design cycle, the analysis of specification needs and usability specification can be highly detailed at this phase.

2.3.3 Conclusions and Further Work

Much ink (and many bits!) have been spilled already in creating ideas for how people will behave with 'beyond 3G' telecommunication concepts like ambient awareness, ubiquitous computing and autonomous agents. We selected a subset of the plethora of 'beyond 3G' scenarios and attempted to analyse them for commonalities, common trends and patterns. We also proposed several ways to analyse specific scenarios in detail in order to derive requirements. Finally, recognizing that key user research questions must be addressed in specifying technology for the wireless world, we proposed a number of high-level issues that should be addressed via user research.

Only by starting early and focusing on the user we can arrive at a wireless world future that delivers on the bright promises of many of the scenarios. Otherwise, we risk a repeat of 3G, where no obvious compelling value will emerge, or worse yet, a dystopian future where the tools of the wireless world may be misused. There is the need to set common methodological grounds to assess and validate scenarios and an agreed research agenda that should help to clarify better the emerging visions of the wireless world.

2.4 User Interface Technologies and Techniques

2.4.1 Introduction

In the field of mobile communication two conflicting trends can be observed: the variety and complexity of mobile devices increases, while the spectrum of users broadens. This means that more and more technologically inexperienced users must deal with devices the operation of which becomes more and more difficult due to their shrinking size and increasing number of features. One feasible way to resolve this conflict is the improvement of the user interface (UI).

Browsing web pages on a PDA or a smartphone is already inconvenient due to the small screen. Entering text by means of typing a stylus on a virtual keyboard or operating a ten digit keypad with three alphanumeric layers is nearly impossible. The situation becomes even worse when having to operate the device out in the street or while driving a car. In these situations, robust communication channels, other than keyboard input, would be advantageous. But even under less extreme circumstances, natural input methods such as speech, gesture, mimics and gaze could add extra value to the communication between man and machine [26]: switching TV channels by hand-gestures, or having a talking virtual assistant in the PDA which autonomously collects and reads out information from the internet according to the user's personal preferences, are just two examples for which new user interfaces are required in the future.

As described in [27], given the bandwidth of future 3G/4G/broadband networks, applications will gradually start to exploit new ways of interaction for which the bottleneck will not be in the transport of data, but will rather be determined by the 'bandwidth to the brain'. Therefore, communication channels are needed which increase the information throughput between man and machine, while simultaneously maximizing the reliability and robustness of this information transfer.

While the underlying technology is becoming increasingly complex, the need for simple and standardized ways to configure, personalize and operate the various devices is growing. Along with the development of new physical UI-devices and natural communication channels, it is necessary to harmonize the 'look-and-feel' of devices and services. As described in [28], the availability of common, basic interactive elements increases the transfer of learning between devices and services and improves the overall usability of the entire interactive mobile environment. Such a transfer becomes even more important in a world of ubiquitous devices and services. Simplifying the learning procedure for end users will allow the reuse of basic knowledge between different terminal devices and services and will lead to a faster and easier adoption of new technologies, fully benefiting the end user without restricting the manufacturer's wish to implement user interfaces based on a corporate look-and-feel and the overall user experience as a competitive edge.

New UIs are also required to improve the usability of modern technology for impaired or disabled people. In particular, means for speech input and output must find their way into future devices. Developing mainstream devices that can be operated equally well by all user groups, from infants to elderly people, opens new opportunities on the mass market.

To predict the chances of new UI techniques on the mobile communication market in the future, it is necessary to investigate the profile of today's customers with respect to their age, profession and financial potential. While some device characteristics, such as ease of use and robustness, are equally appreciated by all user groups, the importance of many issues is rated differently depending on the individual social or financial situation.

Summarizing, the following UI requirements can be identified for user friendly mobile devices of the future:

1. Robust natural communication channels such as speech and mimics that work on any mobile device.
2. A harmonized configuration and look and feel based on meaningful interaction elements such as standardized icons and sounds.
3. The design and functionality of UIs must match the individual preferences of each user group on the market.

In the following sections these technical, social and market-related topics will be investigated in more detail.

2.4.2 Current User Interfaces for Man–Machine Interaction

The most common man–machine interface is the keyboard, the mouse and the monitor. Furthermore, webcams for visual input and headsets and speakers for sound in- and output are in use. For some very specific applications, graphics tablets, data-gloves, force-feedback devices and 3D screens are in operation.

The in- and output devices are designed such that they make the operation of the machine easier for the user. The same is true for the software that utilizes these input devices. For speech input, for instance, dictation software, like IBM's ViaVoice®, aims to ease the job of a secretary. Utilizing the speech channel is only one step towards more natural man–machine interaction. Devices like pen-shaped scanners that employ a small camera to read in printed lines, store the text, translate it into another language and transmit it wirelessly to the PC are good examples of advanced user interfaces. Scanning in lines is similar to the operation of a text marker. Analogously, drawing directly on a graphics tablet is much more natural than using the mouse to draw a picture that is not visible on the mouse pad but on the screen only. Similarly, the operation of a touch screen is more natural than the somewhat indirect way of specifying screen elements by means of the mouse: tapping directly on the object is a natural way of addressing an item.

The touch screen is one of the few examples of currently existing natural user interfaces which have been implemented in both stationary and mobile devices. Most of the currently available user-friendly interfaces are bound to the desktop computer rather than being appropriate for mobile devices. Therefore, it is necessary to investigate new concepts for mobile appliances like mobile phones and PDAs.

2.4.3 Future User Interfaces for Mobile Devices

2.4.3.1 Access to Information

The form of mobile devices and the conditions under which they are used put heavy constraints on the user interface. The user interfaces for pure voice communication between two human partners over cell phones are already relatively user friendly (e.g. Bluetooth headsets, hands-free sets for car phones etc.). In contrast, the small screens of PDAs and cell phones are not appropriate for text output or web pages. However, as the trend in mobile communication shifts from pure voice communication to integrated voice and data transfer, the easy access of data services becomes more and more important. There are two ways to overcome this problem: first, the size of the screen can be enlarged physically or virtually. A physical enlargement is possible with new types of screen such as flexible OLED displays which roll up or fold (see, e.g., [29]). While the screen is stowed away, the devices are very small. During operation, however, the large screen is cumbersome in some situations. The same is true for current head-mounted displays, which provide a large subjective screen size but are still displeasingly large and fragile. However, this drawback may be overcome one day by laser-based systems which project the image directly onto the retina [30].

Another way to enlarge the screen is based on the idea of a virtual display: building a micromachine acceleration- or tilt-sensor [31, 32] into a PDA enables a window functionality. Movements of the PDA towards the eyes, for instance, can be interpreted as a zooming command to enlarge the central part of the screen's contents. Moving the PDA away from the face displays larger parts of the image in lower resolution. This principle can be extended towards the inspection of virtual 3D objects: moving the PDA in 3D-space is detected by the acceleration sensor and used to calculate and display the corresponding views of the virtual object.

A second way to solve the problem of accessing symbolic information on mobile devices is to implement a text-to-speech system which reads out the information to the user. While this works nicely for email and e-books, web pages and other contents with a mixture of text and graphics cause problems. In this case, an intelligent assistant who can autonomously filter out the information which is relevant for the current situation would be advantageous. Consideration of the specific situation's context and interactive communication with the user are required to achieve this goal.

2.4.3.2 Input of Information

New user interfaces for the input of information into mobile devices are based on speech and vision processing. Many current mobile phones and PDAs are equipped with a digital camera (either built-in or in the form of a CF card) and a microphone. Simple speech recognition is already possible for entering phone numbers or contact names in mobile phones. A lot of progress has been made for recognizing whole sentences speaker-independently and in noisy environments (e.g. in a car). Based on this research, a natural dialogue between a user and a virtual personal assistant might be feasible in the future. One step towards this goal is the integration of visual information, e.g. in the form of lip-reading [33]. First experiments show that it is possible to facilitate speech recognition by integrating information from tracking lip motion of the speaker. Simultaneously, this solves the problem of sound interpretation: when the speaker is silent, a speech recognizer, which is solely based on sound, often tries to interpret the background noise as speech. This effect is prevented by activating the recognizer only when lip motion is visually recognized. Another visual information channel is the interpretation and integration of the user's gaze direction. One application would be the automatic movement of the screen focus or the automatic turn of e-book pages when the reader has reached the end of the page.

By means of face recognition software, it is possible to identify the user and to use this information as context for the dialogue. A simple form of this identification is the integration of a fingerprint sensor in the current HP iPAQ® 5400 series. While currently this system is used for access control only, the identification of the user will also allow multiple persons to use the same device: for every person the identification system uploads the personal preferences for the session.

A built-in camera could also be used to recognize head gestures such as nodding or shaking the head. This natural form of input is interpreted as affirmation or denial. Less precise information about the user's satisfaction can be obtained from interpreting mimics.

Hand gestures, such as pointing, do not make a lot of sense for mobile devices, as the distance between user and device is usually very small. For specifying positions on the screen, the touch screen functionality is much more appropriate.

Finally, by pointing the camera away from the user into the current environment, certain other functions are possible: comparing images of objects with an internal or external database enables new features, such as a portable travel guide, for instance (see, e.g., [34]).

As described above, an acceleration- or tilt-sensor allows for new forms of user input, in particular in the field of entertainment. 3D computer games are a lot easier to play by means of natural movements in 3D-space than by the currently existing miniature joysticks for mobile phones.

It is important to note that most of the user interface techniques mentioned above do not require any visionary hardware but appliances that are available already today. However, they require a lot of progress in the robustness and performance of the software. While the robustness can be achieved by a smart fusion of multiple communication channels, the performance issue calls for more powerful and application-specific processors. Due to the cost and energy constraints for mass-market products like mobile phones or PDAs, it will not be feasible to equip the devices with high-performance multipurpose PC processors. Rather, the way to go is to develop processors dedicated to specific tasks such as 3D image rendering, speech and image processing, etc. One example is the ATI graphics chip in the Toshiba PocketPC e740. The concept of dedicated task processors can be developed further towards a parallel platform consisting of a number of programmable task processors. First steps in that direction have already been taken (see, e.g., [35]).

In addition to the development of the image and speech recognition software, researchers have to solve the problem of how to distribute the different tasks across the various task processors and how to organize their interplay.

Another very interesting field of research in the domain of UI technology is the virtual personal assistant (VPA) [36]. The basic idea is to implement an artificial agent in the mobile device, who communicates with the user by means of multimodal natural communication channels. The assistant carries out routine tasks such as dialing phone numbers or reading out emails. However, there are also advanced functionalities imaginable. The VPA could accept requests for information from the user, autonomously search the internet and read back the results once the required data has been found. The advantage for the user would be the possibility to access all functions through a standardized interface using natural communication channels. While the implementation of a virtual assistant is still a future vision, the standardization of UIs and interface elements in today's mobile devices could already significantly increase the user friendliness.

2.4.4 Harmonized UI Elements

The tremendous growth of the telecommunication market is driven to a large extent by the frequent introduction of new services, new devices and new features of the devices. Along with the beneficial aspects of this development goes an increase in complexity in the operation of the new technologies. Most people are overwhelmed with the multitude of new functions and use only a small subset of the features implemented. If this evolution is not stopped, new mobile technologies might soon degenerate to gadgets for a small group of technology geeks. Therefore, the focus must be put on how users can easily learn to use the new features and functions. An example is the similarity between the desktop versions of Microsoft Windows® and the version for PocketPCs and handhelds: although the devices and methods of input differ very much, users can quickly adapt to the PocketPC because they find labels and icons of similar look and arrangement. To harmonize UI elements between different devices and services, a general understanding of the users' basic preferences when operating mobile devices is required. In [28] some common user needs have been identified.

A most basic need, regardless of the user's age, is gaining access and being able to perform the basic interaction in order to place any kind of call and answer or reject a call. Emergency calls from any available device, without the necessity to consult a manual or even having to think about how to achieve this, is another important action, often performed under high stress, where failure is not an option.

Other user requirements include the wish not to relearn or redo complex procedures like inputting text on a 12-key pad, setting up and configuring access to network based services, configuring and connecting wireless accessories, adapting procedures to access network-based content from different devices in different networks, performing financial transactions or accessing speech controlled applications.

Another important developing user requirement is the continuity of data. This means that personal content, such as address books and agendas or configuration data such as access to the mobile internet, should be made easily transferable and compatible in a device-independent way.

With users having an increasing ability to personalize their communication environment, there is a risk that the information required to personalize the product or service may be abused. To ensure that users are confident that their information is not used in ways of which they disapprove, they need standardized ways to control both how and to whom their information is made available. They also need status visibility, e.g. to be warned when there is a risk that information is about to be shared in ways which they may find unacceptable. Users also need confidence that information that they make available cannot be seen as a result of insecure communication links.

Accessibility requirements cannot always be satisfied for all users. Therefore, it is important to support assistive devices in order to be able to display larger font sizes or provide higher volumes of speech output to those who require it. Similar user requirements are to be found when accessing devices and services with multimodal user interface capabilities that need to be adapted to the user's needs, e.g. a blind person retrieving a written short message or a nonliterate child wanting to call their parents by speaking their name instead of entering the digits of their telephone numbers.

When investigating the possibility of harmonizing UI elements, one important fact is the ever-increasing functionality of today's communication devices. Since it cannot be foreseen which new functions need to be implemented on future devices, any harmonization or standardization of user interface concepts or components could be obsolete in the very near future. Any harmonization effort that restricts the implementation of user interfaces for novel functionality is doomed to fail. Harmonization results must therefore be able to coexist with these still to-be-developed UI components.

In the following, we will present some examples of UI elements, the harmonization of which would reduce the amount of adaptation required from the user.

2.4.4.1 Basic Elements and Functions

International Access Code Dialing
As the access code for the international call format has not been fully harmonized, it may be advantageous for the user to harmonize the input of this access code in communication devices in order to fully overcome this technical gap. One solution is a uniform input method for the '+' sign which is used as international code in all GSM networks.

Emergency Functionality and Services
The user procedure to start an emergency call (keys to be pressed, dialogues for user confirmation) has already been subject to harmonization efforts. Especially in this area, a harmonized user interface seems to be important. Everyone has to be able to start an emergency call from every communication device, while keeping the number of unnecessary emergency calls at the lowest possible limit.

Symbols, Icons and Pictograms
For the user without disabilities, icons used in communication devices are mostly for illustration or provision of status-related feedback. Exceptions to this rule are found for graphically displayed menus and functions in a number of devices. While it is impossible to foresee the functionality to be represented by symbols and icons in the future, it is nevertheless important to harmonize the symbols and icons depicting basic communications functionality. In particular for users with cognitive disabilities (e.g. dyslexia), the harmonization of symbols and icons is desirable.

Earcons
Earcons are acoustical interface elements to provide cues about content and the organization of information. They show properties very similar to icons and symbols. Due to the fact that their use is currently quite restricted, it seems to be much easier to reach consensus on earcon harmonization. To a user with limited, or no, eyesight, earcons are crucial for using telecommunication devices. For these users, harmonization of earcons (across devices AND services) offers important advantages. Another improvement area for these users is harmonized master text for announcements, typically used in text-to-speech prompts.

Access to Basic Voice Services Via Shortcuts

For a number of basic voice services, like 'Call set-up to the voice mailbox', de facto standards have evolved during recent years, e.g. most mobile phones accept a long press on '1' as a shortcut to access the mailbox. The effort to harmonize these functions should be minimal and a uniform interface would certainly be beneficial to the average user. Especially for people with visual impairments, who have problems locating the respective function in a menu, these shortcuts are extremely helpful. To achieve maximum benefit for these users, it would be advantageous for the harmonization team to consult with blind UI experts.

Basic Terminology

Different companies producing communication devices use different words for describing telephone functions and services to their users, both in the user interface of the device itself and in the user guide describing the functionality. An agreed upon set of function names would ease the use of these functions for the users of communication devices. For this set of function names, it is important that the average user, not only communications experts, understands the words.

Text Entry and Retrieval

Efficient and intuitive text entry and retrieval are one of the basic, key requirements – and stumbling blocks – in contemporary mobile devices. It would be beneficial to the end users and operators to see efficient, intuitive, and also common solutions to text entry. However, the technologies are continuously evolving, so we must be careful not to stop innovation. The current keypad mappings and predictive text input solutions are still probably far from perfect. Additional complexity arises through the necessary control functionality for predictive text entry systems. Turning these systems on or off, input of new, unknown words and the selection between prediction alternatives are major obstacles to using these systems for many users. Easy-to-use command shortcuts, harmonized over many different devices, might broaden the possible user group of these predictive text entry systems.

2.4.4.2 Configuration for Service and Application Access

A large number of applications, many of them being defined at the moment, might be subject to some form of standardization on the user interface level. Differences in using these features in different devices are very often just disturbing to the user. For many manufacturers, their style of controlling these applications is considered a part of their corporate brand building, so the selection of appropriate harmonization areas, and the identification of ergonomically optimal solutions, is a fairly tedious process.

Examples of applications with user interfaces that could become subject to standardization are: control of audio and video transmission, data enquiry on standard databases and directory services, mobility services (navigation, localization), messaging, remote services, video monitoring, multimedia and interactive multimedia, personal mobility and group communication. It is important to have the service providers and the operators included if trying to achieve any level of harmonization in these application areas.

Another area where harmonization would be beneficial for the user is the task of configuring services and applications. Examples are complex set-up procedures for network access, service logon and availability of WAP and GPRS or Bluetooth accessories inter-working. The problems caused by lower level standards could be, at least partly, overcome by offering configuration support solutions that work with most manufacturers' phones.

Harmonization is also required in the field of service and application access, interworking and portability, as user trust and reliability of terminals, applications and services working across networks is of paramount importance. During the past few months, an industry-wide initiative has been taken in the Open Mobile Forum, leading to the set-up of the Open Mobile Alliance, OMA. Its primary mission is to make the mobile internet work by ensuring interoperability between the components of telecommunication systems, based on a user requirement driven development.

Finally, the terminology for services and applications needs to be harmonized. While, for instance, Bluetooth and SyncML as wireless data transmission technologies have been standardized, in order to allow the devices of many manufacturers to cooperate, the user interface for setting up Bluetooth connections has not been worked out on the same, fine level of detail. Using a common terminology may ease the set-up of communication channels between devices by the average user.

2.4.4.3 Advanced Functionality-related Interaction Elements

Among the more advanced issues which are directly or indirectly linked to the UI is the harmonization of the structure and vocabulary of spoken commands in future verbal user interfaces. Since it is necessary to learn the vocabulary of spoken commands, switching between two nonharmonized devices would raise the learning effort of each user considerably.

Another, more obvious but nevertheless advanced, task is the portability and the harmonization of personal information management (PIM) data across multiple different mobile devices. With such harmonized data formats, PIM data can be transferred without additional obstacles for the user. Given that most communication devices are able to synchronize their data with computer programs such as Microsoft Outlook and Lotus Notes, there should not be too much opposition to such an undertaking. The data formats are quite obvious areas for standardization.

Other harmonization issues which are important, although only indirectly linked to the UI, are [28]: universal addressing and identification of users, detection of the user's position for location based services, ubiquitous presence of services and content within converging networks, access of public ad hoc services (e.g. a museum guide), codecs and formats of streamed media content and privacy/security/safety issues (e.g. encryption functions).

Summarizing, the availability of common, basic interactive elements increases the transfer of learning between devices and services and improves the overall usability of the entire interactive mobile environment. Such a transfer becomes even more important in a world of ubiquitous mobile telecommunication devices and services. Simplifying the learning procedure for end users will allow the reuse of basic knowledge between different terminal devices and services and lead to a faster and easier adoption of new technologies. From the manufacturers' and service providers' point of view, this implies that development costs can be reduced, time to volume markets decreased, larger user segments reached more easily and quickly, thereby ensuring quicker uptakes of key technologies and access to all users, without restricting their competitive edge made possible by the user experience and specific UI implementations offered.

2.4.5 User Interfaces for Impaired or Disabled People: 'Design for All'

Improved user interfaces for in- and output of information are particularly important for impaired or disabled people. Following the argument in [37], the population of most countries is ageing and the number of people with impairments and disabilities is also increasing. At the same time, there is a growing recognition of the need to integrate older people and people with disabilities into society by enabling them to sustain independence for as long as possible. Information and communication technology plays an important part in this integration process.

Taking the needs of a broader spectrum of people into account in the design process is called 'Design for All', 'Barrier Free Design', or 'Universal Design' – the term more often used in the US. The philosophy behind 'Design for All' is to ensure that mainstream equipment and services can be used by a wide range of users, including older people and those with disabilities.

While there are products that are specifically tailored to the needs of disabled users, the mass market of the future requires mainstream information and telecommunication technology products that are designed according to good human factors practice, incorporating considerations for people with impairments, that can be used by a broad range of users.

Only when 'Design for All' is adopted from the start of the design process is it possible to design products that are accessible to a significant number of disabled and elderly people, with minimum effort and cost. The increased usability of 'Design for All' may even bring benefits to all users: providing volume control on a phone, for instance, not only assists people with hearing difficulties but also helps others operating in a noisy environment, or taking account of the needs of people with visual impairment helps all users trying to read a display in poor lighting conditions, or without their reading glasses at hand.

In practice, adopting the 'Design for All' approach means considering the needs and requirements of people at the ends of the population continuum, rather than just those in the middle. This means that, in addition to the general principles of ease of use and harmonization mentioned in the previous sections, in particular the following requirements must be met [38]:

1. *Equitable use.* The design must be useful and marketable to any group of users – avoiding segregation or stigmatization of any users.
2. *Flexible in use.* The design must accommodate a wide range of individual preferences and abilities.
3. *Simple and intuitive to use.* The design must be easy to use and understand, regardless of the user's experience, knowledge, skills or concentration level.
4. *Perceivable information.* The design must communicate necessary information effectively to the user, regardless of ambient conditions or the user's sensory abilities.
5. *Tolerance for error.* The design must minimize hazards and the adverse consequences of accidental or unintended actions.

6. *Low physical effort.* The design must be usable efficiently and comfortably and with minimum fatigue.
7. *Size and space for approach and use.* Appropriate size and space must be provided for approach, reach, manipulation and use, regardless of the user's body size, posture or mobility.

Summarizing, the concept of 'Design for All' is a way to both take the needs of impaired or disabled people into account, and increase the usability and robustness of future information and telecommunication devices for everyone.

2.4.6 Market Aspects

In the last few years, new communication needs have emerged and people worldwide are showing similar attitudes towards technologies. Customers have high expectations for ubiquity of access points, transparency of information, personalization of offerings, relevance of promotions, neutrality of advice, connectivity with peers and elimination of interruptions. All these features are considered a common frameset among different technologies. Also of high importance for customers are mobility issues, user friendly customer interfaces, interaction models free of particular and boring training processes, and fast, goal-oriented actions with explicit consequences.

The number of wireless market users exceeds the number of internet users by far and, moreover, the two user groups show different approaches to the use of technologies. In Italy, for instance, there are approximately 45 million mobile phone users against 'only' 11 million internet users. In Europe, the ratio of mobile to internet users is not far from 3 to 1 (62% and 27% of the population respectively) and in Japan – where the mobile market growth started later – the ratio is 2 to 1. Considering these numbers helps us in trying to identify appropriate market strategies in the field of user interfaces.

In order to be able to provide users with a positive user experience, new design theories, services and applications, as well as evaluation techniques for interaction with mobile devices, must be developed. Special attention must be given to the phenomena of usability and creativity. Young users especially overcome technical and physical limitations by creating new languages through a system of emotions, acronyms and abbreviations. At the same time, teenagers have also developed new interaction models, as well as new uses for the traditional ones (e.g. SMS).

This section aims to identify some crucial drivers of market change that will accompany the evolution beyond the development of mobile interface technologies. The basis is to focus both on technological innovative features and on social communication changes. The social impact of innovative user interfaces is part of the market leadership challenge. To build coherent market scenarios and to develop adequate business models will also be fundamental for the survival of the enterprises involved in such changes.

The debate in WWRF 2002 evidenced that the development of a purely 'technical' vision, focusing, for example, on new network concepts or radio interfaces, will not be sufficient. Such a technical view must be put into a much wider context:

- a user-centred approach that investigates the new ways users will interact with the wireless systems;

- new services and applications that become possible with the new technologies; and
- new business models that could prevail in the future, overcoming the traditional hierarchy represented by user–service provider–network provider.

A necessary prerequisite for success on the mobile market is the investigation of the end user's expectations of these new applications and services. However, different groups of customers have different preferences, needs and desires. Therefore, a customer analysis of the current mobile market is necessary. Such an analysis of the 'Y-Gen', the wireless generation, has been carried out in [39] and [40].

2.4.6.1 Market Analysis and Segmentation

The large majority of users is not interested in the technical details of the mobile devices, but in their usability, efficiency, robustness, design and entertainment qualities. Although these aspects are considered mandatory by most users, their relative importance varies among customer groups. As the UI defines the user's experience with the device, the group-specific preferences must be taken into account during the design process. Considering the individual preferences and the purchasing power, the market can be roughly segmented into three large clusters [41]: (a) teenage users; (b) parental users; and (c) professional users.

In the first group, the usage of wireless services, voice and data is very intense, in particular for entertainment and communication services like SMS, downloadable games, chat, email and other infotainment environments. This group is specifically interesting for usability studies about UI technologies, as teenagers are generally open to new features and innovative designs. However, the teenagers' preferences change quickly, which makes a prediction of successful UI designs of the future very difficult. In addition, the financial capabilities of this user group are very limited. A successful market strategy would require offering cheap mass-market solutions which can be flexibly adapted to the varying user preferences. In particular, it is important to keep teenagers interested in the devices and their design. Skinable UIs, multimedia offerings and device designs which are aligned with the current fashion generate a demand within this group of customers.

The second group, consisting of more rationally allocated mature people, puts the emphasis on seamless communication, the balance between home and job, the simplification of routine processes (e.g. through wirelessly controlled 'intelligent houses'), family management and on self-realization. Within this customer segment, there is no place for gizmos or experimental features. Rather, robustly working devices and services are needed which do not require a long learning process to be operated. Usually, members of this user group put the emphasis on basic voice and data functions, and not on multimedia. User interfaces should reflect these preferences by offering clear, robust and rational operation of the devices. These requirements, together with the generally high financial independence of this group's customers, call for higher class long-life devices with advanced and standardized UIs.

Finally, the financially most powerful group of professional users appreciates everything that facilitates their professional goals. Members of this group require personal information and time management, email, internet access, seamless roaming, location-based services, online financial, travel and weather information, video conferencing, etc. This group of customers is generally literate in modern technology and open to innovative features. Their demands are maximal in terms of performance. Gaming or other gizmos play a relatively low role within this group of users, however, the whole palette of multimedia, voice and data services is required by these customers. The UI is crucial for the task to handle all the advanced functions efficiently. It is this user group where a UI in the form of a virtual personal assistant, described earlier, would pay off most.

To gain success on the market, companies will have to design UIs which can either adapt to the different needs and preferences of the customer groups, or which can serve all requirements simultaneously.

2.4.7 Research Issues

Extended research activities are required to realize user-friendly interfaces that are easy to use, do not need too much adaptation, are harmonized and unified across different devices, services and networks, offer advanced functions for impaired people and take into account the different preferences of various user groups. Many different fields of research must be covered: psychology, artificial intelligence, information processing, hardware and software design, market research, etc.

The task for researchers in hard- and software design is obvious: how can new sensors, applications and functions be built into the ever shrinking mobile devices, without going beyond the scope of energy consumption, robustness and usability? In this respect, the evolution of PDAs and smart phones shows the tremendous development. While only two years ago PDAs were not able to do much more than personal information management, the current models have built-in WLAN, Bluetooth and GSM/GPRS modules, cameras, and fingerprint sensors, are extendable by memory cards and GPS systems and they can display full-screen/full-frame-rate video, for instance. However, the in- and output channel is still predominantly the touch screen. To find new man–machine interaction channels is, therefore, an important research issue, which is linked with the task of representing and integrating multimodal information channels based on vision and speech.

For the development of natural communication channels, such as gaze, gesture and mimics, a lot of research is necessary. In particular, the heavy constraints of mobile devices, such as low power, low cost and small outline, require algorithms that are simultaneously robust and computationally cheap. Exploiting the fact that vision and speech processing algorithms can easily be expressed in the form of highly parallel code, processors with a simple but parallel architecture can show high performance while consuming only low power.

The distribution of multiple different tasks over a number of application-specific task processors is an issue which is not specific to mobile applications. Rather, it is a problem that has to be solved for any future electronic mass-market product which integrates many advanced functions in a low-power small-outline design. To enable a fast adaptation of the programmable task processors to specific applications, even for complicated architectures, a system has to be developed which frees the programmer from routine tasks such as parallelizing the software.

Even with currently existing UIs there is a lot of research to be done. Harmonization and standardization of UI elements, configuration procedures, applications and services not only requires common sense between the device manufacturers but also the incorporation of the customers' preferences. Here, results from usability' studies, as well as psychological studies, are needed.

Harmonization is closely linked with the topic of seamless and ubiquitous mobile access to networks which are converging but nevertheless distinct. Roaming across networks globally, while keeping a unique identity, is an issue that strongly affects the usability of mobile devices and services. Here, international efforts for standardization and harmonization are required.

Finally, social and economical studies play an important role when deciding on new features and services to incorporate into mobile devices. The analysis of different user groups and their individual preferences is vitally important for device manufacturers and service providers.

2.4.8 Conclusions

Within this section we have investigated the technical, economical and scientific issues of user interfaces for mobile devices. We have shown the multitude of problems to be solved and research areas to be touched on when aiming for user-friendly devices that have success in diverse markets.

Due to the limited space, we were only able to give a brief overview of currently existing and envisioned user interface technologies. As this topic is definitely a key issue for upcoming mobile devices and services, the field of research is rapidly changing. By putting the focus on natural multimodal interaction and ergonomically designed UI elements and services, we tried to express our opinion that new user interfaces will only have success on the mass-market if the users have the impression that the new appliances provide added value, such as ease of use, robustness, safety, and that the operation of the devices and services is easy to learn.

As mobile communication is an issue addressing the entire community, and not only IT experts, it is worthwhile to investigate ways of how to provide new features without making the devices and services more difficult to operate. The development of natural communication channels and the harmonization of UI elements, services, configuration processes and roaming is based on the idea that technology must adapt to the user and not vice versa.

References

[1] Jokela, T., 'Assessment of user centered design process as a basis for improvement actions,' Department of Information Science, University of Oulu, 2001.
[2] ISO 13407, 'Human-centered Design Processes for Interactive Systems,' 1st edition, 1999.
[3] ISO 18529, 'Human Centered Life Cycle Processes Descriptions,' Technical Report, 2000.
[4] WWRF, 'The Book of Visions: Visions of the Wireless World,' Version 1.0, December 2001.
[5] Beyer, H. and Holtzblatt, K., 'Contextual Design: Defining Customer-Centered Systems,' Morgan Kaufmann, 1998.
[6] Friedman, B., 'Network Browser Security and Human Values: Theory and Practice,' National Science Foundation Project at the University of Washington. http://www.ischool.washington.edu/research/project-details.cfm?ID=4.

[7] Maslow, A., *'Toward a Psychology of Being,'* 2nd edition, Van Nostrand Reinhold, New York, 1968.

[8] Jordan, P.W., *'Designing Pleasurable Products: The New Human Factors,'* Taylor & Francis, 2000.

[9] Shneiderman, B., *'Leonardo's Laptop: Human Needs and the New Computing Technologies,'* MIT Press, Cambridge, MA, 2002.

[10] Meier, R.L., *'A Communications Theory of Urban Growth,'* MIT Press, Boston, 1962.

[11] Adams, A. and Sasse, M.A., 'Privacy in Multimedia Communications: Protecting Users, Not Just Data,' in A. Blandford, J. Vanderdonkt and P. Gray (Eds): *'People and Computers XV – Interaction without Frontiers,'* Joint Proceedings of HCI2001 and ICM2001, Lille, September 2001, pp. 49–64.

[12] Adams, A. and Sasse, M.A., 'Privacy Issues in Ubiquitous Multimedia Environments: Wake Sleeping Dogs, Or Let Them Lie?,' in M.A. Sasse and C. Johnson (Eds): *'Human–Computer Interaction INTERACT '99,'* Proceedings of IFIP TC. 13 International Conference, 30th August–3rd September 1999, Edinburgh, pp. 214–221.

[13] Dertouzos, M., *'The Unfinished Revolution: Human-centred Computers and What They Can Do For Us,'* HarperCollins, New York, 2001.

[14] Communication of the ACM, **43** (8), August 2000.

[15] Lombard, M. and Ditton, T., 'At the Heart of It All: The Concept of Presence,' *Journal of Computer Mediate Communication,* **3** (2). Available at: http://www.presence-research.org/(2002-12-03).

[16] A. Steinhage, "Tracking Human Hand Movements by Fusing Early Visual Cues", published in the Proceedings of the 4th Workshop on Dynamic Perception, Bochum, Germany, November 2002.

[17] M. Rantzer, "Foresight Paper – All Senses Communication", No. ERA/SVZ/R-01 : 029 Uen, Ericsson, 2001

[18] "Human Factors (HF): Potential harmonized UI elements for mobile terminals and services", Technical Report DTR/HF-00051 of the European Telecommunications Standards Institute, 2002

[19] Universal Display Corporation, www.universaldisplay.com

[20] "Double Vision: Retinal Displays Add Data Layer", published in The New York Times, April 26, 2001.

[21] "The World's Smallest Piezo-Resistance 3-Axis Acceleration Sensor", Hokuriku Electronic Industry Co., LTD, www.hdk.co.jp/english/topics_e/tpc007_3.htm, Feb. 5, 2002.

[22] M. Ward, "3D images in your hand", BBC News online report, news.bbc.co.uk/1/hi/sci/tech/1939181.stm, 23. Apr. 2002.

[23] J.P. de la Cruz, "Motion based lipreading" (Master's thesis), published at the Department of Electronic Engineering, University of Granada, Spain, June 2002.

[24] *'France Telecom Trials Interactive Tourist Guides on PDAs,'* http://www.francetelecom.com/en/financials/journalists/press_releases/CP_old/cp021008.htm.

[25] Raab, W. *et al.,* 'A Development System for Heterogeneous Multiprocessor Architectures,' in *Proceedings of the IEEE Workshop on Heterogeneous Reconfigurable Systems-on-Chip (SoC),* Hamburg, Germany, April 2002.

[26] Steinhage, A., 'User Representation in Cyberspace: Ideas about Key Applications for 3G and Beyond,' in *Proceedings of the 6th Wireless World Research Forum,* London, UK, March 2002.

[27] ETSI, *'Human Factors (HF): Guidelines for ICT Products and Services – Design for All,'* ETSI Technical Report DEG/HF-00031, 2002.

[28] 'The Center for Universal Design,' http://www.ncsudesign.org/content/.

[29] Dainesi, E., 'The Swiss Knife Strategy,' *On Web Marketing Tools,* **35**, November 2002.

[30] Rheingold, H., *'Smart Mobs: The Next Social Revolution Transforming Cultures and Communities in the Age of Digital Access,'* Perseus Books Press, USA, 2002.

[31] Dainesi, E. and Zucchella, A., 'Wireless Marketing: Remodelling the Wireless Future from Technological to Generational Approach,' in *the 7th Wireless World Research Forum Meeting,* Eindhoven, The Netherlands, December 2002.

2.6 Credits

The following individuals have contributed to the contents of this chapter: Ken Crisler (Motorola, US); Andrew Aftelak (Motorola Labs, UK); Satu Kalliokulju, Tea Liukkonen-Olmiala,

Juha Matero, Minna Asikainen and Harri Mansikkamäki (Nokia, Finland); Manfred Tscheligi (CURE, Austria); Angela Sasse (University College London, UK); Mikael Anneroth (Ericsson, Sweden); Petri Pulli (University of Oulu, Finland); Michele Visciola (Polytechnic of Milan, Italy); Thea Turner (Motorola, US); Peter Excell (University of Bradford, UK); Axel Steinhage (Infineon Technologies AG, Germany); Martin Rantzer and Bruno von Niman (Ericsson, Sweden); Lele Dainesi and Antonella Zucchella (University of Pavia, Italy).

3

Service Infrastructures

Edited by Radu Popescu-Zeletin and Stefan Arbanowski
(Fraunhofer FOKUS)

3G Beyond Service Architectures

Since September 2001, Working Group 2 has worked in the area of service architectures
for future wireless systems. A vision of I-centric communication has been developed,
which puts the individual ('I') user in the centre of all activities a communication system has to perform. WG2 is looking for communication systems which, in the future,
are able to model each individual, his preferences, and adapt to different situations and
resources in time. The developed reference model for I-centric communications addresses
the individual user, interacting in his personal communication space.

The reference model for I-centric communications follows a top-down approach, starting with the introduction of individual communication spaces, related contexts and objects
(for more details see next section). It is commonly understood that I-centric services have
to support at least three different features, namely ambient awareness, personalization,
and adaptability. These are discussed in detail in the following sections.

The service platform for I-centric communications is responsible for shaping the communication system, based on individual communication spaces, contexts, preferences and
ambient information. Preferences will be provided by personalization, whereas ambient
information has to be provided by ambient awareness.

The IP-based communication subsystem is responsible for providing the linkage
between different objects in the communication spaces. IP communication is seen as
the common denominator by which to harmonize heterogeneous network infrastructures.

The wired or wireless networks layer implements all aspects of the physical connec-
tion(s) between different end points. Due to the hierarchical structure of the reference
model, a connection in the IP based communication subsystem might use different physical
connections in the underlying networks. Devices and communication end systems provide
the physical end system infrastructure that hosts all other layers. The main features of
I-centric communications (ambient awareness, personalization and adaptability) affect all
layers. Therefore, supporting functions have to be provided as a vertical solution. The reference model introduces the concept of generic service elements that implement common

Technologies for the Wireless Future.   Edited by R. Tafazolli
© 2005 John Wiley & Sons, Ltd   ISBN: 0-470-01235-8

functionalities on all layers. Generic service elements can be seen as a toolbox from which complex services can be assembled and executed dynamically. The vertical approach allows I-centricity on all layers, i.e. it establishes I-centric private virtual networks.

Accompanying all of these technical issues, the business model for I-centric communication identifies the relationships and information flows between all active roles within an I-centric system. The business model will help to identify reference points between all involved entities and will assist the assessment of the applicability of I-centric services for different business domains.

The following sections introduce the results of the Wireless World Research Forum Working Group 2. They explain in detail how personalization, ambient awareness and adaptability can be provided in future mobile communication systems. The papers have been worked out as 'white papers' within Working Group 2 and represent a common understanding between academia and industry.

3.1 I-centric Communications – Basic Terminology

3.1.1 Vision

The communication behaviour of human beings is characterized by frequent interactions with a set of objects in their environment. Humans solve problems in their daily lives, e.g. money and bank accounts need to be managed, food has to be bought and to be prepared for eating, movies are watched for entertainment, places are visited and news consumed to improve education, and other people are met for discussions. The set of objects, controlled by each individual human, defines its individual communication space, as shown in Figure 3.1.

A communication space of an individual is limited: I do not know everybody in the world, I am not interested in everything, and I do not have all necessary devices required by all communication services everywhere at all times.

Furthermore, individuals are interested in semantics and not necessarily in the presentation of a specific service. Services in an individual communication space have to support the quality of the human senses, and since quality of senses is individual, they have to

Figure 3.1 Individual communication space

adapt their presentation to each individual automatically. Services have to adapt to the life stage and the environment of each individual.

In former days, the communication space of human beings has been limited to the actual surrounding physical environment (like a home or an office) because of the spatial range of human senses. With the introduction of telephony, the range was expanded. It became possible to talk with people regardless of their location. With asynchronous services like email and short message service (SMS), the dimension of time expanded. People can send emails and do not need to care whether the addressee is ready to receive the message or not. That is, technology has eliminated distances in time and space, or at least made these boundaries almost imperceptible. By this means, today's communication services act as a prolongation of human senses and extend the individual communication space.

Following this view, a new approach is to build communication systems not based on specific technologies, but on the analysis of the individual communication space. The result is a communication system that adapts to the specific demands of each individual (I-centric). Such a communication system will act on behalf of a human's demands, reflecting recent actions to enable profiling and self-adaptation. I-centric services adapt to individual communication spaces and situations. In this context, 'I' means I, or individual, 'centric' means adaptable to individual demands and a certain environment.

The rationales above require intelligence in service provisioning in order to personalize, adapt to situational and environmental conditions, to monitor and to control the individual communication space. An I-centric communications system will provide the intelligence that is required for modelling the individual communication space of each individual by adapting to its interests, environment and preferences.

The multitude of devices, wearables, different telecommunication technologies, positioning and sensing systems are considered as enabling technologies for I-centric communications. Universal information access (including service interworking and media conversion), flexible management of equipment and facilities (e.g. smart homes [1]), and personal communications (supporting personal mobility and terminal mobility [2]) form the basis of such systems.

3.1.2 I-centric Communications – Basic Terminology

I-centric means to take a bottomless look at human behaviour and to adapt the activities of communication systems to it. Human beings do not want to employ technology. They rather want to communicate, acting in their individual communication space. They meet with others to talk and to celebrate, they read and travel, they listen to news or to music, they make decisions, etc. This abstract description of humans' communication activities requires a set of definitions that allow the mapping of abstract requirements (or wishes) to the physical communication environment later on. In the following, the basic terminology to describe I-centric communications systems is introduced.

As discussed above, human communication behaviour is characterized by frequent interaction with a set of other humans, information sources, and devices within or outside the human's vicinity. All entities that humans interact with will henceforth be called objects. They can be activated or deactivated by an individual, or environmental conditions, to perform an action according to the specific needs of an individual. Objects are directly addressable, and can represent one or more physical entities performing a certain service.

Object: An object is a logical representation of a hardware or software entity, or even a representation of a certain individual, and provides well-defined services from the perspective of an (other) individual.

To model the interaction of human beings with objects of their individual communication space, a general model applicable for all kinds of objects is needed. Objects will be used differently in different contexts by different individuals. In addition, a mechanism is needed for managing the use of objects (e.g. monitoring the activities of each object can be used to profile, or to bill for service provisioning). The focus regarding individual–object interaction is to provide services of objects in a generic way.

To enable ad hoc interaction of beforehand unrelated objects, an interaction model between objects is needed. This will allow the dynamic collaboration of objects for a specific purpose. Together with an organizational model, which describes relations between objects, such as ownership issues, such kinds of interaction can be used to stimulate social behaviour of objects, like multiagent systems [3, 4] do, to perform a specific task.

The basic idea of objects, to model real-world entities, has to be reflected by the I-centric service architecture. General procedures for wrapping legacy technology have to be developed, bearing in mind the fact that environmental constraints can affect the design of a distributed system [5–9]. Middleware concepts have to be selected that on one hand hide the heterogeneity of physical resources (represented by objects of individual communication spaces), but on the other hand support vertical integration, if necessary.

A general methodology on how to select/design objects of individual communication spaces has to be developed. An I-centric system should support massive numbers of objects and should be tolerant against object failures. The population of objects is always changing because they spontaneously enter/leave/roam the environment, and hence the communication space. Already standardized mechanisms for naming, lifecycle, monitoring, fault tolerance, etc. have to be taken into account to determine whether they suit the requirements of I-centric communications.

Due to changes in human beings' daily lives, the amount, or the concrete instances, of objects are changing over time. Nevertheless, the sum of objects an individual might interact with form his individual communication space. Objects may pertain to different communication spaces. They can be controlled by individuals, other objects, or services. Individuals can directly ask for a service to be performed by an object, whereas environmental condition may influence the status of objects indirectly. Communication between different individuals takes place by sharing objects of their communication spaces. In this case, objects representing communication facilities in the different communication spaces will be connected to establish a physical connection between two individuals. What kind of physical resources are used for the communication is decided dynamically and depends on the individual preferences of involved parties, their available communication facilities, and additional ambient information. The process of how to select and activate objects and physical resources underneath is one of the main activities of an I-centric communications system and will be introduced in the following sections.

Individual communication spaces grow and shrink on the time axes, based on the individual life stage, personal interests, working and living environments, and the availability of new kinds of telecommunication services and devices. Meeting new friends or entering a room, which provides certain telecommunication devices, enlarges the communication space, whereas leaving the same room will reduce the size of the communication space dynamically. This process of growing/shrinking the communication space has to be supported by an I-centric communications system in two ways. On one hand, the system must perform this process automatically, but on the other hand, the individual must have the possibility to include and exclude objects explicitly.

> *Individual communication space*: The individual communication space of a certain individual is defined by a set of objects this individual might want to interact with. The size of the individual communication space varies over time due to the appearance or disappearance of objects.

Each individual has only one individual communication space. It contains all objects this individual might want to perform requests on. Objects that pertain to individual communication spaces of different individuals must handle concurrent access from different individuals, or must delegate the concurrency control to the I-centric communications system.

Although the individual communication space provides a repository of objects for an individual, there is no partitioning of objects in the communication space itself. The concepts of context, personalization and ambient awareness provide necessary mechanisms to structure, classify, group, and even to partition objects by several means, i.e. relations between objects to define dependencies, or relations between objects and individuals to define ownership issues or usage rights.

3.1.2.1 Context and Active Context

An individual communication space provides a set of objects whose services an individual can use to achieve his goals. In addition to this, the concept of *context* provides the definition of relationships and causalities between different objects of an individual communication space and the individual. A context represents a 'universe of discourse' in an individual communication space. Individuals communicate with objects in their environment in a certain context.

A context can be seen as a metaphor for:

- those objects of an individual communication space that an individual wants to use in parallel;
- the objects that have to be taken into account to fulfil a certain wish of the individual; or
- whether and how objects are affected by activities of other objects.

Objects may pertain to different contexts (even to contexts of different individuals), because individuals might want to have a certain object involved in different activities.

> *Context*: A context represents a certain 'universe of discourse'. It defines relationships and causalities of an individual to, and between, particular numbers of objects in his communication space.

Contexts are independent from any concrete environment. If an individual wants to act in a certain context this context has to be activated. An *active context* defines the relationship of an individual to, and between, particular numbers of physical resources (see Figure 3.1) at a certain moment in time, in a certain environment. The activation and deactivation of a context should usually be done automatically just by analysing individuals' activities, but an individual should also have the possibility to do this explicitly.

Activating a context means:

- the identification of objects that are required by the context;
- the evaluation of the relationships and causalities between objects defined by the context;
- the discovery of the actual vicinity of the individual to identify physical resources that provide the functionality required by the identified objects;
- the activation/configuration of these physical resources to perform the task required by the context.

The difference between context and active context is characterized by the entities that are considered in relations and causalities. Context only refers to objects as an abstract model of the kind of objects that have to be taken into account in a certain context, whereas an active context refers to physical resources that have been identified during the activation process. Active contexts are of a dynamic nature, reflecting the current environment an individual resides in.

The activation and deactivation of contexts is one task of I-centric services. To activate a context, the I-centric service performs the activities described above. In addition, the I-centric communications system has to manage concurrent access to objects and conflicts caused by contrary wishes, expressed by individuals.

Active context: A context is active when it is adapted to a certain environment at a certain moment in time. It defines the relationships and causalities of an individual to a particular number of physical resources at certain a moment in time, in a certain environment.

Deactivating a context means:

- the identification of objects that have been activated/controlled in the context;
- the identification of effects that have been caused on other objects by the context;
- the identification of activities (e.g. deactivation, reconfiguration) that have to be performed on all of these objects and physical resources underneath for deactivating the context;
- performing these activities on the objects and physical resources.

Acting in a context means using only services that are provided by objects which are part of that very context. Starting to interact with objects that are not part of an active context will cause the activation of another context. This means, on one hand, individuals are allowed to act in several contexts in parallel, and on the other hand, the I-centric communications system must handle conflicts that might occur due to contrary causalities defined in the different contexts.

To handle each individual communication space and associated contexts, a general model of domain information and relationships to objects and physical resources is needed. This model must be open to being enhanced, due to the introduction of new locations, devices, etc. Furthermore, the model has to provide mechanisms to map objects to physical devices to support the activation and deactivation of contexts.

3.1.2.2 Preferences and Ambient Information

Individuals have different preferences in different situations. Sitting alone in a silent room might indicate that an individual is willing to receive incoming phone calls. However, the same individual can perceive this as a disturbance when involved in a conversation with others. With preferences, individuals express their choices of service characteristics in certain contexts. Therefore, preferences provide a powerful mechanism to influence the behaviour of I-centric services by giving them explicit instructions. Considering the example above again, the individual might have the preference not to be disturbed by incoming phone calls in such situations.

> *Preferences*: Preferences are conditional choices of service characteristics of an object, depending on context and ambient information. Preferences are applied to objects during the activation of a context.

I-centric services evaluate preferences to adapt their behaviour to what is 'really wanted' by an individual in a certain environment at certain moment in time. Therefore, preferences have to be either gathered from individuals interactively or automated by monitoring, and they have to be expressed in a machine computable form.

Gathering preferences interactively causes a lot of interaction between the I-centric system and an individual, who could feel bothered after a short period. A more desirable approach is the collection of preferences based on monitoring the behaviour of an individual over time. Inference mechanisms can be applied then to the collected data to predict what preferences a certain individual might have. This process has to run continuously to keep the preferences up-to-date or to make them more precise, step-by-step.

The description of preferences that can be processed automatically is another challenging task. Preferences can capture many aspects, such as mood, interests, life stage, etc. that are even hard to describe in words. Furthermore, the kind of preferences that are relevant to different individuals may differ completely. A model for describing preferences must be as generic as possible to avoid restrictions that might prevent the expression of a certain preference. On the other hand, the model has to provide some structuring, or categories, to allow the assignment of preferences to a certain I-centric service.

Preferences should be as precise as possible. One approach to teach this is to relate preferences to contexts and to information describing environmental conditions. Continuing the example given at the beginning of this section, only the existence of environmental conditions allows the specification of 'a silent room'. Besides the noise level, many other environmental conditions might be sensed by I-centric communications systems. To emphasize the variety of such kinds of information, the term *ambient information* has been coined by the ISTAG group in its 'Scenarios for Ambient Intelligence for 2010' [10].

In general, ambient information is information that can be collected, gathered, or sensed from the environment. Ambient information comprises temporal and spatial characteristics such as any user input, temperature, noise level, light intensity and presence of other people, just to give a few examples. Ambient information is sensed by sensing facilities, like motion detectors or microphones, and transmitted through sensor networks. Ambient information may also include geographical information (e.g. location), environmental information (e.g. temperature), and life conditions (e.g. blood pressure).

> *Ambient information*: Ambient information is information that can be collected, gathered or sensed from the physical environment using the objects of the individual communication space of a certain individual.

Furthermore, ambient information is the basis for ambient awareness. A semantic model is needed to describe preferences and ambient information. Such a model incorporates knowledge representation to qualify available information and ontology languages to relate syntax and semantics to each other (e.g. Semantic web, Agent ontology service). The focus for I-centric communications here is to define a harmonized semantic model that includes human aspects, as well as the process to gather, store, evaluate and exchange preferences and ambient information.

3.1.2.3 I-centric Services

I-centric services define, manage, and (de)activate contexts in an individual communication space, taking the preferences of individuals and ambient information into account. They support an individual (I-centric), adaptive, personalized and ambient-aware way to interact with objects in individual communication spaces. Based on the evaluation of 'profiles' that describe preferences, service capabilities, and sensed information about his actual environment, the individual can be provided with personalized services for his actual demands.

> *I-centric service*: I-centric services define, manage, and (de)activate contexts in an individual communication space, taking ambient information and the preferences of an individual into account. They support an adaptive, personalized and ambient-aware way to interact with objects in individual communication spaces.

Self-learning capabilities are used to profile the behaviour of individuals. Numerous services or several features of different services are combined on demand.

I-centric services need ambient information in order to adapt to the environment. Temporal and spatial characteristics are only two examples of information that may affect the service behaviour. Note, that a certain environment can restrict the functionality requested in a certain context. Interacting in a 'TV context' while driving a car may reduce the available functionality to 'record the movie for later viewing' or to listen just to the audio part.

I-centric services activate contexts by choosing the equipment to be controlled (presentation terminals, handhelds, microelectronic controlled devices) and their quality of service (volume, brightness, etc.) before being finally connected via heterogeneous networks (BANs, PANs, LANs, WANs) to create an I-virtual private network.

The process of choosing and controlling the equipment of the physical environment is supervised by the service logic of I-centric services. The service logic controls the activation of contexts by combining multiple objects dynamically. It parameterizes objects by defining what, when, and how one or more objects behave in a given condition. The service logic decides, based on profiles and on the status of the objects, how those objects should behave in a certain situation. This enables sensitive services that adapt to the environment dynamically. Therefore, mechanisms have to be developed, which enable one to gather and to evaluate profiles. The profiles and the status will be evaluated by a domain-specific service logic that also controls the object(s).

Nowadays, service logic is, in most cases, 'hard-coded'. Once implemented, it cannot be changed afterwards. The basic idea of I-centric communications, to provide individuals with their own services that might change over time, requires a more flexible approach. A generic model for describing service logic independent from any execution environment is needed. The description of service logic has to be an input parameter, like ambient information, during the execution of an I-centric service. This will allow altering the service behaviour according to changing situations. Moreover, creating new, or altering existing, service logic can become an interactive or automated process. Self-learning systems with automatic reasoning engines or interactive tools can then be applied.

The process of creating or modifying I-centric services has to be accompanied by ontology definitions that describe what services an object is providing. Interactive applications are envisaged that allow individuals to assemble their service by a simple 'drag and drop' mechanism. Like a LEGO™ toolbox, the individual should be able to create and to deploy his I-centric services.

3.1.3 Reference Model for I-centric Communications

Following the vision explained earlier, this section introduces the reference model for I-centric communications, as shown in Figure 3.2. It follows a top-down approach starting

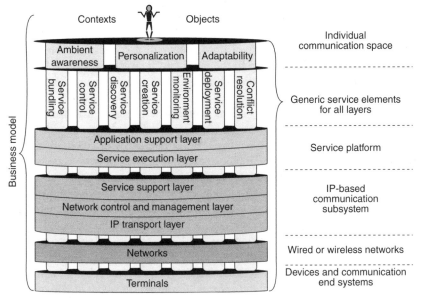

Figure 3.2   Reference model for I-centric communications

with the introduction of individual communication spaces, related contexts and objects. In general, the topmost layer recalls the I-centric vision that human beings interact with objects of their communication space in a certain context. Furthermore, it is common understanding that I-centric services have to support at least three different features, namely ambient awareness, personalization and adaptability. To emphasize that these features are needed for I-centric communications, they have been assigned to the individual communication space in the reference model.

The service platform for I-centric communications is responsible for shaping the communication system, based on individual communication spaces, contexts, preferences and ambient information. Preferences will be provided by the personalization feature, whereas ambient information has to be provided by the ambient awareness feature. More details are given in the section about the service platform for I-centric communications.

The IP based communication subsystem is responsible for providing the linkage between different objects in the communication spaces. These links have to be maintained and managed, even when they are subject to change because of roaming between different network topologies or access networks. There might be non IP-based communication networks underneath the IP-based communication subsystem. These have to be wrapped by bridging facilities (e.g. by SmartIP devices) to include them in I-centric communication systems. IP communication is seen as the common denominator by which to harmonize heterogeneous network infrastructures. The IP based communication subsystem consists of three layers:

- *Service support layer.* This provides well defined APIs for the service platform to access the IP based communication subsystem.
- *Network control and management layer.* This combines the traditional concepts of network management with required real-time aspects needed for system wide control functions.
- *IP transport layer.* This basically represents OSI layer four.

The wired or wireless networks layer implements all aspects of the physical connection(s) between different objects. Due to the hierarchical structure of the reference model, a connection in the IP-based communication subsystem might use multiple connections in the underlying network.

Devices and communication end systems provide the physical infrastructure that hosts all other layers. It can be instantiated as a switch responsible for connecting different networks, or even as a multimodal terminal able to interact with a certain individual.

The main features of I-centric communications (ambient awareness, personalization and adaptability) affect all layers. Therefore, supporting functions have to be provided as a vertical solution. The reference model introduces the concept of *generic service elements* that implement common functionalities on all layers. Generic service elements can be seen as a toolbox from which complex services can be assembled and executed dynamically. The vertical approach allows I-centricity on all layers, i.e. to establish I-centric private virtual networks.

Accompanying all of these technical issues, the business model for I-centric communication identifies the relationships and information flows between all active roles within an I-centric system. The business model helps to identify reference points between all involved entities and assists with the assessment of the applicability of I-centric services for different business domains.

3.1.3.1 Business Model

The vision of I-centric communications requires new business models. The borders between traditional roles and administrative domains: network provider, content provider, service provider and retailer are blurring. An individual may become service provider (ad hoc and peer-to-peer), or content provider, or retailer. Additionally, roles may change depending on context, which implies a very flexible business model.

> *Business model*: A business model is a description of how an entity, or a set of entities, intends to create value with a product or service. It defines the product or service, the roles and relations of the entity, its customers, partners and suppliers, and the physical, virtual and financial flows between them.

The business model for I-centric communications has to cover:

- roles, relationships and reference points;
- business topologies;
- service lifecycle (creation, deployment, management and billing);
- benefits for parties involved in the market value network.

The first objective of a business model for I-centric communication is a model for describing relationships between involved parties in a global business community. Based on these relationships, roles and reference points will be defined. This allows the participation of each business partner in a global business on one side, and provides freedom of development and integration on the other side. Reference points provide standardized points of contact and information exchange between business partners. Such a business model for I-centric communications is a prerequisite for the definition of a service architecture that supports the required functionalities of the whole business lifecycle.

3.1.3.2 Personalization

Personalization is considered to be the key factor for I-centric communications. Information and services must become increasingly tailored to individual preferences to make the usage of services easier and the perception of the individual communication space richer.

An extended personalization concept is needed that enables value networks (e.g. value chains) of content providers, network providers and service providers to offer personalized services to mobile individuals in a way that suits their needs, at a specific place and time.

Personalization aspects, like preferences, role and task, have to be integrated, and the relationship between the aspects must be studied. Investigations into benefits according to the user perception are necessary and need to be considered in the design of personalized services and personalization supporting frameworks.

For I-centric communications, this means that objects available in an individual communications space have to adapt to the preferences of individuals. Personalization models each individual in the I-centric service platform by managing his preferences and providing these preferences to I-centric services.

> *Personalization*: Personalization provides the information for modelling preferences for an individual communication space in the I-centric system.

To reach this goal, personalization federates profile information (containing preferences). Personalization incorporates dynamic behaviour to enrich stored and federated information to enable proactive I-centric services. This leads to an overall profiling infrastructure managing the individual preferences.

The main research issues behind personalization are:

- how to gain personal preferences from individuals (interactively or automated);
- how to store these preferences in profiles (profile format and categories);
- standards for exchanging profiles;
- how to secure privacy of sensitive parts of profiles; and
- how to describe active contexts using individual preferences.

3.1.3.3 Ambient Awareness

In I-centric systems, services will be tailored to contexts of the individual communication spaces. The services will automatically adapt themselves to changes in the environment. The services have to deal with a dynamic environment of nomadic individuals. The adaptation to the situation (in a certain context) is hidden from the individual, who is provided with optimal experiences and benefit. Conversely, the environment can be influenced by the presence and activities of the individual and adapt itself accordingly.

Therefore, ambient awareness deals with gathering ambient information from network, application, individual, terminal and contexts. The federation of ambient information from various sources, according to an individual's mobility and roaming, is an integral part of ambient awareness. Intelligent inference systems for missing information are needed in order to incorporate as much information as possible to provide an automatic, ambient-aware environment to the individual.

> *Ambient awareness*: Ambient awareness is the functionality provided by an I-centric system to sense and exchange information about the current environment an individual is in at a certain moment in time.

Sensor networks will play a major role in providing ambient information. Sensor technologies will be embedded in mobile equipment, communication networks, living and working environments to sense who the user is, where he is, what he is doing and what the environmental conditions are, in order to provide this ambient information to I-centric services.

Advances in sensor technology are necessary to enable further adaptation of services to the environment of individuals. Many devices in our current world already adapt in some form to their operating environment. One example is the television set that adjusts its image contrast to the ambient lighting level. 'Intelligent I/O behaviour' can be used to adapt the output characteristics depending on the context, but also the input characteristics should change in many situations (do not bother the individual with long lasting interactions, but combine the context information with the basic intention that was previously communicated by a few interactions). This intelligent I/O behaviour demands the development of multimode user interfaces and corresponding support functions in the various consumer devices, as well as in the service adaptation network.

3.1.3.4 Adaptability

Adaptability is mainly based on information provided by personalization and ambient awareness. It provides the functionality to adapt I-centric services to personal preferences and environmental conditions. Therefore, adaptability can be seen as a function that activates a context, based on whatever information is provided by ambient awareness and personalization.

In general, I-centric adaptability translates the wishes of individuals, which are usually inaccurate, incomplete and sometimes even contradictory, into a set of rules precise enough to be automated with sufficient reliability. It is the engine that activates a context at a certain moment in time in a certain environment.

> *Adaptability*: Adaptability is the functionality provided by an I-centric service to activate a context, taking ambient information and individual preferences into account. An active context might be adapted continuously to reflect changes of individual preferences or environmental conditions over time.

Typical situations when adaptation takes place include a substantial change in characteristics of connectivity, entering into a new service domain, or changing terminal device in the service session.

By technical means, adaptability requires the adaptation of media, content and service behaviour. During recent years, a variety of concepts for adaptation has been developed [11]:

- communication streams can be altered during transmission (e.g. bit rate adaptation);
- media types can be changed (e.g. text-to-speech conversion);
- the type of presentation can be adapted (e.g. downscaling an image to fit a PDA screen);
- the content of a message can be altered (e.g. adding or stripping off information); or
- the service behaviour can be modified (e.g. by customer service control functions).

Adaptability cannot be only reactive. When the battery of a mobile device dies or the connectivity breaks, many actions become impossible. However, something could have been done beforehand. Therefore, adaptation must also be proactive, which in turn requires predictability of the near future. An important area in prediction is to distinguish between the situations in which the human behaviour seems to be predictable and those in which it becomes unpredictable.

3.1.3.5 Service Platform for I-centric Communications

A service platform for I-centric communications is responsible for shaping the communication system, based on individual communication spaces, contexts, preferences and ambient information. Finally, it (de)activates objects (advised by the I-centric service), identifies causalities between them, based on sensed environmental data, controls the services offered by these objects, and converts data structures and operations for inter-working between services. The equipment is configured dynamically, its state is profiled, distributed objects are controlled, service creation and deployment are supervised, and the interworking among domains is enabled by the platform.

Service platform: A service platform is an infrastructure that supports the development and operation of I-centric services by providing a set of service features:

- execution environment for services and objects;
- supports (re-)deployment (hot-plugging) of services and objects;
- supports the (re-)binding (configuration) of services and objects;
- supports the interworking of services and objects.

The service platform consists of two layers:

- *The application support layer.* This provides well-defined APIs to applications, ser-vices and objects. It offers generic service elements that can be used by developers of these entities to ease and speed up the process of design, implementation, deployment and management.
- *The service execution layer.* This provides the actual run-time environment of applica-tions, services and objects. It supports their secure, QoS-aware and managed execution.

Moreover, the service platform provides functional blocks that directly support the I-centric approach. These functional blocks manage ambient information, preferences and adaptability to be offered to I-centric services. To fulfil the functionalities requested by I-centric communications, I-centric service platforms have requirements on the underlying communication subsystem. This is caused mainly by the empowerment of any individual to act as a service provider or network provider in a paradigm shift from a provider centric paradigm to a decentralized I-centric paradigm. From an I-centric perspective, this is done by sharing objects between different individuals, or by allowing another individual to use objects out of 'my' individual communication space.

On the other hand, the requirements are based on information that has to be provided by lower layers to the service platform. Traditional platform approaches (e.g. object-oriented middleware platforms) try to hide, as much as possible, technical parameters between the different layers. I-centric services have to be provided with ambient information. Therefore, the traditional concept of transparency [5–9] vanishes. Ambient information has to be provided through all layers. Only the I-centric service itself can decide what information is useful or not. Intelligent filtering mechanisms and translation rules will help the I-centric services to 'understand' the ambient information.

In technical terms, the requirements of the service platform to the underlying net-work are:

- peer-to-peer, ad hoc communication (single or multihop, ad hoc relays);
- bearer technology-aware stacks and applications;
- QoS in various areas (guaranteed streaming, data rate, real-time traffic);

- global roaming by integrating access networks supporting various wireless and wire-line bearers, various radio interfaces, and mobile-IP;
- ubiquitous addressing (address mapping, routing over addressless networks);
- multicast services (including advanced addressing methods like geo-cast);
- personal identifiers to overcome address schemes of traditional communication networks;
- sensor networks for gathering ambient information;
- privacy, assurance of the confidentiality of data;
- integrity, assurance that the content of a transmission cannot be altered during transmission;
- confidentiality, assurance of the confidentiality of information;
- nonrepudiation, assurance that the sender (receiver) cannot deny having sent (received) information;
- authentication, assurance of the identity of the sender and the receiver of information;
- access control/authorization, assurance that only authorized people can access information;
- accounting/billing, transparent billing for service usage and integrated network access.

However, the paradigm shift addressed here not only concerns the individual as a provider of network related services. A service platform allowing global mobility and transparent access to any kind of service over a common IP platform is the basis for allowing everyone to provide a wide range of services.

3.1.3.6 Generic Service Elements

I-centric communications systems will have to cope with issues like numerous service providers, always-connected individuals, automatic service adaptation and ambient awareness. Aspects like dynamic service discovery and service provisioning in (for individuals and services) unknown environments, and personalized services usage, requires new mechanisms to support I-centric communication systems.

To simplify the definition and realization of I-centric services and applications, a set of reusable software components will support functionalities common for different services and applications. These components are called *generic service elements* to emphasize their general applicability for all kinds of services.

Generic service elements: A GSE is a functional software component that can be used by other GSEs, services, or applications and it is hosted by the I-centric service platform. GSEs provide functionalities common to different services and applications to ease and shorten their development process.

Because I-centric services should work under changing environmental conditions, serving changing individual preferences, the most promising candidates for common functions are:

- *Service discovery.* A mechanism to discover service features dynamically that are provided within a certain environment or by a certain physical resource.

- *Service management.* How to manage context (dynamic relationships and causalities between individuals and their environment).
- *Service deployment.* How to deploy services in distributed environments.
- *Service composition.* Dynamic interworking of services will help to create and operate contexts.
- *Service logic.* The evaluation of the preferences and ambient information leading to a decision on what has to be done by an I-centric service.
- *Service control.* The process to control all the resources needed for a specific purpose.
- *Environment monitoring.* How to gain ambient information.
- *Reservation.* To manage exclusive usage of objects.

In technical terms, objects and generic service elements are quite similar. They have to provide well-defined interfaces to be used for service developments. Generic service elements can also be seen as enablers for objects, by providing functionalities that are common for all objects.

Consequently, well-defined collections of interface specifications designed for certain business domains are needed. The idea is to equip the same kind of objects with standardized interfaces for functional (usage) and nonfunctional (management) interfaces. In particular, from the area of telecommunications, open service APIs, like OSA/Parlay [W-OSA], build the basis for such interfaces. As the vision for future mobile systems does not cover communication scenarios only, new open service APIs have to be developed. On the one hand, the service infrastructure should be opened, by means of open APIs, to enable ease of use service creation. Such APIs must provide framework functionality, like hot-plugging of services, dynamic (re-)binding (e.g. in ad hoc environments), service mobility, AAA services, support for automated SLA negotiation [12], contracting, and so on. This generative approach may also require well-defined internal APIs and call-backs.

3.2 Business Models in the Future Wireless World

3.2.1 Introduction

Interest in the concept of business models has been closely linked to the rise of Internet-based e-commerce [13]. The (additional or alternative) channels offered by the online environment spurred firms to devise new ways of interacting with their customers, be it end users or other companies. Through various forms of disintermediation and reintermediation, more direct or value-added ways of interaction between firms and customers seemed to become possible. The expectation was that by experimenting in this way, a range of entirely new business models would be found and implemented, creating more value for the customer, and leading to major competitive advantages over firms with more traditional ways of doing business. Although these supposedly new internet business models have in the meantime, mostly proven to be existing business models set in a new environment, the fact remains that the internet has broadened the range of choices for firms on how to interact with their customers and end users.

In the telecommunications sector, interest in the concept of business models has been fuelled by the (partial) unbundling of technical functions and economic roles, caused mainly by technological developments and regulatory pressure, and the expectation of a range of new value-added telecommunications services. For telecommunications firms, the main questions to be answered by new business models are those connected with shifting firm boundaries and the complex provision of new services. The growing notion of a telecommunication system as a complex structure of cultural, process and technology components engineered to accomplish organizational goals, creates the need to analyse just what happens in such systems. This integrated view is what sets business models apart from other models for business design, i.e. process models, business cases, etc. In a business model, technical, organizational and financial aspects are combined in order to capture the complexity of new business structures.

This section presents the preliminary results of the discussions within the Wireless World Research Forum (WWRF) on possible business models in the future wireless world. Most of these discussions have been going on within the WWRF's Working Group 2 (WG2), dealing with the service architecture of the wireless world. The section first addresses different definitions of, and approaches to, business models, and attempts to formulate an integrated view. The later part of the section deals with the business models' underlying current, and emerging, internet services and mobile services, and explores some potential impacts of a future I-centric environment on wireless business models.

3.2.2 Business Models: An Integrated View

3.2.2.1 Definition

Despite widespread interest in the concept of business models, there is still no clear definition of the term. Different definitions emphasize different aspects, such as the architecture for a product or service, a description of roles and relations of a company, the way to do business, how a company goes to market, how value is added, how to make a business viable, etc. (see [14, 15]). The definition we propose here captures the main elements of the definitions mentioned above. It defines a business model as:

> . . . a description of how a set of companies intend to create value with a product or service. It defines the architecture of the product or service, the roles and relations of the actors involved, and the physical, virtual and financial flows between them.

This definition highlights three different levels of any business model: a functional level, an organizational level and a financial level. On a functional level, a business model describes the architecture of a product or service. On an organizational level, a business model demonstrates the roles and relations between actors, and the physical and virtual flows between them. On a financial level, a business model describes the revenue sources and financial flows for the actors involved.

Another feature of this definition is that it puts the emphasis not just on one firm but on the entire network of actors involved in the production, distribution and consumption of a product or service. This reflects the increasing complexity of innovation processes referred to in the introduction. From a financial point of view, it means that the emphasis is on structuring various revenue streams and constructing revenue sharing models.

In terms of the value chain, a concept coined by [16] to describe the primary value-adding activities of a firm or of a set of firms, this means looking at the whole chain. In fact, most scholars agree that the increasing complexity and flexibility of business design means that the representation of business processes by a linear value chain has to be replaced by more fluid value networks, in which roles and functions can be combined in different ways by different actors. Business design is therefore increasingly about defining firms' boundaries and the level of horizontal and vertical integration [17].

All of this implies that the success of a business model is dependent on it finding a 'fit' between different interests, on different levels [18]. Not only the fit between the firm's business model and the end customer is important, but also the fit between the business models of the different actors involved in manufacturing a product or producing a service. This fit will have to take place on all three levels described previously: the functional, organizational and financial levels.

3.2.2.2 Methodology

If the existing definitions of business models are not always very clear, the previous section showed that a common definition of business models can be constructed without too many problems. This is not the case when the methodology of business modelling is concerned. Therefore, the remaining part of this chapter will be devoted to this subject.

Modelling is an expression of concepts that allows each part of an organization or a system to understand it and contribute to its development. Models also promote understanding across different parts of an organization or a system. A good model can communicate much of a system's purpose to its stakeholders, whether they are employees, shareholders, or customers. Modelling can be applied to all stages of systems development.

What is a model? A model is a representation of concepts that:

- can be validated;
- can be checked for rigour and robustness;
- captures and communicates ideas;
- can be changed;
- can provide scenarios ('what if' and 'where do I fit in?').

What is not a model?

- a collection of drawings and documents;
- anything that cannot reflect the business changes;
- anything that cannot derive the impact of change;
- anything that cannot be navigated.

There are many modelling methods available. Some of these methods are designed to serve a specific purpose (modelling object oriented systems for instance) but are well-suited to performing outside their intended role (e.g. modelling business processes). Business modelling is the act of providing a set of descriptive representations (e.g. models) that are relevant for describing a system, such that it can be produced to management's requirements and maintained (or changed) over the period of its useful life.

When building a business model a common question that arises is what to model and when. There are different models that a business analyst might use in order to complete his or her part of the picture. These include organizational models, process maps, process flows, functional decomposition, requirement definitions, technology models, use case models and revenue models. In a business model, these views are combined according to the purpose for which the business model is used. This can be to articulate the value proposition, to define the position of the firm within the value network, to identify the market segment, to estimate the cost structure of a product or service, to formulate a competitive strategy, and so on [14]. Usually, depending on its specific purpose, the business model will describe business processes on one of the three levels – functional, strategic/organizational, financial – we described earlier. Each of these levels has its own specific subset of models, methodologies and terminology.

In telecommunications, it is usually the case that the technological infrastructure is set up before the actual business model [19]. As a result, functional models have prevailed in the telecommunications sector up to this point. However, this is changing. Liberalization has shaken up the structure of the sector over the last decade. Organizational and financial aspects are more and more becoming leading factors in new systems development. In the case of the development of wireless systems beyond 3G, which is still characterized by a large amount of uncertainty over technological as well as other factors, this creates the need to look for a generic business model containing all three levels, and outlining explicitly the way in which they interact with each other.

The next sections will look into some of the modelling approaches at the functional, organizational and financial levels, and into a possible 'fit' between them.

Functional Level

On a functional level, business models describe the architecture of a product or service. The ISO Reference Model for Open Distributed Processing (RM-ODP) is an international standard that can serve as a framework for describing architectures of open distributed systems.

An RM-ODP specification of a system consists of five different specifications, corresponding to five different, but related, viewpoints of a system. They are the enterprise, information, computational, engineering and technology viewpoints. Each viewpoint is an abstraction of the whole system focusing on a specific area of concern. The enterprise viewpoint is concerned with the purpose, scope and policies of the ODP system. When describing a distributed system using the RM-ODP framework, the RM-ODP enterprise viewpoint is typically used as a first step in the process.

The enterprise specification for an ODP system specifies the purpose, scope and policies for that system in terms that are meaningful for the stakeholders for that system. This specification is expressed in terms of the behaviour expected of the system by other entities within the enterprise. The term 'enterprise' is used here in its widest meaning, i.e. a configuration of entities that may be very large (e.g., a group of cooperating companies), more limited (e.g., a particular service inside a company), or much smaller (e.g., a service provided by an IT system).

The enterprise language introduces the concepts necessary to represent the behaviour expected of an ODP system by other entities within the enterprise and the structuring rules for using those concepts to produce an enterprise specification.

An enterprise specification is an abstraction of the enterprise, describing those aspects of the enterprise that are relevant to specifying what an ODP system, or set of ODP systems, is expected to do in the context of the purpose, scope and policies of this enterprise. In an enterprise specification, an ODP system and the environment in which it operates are represented as a community. At some level of description the ODP system is represented as an enterprise object in the community. The objectives and scope of the ODP system are defined in terms of the roles it fulfils within the community of which it is part, and policy statements about those roles. A community is defined in terms of each of the following elements:

- the enterprise objects comprising the community;
- the roles fulfilled by each of those objects;
- policies governing interactions between enterprise objects fulfilling roles;
- policies governing the creation, usage and deletion of resources by enterprise objects fulfilling roles;
- policies governing the configuration of enterprise objects and assignment of roles to enterprise objects;
- policies relating to environment contracts governing the system.

Community is the key enterprise concept in RM-ODP. A community is defined as a configuration of objects formed to meet an objective. The objective is expressed as a contract that specifies how the objective can be met. A contract is a generic RM-ODP concept that specifies an agreement governing part of the collective behaviour of a set of objects. A contract specifies obligations, permissions and prohibitions for the objects involved. Therefore, a community contract specifies the roles within a community, the relationships between the roles and various policy statements about the roles and about the enterprise objects that fill them. In situations when two or more groups of objects, under control of different authorities, engage in cooperation to meet a mutual objective, they form a special kind of community, called a federation.

A role is a specification concept describing behaviour. A role may be composed of several roles. A role serves as a placeholder for an object. An enterprise role is defined in terms of the permissions, obligations, prohibitions and behaviour of the enterprise object fulfilling the role. A role thus identifies behaviours to be fulfilled by the enterprise objects comprising the community. An enterprise object is an object that fills one or more roles in a community.

A policy statement provides an additional behavioural specification. It may be either a constraint or an empowerment. Examples of policy statements are the statements of permissions, prohibitions and obligations related to the roles or to the enterprise objects. In particular, these statements relate to:

- interactions between enterprise objects fulfilling roles;
- the creation, usage and deletion of resources by the enterprise objects fulfilling roles;
- the configuration of enterprise objects and the allocation of enterprise objects to the roles; and
- the environment contracts governing the system.

A model of the key enterprise language concepts is shown in Figure 3.3 using UML (Unified Modelling Language) notation. From this model, specific enterprise models can

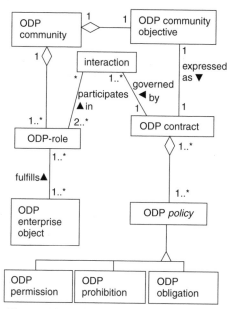

Figure 3.3   Model of community concepts

be derived. Note that ODP policy is an abstract class, so only its subclasses can be instantiated.

The specification parts discussed incorporate both the functional and nonfunctional aspects of the system to be built. The functional aspects relate to the description of what the system does. Nonfunctional descriptions address how well this functionality is (or should be) performed if it is realized.

So, although the environment is liberalized and competitive, the cooperative relationships between service providers, with network operators and with customers, is vital when providing end-to-end services, especially in order to react flexibly to the customer's requirements for tailored service delivery. The introduction of advanced services should not be hindered by the lack of coordination between the stakeholders regarding management of these services. There must therefore be a consistent management understanding and functionality across all domains involved in managing such services.

For the last decade, the Telecommunications Information Networking Architecture (TINA) Consortium has been working towards creating general ODP models. TINA is intended to be applied to all parts of telecommunications and information systems. It is divided into three subarchitectures: the computing architecture, the service architecture and the network architecture.

The complexity of a telecommunications and information system is captured in three types of model: the business model, the information model (modelling information-bearing entities and their relations to each other) and the computational model (describing computational objects and their relations). TINA is task oriented: relations and roles are described per session.

The business model describes the different parties involved in service provisioning and their relationships to each other. A small number of roles are defined, which reflect the major business separations of a complex telecommunications and information market: consumer, retailer, broker, third party service provider, content provider and connectivity provider.

Two important notions in TINA business models are *administrative domains* and *reference points*. An administrative domain is the unit which owns the informational, computational and engineering objects of the system. Reference points are interfaces describing the interactions taking place between the different roles. For example, the reference point 'retailer' describes the relationship between the consumer and the retailer.

Organizational Level

At the organizational level, a business model describes the roles and relations of the different actors and the physical, information and financial streams between them. There are two kinds of model which are commonly used to describe roles of actors in the telecommunications sector, and which we will deal with here. The first type is a layered model corresponding with the different layers in telecommunications service provisioning, the second type is a more general value chain/value network type of model.

Layered Models

Layered models take into account the different layers in telecommunications service provisioning, such as infrastructure, transport or end user services. An example of such a layered model is the simple enterprise model, as shown in Figure 3.4.

In the simple enterprise model, the primary purpose of a business model is to identify interfaces that are likely to be of general commercial importance. In order to do this, a number of roles are identified which describe a reasonably well-defined business activity that is unlikely to be subdivided between a number of players. In addition, it should be anticipated that the roles would have a reasonably long existence. The interfaces surrounding the role must persist for some time in order that customers and suppliers of the role can successfully interact with it. In addition, many players may choose to take on the same role, in which case the role becomes a competitive activity. The role also needs to be reasonably stable for a successful competitive marketplace to emerge.

According to the simple enterprise model, services flow through various functions or organizations, such as the service creation function or service provider, brokerage function or broker, etc. Based on this value flow, it is possible to classify interconnection types.

In understanding the roles and the nature of the services flowing between them, it is possible to identify more generic features that are needed to support the roles and the

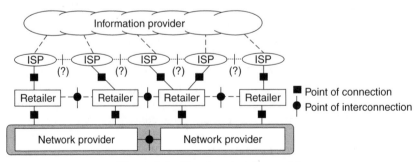

Figure 3.4   Simple enterprise model

interfaces between them. As well as the services, other entities must pass between roles, including information and contractual and legal obligations.

The first distinction made in the simple enterprise model is between *roles* and *players*. A *role* is a reasonably well-defined business activity and will form, along with other roles, a value chain by which services can be produced for end users. A *player* is a defined organization, for example a company or government body, which undertakes one or more roles.

The simple enterprise model is concerned only with roles. This method of definition allows a clear translation from business activity to the relationships which may be the subject of standardization. The information industry has a set of structural roles which form a primary value chain, including:

- *Information ownership role.* There is frequently value in owning the source of information; for example, a gallery owns pictures, images of which can be sold.
- *Provision of information and related content role.* This role takes raw information and makes it available for inclusion in information services and information-based services; for example, it could provide a library of photographs which could be used by travel agents in describing their services.
- *Provision of information-based services role.* The role builds information and information-based services which it makes available to end users. This includes services that deliver information – for example, a yellow pages service – and also includes services where information is only part of the service – for example, home shopping – where, in addition to the information transactions, physical goods must be delivered to the end user.
- *End user role.* The end user can be either a private individual or a role in another industry which requires information-based services.
- *The structural roles.* These, as well as the relationships between them, require support services and these are supplied by infrastructural roles. The set of infrastructural roles includes:
  - *Communication and networking of information.* This role provides a general, distributed platform on which information applications and services can be supported. It provides the means by which an application and its users can be distributed without being aware of the nature of the distribution. It will invoke the application when requests are sent to it, support messaging to and from the application and also between components of an application. It will include capabilities to support directory services, navigation, security, payments, browsing and searching, etc.
  - *Distributed information processing and storage service provision.* This role provides a processing platform on which applications can run and can store data. This role will involve the use of an information appliance.
  - *Generic communications service provision.* This role provides telecommunications services for the transport of data, voice and video.
  - *Service and application creation support.* This role supplies features which facilitate the production of services and applications which can use communication and information networking.
  - *Terminal equipment supply.* This role supplies information appliances to end users which can form an integral part of the distributed platform operated by communication and information networking.

The relationship between the roles is commercial and so there is a customer side and a supplier side to the relationship. In addition, this relationship needs to be supported by brokerage and management services. These enable a customer to find a supplier and a commercial transaction to be negotiated and completed, respectively.

Since there can be more than one player undertaking the customer role and more than one player undertaking the supplier role, a relationship between roles also represents a competitive marketplace. This marketplace trades the intermediate services that are passed between the roles, and establishes the price at which they are traded.

The simple enterprise model also provides a semi-formal definition of business modelling terms:

- *Role:* The role is a business activity that fits in a value chain. The role is constrained by the smallest scale of business activity that could exist independently in the industry, and so a marketplace will exist for every relationship between roles.
- *Value chain, complete value chain and primary value chain:* A 'tree' of roles are connected together to make a service. The total set of roles involved in producing a service, and the way they pass intermediate services between the roles, is called the complete value chain. The set of roles which form the only principal activity of a generally recognized industry that produces the service are the primary value chain. All the other roles in the complete value chain will be providing support services for roles in the primary value chain.
- *Structural role:* A structural role is a role in the primary value chain of an industry. A structural role will therefore involve a business activity that is directed towards that industry and, in general, only towards that industry, and the output services of a structural role will be directed, in general, only to the next structural role in the primary value chain.
- *Infrastructural role:* An infrastructural role is not in the primary value chain of the industry which is under consideration, but supplies services for one or more roles to the primary value chain of that industry. The business activity of an infrastructural role will normally be directed towards many other roles, even roles belonging to more than one primary value chain. The output services of an infrastructural role are likely to be based on reusable components in order to meet the requirements of its many customer roles. However, an infrastructural role may itself belong to a primary value chain within its own industry, and therefore be a structural role from the perspective of its own industry. From the perspective of the roles that it is supplying, it is an infrastructural role.
- *Player:* A player is an organization, or individual, which undertakes one or more roles. The player can be a commercial company, a government agency, a nongovernmental organization, a charity or an individual.
- *Relationship between roles:* When intermediate services are passed from one role to the next in a value chain, there is a relationship between the roles. The relationship implies that a marketplace exists which can match players who undertake the role on one side of the relationship with players who undertake the role on the other side of the relationship. Some players may choose to be integrated and undertake both roles, in which case the relationship is within the players' domain.

- *Horizontal relationship:* This is the relationship between two adjacent roles that are in the same primary value chain. The roles are therefore in the same industry. A horizontal relationship can also be called a structural relationship.
- *Vertical relationship:* This is a relationship between two roles that are not in the same primary value chain. One role will be supplying services in order to provide some of the infrastructure required by the structural role.
- *Segment:* A segment is the entity which is common to the business modelling, the structural modelling and the functional modelling. A segment is part of one role, owned and operated by one player, part of one (and only one) service provisioning platform, and part of one domain, and is composed of a well-defined set of functions.

Value Network Models

The simple enterprise model has already made a number of references to value chains. Just like traditional value chain models, these types of layered model are characterized by a rather strict hierarchy and linearity, dictated by the different layers in traditional telecommunications systems. As the linear hierarchy and sequential dependencies associated with telecommunications systems are decreasing, these models are increasingly starting to resemble the more general value systems or value network models.

In recent years, the concept of value networks has gradually replaced that of value chains to describe sequences of value-adding activities. Authors such as [20] and [21] have argued that Porter's value chain [16] and value system analysis fail to include a number of activities that contribute to what is considered of value for the end customer. The value network concept attempts to broaden the value chain perspective by considering links with both upstream and downstream value chains. In this view, a value network is a set of relevant activities behind a product or service offering. These activities, and not the actors performing them, are the primary nodes in the network. The relationship between the nodes may consist of flows of materials, services, information or financial resources.

Actors in a value network may be responsible for one or more activities and may participate in more than one value network. They may be suppliers, producers, intermediaries or users of a certain good or service. Among the actors is the end customer, who not only receives and (usually) consumes the value created, but may also participate in value-creating activities. Håkansson and Johansson [22] name the following characteristics of actors in networks:

- actors control and execute activities. They develop relationships through repeated interactions;
- each actor is embedded in a network of strong and weak relationships;
- each actor controls a number of assets. This control can be direct or indirect, i.e. through intermediaries;
- each actor has specific objectives. Changes in the network occur as a result of the pursuit of these individual objectives;
- actors have specific knowledge about specific parts of the network. This knowledge is mainly developed on the basis of past experience.

How do actors position themselves in the network? One implication of the value network concept is that company strategy is no longer a matter of positioning a fixed set of activities along a value chain [23]. The focus of strategy is shifting from the company, or even the industry, to the value-creating system itself, within which different economic actors (suppliers, partners, customers) coproduce value. The key strategic task is the reconfiguration of roles and relationships between this constellation of actors in order to create an improved fit between a firm's competencies and its customers' needs.

The various activities may be coordinated by the market, a hierarchy or intermediate forms of coordination. This raises a crucial question regarding value networks, i.e. that of control over the network [20]. The control an actor can exert over a network is dependent on:

- an actor's resources (e.g. knowledge or financial resources);
- functional structures (e.g. the structural importance of specific nodes);
- time-dependent structures (e.g. trust relationships, established practices).

The fact that time-dependent structures play an important role in the control over, and thus the shaping of, a value network, means that any business model design will have to take into account the present-day situation, evolutions and trends. Value networks are not formed overnight but are evolving. Being in a network is not merely a rational choice backed by a business model, but also the result of trust relationships through repeated transactions, partnerships, personal networks, etc.

Financial Level

Financial models can be used to sustain a business case. A business case is a specific business model, (usually) aimed at demonstrating the profitability of a specific product or service. In sectors such as telecommunications, the field of techno-economics deals with the financial impact of different technically feasible service options and provides a number of tools and models for constructing business cases. One example is the model developed in the European TONIC project, which is a techno-economic tool to model market conditions, services and technical architectures, and the resulting cash flows [24].

The most important elements of financial models are investment decisions, projected revenues, estimated costs and potential risks. Traditionally, investment decisions are based on a calculation of the net present value (NPV) of a project. Other important parameters are the internal rate of return (IRR) and the payback period of a project. These can be defined as follows:

- *NPV.* The discounted value of future cash flows, which consist of cost and revenue streams, plus the discounted rest value of the project. If the NPV of a project is positive, the project is usually deemed profitable.
- *IRR.* The discount rate at which the NPV is zero. If the IRR is higher than the opportunity cost of the necessary investment funds, the project is viable.
- *The payback period.* The amount of time until the cash balance of a project becomes positive. This is when return on investments takes place.

The costs to be taken into account in order to calculate the NPV can be broken down into fixed costs (in telecommunications these are often sunk costs), operational costs, marketing costs, licensing costs etc. Revenues depend on market penetration, usage, average return per user, etc.

Starting from a basic NPV model, sensitivity analysis can be conducted to see how certain elements influence the NPV, as well as risk analysis in order to provide insight into major risk factors. These are also quantitative calculations.

However, a number of enhancements to NPV analysis are necessary. First of all, the NPV can be enhanced by real options theory, which takes into account elements such as sunk costs, timing and uncertainty [25]. Secondly, the NPV can be enhanced by the use of game models simulating market dynamics and taking into account network externalities. Thirdly, the use of other types of indicators and approaches is often necessary in order to do justice to the high level of uncertainty in the future that is to be modelled. A classic quantification problem is that of the expected revenues. To estimate revenues requires the modelling of demographics, user preferences, demand, product or service lifecycles, etc. This is a very difficult task, especially when services are concerned [18]. The value proposition of services, and the relationship with revenues, requires the introduction of qualitative elements and scenario-type approaches in techno-economic modelling.

There are also more structural elements to financial business models, which relate to the network character of the sector. This is reflected in the introduction of sharing models, i.e. revenue sharing models, investment sharing and cost sharing models, and risk sharing models.

The most important structural element is the commercial relationship with the customer. This relationship between the originator and the user (organization or consumer) can be conceptualized as having three modes [13]:

- *The transaction mode:* The basic relationship between the originator and the user of specific goods or services. This can be a direct transaction or an indirect transaction (by means of intermediation);
- *The revenue mode:* How income is generated. This can be by procurement, by retail, by wholesale or by brokerage;
- *The exchange mode:* The way value is established. This can be by a fixed price transaction or by a negotiated price transaction.

Fitting the Models

As business practices become increasingly diverse and complex, the focus of business models shifts from the enterprise architecture to a specification of interactions [26]. The previous sections demonstrated that a business model should concern itself with the interactions between activities (or roles) and between actors, but also with the interactions between the different levels of a business model.

In an overview of the current understanding and use of business models in literature and practice, Alt and Zimmermann [27] identify six different dimensions to business models. These are:

- *Mission:* A high-level understanding of the overall goals, vision, and the value proposition, including the basic product or service features.
- *Structure:* Determines which roles and agents constitute and comprise a specific business community, as well as the focus on industry, customers and products.
- *Processes:* Provide a more detailed view showing the elements of the value creation process and which requirements they address in the customer process.

- *Revenues:* Are the 'bottom line' of a business model. This dimension deals with sources of revenue and necessary investments.
- *Legal issues:* Have to be considered with all dimensions of business models.
- *Technology:* Both an enabler and a constraint for the other dimensions.

The authors emphasize that not only these six levels, but also the dynamics and the interdependence of the levels, have to be considered. From the previous sections, a simplified framework with three dimensions can be constructed, i.e. functional, organizational and financial. This is illustrated in Figure 3.5.

As stated previously, in mobile telecommunications the technological infrastructure is usually more or less determined before the business model. Compared to most other businesses, this has led to a predominance of functional models. It is an open question whether this will remain the case in the future wireless world. However, it seems likely that technology will retain a role more important than that of an enabler (as opposed to, for instance, legal issues), and that it will remain strongly connected to the design of business processes. This is the functional level that was referred to earlier.

On the other hand, process design, increasingly a matter extending across different actors, has become, in part, a structural and strategic element. The overall mission of a business model cannot be seen as independent from this structural level either. This overlap is referred to in the organizational level. As was stated above, the level of control exerted by the different actors involved, i.e. the measure in which partners play a structural or merely a supporting role, is one of the most crucial aspects to be considered.

Finally, the financial level is in effect the 'bottom line' of the model, although elements such as revenue sharing and the design of the commercial relationship between originator and consumer, have also become defining features of the business structure.

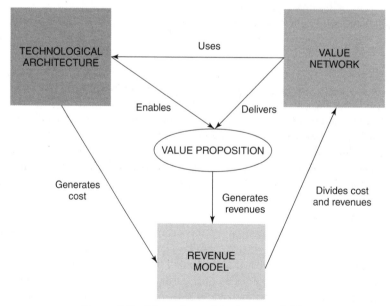

Figure 3.5 Dimensions of business modelling

These three different levels can therefore be seen as three layers of the same structural 'map'. At a high abstraction level, a generic business model is comprised of actors, i.e. persons or organizations acting in a product or service market, the various roles these actors play, and the relations (contractual or not) between these roles in the realm of a specific business case.

In an increasingly unbundled telecommunications sector, different functions in the architecture of a product or service lead to different economic roles. These roles can be combined by actors in a number of constellations, influenced by the dependencies and hierarchies in the network. It seems unlikely that it will be possible to describe wireless systems beyond 3G with strictly hierarchical layered models. There will, however, remain primary roles or activities that can be referred to as central nodes within the value network, and less central, dependent or enabling ones. Revenue sharing and the design of the commercial relationship between originator and customer of a product or service are important parts of the relations described in the business model. These relations, or interactions, and the interfaces that make them possible, form the core of the business model.

3.2.2.3 I-centric Business Models

The Wireless World Research Forum's Book of Visions (2001) [28] provides a conceptualization of future wireless, internet-based services as I-centric services. However, the outlines of any associated I-centric business models are not yet clear. This section briefly describes the current and emerging business models underlying internet services and mobile services, and explores some potential impacts of the I-centric environment on future wireless business models.

Current and Emerging Internet and Mobile Business Models
There is extensive literature from both academia and practice on so-called internet or e-business models. In fact, the whole business model debate, as developed over the last couple of years, has been largely centred around the anticipation that the internet would thoroughly alter the overall commercial environment, and make new business models a necessity.

There are a number of taxonomies of such internet business models available (see, e.g., [29]), although nearly all lack an underlying systematic framework, and therefore often seem quite arbitrary. As Hawkins [13] observes, most of such online business models emulate offline models. In these cases, the novelty of e-commerce or e-business is simply that commerce or business is taking place in a new, electronic environment. However, there are some structural changes caused by the advent of the internet that can be observed, and which are relevant for future wireless services and business models. The most important ones are:

- *New channels:* The online environment offers companies new channels to reach the customer. 'New' business models are formed when these channels are used as alternatives to, or in combination with, traditional channels.
- *Disintermediation and/or reintermediation:* A lot of the 'new' internet business models are based on disintermediation or new forms of intermediation. As the internet makes a direct relationship with the customer possible, it may serve to lower transaction

costs and thus lead to disintermediation. Just as well, the internet has made a new range of value-added reintermediation services possible, thus stimulating changes in business models.

- *New role of the user:* The user has become an integral part of e-business models, in the sense that the internet provides the user with opportunities to take on a variety of new roles extending beyond that of mere consumption, and to be involved in the business process as an individual, rather than as a generic consumer.

As the success of the Japanese I-mode services has been in part attributed to I-mode's supposedly superior business model, the particularities of this model have been well documented (see, e.g., [30]). However, as far as the whole field of mobile services is concerned, a systematic taxonomy and comparison of mobile business models on the value network level is still lacking. This constitutes an important challenge for any research into current and future wireless business models.

A report by the Yankee Group [31] provides an attempt at 'deconstructing' the emerging mobile value network (see also [32]). The most striking difference with the traditional mobile value chain is that the latter is characterized by linear sequential dependencies, while the former is organized in the form of parallel, but interlinked, tracks of different chains and systems. The Yankee Group describes a mobile value network existing of five major value chains. They refer to:

- *Network transport.* Network operators have traditionally integrated the whole network operating value chain, consisting of spectrum brokerage, mobile network transport, and mobile service provisioning. They are often labelled as gatekeepers, both in terms of customer ownership and in terms of ownership of limited resources such as spectrum and operating licences. With the subdivision of telecom groups into fixed and wireless operators, and the advent of so-called mobile virtual network operators (MVNOs), some fragmentation of this value chain can be expected.
- *Applications operation.* The application environment includes application developers, systems integrators, and applications operators. Companies that bundle these activities are also labelled wireless application service providers (WASPs). WASPs may develop and host applications for end users, but they may also concentrate on providing solutions for mobile network operators. This means that there are strong links with middleware/platform providers (see below).
- *Content provisioning.* This value chain consists of content providers, content aggregators and portals. Portals also serve as wireless internet service providers (WISPs), as they become the gateway to internet content.
- *Payment processing.* Traditionally, network operators have had the only billing relationship with the client. With the possible advent of mobile commerce, requiring a number of mobile financial services, other parties, such as banks, specialized billing companies, and mobile commerce platform vendors, have opportunities to get involved in this activity.
- *Providing device solutions.* Handset vendors are a well-established part of the mobile value system. As they provide hardware, as well as software, solutions, they not only have access to the user because of the direct buying relationship, but they can also preset the operating and browser systems running on the handsets to their own advantage.

In addition, there are two 'enabling' value chains involved:

- *Network equipment provisioning.* Companies providing network equipment are, e.g., Ericsson, Nokia, Motorola, Alcatel, Nortel. Traditionally, infrastructure vendors provided a relatively standardized product. However, this is changing as new applications and middleware (see next bullet) are being developed by these companies.
- *Middleware/platform provisioning.* This is becoming an ever more important part of the wireless value system. Examples are WAP gateways, SMS gateways, mobile portal platforms, mobile commerce platforms, and other applications platforms.

A lot of speculation has been made about the precise configuration of these interdependent chains in the future wireless value network. In general, it can be argued that business models for mobile services have traditionally been characterized by an important dependency on the underlying technological infrastructure, resulting in a rather closed model with a central 'gatekeeping' role for the mobile network operator. Recent research [33, 34] shows that this constellation is still valid with the advent of new services over GPRS systems, although there are a number of profound underlying changes that are becoming visible.

Handset and network vendors are increasingly central in the value network, even more so as they are providing more and more of the platform and middleware functionalities;

- the billing relationship with the customer, however, is still largely held by the mobile operator, although it is no longer restricted to the operator;
- there is no well-defined content provisioning model yet, with the I-mode model and the messaging model being, at this stage, the most successful ones.

In terms of near-future UMTS and WLAN systems and services, some of the decisive factors shaping future wireless business models are starting to appear:

- the emerging network environment will be characterized by heterogeneity (2G, 2.5G, 3G, WiFi, etc.). Interoperability, rather than integration of these networks, will become a crucial factor determining new wireless business models (see, e.g., [35]);
- a central question will be whether new solutions will be network-centric or terminal-centric. Up until now, three different, partly conflicting approaches have been observed: 'service-based approaches' (e.g. I-mode, Vodafone live), 'platform-based approaches' (e.g. MS Smartphone, Symbian Series 60), and 'protocol-based approaches' (e.g. SMS, MMS) [36];
- Ad hoc networks might be a disruptive technology in the mobile telecommunications industry and may threaten the position of (some of) the mobile network operators [37];
- Capturing value from information goods, which are widely perceived as being public goods, will remain an important challenge for any business model for mobile services [38].

The outcome of these factors and their impact on the future wireless value system is still very much under debate. In any case, the actors involved will continue to determine their business model by bundling different elements of these value chains (e.g.

network equipment vendors that are also developing middleware platforms and turning into application providers, or that are even turning to content provisioning and aggregation) and by (re)designing their interactions with other parties in the wireless access value system, including the customers (e.g. MVNOs or WASPs that have a direct billing relation with the user, and also mobile banking services involving financial institutions).

3.2.2.4 Towards an I-centric Business Model

The I-centric Vision

Following its aim to conceptualize future wireless developments that go beyond the current and emerging systems, the WWRF's Book of Visions (2001) [28] stated that future services will adapt to individual requirements, i.e. they will be I-centric. The communication system will provide the intelligence required for modelling the communication space of each individual, adapting to his interests, environment and life stage.

I-centric communication considers human behaviour as a starting point by which to adapt the activities of communication systems. Human beings do not want to employ technology, they rather want to communicate and act in their individual communication space. They may meet with others to talk and to celebrate, they read and travel, they listen to news or to music, they make decisions, etc. The individual communication space grows and shrinks on the time axes, based on the individual life stage.

Human beings are interacting with objects in their communication space in certain contexts to solve the problems of daily life: money and bank accounts need to be managed, food has to be bought and to be prepared for eating, movies can be watched for entertainment, places are visited and news consumed to increase knowledge, other people are met for discussions.

A context represents a certain 'universe of discourse' in an individual communication space at a certain point in time. Contexts and the related objects define the communication space of a human being. In general, human beings communicate with 'objects' in their environment in a certain context.

Note that the same objects may pertain to different contexts and to different communication spaces. Objects pertaining to a certain context can be active or passive at a certain moment in time depending on the situation of the user. They can be activated or deactivated by the user or environmental conditions. They can be directly addressable or represent a set of physical entities performing a certain service as a whole.

A user might have different preferences in different situations. Sitting alone in a silent room might indicate that the user is willing to receive incoming phone calls. However, the same user can take it as a disturbance when involved in a conversation with other people. To be I-centric requires knowledge of the actual situation of a user. An active context defines the relationship of a human being to a particular number of objects in its communication space at a moment in time, in a certain environment. I-centric communication systems have to be aware of the context a user is in, and have to adapt their service provisioning to that very context in a certain temporal and environmental situation.

The framework consisting of communication space, contexts and objects provides a powerful model for describing the I-centric paradigm, not only from a service architecture point of view but also for modelling administrative domains, security and trust, resource and traffic engineering of the communication subsystem, and encapsulation of heterogeneous end systems, etc.

I-centric services describe the ability to define and to manage contexts that are tailored to the preferences of single users. Based on the evaluation of 'profiles' that describe user preferences, service capabilities, and on sensing information about its actual environment, the user can be provided with individualized services adapted to his present environment. Self-learning capabilities can be used to profile the behaviour of users, services can then be assembled on demand and appropriate terminals and conversion strategies evaluated.

Interaction with the environment is required for a certain context in order to be able to adapt to it. Temporal and spatial characteristics are only two examples of information that may affect the context. Temperature, noise level, light intensity, presence of other people and objects in the vicinity are additional parameters which may help to adapt the applications to the user needs and profiles. Note that a certain environment can restrict the functionality provided by a certain context. Interacting in a 'TV context' while driving a car may reduce the functionality to 'record the movie for later viewing'. Sensing the environment provides the information necessary for the choice from several types of equipment which have to be controlled (presentation terminals, handhelds, microelectronic-controlled devices) and their quality of service (volume, brightness, etc.). These types of equipment are connected via an access network (infranets, intranets, internet, extranets, telecommunication networks) to create a dynamic, I-centric virtual private network.

A service platform will be responsible for shaping the communication system, based on the contexts identified and the actual sensed environmental information. It activates the objects involved in the context, identifies causalities between them, based on sensed environmental data, controls the services offered by these objects, and converts data structures and operations for interworking between (possibly unrelated) services. The equipment is configured dynamically, its state is profiled, distributed objects are controlled, service creation and deployment is supervised, and the interworking between domains is enabled by the platform.

The multitude of devices, wearables, different telecommunication technologies, positioning and sensing systems, location-aware or context-aware applications, etc. can be seen as enabling technologies for I-centric communications. Universal information access (including service interworking, media conversion), flexible control of equipment and facilities (e.g. smart homes), and personal communications (supporting personal mobility and terminal mobility) form the basis of such systems.

The aim of the service architectures in the wireless world is to devise a framework that models the communication behaviour of human beings. This will lead to an expandable system that is almost invisible to the user, that requires no time-consuming configuration, and provides customized interfaces to each single user, based on his own preferences and situations in time:

> I-centric services define additional requirements on the communication subsystem and smart IP devices. A further aspect of I-centric technology is the empowerment of any user to act as a service provider or network provider (ad hoc networks) in a paradigm shift from a provider – centric paradigm to a decentralized peer-to-peer paradigm.

Significant technological advances in recent years in the areas of palm-sized computers and wireless communications, accompanied by an infiltration of the internet in all aspects of our lives, have given reason to analysts worldwide to forecast that, in the course of 2003, almost half of the internet population will consist of mobile access devices. By that time, a variety of different wireless network technologies with different properties, capable of transporting internet traffic will be available. IP-based technologies that allow the integration of available heterogeneous and homogenous networks into a single platform capable of supporting user roaming between them, while not interrupting active communications, are already gaining importance. This development will be assisted by the rise of new mobile devices capable of flexible loading and maintaining various access interfaces (e.g. using SDR technologies) that will allow connectivity over a range of providers and technologies. Finally, the emergence of a plethora of access devices will dictate the liberation of users from a single device (i.e. mobile phone) and allow mobility between devices, even as the user is communicating.

However, the paradigm shift addressed here not only concerns the user as a provider of network-related services. An integrated network platform allowing true global mobility and transparent access to other nodes over a common IP platform can function as the basis for a technology that will allow everyone to provide a wide range of services, ranging from typical internet services to 'traditional goods', like physical products, as well as conventional services, to other customers. The main characteristics of peer-to-peer systems as they are starting to emerge today are:

- all nodes are equal – there is no distinction between client and server nodes;
- there is no central element of control (and thus no central authority);
- There is a strong sense of 'locality', 'proximity' or 'community' in which a node operates.

The technology needed to support the aforementioned shift to an I-centric model where every node can function as provider or customer is a dynamic service discovery mechanism. This can locate any available service according to criteria defined by the user, regardless of availability of internet connectivity, physical position of the user or the service provider, virtual position in respect to network bearer technology or segment of the network.

Implications for Business Models

The vision for I-centric communications has two major implications for future business models:

1. Changing constellation and increased flexibility of roles and actors.
2. Influence of feasible business models on functional architecture.

The increased flexibility of roles and actors refers to the increasing unbundling of functions and roles and to their peer-to-peer characteristics. The introduced vision implies the convergence of traditional telecommunication systems, internet-based systems, and the emergence of new applications. The borders between traditional roles and administrative domains, such as network provider, content provider, service provider and retailer are blurring.

An individual user may become service provider (ad hoc networking), or content provider (e.g. music sharing), or service provider (peer-to-peer), or retailer. Additionally, the roles may change in the same active context, implying a very flexible business model.

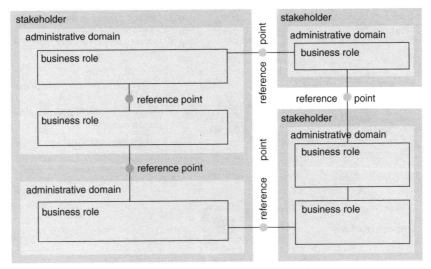

Figure 3.6   Generic business model

Different scenarios on the combination of roles and functions, which in turn influence, and are influenced by, the business topologies can be envisaged here. The business topology determines where intelligence (service/business logic) is realized (on the terminal or inside the network) and whether the service or the individual user is mobile.

Figure 3.6 shows a general business model which has been introduced by TINA-C. This model is general enough to be applied to any business domain or application field, such as I-centric communication systems. It identifies administrative domains as working environments for dedicated business roles, Administrative domains are separated from each other. The communication between administrative domains is done via well-defined reference points. A reference point specification does not only contain an API or protocol specification. Moreover, a reference point is characterized by constraints, contracts, QoS requirements and security definitions assigned to the communication exchange between two different administrative domains. Such information is also called a service level agreement (SLA). Contracts define the relationship between different business roles. Constraints define additional requirements between different administrative domains or business roles. For example, one constraint for the communication between two business roles can be the existence of a third one that acts as a trust centre for their communication. Stakeholders represent the highest aggregation level. They compose different administrative domains.

To illustrate the flexibility of such a business model, a short example for a business interaction is given in the following. The example uses four different business roles: customer, retailer, provider and user.

From step 1 to step 16 (see Figure 3.7), the complete process of offering services at the provider side up to the service usage at the user side is shown. A clear separation of responsibilities and relationships is required to ensure a trustful business interaction. The scenarios outlined within the vision for I-centric communications require exactly such interactions by again placing the individual user as the central element. Individuals will act in distributed ad hoc environments, changing roles dynamically, providing and consuming information from other individuals, etc.

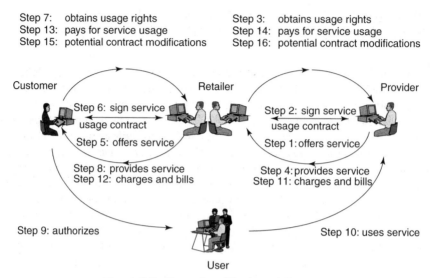

Step 7: obtains usage rights
Step 13: pays for service usage
Step 15: potential contract modifications

Step 3: obtains usage rights
Step 14: pays for service usage
Step 16: potential contract modifications

Customer

Retailer

Provider

Step 6: sign service
usage contract
Step 5: offers service

Step 2: sign service
usage contract
Step 1: offers service

Step 8: provides service
Step 12: charges and bills

Step 4: provides service
Step 11: charges and bills

Step 9: authorizes

Step 10: uses service

User

Figure 3.7   Examples of business interaction

As was stated previously, the influence of feasible business models on functional architecture will increase. Rather than being the result of a research and development trajectory towards a coherent technological system, the I-centric communications system will be made up of various technologies configured around specific user requirements, as identified in the market. One of the most difficult problems for information and communication technology suppliers is linking product capabilities to evolving demand in the market. Historically, this has led to a complex relationship between the research and development, marketing and financial sides of the information and communication technology supply companies. The problem is now even more complex in that much of the technology development process is now distributed among many firms and some major players in key information and communication technology segments are downsizing their research and development commitments. Information and communication technology now typically involves a substantial design component, in which various technologies from different suppliers are configured together. Relationships between the installed technology base and the planning of future systems, products and services were once managed via nonproprietary standards that were linked to the research and development cycles of dominant firms in various information and communication technology industry segments. As the supply environment becomes much more fluid and dynamic, the design function can assume much of this coordination role, potentially creating new markets for new technologies.

How this design of future systems and services will materialize is not yet clear. An important research question raised is how reference points are affected by the business environment. In order to answer such questions, dynamic modelling techniques are required.

The first major research task is the development of a generic business model for I-centric communications. This generic business model contains a set of common terminology as well as a generic framework of business roles, stakeholders, administrative domains,

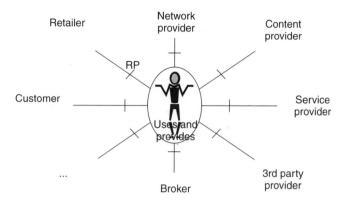

Figure 3.8   The individual as the centre of the I-centric business model

relations, reference points and revenue streams. The individual can act as any available communications role with any other role (see Figure 3.8).

The generic model has to support the vision for I-centric communications featuring ambient awareness, personalization and adaptability. The function of the model is to:

- specify a common frame for all roles in a system;
- allow flexibility against business, regulation and technical change;
- allow one to define the reference points in a multidomain environment;
- allow the derivation of requirements for systems of different stakeholders.

Again, the generic business model introduced in Figure 3.6 provides enough flexibility to be applied to I-centric communications. The remaining question is how to integrate the individual user and the dynamic character due to changing business roles, into such a model.

Relationships Between I-centric Communication Spaces

The vision for I-centric communications has introduced the concept of individual communication spaces. An individual user is interacting with objects in his communication space. Communication between different individuals is done by sharing objects of their individual communication spaces, as shown in Figure 3.9.

The particular physical resources that are used for a certain communication request, are determined in the activation process of contexts and objects. These physical resources belong to administrative domains where certain roles are assigned. As users act in different contexts, the relationships to the administrative domains involved in a communication process have to be managed. This causes the dynamic assignment of roles and employment of different reference points from the viewpoint of an individual (Figure 3.10). This fact is also amplified, as the environments themselves are highly dynamic and characterized by ad hoc and peer-to-peer communication, maybe without any centralized control.

The dynamic relationships between individual users and objects require new concepts like online subscription and accounting, micropayment and federation between ad hoc communication environments. The temporal unavailability of objects and services, and the question: who pays whom for providing or using any physical resource? is to be answered.

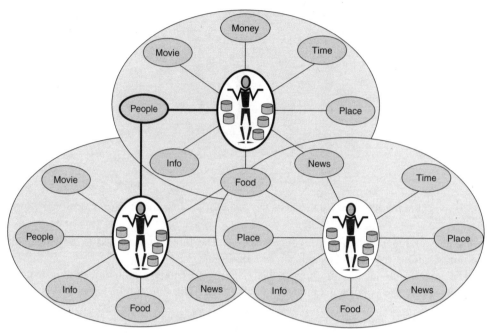

Figure 3.9   Interconnected communication spaces

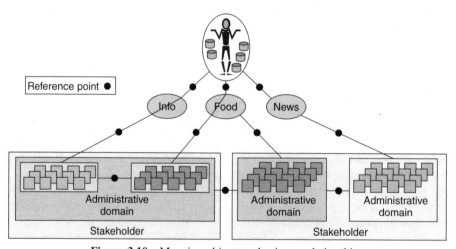

Figure 3.10   Mapping objects to business relationships

This can be done by specifying reference points between all involved instances. The reference points have to support the functions requested above. For example, micropayment between two objects has to be reflected by specifying the information to be exchanged and the relationships to other objects, which might act as certifying instances, billing or accounting server.

3.3 Personalization for the Future Wireless World

3.3.1 Introduction

The phenomenon of information flooding applies not only to the entire internet, but also to individual internet sites and persons. Personalization aims to increase the acceptance of a set of information, or an internet portal, as the user can manage his own individual range, so that he can select, configure and arrange individually presented information and control of systems. Here, an important service achievement is that the customer recognizes 'his side' with a renewed attendance because this facilitates fast orientation and use. Besides the user's explicit personalization, there is the indirect, usually unnoticed form. With this form, the set of offered information is adapted to the user by learned schemes or techniques.

The basic element used for personalization is the context a user is in. A context may consist of many aspects like the needs, preferences, history and behaviour of the user, location-related aspects like physical coordinates and velocity, and also ambient conditions, technical aspects like bandwidth of the network and capabilities of the terminal, business rules that apply, etc. Context information can thus be defined as any information that can be used to characterize the situation of an entity.

The optimal situation is when the wireless services are tailored to the integrated aspects that form the user context. Preferably, the services should automatically, in real time, adapt themselves to changes in the context. The mobile user moves around, and thus the services have to deal with a dynamic user environment. Furthermore, the adaptation to the context should be hidden from the user, provide him with optimal experiences and add value. On the other hand, the environment of the user can, and should, be influenced by the presence and activities of the user and adjust itself accordingly. The basic element of these requirements is the technique of personalization.

This section discusses different forms of personalization, it describes the technologies used and outlines several aspects where significant research is required.

3.3.2 Terms and Definitions

There is no unique definition of personalization or context out there. Researchers from different application domains have numerous approaches to these terms. Depending on their specific requirements, they develop descriptions that are mostly feasible for their own needs in the respective application domain.

3.3.2.1 Definition of Personalization

Personalization means the appliance of actors' needs or interests to the problem of offered sets of information and control. Each user acts in several communication spaces with different preferences concerning control, information and navigation. Here, the communication space implies the context in which the actor is situated, along with the objects surrounding him, with which he is communicating.

Definition of Objects and Contexts

An *object* is a logical representation of anything like a hard- or software entity, which provides well-known services in the individual communication space. Objects may pertain to different contexts and communication spaces. They can be activated or deactivated by the user or environmental conditions. Communication between different individuals or objects takes place by sharing different objects out of their communication spaces.

The recent work of [39] and [40] offer a good approach for the definition of *context*, since they combine the common ground of several domain specific definitions:

> *Definition 1*: Context (from an entity's viewpoint) is information that can be used to characterize the situation of an entity and can be obtained by the entity, where an entity can be a person, place, physical or computational object. Context awareness allows an entity to adapt its behaviour to the circumstances in which it finds itself [39].

> *Definition 2*: Context is any information that can be used to characterize the situation of an entity. An entity is a person, place or object that is considered relevant to the interaction between a user and an application, including the user and the applications themselves [39].

These two definitions mark a very general attitude towards context that can be applied to a wide field of intended application environments. It can be helpful for the identification of relevant aspects or objects that are present in the environment of a user for a given moment in time.

There are several applications in the wide field of communication technologies, applications and services that can be enhanced through the use of context. Therefore, two examples will be given that illustrate the role of context in the area of telecommunications.

The first example is concerned with a problem that can be described as the 'internet effect'. A typical user has access to several useful telecommunication services and internet portals, all of which can be personalized to his individual demands. Each of them has to be configured in a specific way which is totally incompatible with any other [41]. Leaving for a vacation for instance, a user has to switch on his answering machine at home, configure his automatic email response, and redirect his telephone at work to his colleague.

In this example, contextual information can be used to provide a unified model to feed context-aware services. The user resides in a context that states him to be on vacation and to be unreachable. Processing this context, each context-aware service can initiate appropriate actions in case of incoming communication requests. They are able to use information from contexts related to the user, like the telephone number of his colleague at work, or the addresses of his personal email accounts.

The second example is related to business environments. There are several occasions in which incoming communication requests like phone calls are disturbances. These comprise meetings, lunch breaks, discussions, etc. A telecommunication service that uses contextual information is able to detect a situation in which the user does not want to be interrupted and can refuse or redirect requests. It is therefore capable of reacting appropriately to the user's current communication context [42]. For example, if the user is currently participating in a meeting, it is very likely that he does not want to receive telephone calls. Therefore, the context-aware service forwards all incoming telephone calls to the user's mailbox. In order to avoid disturbance, each participant of a meeting or talk has to switch off, or at least mute, his personal communication devices like mobile phones and pagers.

The aim of personalization is to adopt the actual communication context in a way that the user needs. This applies to the presentation of information, the style of navigation and control, as well as to the parameters of the environment where the user is. Each object of the communication space should be adopted in the way that is needed.

I-centric Communication

I-centric communication describes the communication relations around the user. It is driven by the user and his needs, and uses the technology and environment in a well-defined way.

In I-centric computing, the human user is placed in a central position [43]. Looking at humans' communication behaviour, it is obvious that human beings interact habitually with their everyday environment. The interaction is normally limited to a small set of objects that are used frequently. The multitude of information, technologies and content provided by these objects define the personal communication space of a human being. Following this view, a new approach is to build communication systems, not on the basis of specific technologies, but on the analysis of individual communication spaces. The resulting next generation communication systems will be able to fit each user's demands individually. The systems will be able to reflect recent actions for automatic profiling and adapt themselves to the users environment.

An interaction between a human being and an object in his environment always takes place in a certain context. A context represents a certain 'universe of discourse' in a human communication space. It comprises objects and their relationship to the human being at a fixed moment in time. Depending on the user's communication space, objects can belong to several different contexts or to none at all. In a certain context, an object can have an active or a passive role, according to its offered functionality [43].

Since I-centric computing is concerned with humans' communication spaces and individual demands, terms like object, user, person, human or thing are very often used. To state clearly what is meant when using such terms, the following definitions are provided:

Definition 1: Following the view of I-centric computing, human beings can interact with 'things' in the real world, like objects, devices, applications or gadgets. To avoid any misunderstanding, the term 'artefact' will be used in the following to describe any object, gadget or service a human being can interact with, or to which he can have a relationship [44].

Definition 2: When speaking of the person who is actually using a computing system, the term 'user' is often applied. This relates to the fact that the person has an interaction with an artefact: he is using a service or application. But why should a person still be called user when he is not using the artefact anymore? To avoid this shortcoming, each person (using an artefact or not) will be called an 'actor' here after, emphasizing his active role in a system, and leaving any relationship to other actors or certain artefacts open [44]. Knowledge of a recent, and the present, state of the actor's communication space, his context, is the important factor for a system that focuses on an actor's individual demands.

Definition 3: Context in I-centric computing defines a certain relationship between the actor and a particular number of artefacts of his communication space at a fixed moment in time. Context enables the system to act or react appropriately to an actor's demand whenever changes in the actor's communication space make it necessary.

3.3.2.2 Different Categories of Context

In general, the context of an actor consists of properties of the actor himself (human factors) and/or one or more entities in the actor's physical environment. Gathered information that characterizes the situation/context of this entity has to be categorized for reasons of a structured and organized processing methodology.

Actor-related properties could be structured into different categories [45]: information on the actor (e.g. knowledge of habits, emotional state, biophysical conditions), the actor's social environment (like colocation of others, social interaction, group dynamics) and the actor's tasks (including, e.g., spontaneous activity, engaged tasks, general goals). A possible specialization of the physical environment could be [45, 46]: location (e.g. absolute position, relative position, colocation, time, speed), infrastructure (e.g. devices, surrounding resources for computation, communication, task performance) and physical conditions (like noise level, light, pressure, temperature). Further, the business rules and relations that apply to the actor, as well as information on, for example, how much the actor is willing to pay for a certain service, give part of the context.

An actor's overall situation, which is relevant for adaptation of a service, is composed of one or more contexts from one or more of the above properties. The context can be either direct (measurable) or indirect (derived from one or more direct contexts). A special category of the latter is the context information that is deduced from the usage history, which will be further denoted as context history.

Another classification is into primary and secondary context types. Following the work of [39], four types of contextual information that are more important than others for practical reasons can be identified. These primary context types are: location, identity, activity and time [39]. A primary context type has two purposes: first of all it can be used to answer questions like who, what, when and where, about a situation of a particular entity. Secondly, it acts as an index into other sources of contextual information. With given information about an entity's identity, additional information can be deduced that belongs to this entity. For example, given a person's identity, many pieces of related information, such as phone numbers, addresses, email addresses and birth date can be acquired too. With an entity's location, other objects or persons that are nearby can be determined, as well as other activity occurring near to that entity at the moment. Information about an entity's primary context can be used as indices to find secondary context (e.g. email address) for that same entity, as well as primary context for other related entities (e.g. objects in the same location) [39]. The term secondary context acts as a placeholder for all information that belongs to an entity from one of the four primary entity types, but itself is not part of these types. All pieces of information share a common characteristic: they all are an attribute of an entity with primary context, and therefore can be indexed by this context. For example, a person's phone number is a piece of secondary context and it can be obtained by using the information about his primary context as an index into an information space, like a phone directory. Although, there can be information of a secondary context that needs to be indexed by information from multiple primary contexts. A weather forecast, for instance, belongs to a specific location context and a time/date context at the same time [39].

To adapt successfully a (content) service to the context, an integrated approach that takes into account all aspects of the context categories in relation to the communication and computation capabilities of the environment is needed.

Structuring the User Context

A user context describes different aspects of a recent situation that the user has been, is, or will be, in. A context describes important aspects of a complete situation, and typically information for the user context is gathered over a period of time, which can be short or long. The context information will often change, especially for mobile users.

The main intention of a context-aware application should be the provision of intelligent, situation-dependent, customized, personalized and self-adapting systems for different application domains, including communication, control, information, etc.

The *user context* forms the basis for personalization and context-based adaptations. Figure 3.11 shows the proposed components of the user context.

A generic user context consists of five parts:

1. *Environment context.* This part of the user context captures the entities that surround the user. These entities can, for instance, be things, services, temperature, light, humidity, noise and people. Information (e.g. text, images, movies, sounds) which is accessed by the user in the current user context is all part of the environment context. The various networks in the surroundings can also be described in the user's environment context.
2. *Personal context.* This part of the user context consists of two subparts: the physiological context and the mental context. The first part can contain information like pulse, blood pressure, weight, glucose level, retinal pattern and hair colour. The latter part can contain information like mood, expertise, angriness, stress, etc. Some contextual information is quite static, while some is, rather, dynamic in time.
3. *Task context.* This context describes what the user is doing in this user context. The task context can be described with explicit goals, tasks and actions.
4. *Social context.* This describes the social aspects of the current user context. It can contain information about friends, neutrals, enemies, neighbours, co-workers and relatives for instance. One important aspect in a social context is the role that the user plays in the context. A role can be described with a name, the user's status in this role, the tasks that the user can perform in this role, and the various subroles that the role can have. A role can, in addition, be played at a social arena. A social arena has a name like 'at work' and a geographical area.

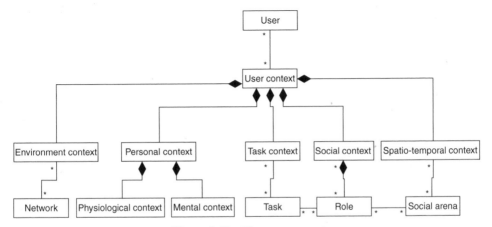

Figure 3.11   The user context

5. *Spatio-temporal context.* This context type describes aspects of the user context relating to the time and spatial extent. It can contain attributes like: time, location, direction, speed, shape, track, place, clothes and the social arena. The social arena is related to the role that is a part of the user's social context.

In other words, a user context is a data structure describing:

- the user's interest and his/her state;
- ongoing activities in the situation;
- the social setting (e.g. social arena and the role);
- spatio-temporal aspects of the situation;
- entities in the environment of the user;
- information that the user has perceived in this context.

For each user a specific context space exists containing her/his context history, the current context and possible future contexts the user may enter, as shown in Figure 3.12.

The context history contains all the user's previous user contexts. A context trace is a subset of user contexts in the context history. This means that the order of the context trace does not need to be the same as in the context history. A context trace regarding your visits at the shopping centre and a context trace regarding your role as a professor will only contain the contexts that are related to such situations. The current context is a user context that consists of environment context, personal context, task context, social context and spatio-temporal context. The future contexts can be user contexts that either the systems or the users can predict and describe. Examples of this can be planned activities in your planner. The system can, for instance, also predict some user contexts when you begin subscribing to a service.

Direct Context
Direct context information is that which is directly measurable, or provided manually, by the actor, e.g., the location of the actor, the current time and/or the blood pressure

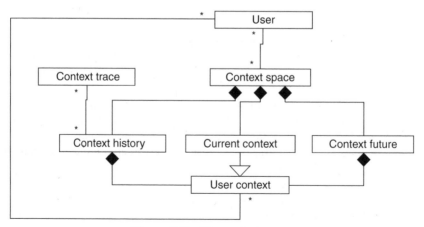

Figure 3.12   The context space

of the actor. This type of context is easy to determine, e.g. through 'sensors' or through information provided by the actor himself (the actor is like a sensor). The context can be easily used to adapt applications according to it. The main advantage of direct context is that measurement usually happens automatically and does not require a high level of interconnection with other systems or with the actor. On the other hand, direct context information alone only captures a part of the actual context, which might not be sufficient. For example, a system may detect that the actor is currently in a foreign city, but does not sense if the actor is there for recreation or business and offers the wrong kind of information (closest museum vs. closest hotel).

Indirect Context

Contrary to direct context, indirect context cannot be measured directly; rather it is derived from one or more direct contexts. As an example, if the actor were located in a conference room where, at that very moment, a conference about context awareness is scheduled, the indirect context would be that the actor is interested in context awareness topics. The more direct context properties are taken into account, the more exact the actual context is. Obviously indirect context allows exact determination of the actor's current situation if many direct context values are used. But this might require a high level of interconnection with other systems to allow the retrieval of the direct context values, which might raise problems about security and privacy.

Context History

The context information that is generated from the actor's behaviour over time is a special category of indirect or implicit context. The historical actor context can originate from simple, straightforward interactions, like clicking/surfing behaviour on the internet, but also from more advanced methods where the system 'learns' aspects of the actor's behaviour and preferences, and adapts to them.

When dealing with context history, additional problems arise. Which context history should be remembered, and for how long? Intelligent mechanisms, possibly agents, have to decide which special context information is, or could be, relevant in the future. The actor must have the option to define if, and which, context information should become a part of his history. This raises a very severe privacy issue as it could allow tracking down the actor's entire life. But on the other hand, context history allows retrieval of information by a 'life period', in the sense of 'the painting I saw when I was in Paris last year', and the adaptation of content based on 'learning' from the actor's behaviour.

3.3.3 Managing Context Information

The context of an actor can consist of several different properties representing actors and artefacts of the actor's current communication space. For the provision of intelligent services, it is necessary to cope with the management (gathering, categorizing, storing, updating and erasing) of data and the evaluation (interpretation of contexts) of that information. This process is called *profiling*.

Since the term management could have different meanings in the context of information technology, it is appropriate to clarify the meaning of the term from the perspective under discussion.

3.3.3.1 Definition

In this chapter the term management is related to the concept of managing information in a wide sense. The area of information and data management comprises the gathering, representing, categorizing, storing, updating and localizing of data. It has no direct relationship to the classical network management philosophy that can be found in management architectures like SNMP or the telecommunication management network (TMN).

All information that can be gathered about actors and artefacts has to be described and stored in profiles. A profile is a data record that electronically represents/contains information on the context of the entity it reflects upon. As such, are profile data can be regarded as a set of metadata describing the user (actor) with his abilities and needs. Presence-related information, terminal-related settings, home environment-specific data, application-specific data, service provider-specific data, user-related preferences, location-related data, and so on – all can be specified in the profile. Sometimes certain parts of the profile data will be specified. At other times it will not be present. Different stakeholders will have different parts of the profile data. The profile data gathered, used and stored by different stakeholders will, in general, not be accessible to the actor (e.g. the Amazon profile records are not accessible for customers). The profile data will never be complete. Furthermore, it will not be stored in one location. It can reside partly in the network (distributed) and partly on (multiple) terminals. It could be updated, implying synchronization of the profile data parts needs to be addressed.

The process of ongoing developments and day-by-day introduction of new end devices make it impossible to define a fixed set of artefacts and actors that can describe all possible contexts. The structure, as well as the meaning of a property in a profile, can change over time. Therefore, an open profiling mechanism is needed that is flexible enough to cope with changing parameters inside profiles. With regard to the implementation of profile data, these characteristics lead to a preferred format that supports:

- semi-structured data;
- extendibility;
- distributed storage and management;
- interoperability; and
- security of privacy-sensitive parts.

Furthermore, the profile data needs to be dynamically adjusted according to the changing context of the mobile end user. A mobile user is typically able to move around and, while moving, not only are his context and his physical location changing, but also the infrastructure, like the network he has access to. For instance, think about a movement from outside with a GPRS network, to within a business environment with a WLAN. For example, a streaming video as a mobile service, this will have a huge impact on what picture quality the user can obtain. Services that are not specific for just mobile, but are also applicable to a wire-line service, provide a changing task or role of the actor, which also influences part of the context information.

An I-centric computing system has to have a profound knowledge about actors for realizing the introduced functionality. That knowledge comprises sensitive information (e.g. content description of information to be delivered), which should be taken into account by the system, but simultaneously has to be inaccessible for other actors. On the other hand, this kind of information has to be partially accessible for a trusted group of actors. Furthermore, the information/personal data present in the profiles needs to be accessible by different stakeholders/interested parties with a different amount of detail. This will require multiple techniques, like password protection, encryption and user-defined authorization, but is also partly covered by legislation. It is essential to provide trust and security mechanisms, which ensure actor-defined policies for information access. In Section 3.3.8 these privacy-related issues are further discussed.

3.3.3.2 Time-critical Aspect

The usage of the context data determines whether gathering and managing this data is time critical. From a service perspective, one may ask if the service is in real-time (live), if it is time critical (in-time), anytime (semi-live), or if it is on demand (as fast as possible). And this, in turn, determines updating context information – how time critical it is. In most information retrieval applications, a delay of a few microseconds will not be noticed, whereas this is a noticeable interval in speech. The application of the data and its intended use, together with the circumstances of the actor, also determine the required freshness of the data – the relationship of this to the data gathering process determines how critical the data is.

Most of the current mobile data services are either time- or location-critical (e.g. file transfer services), whereas the value-added services of tomorrow may migrate from these kinds of service via location-based services (e.g. location-based advertising) and/or time-critical services (e.g. real-time monitoring) towards location-based time-critical services (e.g. tracing and tracking services).

3.3.3.3 Storage Management of Contexts and User Profiles

Due to the inherent distributed nature of profiles (they are gathered, used, stored and managed by multiple stakeholders and at several locations) a generic support for distributed context information is needed for a proper facilitation of service adaptation to the context. In general, the central idea of importing (parts of) different metadata standards yields the so-called application profiles.

Application profiles, as defined by Heery and Patel, are metadata schemes that consist of metadata elements drawn from one or more namespaces, combined together by implementers and optimized for a particular local application [47]. The main issue is that the profiles must be stored in an adequate way in an appropriate place to be able to use the information at the right moment. The profiles are the very basic set of information for personalization and adaptability of the actor's communication contexts. Therefore, these sets of information need to be available in a ubiquitous way. The place to store the profiles and contexts is difficult to determine because it can be used at different places at the same time. On the one hand, the simplest way is to store it on the mobile terminal of the owning actor, on the other hand, it should be located in the network for processing reasons. To solve such problems, modern mechanisms of distributed storage management and caching systems should be investigated. This issue is another major research topic for further investigations.

3.3.4 Personalization Criteria and Systems

Similar to the different types of definition for personalization, different approaches can be identified for how to personalize sets of data.

3.3.4.1 Server Side Personalization – Implicit Personalization

Server side personalization can happen without the knowledge of the actor. The actor does not have influence on the kind of personalization, and in many cases he is unconscious of the fact that personalized sites are presented in a somehow user-unaware fashion. An example of implicit personalization is the adjustment of the navigation paths by learning the actor's behaviour and use of internet navigation. Actors can be categorized in accordance with their behaviour, and can be assigned to a certain class, i.e., to a common profile (beginner, professional, technician, lecturer, tutor, student). Rules define ('Business Rules' [48]) the manner and the extent of personalization [49].

The server side kind of personalization can be extended so that the actor can influence the personalization. It is controlled by user profiles that require the storage of personal data (the actor is usually informed of the procedure of personalization for legal reasons).

3.3.4.2 User-defined Personalization – Explicit Personalization

If the possibilities of personalization go beyond indirect controlling of systems internally, personalization becomes actor-steered or controlled, i.e. it becomes explicit. The actor can affect the appearance of objects in his communication space. Contents could be selected, bookmarks could be set, colour patterns (so-called 'themes' or 'skins') can be adjusted at will, or certain function extensions, like calendar, email and the like, could be adopted. Usually this happens with the help of so-called 'portlets', which offer defined access to excellent information. They represent information channels to different press agencies, weather forecasts, documents or applications, and are generally regarded as units of access and personalization. An actor is able to select a relevant subset for himself from the multiplicity of offered portlets of a portal, and access their contents if he is authorized to do so. Admittedly, portlets are predefined components, which can be adopted flexibly within a common view, but no comprehensive personalization is possible because portlets cannot be combined.

3.3.4.3 Personalization of Products

Personalization of products enhances the individual's configuration of products belonging to the actor's needs. The actor and/or his preferences are included more and more into the product creation process.

3.3.4.4 Personalization of Services

Personalization of services follows a similar course to the personalization of products. The actor can adapt a service according to his individual needs, desires and conceptions. So a mobile operator can offer, for example, its (future) customer the possibility to arrange his individual tariff online. In doing so, the actor could find for himself a tariff-dealing service with which he could arrange his individual mobile phone tariff.

3.3.4.5 Personalization of Data

The personalization of data could takes place anywhere within the actor's communication space. The simplest example is an email client that fetches and indicates the relevant reports for the actor.

3.3.5 Systems Used for Personalization ('How to Learn . . . ')

In order to apply the personalization methods in a well-defined manner, the actor's profile must be learned first. Regarding the depth and complexity of personalization, four systems for gathering information can be differentiated, as shown in Figure 3.13.

3.3.5.1 Rule-based Systems

To use rule-based systems for personalization, the operator of the personalization system determines the rules to offer an actor contents and products. For this, the operator must develop itself an information basis, which is based on empirical values or current trends and is suited to the actor's behaviour. With the help of this information basis, forecasts

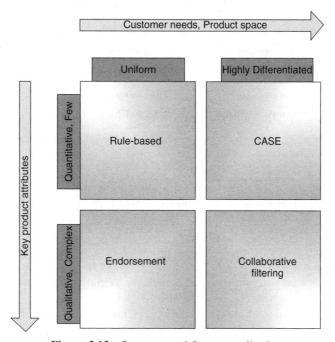

Figure 3.13   Systems used for personalization

Figure 3.14   Rule: persons with green clothing will be offered green hats

and prognoses can be met by possible future preferences of the actor (see Figure 3.14). Updating the information basis could be attained through refinement and improvement for improved accuracy. For example, recommending a table battery charger for customers who bought a mobile phone lately.

3.3.5.2 CASE Systems

Rule-based systems are based on behaviour observations with regard to geographical, demographical, psychological and other user profiles and their use with defined rules. This delivers a small differentiation of the products or services and a smaller number of product or service attributes in itself.

In a CASE system (computer-assisted self explication), an actor is interactively included to specify his preferences online. The importance of attributes is to form the basis for product or service recommendations of the personalization system.

CASE systems are particularly suitable for products and services that have few, but clearly identifiable, attributes and which can be simply evaluated by the customer.

3.3.5.3 Filtering Systems (Simple, Content-based and Collaborative)

On product attributes, different filter procedures can be used to analyse the actor's profile. We can differentiate between simple filtering, content-based filtering and collaborative filtering.

In the case of *simple filtering*, an actor is assigned to a predefined group of persons (see Figure 3.15). Depending on the kind of group, content or products are adopted to suit the actor. A person in the communication university can be, for example, assigned to groups of workers, students or guests.

Content-based filtering relies on a comparison of one or more attributes of different objects (see Figure 3.16). Examples for this could be searching machines that search through the content of different documents with the help of several keywords.

Collaborative filtering pursues another strategy for personalization. The opinions and recommendations of the actor are here from special interest, and there is no longer a comparison of, or search for, suitable attributes. The actor's recommendations are compared and combined into groups with similar profiles (see Figure 3.17). These recommendations

Figure 3.15   Assignment to the group of persons with green clothing and brown hats

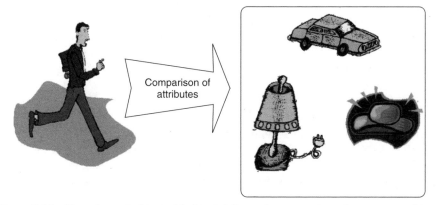

Figure 3.16   From 'green' objects (clothing) follows the suggestion of more green objects

Figure 3.17   The right group recommended because of their opinion about brown hats

are then submitted to new actors with similar groups of interest. For instance, in applying this kind of filtering, the actor could have chosen a book he wanted at an online bookshop. At the same time, the reviews of other buyers that were indicated beforehand could be offered to him so that he could make his best decision. These purchase recommendations can deviate from the filter results of a content-based system. For instance, the book could be recommended because of its content, and not because of the actor's opinion, in which case this book would not be found by collaborative filters.

Product attributes in terms of quality and complexity are very different and difficult to ascertain.

3.3.5.4 Endorsement Systems

Endorsement systems are recommendation systems. They are suitable if the product or service attributes are not clear, or cannot be separated using clear strategies. They ensure a subjective minimum quality of the suggested products or services.

Contrary to collaborative filtering, the endorsement system proceeds from the uniform preferences of the actor and offers an appropriate recommendation. For these systems, a database can be built by direct representative questioning, using selections the actor has made, or by expert analysis.

3.3.6 User Profiles and Context Representation

In order to be able to personalize the objects in the actor's context it is important to know the actor's profile. This profile is learned in different ways and via different mechanisms.

It is, therefore, absolutely essential to define the semantic meaning of the different information items within a profile. The provided sets of information can be sharply defined or specified in a fuzzy way. It is more than necessary to define the meaning of these information sets formally. According to the modern use of software design methods, UML or XML technologies can be used to define the semantics of the different sets of information within a profile.

3.3.6.1 Ontology

The possibility of changing profile structures creates additional problems in an I-centric computing system, because all participating components have to ascribe the same meaning to values and structures stored in the profile. The semantics of a fixed structure can be previously described, but changing structures make a dynamically accessible semantic definition, a so-called *ontology*, necessary. The Foundation for Intelligent Physical Agents (FIPA) has developed several specifications for communicating intelligent agents, which can serve as basis for such an ontology service [50].

Modelling, representing and supporting ontology has been a long-standing research area and is being actively investigated in the areas of artificial intelligence, developing an area of research called knowledge representation.

As the digital society grows, many developers and users of this space realize that our current keyword search is insufficient, and have moved towards ways of improving representing services and information in a more structured way. Some results of this are currently seen in the well-published Semantic Web and in the W3C [51]. There are a number of standard initiatives and tool developments that specialize in modelling ontology services for the web, such as RDFS, DAML + OIL and DAML − S [52] and FIPA Ontology TC [50].

The concept, and some degree of standardization, of ontology are used in many areas of software design, development and deployment. It is clear that this modularization is a heterogeneous approach being applied to:

- Databases in the form of schemas to create collection of knowledge. Capability for certain useful transformations is generally provided; selection, projection, and joining, for example, are common. These can be used for access to, and to a limited degree, for inference on the database.
- The internet through RDF schema, which is currently a descriptive approach. How concepts are shared and matched is left to the application developer. Although the future vision of web-based services fits in well with the concept of context-aware systems, there are currently some deficiencies that need to be addressed. The deficiencies in the current web model can be seen by considering two common requirements in context-aware systems that maintain both scalability and dynamism: those of openness and autonomy.
- All using knowledge representation languages and DAI, such as open knowledge-based connectivity (OKBC).
- Intelligent agent technologies, such as FIPA standard, for multiagent systems.
- Natural language systems.

The purpose of the ontology work is to remove human intervention, where possible, for the task of adding knowledge context, and to allow easier integration when intervention is necessary – i.e., the 'a priori knowledge' is reduced. Efforts in ontology design are principally dedicated to sharing and reusing ontology. Current ontological developments and standards address some classical basic requirements:

- *Knowledge reuse by multiple systems:* Design a particular domain once and reuse it many times.
- *Knowledge exchange and understanding between systems:* Design communication of content between entities (often these entities are loosely classed as agents) to enable coordination on solving a particular problem, answering common queries, etc.
- *Knowledge maintenance without changes to the core reasoning system:* Enable basic changes to a domain without having to change fundamental aspects of the agent itself.

The underpinning of context aware systems requires the use of ontology. The form of which will depend on the environment, services, applications, etc. One of the most important research tasks in upcoming investigations should be the definition of the semantic meaning of profile and its elements, or ontology.

3.3.7 Responsibilities for Personalization

The usage of systems that are based on the research of 'intelligent software agents' is very helpful, in that they deploy precise personalization with its different aspects. So the actor is able to comfortably handle the amount of diverse personalized data. The area of intelligent agents is very broad and many-sided. It arises from research within the ranges of artificial intelligence, software engineering and user interfaces, and represents a man–machine system.

A simple definition of the term 'agent' is: a person or a thing, who is able to actor, or implement, an order of a third party. This deduces the attributes that an agent, on the one hand, achieves trivial things, and on the other hand, he acts by order of a person or a thing. An application, or a special software which settles autonomously the tasks of an actor (e.g. sorts and filters emails or administrates entries in the calendar), is called a *software agent*.

In [53] intelligent software agents are defined more exactly and are described by the following characteristics:

- *Delegation:* An agent implements various tasks, to which he was explicitly authorized by the actor, on behalf of the actor (or other agents).
- *Comware:* An agent has to interact with the actor (or other agents). The agent has to receive the instructions of the actor and inform the actor about the current status and the completion of the task. For this purpose the agent uses either a user interface or an appropriate communication language.
- *Autonomy:* An agent implements his tasks without direct intervention (e.g. in the background). The autonomy of an agent passes from the initialization of a nightly covering of data, up to price negotiation over a product.
- *Control:* An agent must be able to supervise its direct environment for the independent execution of a task.
- *Action:* An agent must be able to affect its direct environment over an appropriate mechanism.
- *Intelligence:* An agent must interpret the supervised events in order to be able to make all necessary decisions in terms of an autonomous enterprise.

Additionally, an intelligent agent can have the following characteristics:

- mobility;
- safety;
- personality.

The agent model depicted in Figure 3.18 [53] is based on the view of the agent in three layers, but the layers are not in a hierarchical structure. The first layer, the task layer, consists of 'individual talents', those that are needed by the agent to reach its goals. For example, searching and filtering of information belong to those goals. With sensors, the agent seizes its environment and implements its tasks in this communication space. The training of the actor to use the agent is also a task of this layer.

In the knowledge layer, the agent gets the rules for the tasks it has to fulfil. The knowledge can be determined by different methods. A part of the knowledge is implemented by the developer, this sometimes could be difficult to adapt later. A second possibility for gaining knowledge is the use of rule-based systems. Further, agents can use the base of knowledge of other agents to extend their own knowledge. The fourth possibility to

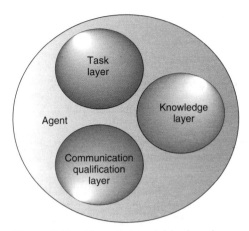

Figure 3.18 The agent model in three layers

'learn' knowledge is to deduce knowledge from the actor, his attributes and characteristics and his communication spaces.

The third layer is called the communication qualification layer and represents the interface to the actor. This causes a suitable choice of language and also contains social aspects on how the actor is communicating. Further, the communication for interacting between the agents is integrated in this layer. For this, a special communication language for use between agents is necessary.

The different types of agent are described by their task, environment and architecture. So, task-specific agents are clearly described by a name containing their talents (e.g. searching agents (machines)). Other agents are described by the environment in which they act and so are called operating system, application or network agents. Network agents are differentiated still further into internet and intranet agents. The architecture agents are designated by their internal knowledge architecture. Examples for this are learning agents, who are learning directly from the actor, and neural agents, which use a neural network to acquire knowledge sensitively. In the following the environment agents are described in more detail.

3.3.7.1 Desktop Agents (Central Agents)

Operating system agents and application agents belong to the set of desktop agents. Here, both pertain to the interface agents, those that assist the actor during the operation of the desktop operating system on the one hand, and on the other hand, of different matching applications.

3.3.7.2 Intranet Agents

Examples of intranet agents are (among others):

- adaptable-collaborative agents, who automate operational sequences within a division;
- process automation agents with these, operational sequences can be automated;

- database agents, who offer a service for the company-owned database to the actor;
- resources switching agents, who assign resources in a client/server architecture.

3.3.7.3 Internet Agents (Distributed Agents)

Internet agents are classified by the following examples:

- web search agents, who offer certain search services to the actor;
- web server agents, who furnish selected agent services on a special web server;
- filter agents, who filter electronic information due to the actor's defaults;
- search agents, who send personalized information to a desktop;
- memory agents, who inform the actor about certain important and interesting events;
- service agents, who put a specialized service to the actor at his disposal;
- mobile agents, who travel from one computer to another, thereby implementing certain tasks.

3.3.8 Privacy and Trust

Context information and the actor's profile must be subject to privacy and security considerations. The most obvious examples for sensitive context information might be a credit card number or an individual's location at a certain point in time, but virtually any context information, such as terminal preferences, actor's interests, user credentials, etc., should be considered as sensitive with respect to privacy and security.

It is likely that the end user's awareness of privacy issues will soon rise to a much higher level than is currently the case, in the internet world, and users will more carefully decide which service providers they trust to adequately handle certain context-related information. The current introduction of location-dependent services, as well as the rapid growth in email spamming, have become triggers for an increasing awareness of privacy concerns.

Both globalization and virtualization of society contribute to a greater privacy risk. The increasing use of automated processing of personal data over the past few decades has exacerbated the risk of illegal use of personal data, and facilitated their transfer across frontiers between countries with great differences in the level of protection provided to personal data. Especially in the case of roaming mobile services through different countries, with their respective privacy laws, additional challenges are posed on the management of those services, as well as on the security and privacy of personal information. It is therefore becoming increasingly clear that privacy and international law are important issues that need to be considered for personalized mobile services. At the point where information regarding the actor's behaviour, history (online profiling information), services rendered, etc. is linked to the user's identity, a serious privacy threat arises [54].

Clearly, in different places and by different players, different types of context information will be collected, as discussed in the previous sections. For example, an actor's location could originate either from the terminal or the access network, whereas service-related profile data, such as the set of books an actor has recently purchased, is more likely to be collected by a service provider such as an online bookstore. It can therefore be expected that there will not be one single place for the storage of an actor's context information, such as the terminal. Further, context information will be processed and used by different players, and potentially exchanged between players.

Correct handling of privacy is especially an issue in situations where fragments of a user's profile may be stored and used in different nations and when, in addition, the actor is roaming between countries. Furthermore, more restrictive rules often apply for personal information collected from children.

The legal and regulatory frameworks regarding privacy differ between various countries. One can distinguish between:

- *Comprehensive laws:* In most countries there is a general law that prohibits the collection, use and dissemination of personal information, both in the private and public sectors. An overseeing body ensures compliance. Examples of this kind of data protection regime are the countries included in the European Union (EU).
- *Sectoral laws:* In some countries, of which the United States is probably the most important, no general data protection law exists, instead, several specific sectoral laws govern, such as those for video rental records and financial privacy. A major drawback of this approach is that legislation almost always lags behind the technology, since with each new technology, new legislation has to be introduced.
- *Self-regulation:* Various forms of self-regulation are another approach, often seen in the United States and also, e.g., in Japan. Companies and industry bodies establish codes of practice and engage in self-policing. The drawbacks are that there is mostly only weak data protection and that the cases lack enforcement.

While there is a rather high level of harmonization within the European Union, comprehensive legislation is lacking, mainly in Asia and the US. In the case of transborder flow of personal data between the EU and US, to bridge the difference in privacy legislation, the additional agreement known as 'Safe Harbor' was reached in 2000 [55].

The idea of the 'Safe Harbor' was that US companies would voluntarily self-certify to adhere to a set of privacy principles worked out by the US Department of Commerce and the Internal Market Directorate of the European Commission. These companies would then have a presumption of adequacy and they could continue to receive personal data from the European Union. Subject to the agreement are: informing the individuals about what is collected and why, offering the opt-out option, offering the opt-in option in case of sensitive information, only transferring data to a third party after an additional signature for data protection, and providing individuals with access and correction possibilities to any personal information held about them. Drawbacks arise from the self-regulatory basis, there is little enforcement, no individual right to appeal and the telecommunications sector is excluded (since the agreement only applies to companies overseen by the Federal Trade Commission and the Department of Transportation).

While there is no internationally consistent legislation, the US Federal Trade Commission in 1998 issued a report on five 'fair information practice principles' which may serve as a basis for privacy considerations. These practices are:

1. *Notice.* An entity wishing to collect personal information should inform the actor about its policies, i.e. what data will be collected for which purposes, to which third parties is it passed, etc.

2. *Choice.* An entity wishing to collect personal information should give the actor the choice of whether it may be collected and how it may be used. Choice can be realized using 'opt-in', where the actor has to perform some action before the information may be collected and used, or using 'opt-out', where by default it is assumed that the collection and use is permitted.

3. *Access.* An entity that has collected personal information should allow the actor to view the information collected about them and, if the information is incomplete or inaccurate, to correct the information.

4. *Security.* An entity that has collected personal information should take reasonable measures to ensure that the information is protected against loss, unauthorized access or modification, destruction or disclosure.

5. *Enforcement.* The previous four information practices only make sense if there is a mechanism in place (industry self-regulation, private remedies or government enforcement) by which they are effectively enforced.

The last point makes it clear that privacy and security issues can only be supported but not resolved, by means of technology. Once personal information is in the hand's of an external entity, there is no purely technical way of stopping them from abusing this information for illegal or immoral purposes, or from handling it negligently. This has to be ensured by managerial means in conjunction with a proper enforcement mechanism. What technology can provide are essentially frameworks for describing and enforcing privacy policies spelled out by the end user.

Furthermore, even if technology can be successfully used to fully secure privacy-sensitive context data, once an end user chooses to say so, no use can be made of his contextual information. Hence, no personalized service can be offered.

A far more important issue in wireless information and communication services is, therefore, trust. And because of this importance it can be stated that the perceived privacy is a mobile business enabler. Perceived privacy reflects how individuals feel and experience the care that is taken in the use of their personal information. With a satisfactory perception of privacy, users are more likely to give other parties access to their personal data. This, in turn, enables personalized mobile services to exist and become successful. An example: if a user gives access to his preferences with regard to his favourite food and also to his location, a mobile service might alert him around his usual dinner time when he arrives in the neighbourhood of a potentially interesting restaurant. Another example is, e.g., knowledge of the terminal the user has access to. If this terminal is capable of displaying high quality video streaming, then knowing this part of the profile data enables, for instance, trailer-streaming services from the video business. From a business point of view, it is essential to have access to the profile data, and it is of vital importance to have the user's trust with regard to the protection of his privacy [54].

Unless context information is locally stored (and its dissemination locally enforced) in the user's terminal, there needs to be some trust relationship with an entity that is collecting and/or storing context information. Note that 'trust' does not have to be comprehensive, but the user needs to trust a specific entity to handle certain issues correctly and according to principles that the user is aware of. Examples of such trust relationships are that between a user and a mobile phone operator (the user needs to trust the operator to charge correctly for phone calls and not to disseminate the numbers the user has called to third parties), or a bank (the user needs to trust the bank not to execute any unauthorized transfers from the user's account or distribute information about the user's financial situation). Thus, a technical mechanism for privacy is anonymization/pseudonymization, where a trusted entity passes on context information to an untrusted third party in such a way that it is not linkable to an individual (pseudonymization) or not even linkable to previous invocations of a service by the same individual (anonymization).

Two guiding principles should be paramount in designing support architecture for handling context information in a distributed fashion: user control and usability. User control means that the user can make decisions on who may handle what kind of context information for what purposes. It also includes the freedom of the user to decide which context information they consider sensitive. Usability means that the system must be designed in such a way that the user is actually encouraged and enabled to use it for making active choices on privacy options. A prerequisite is transparency, i.e. the user needs to understand in a simple way what is happening to their context information. Transparency can be better achieved using 'opt-in' mechanisms than using 'opt-out', no information whatsoever should be collected and used without explicit user consent. In today's internet, there are many examples of security mechanisms that technically work well but that are not useable enough. For instance, very few consumers using SSL/TLS-enabled browsers ever bother to make active decisions on which certification authorities to trust, nor do they verify certificates their browser warns them about. Rather, default browser settings are almost never changed or even understood. Depending on the implementation, there is a risk that a similar situation will also occur as P3P-enabled browsers become widespread.

Only user control and usability taken together will lead to the user perception of privacy and security. This will be a mandatory prerequisite for the business success of I-centric mobile applications. Therefore, context information and user profiles are the most sensitive set of information in this environment. Unlike confidential communication, it is not only necessary to secure the means of transmission, but also to secure the information itself. Because of the distributed storage management of the data it is highly important to secure the stored information.

With regard to securing this information, a powerful access control scheme is necessary. On the one hand, nearly every object in the actor's communication space should have access to these profiles in order to adapt in the right way, on the other hand, the profile data should be secured as much as possible.

It is absolutely necessary to install well-suited mechanisms securing the data in order to achieve user acceptance for such systems. If the user does not trust such a system he will never send his personal preferences into it. One major example for the importance of security issues concerning the profile is the. NET environment of Microsoft. The first and very basic service of the whole system is the. NET Passport service, which provides security mechanisms for the user's profile.

3.3.9 Standardization and Supporting Technologies

One way to enable distributed service calls and profile accessibility is to make use of loosely coupled internet-based services, remote procedure calls over XML (SOAP (simple object access protocol), UDDI (universal description, discovery and integration) and

WSDL (web service description language)) and PAM API specifications, for coupling IDs to profiles.

In the internet today, personalization and profiling services are mainly provided to actors by portals based on user profiles. The user profile information is obtained by different means, including, e.g., user interaction, tracking the user habits and preferences and sharing information between different portals. Based on this information, mostly only a small part of user-related information, the personalization of the content is made on the content origin web server.

In the following, different methods for the definition and the exchange of context information elaborated by W3C and by the WAP Forum are briefly described. Also, relevant standardization efforts are mentioned that relate to using context information to adapt services. Many more standardization bodies exist that address the problem space of context awareness. A forthcoming report will go into detail and enumerate them. The key standardization bodies are now outlined.

3.3.9.1 Internet Engineering Task Force – IETF

WEB Intermediaries – WEBI
In today's worldwide web, content is served from an origin server to a client. The new concept of intermediaries allows modifying the content stream by computational entities, so-called intermediaries [56].

The simplest example of such an intermediary is a proxy server, where the purpose of the proxy intermediary is to provide a caching service. But an intermediary can do more [39]. It especially becomes useful if the content stream needs to be modified after it has been requested, or if it cannot be modified on the origin server. For example, this enables one to adapt content (e.g. transcoding, distillation) according to context (including ambient) information, which is retrieved from a third party.

Content Distribution Internetworking – CDI
Apart from applications, one of the most important factors for successful service provisioning is the content that is being delivered. Over the past few years, several tools, protocols and mechanisms have been implemented for the creation, management and tracking of content and its delivery. This evolution has led to a new type of network, called content networks. In terms of the OSI model, content networks could be introduced as a new layer above the application layer, as they rely on, and use, methods from the application layer to distribute content (e.g. HTTP). A content distribution (inter) network consists of an origin server, several delivery servers (surrogates) along with content delivery infrastructure, as well as a content delivery, request routing and accounting infrastructure [56]. The request routing infrastructure contains mechanisms that allow content delivery from a certain surrogate to a client, whereas the delivery infrastructure contains mechanisms for synchronizing and delivering content from the origin server to the surrogates. Finally, the accounting infrastructure is for management purposes by collecting data about the request routing and delivery of content within the content distribution network.

Open Pluggable Edge Services – OPES
OPES [57] rely on an underlying content distribution network and provide services that enhance the content transmitted through the content distribution network (CDN). In an

OPES enhanced content network, a client establishes a connection with an edge server (OPES intermediate server) rather than a (very distant) origin server.

An OPES edge service system model consists of different components, like the OPES intermediary, an OPES admin server, a remote call-out server and, of course, the user agent and the content server [56].

The OPES intermediary, which is located on the content path between the user agent and the content server, consists of the OPES engine, a local execution environment and remote execution environment. The OPES engine provides content services (generating added value to the content), with the help of the execution environments. So the main purpose of the OPES model is to provide functions which allow the (value-added) processing of content that is being transmitted within a content distribution network.

Geopriv
The geographic location/privacy (geopriv) working group is currently addressing the standardized description of location information, as well as its transfer, in a controlled and secure fashion. If successful, the results of this group might also be extended to other types of context information.

Virtual Identity on the Internet – IDsec
IDsec provides a generic mechanism for establishing virtual identities on the internet that standardizes protocols and interfaces for exchanging identity information between actors and service providers in a secure manner. An identity is known by a certain profile that contains exactly those attributes that the actor wants to reveal to the requester of his profile. Secure transport of the profile is taken care of by a trusted profile manager. Actors specify the access and adaptation rights regarding the profile attributes. Certificates and public/private key mechanisms ensure that information is exchanged in a secure manner between trusted parties. The profiles use a simple XML format.

3.3.9.2 Worldwide Web Consortium – W3C

XML
The distributed and incomplete character of profile data calls for a language and format that accommodates extensibility and interoperability, features that are provided by the extensible markup language (XML). XML is a markup language for documents containing structured information. It enables operations to be performed on content. XML is often chosen as the language for implementing profile data. This ensures interoperability and offers extensibility. New languages can thus be defined more easily.

RDF
The Resource Description Framework (RDF) integrates a variety of applications (e.g. from personal digital data to library catalogues), using XML as an interchange syntax. The RDF specifications provide a lightweight ontology system to support the exchange of knowledge on the web. Also, domain-specific XML dialects (like ebXML) exist, or are under development. In W3C, there is ongoing work to develop a process for encrypting/decrypting digital content, including (parts of) XML documents and an XML syntax used to represent (1) the encrypted content and (2) information that enables an intended

recipient to decrypt it. They also have a working group (joint with IETF) on signatures, whose mission it is to develop an XML-compliant syntax for representing the signature of web resources and portions of protocol messages (anything referable by a URI) and procedures for computing and verifying such signatures.

Composite Capability and Preference Profiles (CC/PP)

A dedicated working group at the Worldwide Web Consortium (W3C) has designed the composite capability and preference profiles (CC/PP) architecture and framework. Its goal is to define how a client communicates its capabilities (device context) and preferences (the user agent profile) to a remote content server. With respect to ubiquitous web applications, this framework will allow the server to adapt the output according to the needs of the client, making it possible to deliver content suitable for a wide range of different devices, such as desktop computers, personal digital assistants and mobile telephones, taking their capabilities, display size for example, into account. At present, applications need to rely on the HTTP user agent string to get some basic information about the capabilities of a client. CC/PP utilizes the standardized XML and RDF technologies to define a flexible, extensible and decentralized framework. The client capabilities are expressed by sending an XML data stream to the server.

WebOnt

The W3C WebOnt working group, part of the semantic web activity, focuses on the development of a language to extend the semantic reach of current XML and RDF metadata efforts. The ontological layer and the formal underpinnings thereof will be addressed, as this is necessary for developing applications that depend on an understanding of logical content, not just human-readable presentation. Ontology in the context of this work refers to what is sometimes called a 'structural' ontology: a machine-readable set of definitions that create a taxonomy of classes and subclasses and relationships between them.

Such language layers are crucial to the emerging semantic web, as they allow the explicit representation of term vocabularies and the relationships between entities in these vocabularies. In this way, they go beyond XML, RDF and RDF-S in allowing greater machine readable content on the web. A further necessity is for such languages to be based on a clear semantics (denotational and/or axiomatic) to allow tool developers and language designers to unambiguously specify the expected meaning of the semantic content when rendered in the web ontology syntax.

P3P

The most serious standardization effort in the internet world so far to address privacy of personal information has been the Platform for Privacy Preferences (P3P) [58] project within the Worldwide Web Consortium (W3C) [59]. P3P is a framework for describing the privacy policies of a website in a both human- and machine-readable format. It enables the actor to view, in a standardized format, the privacy policies before deciding to use a website or to disclose any personal information to it. Further, the user can configure a P3P enabled browser according to their own privacy preferences, which are then automatically matched against a visited website's policies. It must be reiterated, though, that obviously P3P is no more and no less than a description format – it is not capable of ensuring that the website's operator actually complies with its published privacy policies. The P3P specification uses XML and RDF to capture the syntax, structure and semantics of the personal information.

In the current P3P model the client can only accept or reject the server's proposal. A drawback is that the P3P agreements only cover a single session. By assembling information collected over multiple sessions, facts can be derived that the end user did not necessarily want to disclose. Further, multichannel sessions are also not covered by the P3P agreements.

3.3.9.3 Industrial Initiatives

WAP Forum/UAProf (User Agent Profile)
For the context of wireless terminals, the WAP Forum elaborated the UAProf [44], based on the CC/PP work. User agent profiling used in the WAP architecture is the first real-world implementation of CC/PP. The capability and preference information (CPI) of the user equipment is initially sent whenever a session to either a WAP proxy or content server is established. During the session, the CPI is cached by the server and used to adapt the output according to the profile information. If a server (proxy or content) does support user agent profiling, it responds to the request with a profile-warning header having the value 100 ('OK'). This way, a client can determine if user agent profiling and CPI are supported.

OSA/Parlay
Different solutions have been adopted to achieve the goal of open interfaces, including Parlay, OSA and JAIN-. These solutions are more or less based on an open application programming interface (API) method. Such an API is defined as a set of technology-independent interfaces in terms of procedures, events, parameters and their semantics, and it is based on distributed computing concepts such as CORBA, Java and other technologies. OSA/Parlay specifications define open APIs that link advanced applications to telecom capabilities, including mobility APIs.

PAM
The PAM Forum (www.pamforum.org/) has defined a set of specifications for presence and availability management, with the goal of reaching a standard on digital identities, characteristics and presence status of agents, capabilities and state of entities, and availability of entities for various forms of communication. An identity is a limited electronic representation of entities. Profiles (named sets of attributes and attribute value pairs) are used to associate data with identities. Programmable interfaces to define profiles are part of the specifications.

OPS
OPS (open profiling standard) specifications enable personalization of internet services while protecting users' privacy. OPS provides internet site developers with a uniform architecture for leveraging personal profile information to offer individuals tailored services that match their personal preferences. Netscape, Firefly Network and VeriSign developed OPS, and it has been announced that the OPS specifications will be submitted as a proposed standard to W3C.

UDDI

The Universal Description, Discovery and Integration (UDDI) project is a sweeping industry initiative. The project creates a platform-independent, open framework for describing services, discovering businesses, and integrating business services using the internet, as well as an operational registry that is available today.

WSDL

Web service description language (WSDL) is an XML format for describing network services as a set of endpoints operating on messages containing either document-oriented or procedure-oriented information. The operations and messages are described abstractly, and then bound to a concrete network protocol and message format to define an endpoint. Related concrete endpoints are combined into abstract endpoints (services). WSDL is extensible to allow description of endpoints and their messages, regardless of what message formats or network protocols are used to communicate. However, the only bindings described in this document describe how to use WSDL in conjunction with SOAP 1.1, HTTP GET/POST and MIME.

UpnP

The Universal Plug and Play Forum is an industry initiative designed to enable simple and robust connectivity among stand-alone devices and PCs from many different vendors. As a group, it is leading the way to an interconnected lifestyle. The Forum consists of more than 488 vendors, including industry leaders in consumer electronics, computing, home automation, home security, appliances, printing, photography, computer networking and mobile products.

3.3.10 Conclusions

Personalization is one of the major issues in the I-centric communication space of modern users in order to be able to handle and use the entire universe in a well-suited manner. To be able to use and apply personalization, several major issues needed to be investigated in further research activities. We have mentioned several mission-critical issues for personalization on the technical level, as well as on the social and emotional levels. The first and most important precondition is the trust of users in such systems. Therefore, one major research activity should be concentrated on the necessary elements concerning trust and acceptance of the user. With this text we propose to start an in-depth research activity for personalization of communication contexts and the necessary processes.

3.4 Ambient Awareness in Wireless Information and Communication Services

3.4.1 Introduction

Ambient awareness is particularly relevant for wireless applications, due to the use of positional user context information, such as location and movement, for applications such as location-based services and the limited user interface. This relevancy motivates the current collaborative work on ambient awareness in wireless information and communication services. This collaborative work is the result of a joint effort of several participants within Working Group 2 of the Wireless World Research Forum (WWRF).

Ambient awareness is part of ubiquitous attentiveness and context-aware computing, where a combination of ubiquitous context-aware computing devices, exchanging information using communication, and the proactive responsiveness of the ubiquitous environment is applicable. A (wearable) system is ubiquitously attentive when it is cognisant and alert to meaningfully interpreting the possibilities of the contemporary human communication space, and intentionally induces some (behavioural/system) action(s). Ubiquitous attentiveness comes close to the term context awareness. Within Working Group 2, we split the technological issues of ubiquitous attentiveness and context awareness into personalization, ambient awareness and adaptation. The current section focuses on ambient awareness, which denotes the functionality provided by an I-centric system to sense and exchange the situation in which the individual is at a certain point in time. The ambience, or situation, is a trans-section of the context.

The most appealing and typical mobile services of today are based on derived information using location awareness. Services can be delivered to users everywhere. Moreover, the location of the user influences the delivery of a service (what is delivered, e.g. a map of a user-unknown city that dynamically changes in accordance with the user's movement, or of which the scale dynamically changes depending upon the user's velocity). Location awareness involves more than just physical coordinates and velocity. Ambient conditions, like, for instance, indoors/outdoors, humidity and temperature play a role as well.

In the current section, different aspects that need to be covered in ambient awareness are further discussed in relation to open research issues that need to be solved for their realization. First, the term 'ambient awareness' is further defined.

3.4.2 Definitions

Ambient in the sense of I-centric systems refers to the situational context in which an individual user or actor is. Ambient awareness deals with sensing and exchanging the ambience of a user/actor in the human communication space. The ambience, or situation, includes spatial, environmental and physiological information.

Examples of spatial information are geographical data like location, orientation, speed and acceleration. Environmental information includes temperature, air quality, and light or noise level. Physiological information depicts life conditions, like blood pressure, heart rate, respiration rate, muscle activity and tone of voice.

An important goal of ambient awareness is to acquire and utilize information about the situation of an actor to enable services in an active context, i.e. services that are personalized and adapted to a certain situation at a certain moment in time. In ambient awareness, a key aspect is situation sensing, where a sensing device detects environmental states of different kinds and passes them on.

Following the definitions of context and context awareness, we define:

Actor. When speaking of the person who is actually using a computing system, the term 'user' is often applied. This relates to the fact that the person has an interaction with an object: he is using a service or application. But why should a person still be called user when he is not using the object any more? To avoid this shortcoming, each person (using an object or not) will be called an 'actor' in the following. This emphasizes his active role in a system, leaving any relationship to other actors or certain objects open.

Ambient. Ambient in I-centric computing defines the ambience or situation the actor is in, given his communication space at a fixed moment in time. It includes spatial, environmental and physiological information.

Ambient awareness. This is the functionality provided by an I-centric system to sense and exchange the situation in which the individual is, at a certain moment of time.

Effector. An effector is a device that can modify the physical world in a way that can be detected by actors i.e. by his/her senses directly. Examples of effectors are lights, automatic doors, phones and loudspeakers.

3.4.3 Acquiring Ambient Information

A device is ambient-aware if it can respond to certain situations and stimuli in its environment, representing actors and objects of the actor's current communication space. Sensors or human–machine interfaces can gather information about these actors and objects. Actors can also provide this information themselves. Sensors can, for example, be located in the network, in devices or in the environment of the individual. For actors, an automatic context information gathering is preferable. Advances in sensor technology are needed to further adapt services to–and cooperate with–the environment of actors.

In general, we can distinguish the following categories of context describing the environment:

- *Computing context:* such as network connectivity characteristics, nearby devices, displays and hardware resources;
- *User context:* such as user's location, social conditions, user's profile and history;
- *Physical context:* such as levels of lighting, noise, humidity and temperature;
- *Time context:* such as time of the day, day of the week, time before or after a specific moment.

This categorization provides a guideline for the categorization of the sensors or interfaces needed to acquire ambient information.

A special category of ambience/situation is positional information as part of the user context. Positional information is seen as accurate information (x,y,z coordinates) representing one's location, while location can be seen as a wider geographic area. An actor, network or terminal can provide positional information for a mobile terminal.

Wide-area positioning technologies range from Cell-ID to enhanced GPS. Today, network technology makes use of Cell-ID to calculate the coordinates. Within a year, the Enhanced Observed Time Difference (E-OTD) solution will be ready, while for 3G networks, RTT/IP-DL is foreseen to be capable of providing position information through/from the network. High-end terminals will be GPS-enabled. E-OTD makes use of location information provided by the terminal, while the actual calculation of the position is done within the network (handset-assisted technology, or vice versa, in which case it is called network-assisted technology). Round trip time (RTT) enhances the Cell-ID with propagation time measurements. Idle period downlink (IP-DL) is a cellular signal timing-based method for WCDMA. Global positioning system (GPS) makes use of 24 satellites in orbit around Earth. Differential GPS can be used to determine the 3D position of an entity with an accuracy of 1 m. However, GPS has a major drawback in that it cannot be used indoors. Further, it provides no information on orientation (such as, is the entity oriented to the west?)

If coverage is fulfilled, short-range wireless technologies, like Bluetooth, IEEE 802.11 or Hiperlan-II, can complement the wide-area positioning information. Accuracy and acquisition time can be improved.

Position sensors provide yet another source of coordinate information, with differences in accuracy, range and costs. So-called inertial sensors can sense acceleration and rotation. Motion sensors detect changes of motion.

Body sensors could be used for user context regarding the actor's condition. They gather information by measuring directly from the actor. Heartbeat is an example, as is skin resistance. These might be used to gather physiological information such as emotions and anxiety.

Terminal-activity sensors obtain context information from monitoring the interaction and communication with the actor, like key presses on a keyboard. Several types of visual sensing can be provided by a camera (via light input, or though the use of visual context markers like pictograms). These sensors act on the borderline of human–machine interaction, detecting both user context and computing context.

There is quite some ongoing research on computing context in terms of the request for, and the announcement of, computing and communication services in a networked environment. 'Service discovery' basically describes the mechanisms to get information about the availability of resources in a local area network, e.g., SLP. Note that, in this case, the computing context is already provided as a service and not as pure information.

For physical context, several types of sensing device can be used as general-purpose sensors. For example, cameras that can sense not only the light intensity, but also wavelengths invisible to the human eye, e.g. infrared, and are appropriately calibrated, can be used as temperature sensors. Investigating how multiple readings can be derived from one single device is a challenging research task.

3.4.4 Crunching (Interpreting) Ambient Information

In the previous section, a few examples of sensors were given. The sensing equipment itself can be quite small and distributed, the problem being to acquire the sensor data at an appropriate aggregation level. Knowing that the temperature varied 1 degree at 1 000 000 positions may not be appropriate, if the values cannot be aggregated to describe the area in relation to a known reference, e.g. the geographic location and time of the sensor. These, in turn, have to be aggregated appropriately. Knowing that the temperature has risen one degree over the province of Skåne in March will tell Swedish people that spring is approaching, for instance.

Sensors also interact, as the example above shows. While a clock is, strictly speaking, not a sensor (since it does not retrieve its input from the outside, but from the oscillations of a quartz crystal), it is likely that the more general-purpose sensors become, the more they will trust other accessible information sources to provide them with information about their context. Today, if a thermometer is to communicate temperature readings, the location has to be determined and communicated out-of-band (i.e. by setting it in software, or the thermometer itself). A general-purpose thermometer could sense its own location (and, using GPS, the time), and use this in communicating its values.

The above examples show that an appropriate interpretation of sensor readings is needed. Ambient information is useless until an evaluation takes place that can take advantage of the additional situational information. Evaluation of the individual's situation in the communication space is therefore as important to an I-centric computing system as sensing it.

Different parts of the actor's situation (and of the context of the information, such as publisher-created document information) may have different weights in determining the ambience of the actor. In addition, the ordering of the processing may affect the resulting ambient awareness. In traditional programming languages (including script languages and transformation templates, such as XSLT) there may be an inherent ordering of the processing, whether consciously created or not. This ordering can be formally expressed, and externalized from the program which actually executes the processing.

If the information about the relative weights and the ordering of processing is not communicated in profiles (storing the user's context), it has to be communicated as a separate description, essentially a metaprofile for the processing. This can be expressed in a formal rules language, or as an ontology using the names and/or properties of the profiles as nouns. This, in turn, implies that the processing entity can handle the reception of such metaprofiles. This would determine the processing order and the relative weights. Database selection, transcoding, annotation and other contextual processes could be applied as part of this process.

Profile evaluation takes place in every kind of service logic inside an I-centric computing system for the purpose of decision making. Normally, different profiles of an actor will be combined and evaluated to react to an actor's command, or changes in his environment. The profile evaluation process has to provide several features for enabling mappings of abstract 'wishes' to concrete behaviour of certain objects. If two or more objects are needed to realize a 'wish', the interaction between these objects in an actor's surroundings has to be ensured by some kind of service logic. The outcome of the profile evaluation process, and the resulting actions, have to be stored in a history profile to provide information for self-learning and adaptation capabilities. The result of an evaluation process influences the system's behaviour. For I-centric computing, this means that the result of the evaluation of the current situation of an actor defines what has to be done for him by the system at a certain moment in time.

The actual creation of a multiprofile system is yet to be realized, but promises to bring a number of interesting challenges. Communicating and aggregating the values also present a number of special challenges. For some applications (e.g. 'if the day is Saturday, the month is August, the time is morning, the temperature is above 20 degrees centigrade, and the sun is shining, show me the way to the beach'), the number of values is quite limited, and the aggregation and rule system quite straightforward. However, if there is a large number of uncoordinated sensors, then statistical methods may have to be used to determine the current values for the current position ('what is the temperature near me?' can be determined by averaging a number of fixed sensors on buildings, for instance). The problem then becomes one of communicating these readings, as well as one of privacy. The fact that the sun is shining is well known to anyone in the area, but the fact that a person is stressed may not be immediately obvious, and may be something that the person wishes to hide from others. This will require privacy protection, which is discussed further in other sections in this chapter.

An important aspect of information processing for context-aware applications is the social factor. This addresses an automatic or semi-automatic grouping of relevant context information in order to leverage the complexity of context heterogeneity, before the context is processed by an actor. For example, all information regarding user location could be accessed by addressing the issue 'user location', without having to search for all relevant pieces of information.

3.4.5 Issues Relating to the Tailoring of Input/Output

3.4.5.1 Interaction by Autonomous Software Agents

The continuous nature, and the infinite number, of states and precepts about the environment surrounding a mobile ambient-aware application, provides inherent difficulties for gaining knowledge of the environment. Autonomous multiagent system research related to socialization and interaction of agents, with knowledge of environment states and settings, coupled with investigation diligence, may prove to be a successful strategy for overcoming the inherent difficulties of the continuous problem domain.

Autonomous software agents, or just agents, are capable of autonomous proactive behaviour, reactive behaviour, and goal-driven social behaviour using communication in their environment. Agents can be as simple as subroutines, but are normally larger entities with persistence. In rich multiagent systems, a strong relationship must occur between the communication act of one agent and the necessity of interaction, which is the act of another agent understanding the original communication.

An intelligent interaction framework for ambient awareness must provide knowledge distribution and the principles and processes governing and supporting the exchange of ideas, knowledge, information and data. To address the open research requirements for an intelligent interaction framework, research is needed to provide functions and structures that are commonly employed in multiagent research to enhance knowledge communication using mechanisms, as described in [60]:

- *Intelligence.* This state is formalized by knowledge and interacting using symbolic knowledge.
- *Scalability/community.* Support for community organization and creation of societies. Social participants are normally characterized by the interactive ability to be included in a multiparticipant system.
- *Coordination.* Performance of an activity in a shared environment with other agents. Coordination activities often require negotiated plans, workflows, or some other activity management mechanisms.
- *Cooperation.* The ability to coordinate with other participants to use a common resource.
- *Collaboration.* To be able to coordinate with other participants to achieve a common goal.

- *Competitiveness.* The ability to coordinate with other participants, except that the success of one agent implies the failure of others.
- *Multimodality and intelligent I/O behaviour.*

3.4.5.2 Multimodality

In multimodal communication, humans utilize a combination of different modalities, taking advantage of both the individual strength of each communication mode and the fact that both modes can be employed in parallel. The ability to integrate information across sensory modalities is essential for accurate and robust comprehension by machines, and to enable machines to communicate effectively with people. Sensory modalities include vision, hearing, kinesthesia (muscle sense), somatic sense (touch, pressure, temperature, pain) and chemical sense (taste, smell). The area where the greatest cross-modality interface research has been carried out has been in the disability access area. Multimodality in the context of ambient awareness is also about gathering information (images, sounds, gestures, etc), interpreting them, and using them (for output) in an appropriate manner.

Multimodal systems can be used in two ways: to present the same or complementary information on different output modes (e.g., a phone that rings (hearing), lights up the display (vision), or vibrates (somatic)), and to enable switching between different input modes, depending on the current context and physical environment. Therefore, research on ambient awareness may benefit from research that ensures modality independence.

Many ambient-aware applications make use of sensory information sensed using diverse sensors. The research challenge is to be able to use and integrate that information and understand sensory modelling and identifying requirements for a cross-modality sensory model for ambient awareness.

3.4.5.3 Ambient-aware Intelligent I/O Behaviour

The presence of multimodality and the availability of relevant ambient information provide the need for introducing intelligent human–machine interface management, which enables intelligent I/O behaviour adaptation. This means that the input/output behaviour of a user end system–for example the modality–can change dynamically based on the actual context. For example, the user interface (UI) to an email-system is typically text-based, whereas it would become speech-based if the user is located in a car. The role of the actor for the physical output could move from the laptop/PDA to the audio equipment inside the car. This implies that the end device is not restricted any longer to a single device, but is dynamically reconfigured based on the needs of the application, the surroundings, and, of course, the available modalities (that can be considered as terminal capabilities).

The ontology for an MMI configuration is provided through a set of profiles and policies. To apply ambient-aware adaptation to a multimodal interface (MMI) configuration, an MMI management unit has to deal with the ambient information captured in various profiles, and with policies. Thus, for ambient awareness, overlap and strong interactions with the research issues described in both the personalization and the adaptation sections exists.

The disability access area is also an important application area for ambient-aware intelligent I/O behaviour. Again, suitable policies are needed to optimally serve the physical condition, or even handicap, of the user with his particular surroundings and preferences.

3.4.5.4 Modifying the (Real-world) Ambience by Effectors

In the same way as sensor processing converts real-world information into the digital domain where it can be used for supporting an actor, effectors convert digital information into real-world state changes in a way that can be sensed by actors directly. Thus, using effectors is the ultimate result of any I-centric application, because, in the absence of effectors, no actor support would be possible. Examples of effectors are lights, automatic doors, phones, loudspeakers, projectors, printers, copy machines, faxes, displays, air conditioning, blinds, etc.

Like sensors, effectors can be modelled as objects and share the problems of services in a network:

- they need to be discovered (which can be additionally complex, as some effectors might be reached only by local communications means, e.g. a public area network (PAN));
- they need to be access controlled because not every actor will be able to, for example, control the blinds in a room;
- there need to be means to allow other objects to actually use effectors, i.e. knowing the achievable effect and using the correct parameters;
- there need to be means to handle the case that an effector in use becomes unavailable due to, for example, communication problems; and
- effectors can be stationary or mobile, and mobile effectors might pose special requirements in terms of discovery or usage.

However, unlike generic services (but also like sensors), communicating with effectors might require proprietary protocols and might take place at lower layers (i.e. below Layer 3). Therefore, one aspect to examine is to what extent the mechanisms for dealing with sensors can be applied to effectors. One difference between effectors and sensors is the fact that, in most cases, the usage of an effector will be exclusive and consumes data instead of generating it. So, for example, the efficient usage of a single value by many readers is not an aspect for effectors.

3.4.5.5 Topology-related Issues in Dealing with Sensors and Effectors

An aspect of ambient intelligence is interpreting a variety of sensor information in order to act on effectors in an intelligent way (according to the user surroundings). This becomes more complicated when this 'ambient intelligence' is to be applied in a non-open space, like a house or apartment. In such an environment, the topology has to be taken into account. This requires intelligent reasoning about the ambience. Let's illustrate this complexity with the following example. A person arrives in his apartment that is equipped with quite a dense lattice of BlueTooth tags. Let's imagine the user is located with respect to those BlueTooth tags. While he was absent his daughter called him. Upon arriving in his apartment, the man's context includes the information (incoming event) that his daughter would like him to call her back. The so-called 'ambient intelligence' gives the man notification about that incoming call at a appropriate time, using an effector such as a blinking frame or any kind of 'beeping' device.

The problems to be solved in this example include: which kind of effector to use? How many effectors should be used? Which one(s) should be switched on according to the estimated user's location? Certainly, sensing the man's location using a BlueTooth tag and then acting on the closest effector (lightening a lighting frame for instance) is not sufficient because (a) people might be sensed through a wall, (b) the sensor and the effector can be separated by a wall as well, and (c) the man might take different directions and then be missed by the effector. It appears in this example that the reasoning depends on the topology where it takes place: not necessarily the closest effector is to be used, but rather the one that is 'known' to be visible from the estimated user's location. Maybe two effectors must be used to be sure the man's attention is drawn, as he might take different directions from his current estimated location.

Taking the topology into account can be achieved in some classical ways if this topology is known in the system, e.g. if the 'ambient intelligence' has access to the apartment plan. In most general cases it is desirable that this topology is to be learnt automatically by the system, or at least emphasized automatically in the ambient reasoning. In the first case, the 'ambient intelligence' is going to infer a topology based on sequences of locations gathered during the various user's displacements, with possible reinforcement methods. The latter case can be achieved by learning the user's behaviour (which itself emphasizes the topology) by correlating the user's actions upon effectors with values of sensors (using incremental clustering is a possibility here). For instance, if a person located at (x,y) never reacts to an effector, this being a lamp located at (x',y'), it is likely that there is a wall between (x,y) and (x',y'). In this case the topology is never made explicit as only relationships between sensor–effector–user–action are considered. Of course, getting feedback from the user will help to improve the learning and reasoning (the person not reacting to the lamp could be blind).

Key technologies envisioned here as being in need of research are, among others, reinforcement learning, incremental clustering, rule-based deduction and some not yet identified graph techniques.

3.4.5.6 Intelligent Sharing of Resources: Super Distributed Objects

The increasing propagation of devices, disposing of processing capabilities and having the possibility of connecting to each other in an easy and ad hoc manner, leads to the ubiquitous availability of services and enables the introduction of new computing paradigms, like mobile computing, ubiquitous computing and pervasive computing. In addition, ubiquitous network connectivity allows building global software infrastructures for distributed computing and storage. A goal of these infrastructures is to provide a distributed community of software components that pool their services for solving problems, composing applications and sharing contents. The resources, such as hardware devices and software components, can be abstracted as super distributed objects because both share common assumptions and technical issues, such as a large number of distributed resources, ad hoc application boundary, temporal unavailability of resources, and decentralized organization of resources.

An SDO is a logical representation of hardware, or a software entity that provides well-known functionality and services. Super distribution means incorporating a massive number of objects, each of which performs its own task autonomously or cooperatively with other objects, without centralized control. Examples include abstractions of devices such as mobile phones, PDAs, home appliances and various software components. An SDO may also act as a peer in a peer-to-peer networking system, or a storage node in a global storage network system.

Due to their inherent features, SDOs need interoperable middleware technology that enables uniform access to them in order to support easy and rapid service creation. The middleware shall provide SDOs the functionalities of access control, general management (e.g. configuration, monitoring and reservation), discovery with support of mobility, social networking (cooperative processing) and spontaneous networking in an ad hoc manner. Today, there are several hardware and software interconnection technologies, such as HAVi, BACnet, OSGi and Jini. However, they are restricted to specific platforms, network protocols, programming languages, or they focus on limited application domains. No common model-based standards exist to handle various resources in a unified manner independently of underlying technologies and application domains.

Characteristics

The characteristics of an SDO are defined by its properties and the services it offers. An SDO is dynamically characterized by its status, which is nonambiguous for every point in time. The status of an SDO can be modified solely through the invocation of its services. To summarize:

- services define the capabilities of a super distributed object. A service is represented by a group of operations that can be executed by the SDO. A service execution can modify the configuration or status of an SDO.
- While some SDOs can act fully autonomously, others need to make use of services of other SDOs before they can accomplish their own services. As an example, a dimmer SDO with service providing constant brightness in a room can be considered. It should find an SDO offering a brightness measuring service and SDOs that control light switching devices in the room, such as dimmers, lamps, etc. in order to have the possibility of building and executing its own service.
- An SDO can wrap a device. Devices wrapped by SDOs are called embedded devices. The SDO is responsible for the mapping of device capabilities to the operations of the SDO interface, as well as for representation, monitoring and configuration of device capabilities. By means of its interface, the wrapping SDO also enables access to the device functionality.
- SDOs can gather in groups. The group concept can assist better organization of communication and interaction between SDOs. The group organization principle can be chosen freely. It is possible to group SDOs with similar properties or capabilities, for example, all SDOs placed in a certain location, or all SDOs controlling light-switching devices. An SDO can belong to several groups at the same time.

SDO Interface

In addition to its services, an SDO offers operations intended for resource management. Therefore, all SDO interface operations are divided into two large groups. *Service operations* compose the SDO's operational interface, while *management operations* constitute the management interface. The operations of the management interface can be divided further into *monitoring, configuration, reservation* and *discovery* interfaces. The monitoring interface enables one to check the current state of SDO resource data; through the operations of the configuration interface, the resources can be configured; the reservation interface permits one to reserve SDO utilization; the operations of the Discovery interface allow an SDO to advertise its capabilities in the SDO network and search for services it needs.

SDOs can be composed; the result of composition of several SDO entities is again an SDO. In this approach, an SDO is considered as the smallest addressable unit. Beginning with the moment when several SDOs have built a composed SDO, they can be accessed only through the composed SDO's interfaces; there is no possibility to address any of them individually. The services provided by those former SDOs, are henceforth provided by the composed SDO. In addition to the services of partial SDOs, the latter can offer further services, possibly based on their services. A composed SDO can maintain an internal table with the list of partial SDOs' services. Then, when an invocation request arrives for one of these services, the SDO can forward the request to its immediate addressee.

Required Functionality

To facilitate SDO communication and collaboration, the SDO system architecture should offer several functions. The following scenario can be used as an example of SDO interaction in the system. Several SDOs that control electrical appliances in the office build an SDO group. A new SDO joins this group. The newcomer offers a service that turns off the office appliances (for example, lighting, air conditioning system, etc.) at 7 pm. The SDOs in the group should supply some mandatory functions to enable the newcomer to become integrated into the group so that it will be able to communicate and collaborate with other SDOs. A first collection of these mandatory functions is listed below.

- *Addressing.* To make the communication between SDOs possible, some kind of SDO entities addressing is required. In other words, SDOs should be able to address each other to achieve others' services. In the surveyed scenario, the locations of the SDO entities with required functionality should be known to ensure that the newcomer can contact them directly.
- *Discovery.* An SDO may need other SDOs to accomplish its services. So, when changing its location and joining a completely new SDO group, it must first find SDOs offering services it needs before its services can be executed. Knowing exactly, or at least similarly, what services are required, the SDO should be able to initiate a search for services with the necessary capabilities. Therefore, the discovery facility is required to support SDO collaboration. At the same time, a newcomer SDO should have the possibility to announce offered services to other SDOs in the system. In the above scenario with the 'power down' service, the service would search for SDOs offering, among others, lighting and air conditioning services, probably by asking the system discovery service if there are some entities with specified functionality in the office, or by sending requests to all SDOs.

- *Reservation.* It should be possible for SDOs' to reserve other SDOs' resource utilization, therefore reservation functions should be offered by the SDO interface. The requirement arises from the necessity to get exclusive access to services. For example, if, while the 'power down' service was trying to turn one of the light switches off by invoking the corresponding operation on a light SDO, some other SDO service was trying to turn it on, the attempts of the former would become rather useless. Hence, it can be advantageous if the SDO is able to get exclusive access to the services provided by the condition SDO.

- *Monitoring.* SDOs should provide interface functions enabling monitoring of their services' states described by resource data. Services can need a monitoring possibility to control the service state for some purposes. In the example scenario, the service would have to monitor the state of the light SDO. In fact, if the light was turned off before the service started its execution, there would be no need to turn it off again. On the other hand, if the light was on, it should be turned off.

- *Event communication.* A monitoring mechanism permits an SDO entity to watch the other SDOs' states. While it is based on polling, another mechanism for state information propagation can be taken into consideration, enabling SDOs to communicate changes in their state to other system components. When the SDO state changes, other system objects can be interested in immediate notification of this event, and in receiving the data describing the new SDO state. So, a mechanism for event transmission can be specified, enabling system components to propagate information about their state changes by event notifications. In order to ensure that event notifications need not be sent to all SDOs that are currently present in system, as this could lead to network congestion, definition of a subscription mechanism is required. With the introduction of a subscription concept, SDOs would have the possibility to subscribe to notification of events they are interested in. In the scenario with the 'power down' service, the service could subscribe to event notifications of appliance services some time before its settled start time. By this means, the service would be able to collect information about appliances' states and eventually would know at the launch time if they are on and should therefore be turned off.

- *Configuration.* The SDO interface should provide functions enabling configuration of its resource data and service parameters. This can be required to configure SDO location information, services, alive-notification intervals, etc. In the example scenario, the 'power down' service should offer an interface enabling the configuration of its resources, for example start time. So if somebody would like to work in the office after 7 pm, he would be able to configure the service to start its execution after he leaves.

3.4.6 Modelling the Layers of Ambient-aware Applications

Structuring ambient-aware applications is important for modularizing their implementation. The modularization is necessary for reuse and sharing of software and hardware components. For this purpose, a five-layer model for structuring ambient-aware applications is proposed. The layers are Physical, Data, Semantic, Inference and Application. On the lowest level, the Physical layer, there are sensors and other objects, producing output in raw format. Examples of these data are analogue microphone signals or the strength of an RF signal sent by a WLAN access point. On the second level (Data layer), there are objects producing processed data, for example spectral information of the phonemes in the audio signal or location coordinates computed based on three RF signal strengths. The third level, the Semantic layer, contains objects that transform the data into a form meaningful for inferring context. These objects could analyse the spectral data and state that the speaker is Peter with confidence of 0.9, or they could state that the coordinates computed from the RF signals indicate that the mobile terminal is in Lisa's office. The fourth level, the Inference layer, uses information from the semantic level, earlier information and inference rules, possibly autonomously learned, to proactively formulate and reason as to what the user (either man or machine) is doing, and what kind of services he, she or it might want. For example, if Peter is in Lisa's room and Lisa is his boss, the inference object (agent) might suggest that he is in a meeting. On the uppermost level, the Application layer, his personal mobile agent might then decide that he probably won't want calls to his terminal from football team-mates during the meeting, and will block those calls.

It should be noted that objects at a higher layer may combine input from one or more objects at a lower layer with stored information. This is indicated in Figure 3.19 by input arrows.

In general, context models define a conceptual basis to support the development of context-aware and adaptive systems and applications. They aim to facilitate the gathering of information from sources such as sensors and context providers, performing

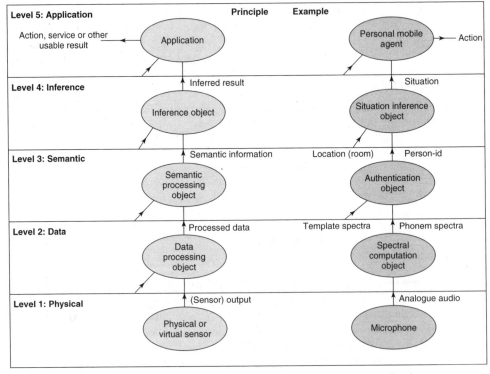

Figure 3.19   Five-layer model for structuring ambient-aware applications

interpretation of data, carrying out dissemination of contextual information to interested parties in a scalable and timely fashion, and providing models for programming context-aware applications. In [61] an overview is given of context models, as well as several context-aware applications.

3.4.7 Interactions of Humans and Machines in Processing Ambient Information

It has been recognized that there is a pressing need to obtain a better understanding of what ambient is in order to facilitate the exploitation of an individual's situation through ambient-aware applications. The bottom line is what situational information is relevant, useful and how should it be used?

All indicators of ambient awareness constitute a parametric space in which information has been gathered about complex situations (e.g. computing context, user context, physical context, time context). More information is not always better, however. Situational information is only useful when it can be usefully interpreted. It has been argued that context-aware computing (and hence ambient-aware computing) redefines the basic notions of interaction, with questions such as 'What role does context play in our everyday experience, and how can we extend this to the technical domain?'

A computational solution that covers a human strategy to deduce information from their environment, use that in interactions and act appropriately, may point to the solution but is tremendously complex. As humans interact with humans they make implicit use of verbal (e.g. tone of voice) and nonverbal (e.g. facial expressions) information. In general, human beings imply knowledge about their current situation in order to act according to the state of the environment. This usually includes perceiving and acting, listening and making conversation, rational and irrational decision making, and common sense reasoning, which are often also based on situation-dependent knowledge. Further, the knowledge of a person's situation can comprise information about objects of his current environment, as well as abstract and nonphysical relationships to other human beings. This kind of information needs to be transparent and needs to be made explicit in order to make mobile applications adaptable.

On the one hand, by way of contrast, human–machine interaction is substantially different to human–human interaction. A typical communication between a human and a computer system consists of input and output data, which is exchanged sequentially. A human being interacts with the computer through the use of devices like a keyboard, mouse or monitor. A movement with the mouse, for instance, is directly followed by a reaction of the computer system on the monitor. But the computer is (in general) not able to 'sense' the current situation like the human does. It is therefore unable to adapt its behaviour to the current state of the environment. On the other hand, the interpersonal interaction rules (how humans interact in the real world) have been applied to how humans interact with computers. The research is based on the idea that one can apply theories and methods from the social sciences directly to users' interactions with computers. Amazingly, users' responses to computers are fundamentally social and natural.

The situational circumstances of an actor can consist of several different properties representing actors and objects of the actor's current communication space. Sensors or human–machine interfaces can gather information about these actors and objects. Logging these human–computer interactions makes a computer system self-knowledgeable, that is, aware of its own context. Profiles have to be created manually by actors, by human–machine interaction, or automatically by self-learning capabilities. It is recognized that human–machine interaction is restricted to direct interaction, which makes it ineffective and slow in comparison to human–human interaction. To improve this kind of communication, it is necessary to adjust it to typical face-to-face communication between humans, who make use of implied situation-dependent information. It is believed that the interpersonal interaction rules and strategies (human–human and human–computer) may determine the coupling of the indicators in order to be really ambient aware in the provision of an active context.

3.4.8 Standardization Efforts and Supporting Technologies

Standards to exchange the situational circumstances of an individual and generic support for this exchange, as well as standards to secure protection of sensitive data are needed. Though ambient information might be easy to acquire (emotions are difficult, absolute position can be easy), there are only a few well-defined mechanisms, frameworks and standards specifying how the ambient (mostly location or position) information can be made available to applications.

3.4.8.1 Geographic Information

In the geospatial community, there are three important international organizations leading the development of industry standards and specifications: ISO/TC211, ISO SQL3/MM and OGC. *ISO/TC211*, a formal international standards workgroup, focuses on an entire family of geospatial standards ranging from exchange formats to metadata to spatial data models. *ISO SQL3/MM*, also a formal international standards workgroup, works on multimedia and spatial extensions to the SQL3 dialect of the SQL standard. The *OGC*, a consortium of industry, government and academic organizations from around the world, develops software specifications (APIs) that enable location-based technologies to interoperate. OGC's interoperability programme started an initiative to integrate mobile services into its GIS (geographical information systems) test bed. The *Open Location Services* (OpenLS) initiative is intended to engage the wireless community in taking location services to the next level.

There are many different ways of expressing location information designed by numerous domains and organizations. They include:

- Expressions standardized for GSM and UMTS (called here '3GPP') to be used internally in the GSM and UMTS mobile networks specified by the Third Generation Partnership Project (3GPP).
- An interface towards mobile networks (e.g. GSM) for providing access to location information of mobile terminals under consideration by the Location-interoperability Forum (LIF, integrated into the Open Mobile Alliance, OMA). The LIF has produced a specification for a mobile location protocol (MLP), which is an application-level protocol for the positioning of mobile terminals. It is independent of the underlying network technology (and, thus, of the positioning method), so it can also use position

data from GPS – if there is a way to get it from the GPS receiver to the location enabling server (LES). The MLP serves as the interface between a location server and a location enabling server, which, in turn, is interfacing to the application server. It defines the core set of operations that a location server (essentially an MPC) should be able to perform.

- The geography markup language (GML) for storing and transporting geographic information specified by the Open GIS Consortium (OGC). The OpenGIS GML is part of an entire framework that includes the means for digitally representing Earth and Earth phenomena, both mathematically and conceptually; a common model for implementing services for access, management, manipulation, representation and sharing of geodata between communities; and a framework for using the Open Geodata Model and the OpenGIS Services Model. The intention is to solve not just the technical problem, but also to define a vocabulary and a way to extend it, which provides a common ground for information exchange between different information communities (sets of users with different meanings, semantics and syntax for geodata and spatial processing).
- Navigation markup language (NVML) for describing navigation information submitted by the Fujitsu Laboratories to the Worldwide Web Consortium.
- Point of interest exchange language (POIX) for exchange of location-related information over the internet, created by Mobile Information Standard Technical Committee (MOSTEC) and submitted to W3C.
- Geotags for geographic registration and resource discovery of hypertext markup language (HTML) documents.
- National Marine Electronics Association's (NMEA) interface and data protocol NMEA-0183, often used by GPS receivers.
- The electronic business card format VCard and ICalendar for exchanging electronic calendaring and scheduling information in the internet, including elements to specify position.
- A means for expressing location information in the domain name system (DNS-LOC).
- Simple text format for the spatial location protocol (SLoP) proposing a simple text-based format to carry a minimal location data set.
- GMML, XML-based geographical information for navigation with a mobile.
- LanXML, an XML-based data format for exchange of data created during land planning, civil engineering and land survey processes.
- Geospatial-extensible markup language (G-XML) for encoding and exchanging geospatial data, specified by the G-XML committee in Japan.
- The magic services API (Mobile Automotive Geo-Information Services Core) was created by a loose industry group with the goal of creating a web services API for location information. The protocol used between the mobile station and the server is simple object access protocol (SOAP), and the document type is XML. The idea is to create a set of core services, which can be called web services in the emerging W3C architecture, that could be used by application service providers (ASPs) who intend to set up location-dependent services. The scope of the services includes route planning and geocoding (= the conversion of human text or speech defining an address or other location expression to corresponding geographic coordinates).
- The geographic location/privacy (geopriv) working group of the IETF is currently addressing the standardized description of location information, as well as its transfer

in a controlled and secure fashion. If successful, the results of this group might also be extended to other types of context information.

In addition to these, there are several other nonpublic specifications of location information expressions, including those from the WAP Forum Location Drafting Committee, Bluetooth Special Interest Group, ISO/TC211, etc.

3.4.8.2 Agent Technology

Agent technology has grown from roots grounded in the subdomains of Information Science, Artificial Intelligence and Distributed Computing. Agent research has covered a wide range of topics including:

- communication technologies (semantics, expressions of content);
- architecture (internal reasoning models for agents);
- knowledge representation (in communication or internal);
- distributed problem solving (planning, scheduling);
- coordination (coherent joint action);
- self-interestedness (negotiation, coalition formation) and embeddedness (interaction with the environment including humans).

Software agents can be used for supporting functions needed for ambient awareness:

- to solve the knowledge problem in open and dynamically changing environments;
- to handle joint socialization, coordination, negotiations and planning;
- to learn about context and generate detectors and analysers;

A multiagent system, or multiagent society, is a social agent communication environment. A multiagent system must involve [62]:

- the principle of coordination – which agent societies must use to provide information about an agent's role in the society;
- the principle of cooperation – which an agent society will use to perform activities in a shared environment; and
- the principle of competition – which an agent society must address due to that fact that agents can have self-interested goals that may supersede the goals of the group or another single agent.

Of particular importance is the social agent communications environment that provides the processes that allow agents to interact productively. The productive interaction aspects that are particularly noteworthy include, social differentiation, whereby a group from the agent society can play multiple roles in multiple groups; interaction management, or the rules that manage interactions between agents to ensure correct interaction; and social order, which are the rules for production of structured relationships among social agents.

A role is an abstract representation of an agent's function, service or identification within a group.

There are three main concepts of socialization and community that are addressed by the use of an agent social group.

1. *Intra-group associations.* These allow for the partitioning of a larger group into smaller communication domains for specific topic- or domain-oriented interaction.
2. *Group synergy.* This allows the agent members of the social group to use the abilities and services that may be offered, but which are not possible by any single agent.
3. *Inter-group associations.* These serve as a mechanism whereby groups of agents in an agent social group can interact with other sets of agents in another social group.

One proposed solution to the requirement of enhanced agent communication in the social agent communications environment is the use of an agent communication language (ACL). ACLs have grown from roots in the research domains of Artificial Intelligence and Linguistics. Agent communications messages must have well-defined semantics that are computational and visible. Using an ACL-based approach, different parties can build their agents to interoperate using the ACL as the basic element of an interaction protocol between agents. Notable ACL implementations include:

- KQML – the knowledge query and manipulation language defines an 'envelope' format for messages, using which an agent can explicitly state the intended elocutionary force of a message.
- FIPA ACL – the Foundation for Intelligent Physical Agents (FIPA) has released a specification of an 'outer' language for messages. FIPA ACL defines 20 performatives and is superficially similar to KQML.
- KIF – the knowledge interchange format is a language explicitly intended to allow the representation of knowledge about some particular 'domain of discourse'. It was intended to be used with KQML.
- XML – schema-based implementations.

A second proposed solution to the requirement of enhanced agent communication in the social agent communications environment is the use of agent interaction protocols, which can also be known as conversation protocols. Cooperative and collaborative agents work in association and towards joint and common goals. To produce coherent plans, a social agent communications environment must be able to support agents recognizing interactions to plan appropriately. There are three possibilities for multiagent interaction planning in a social communications environment.

1. *Centralized planning for distributed plans –* a centralized planning system develops a plan for a group of agents, in which the division and ordering of labour is defined. This is a master–slave cooperation strategy. An example of this strategy is the Shared Plan Model.
2. *Distributed planning for centralized plans –* a group of agents cooperates to negotiate a centralized plan. This strategy allows agents to be specialists in different aspects

of the overall plan and to contribute part of it. An example of this strategy is joint commitment and partial global planning.

3. *Distributed planning for individual plans* – a group of agents cooperate to form individual plans of action, dynamically coordinating their activities along the way. This strategy may be best suited to self-interested agents, where there may never be a 'global' plan, but where there is only a virtual emergent global plan for aspects of sharing the cost of some common resource. An example of this strategy is self-interested multiagent interactions.

3.4.9 Conclusions

Aspects related to ambient awareness that need to be researched and solved for the next generation I-centric wireless world are described in this section. Research challenges that follow from these aspects are:

- Advances in sensor technology are necessary to further adapt services to – and cooperate with – the environment of actors.
- How to optimally combine and express (semantics) different position information acquiring technologies:
  — wide-area position technologies;
  — short-range wireless technologies;
  — sensory information.
- Dealing with location information in wireline networks.
- How to deal with partial information.
- Investigating how multiple readings can be derived from one single device.
- Rightful evaluation of ambient indicators:
  — proper aggregation in relation to a known reference;
  — communicating and relating to additional situational information;
  — proper weight and ordering of the processing of the relevant ambient indicators;
  — optimal use and integration of sensory information sensed by diverse sensors and understanding sensory modelling and identifying requirements for a cross-modality sensory model for ambient awareness.
- Privacy protection of user-regarded sensitive or private ambient information.
- Policies on ambient information access and exchange.
- Interpersonal interactions rules and strategies (human–human and human–computer) to determine the right coupling of the indicators in order to be really ambient aware in the provision of an active context.
- Effector-related issues, like:
  — modelling of effectors;
  — semantic descriptions of effector abilities;
  — handling disconnection problems of effectors.
- Reinforcement learning, incremental clustering, rule based deduction and graph techniques.
- An intelligent interaction framework for ambient awareness with functions and structures covering aspects such as intelligence, scalability/community, coordination, cooperation, collaboration and competition.

- Extend standardization specifications on denoting and exchanging ambient information on a global scale across heterogeneous networks and technologies conforming to an appropriate open interworking architecture.

3.5 Generic Service Elements for Adaptive Applications

The telecommunications world is moving towards internet protocols and service/application frameworks/platforms. These trends originate in the convergence of data and telecommunications, and in the demand for ever-faster service development and deployment.

Middleware is a widely used term to denote a set of generic services above the operating system. In this text we identify and describe the upper middleware layer, called generic service elements in the Wireless World Research Forum. We have identified eight generic service elements supporting adaptive applications that will be one of the cornerstones for the success of systems beyond 3G. The generic service elements described include environment monitoring, event notification, a distributed application framework, a perception service, modelling services, a mobile distributed information base, on ontology service and a semantic matching engine.

3.5.1 Introduction

The current trend in developing forthcoming telecommunication networks is to utilize internet protocols. An immediate implication is that IP is the OSI Layer 3 protocol and that the addresses are IP addresses. However, this is not sufficient. Other solutions – both above and below the IP protocol – are also needed to meet the requirements for the next generations of telecommunication networks. Issues under study in the internet community and in various standardization bodies, forums and consortia of telecommunications, include mobility, quality of service, security, management of networks and services, discovery, ad hoc networking and dynamic configuration.

Another significant trend is the requirement of ever-faster service development and deployment. The immediate implication has been the introduction of various service/application frameworks/platforms. Middleware is a widely used term to denote a set of generic services above the operating system. In the WWRF service architecture, the traditional middleware layer is further divided into the service execution layer, the application support layer and generic service elements.

Inside the WWRF there is a wide consensus that personalization, ambient awareness and adaptability will be the fundamental characteristics of services and applications for the wireless world of the future. In this text we identify and examine the generic service elements needed to build adaptive applications. By adaptation we mean the ability of services and applications to change their behaviour when the circumstances in the execution environment change. Issues in personalization and ambient awareness have already been addressed in Section 3.3 and 3.4.

Our brief literature survey indicates that the terminology in the field is not unified. People speak about nomadicity, ubiquitous computing, pervasive computing and ambient intelligence. If we take a serious look at different 'definitions', or characterizations, of those concepts, we must conclude that they are at least very close to each other, if not different names for the same thing. There may be some minor differences in emphasis, but the core and major challenges are similar.

Based on the functional requirements for adaptive applications, we have identified eight generic service elements. They include environment monitoring, event notification, a distributed application framework, a perception service, modelling services, a mobile distributed information base, on ontology service and a semantic matching engine. We have not identified service discovery and autoconfiguration as generic service elements. Instead, they are included in the distributed application framework due to their crucial importance.

3.5.2 Related Work

The notion of nomadic computing was introduced in the mid 1990s to launch information processing services accessible at any time and anywhere. According to Professor. Leonard Kleinrock [63]:

> "Nomadic computing and communications will dramatically change the way people access information – and a paradigm shift in thinking about applications of the technologies involved. It exploits the advanced technologies of wireless, the Internet, global positioning systems, portable and distributed computing to provide anytime, anywhere access. It is beginning to happen, but it makes everything much harder for the vendors."

The Cross-Industry Working Team [64] identified the following challenges in nomadicity: location independence, device independence, widespread access, security, adaptability to new technologies, friendly interface and partitioning.

The concept of ubiquitous computing was introduced in the early 1990s [65, 66]. Mark Weiser from Xerox PARC expressed the goal as being to achieve the most efficient kind of technology that is essentially invisible to the user, to make computing as ordinary as electricity. In the beginning, the focus was on small special purpose devices, network protocols, interaction substrates and new styles of application. After the first prototypes, additional research directions were identified: wireless communications, partitioning and disconnected operation, location and resource discovery, privacy and power consumption [67].

The notion of pervasive computing is yet another closely related concept. It is characterized by different attributes by different research groups, but most of them agree that pervasive computing is:

- about mobile data access and the mechanisms needed to support a community of nomadic users;
- about 'smart' or 'active' spaces, context awareness, and the way people use devices to interact with the environment; and/or
- about the best ways to deploy new functions on a device and to exploit interface modalities for specific tasks.

Nevertheless, there are three major focus areas, which are widely agreed, in pervasive computing:

1. The way people view mobile computing devices, and use them within their environments to perform tasks.
2. The way applications are created and deployed to enable such tasks to be performed.
3. The way in which the environment is enhanced by the emergence and ubiquity of new information and functionality.

If we take a serious look at the different 'definitions', or characterizations, of these three concepts, we must conclude that they are – at least – very close to each other, if not different names for the same thing. There may be some minor differences in emphasis but the core and major challenges are similar. As a validation and verification procedure we carried out a literature (and web) review study on current research activities in various research areas of software systems. Below we briefly summarize our major findings.

The Endeavour Expedition at the University of California in Berkeley [68] is a collection of projects that examines various aspects of ubiquitous computing. The goal is to enhance human understanding through the use of information technology, by making it dramatically more convenient for people to interact with information, devices and other people. A revolutionary Information Utility, which is able to operate at planetary scale, will be developed. The underlying applications, which are used to validate the approach, are rapid decision making and learning. In addition, new methodologies will be developed for the construction and administration of systems of this unprecedented scale and complexity.

In the MIT, the corresponding project is called Oxygen [69]. The Oxygen project targets the means of turning a dormant environment into an empowered one that allows the users to shift much of the burden of their tasks to the environment. The project is focusing on eight enabling technologies: new adaptive mobile devices, new embedded distributed computing devices, networking technology needed to support those devices, speech access technology, intelligent knowledge access technology, collaboration software, automation technology for everyday tasks and adaptation methods.

Project Aura at Carnegie Mellon University [70] will fundamentally rethink system design to address the problem of user attention. Aura's goal is to provide each user with an invisible halo of computing and information services that persists, regardless of location. Meeting this goal will require effort at every level: from the hardware and network layers, through the operating system and middleware, to the user interface and applications. The project plans to design, implement, deploy and evaluate a large-scale system demonstrating the concept of a 'personal information aura' that spans wearable, handheld, desktop and infrastructure computers.

The Future Computing Environments (FCE) Group at Georgia Tech [71] has addressed the problems in building intelligent and interactive human-centric systems that support and augment our daily lives, relying on the concepts of ubiquitous and aware computing. The group is attempting to break away from the traditional desktop interaction paradigm and move computational power into the environment that surrounds the user. The research challenge not only involves distributing the computation and networking capabilities, but also includes providing a natural interface to the user. It also aims to provide knowledge about the user and the environment that surrounds the user.

The project in the University of Washington at Seattle is called Portolano [72]. The project has three main areas of interest: infrastructure, distributed services and user interfaces. An essential research area is data-centric routing that facilitates automatic data migration among applications. Context-aware computing, which attempts to coalesce knowledge of the user's task, emotions, location and attention, has been identified as an important aspect of user interfaces. Task-oriented applications encounter infrastructure challenges, including resource discovery, data-centric networking, distributed computing and intermittent connectivity.

In the University of Illinois at Urbana-Champaign, the research project in this area is 2K: A Component-Based, Network-Centric Operating System for the next Millennium [73]. The 2K is an open source, distributed adaptable operating system. The project integrates results from research on adaptable, distributed software systems, mobile agents and agile networks to produce open systems software architecture for accommodating change. The architecture is realized in the 2K operating system. It manages and allocates distributed resources in order to support a user in a distributed environment. The basis for the architecture is a service model in which the distributed system customizes itself. The objective is to achieve a better fulfilment of user and application requirements.

The Mobile Computing Group at Stanford University, MosquitoNet [74], has developed the Mobile People Architecture (MPA) that addresses the challenge of finding people and communicating with them personally, as opposed to communicating merely with their, possibly inaccessible, machines. The main goal of the MPA is to put the person, not the device that the person uses, at the endpoints of a communication session. The architecture introduces the concept of routing between people by using the *personal proxy*. The proxy has a dual role: as a tracking agent, the proxy maintains the list of devices or applications through which a person is currently accessible; as a dispatcher, the proxy directs communications and uses application drivers to massage communication bits into a format that the recipient can see immediately.

3.5.3 Functionality Needed in Adaptive Applications

The basic principle of adaptability, i.e. the behaviour of an application changes when the circumstances change, requires that the system detects changes and notifies of them. Therefore, the generic service elements must include *environment monitoring* and *event notification*.

Adaptive applications need to change their behaviour. In some cases this is achieved by changing an algorithm internal to the application. However, in many cases an alternative approach is much more plausible, through replacing some components of the application, or through relocating them into other network elements or service nodes. Therefore, the generic service elements must include a *distributed application framework* that enables seamless and flexible replacement and relocation of application components, but also allows flexible combination of service elements. A crucial functionality that the framework must provide is autoconfiguration.

Adaptive applications are based on various models: models of user preferences and behaviour, models for QoS of connectivity, models of the environment, etc. These models are implications of extensive usage of artificial intelligence. Due to different needs of applications, behaviours and environments, the models that need to be supported are manifold: Bayesian, probabilistic, regressive, predictive, propositional, logic, etc. Therefore, the generic service elements must provide basic means of modelling – without restricting the actual form of the model – including a *perception service* that provides a distributed information base to store and retrieve pieces of knowledge, a *model builder* that takes care of parameter estimation according to the given estimation criterion, a *model combiner* that builds up a combined model from given submodels, and a *model evaluator* that returns the output of the model.

Data management, including perceptions, user preferences, device capabilities and application requirements, as well as a user's own data, will be of fundamental importance for mobile users. Therefore, the generic service elements must provide a *mobile distributed information base* providing consistent, efficiently accessible, reliable and highly available information, that is, a distributed and replicated worldwide file system or database. It should also provide intelligent synchronization, role-based views and transactional operations with user-defined semantics of correctness.

The models used to describe the state of the world of interest – behaviour of users or characteristics of environment, for example – in a distributed system, need a common set of concepts so that apples and oranges are not intermixed. Therefore, the generic service elements must include an *ontology service* that provides a well-defined basic set of concepts and a way to introduce new concepts with well-defined semantics.

In computing three different worlds meet: user preferences, device capabilities and application requirements. Adaptive applications need to match the three worlds so that the user preferences, device capabilities and application requirements are met. This requires common representation of preferences, capabilities and requirements, as well as matching rules. In the future, matching cannot only be a string comparison, but needs to be based on semantics. Therefore, the generic service elements must include a *semantic matching engine*, selecting the most appropriate of given alternatives for a given criterion.

3.5.4 Generic Service Elements for Adaptation

The identified generic service elements needed in adaptive applications are further elaborated in this section.

3.5.4.1 Environment Monitoring

In order to facilitate the creation of software that adapts to changing conditions, we need to support environment monitoring and event notification. An environment monitoring component or service [75] allows the software to observe its surroundings using rules that have been defined *a priori*. Observed signals and events in the environment are notified to components that have expressed interest in them. These observed changes may be used to build various models of the environment, the middleware system and the user. It is envisaged that an advanced middleware system may build a model of its own operation and latencies, and that this model would be used to fine-tune and optimize the run-time behaviour of the system.

The observable environment consists of sensors, network elements, hardware and software components and user interests. These observable components are ambience elements that surround the system and may be physical or virtual in nature. A software component using the event monitoring service needs to formulate the phenomenon to be observed by specifying a set of rules that uniquely identify its occurrence. A component that observes the environment may do so by using any means of observation available, and may also employ the monitoring service in this process. The observation of a signal may cause an event to be delivered to one or more recipients that use the monitoring service. The reception of these events may cause further events to be triggered.

Comprehensive event detection rules are needed in order to maintain high precision in event delivery. The expressiveness of this rule system, or language, is one of the main factors that define the power and scalability of the event monitoring service. High precision results in fewer numbers of ignored messages, and reduced communication cost and server load. These rules are usually defined using an event filtering language, for example SQL-92 used in the Java Message Service [76]. In essence, components subscribe to events using a filtering language that allows the formulation of constraints on published events in order for the system to decide whether they are of interest to the recipient or not. The filtering language allows the components of the middleware system to explicitly define the parameters and conditions pertaining to the change of which they wish to be aware. The filtering language may also be realized using pluggable code, for instance programs created using the Java language. Examples of expressive filters are statements such as 'temperature at my location larger than 30 degrees Celsius' or 'five professors within a range of 10 m'. In topic-based systems, the component would specify the topic names of interesting events. More expressive systems support content-based filtering or addressing, where the whole content of an event is used in determining the set of recipients.

The basic event filtering language is generally event or notification specific, that is, a filter is matched against one event at a time and no correlation between different events is possible. Complex event processing [77] languages allow the specification of very expressive filters, or patterns, that are temporal in nature and require that a conditional sequence of events is observed in order that a notification is sent to the recipients. Complex events reduce application complexity, because middleware is responsible for managing the detection and notification tasks.

The event monitoring service needs to provide interfaces that allow components to subscribe to the events they wish to receive. The event notification service is responsible for delivering occurred and matched events to the correct recipients that have previously subscribed to them. In many cases, event monitoring and event notification are combined in a single service that provides interfaces for both subscribing and producing events. In addition to basic publish/subscribe operations, the event monitoring and notification services need to address privacy issues, access control and authentication. Information that is declared as private or secret, such as location, may not be notified to other parties unless explicitly mandated by the owner of that information.

Since the event monitoring service needs to support the monitoring of arbitrary signals from any ambience element of the environment, it needs to support distributed operation and, in many cases, also the mobility of observable components and components that observe. Mobility imposes new requirements to the event architecture, because it needs to support disconnected components and frequent changes in the subscription topology of the system.

3.5.4.2 Event Notification

Event notification is responsible for transporting an event that has been previously observed by environment monitoring. The event notification component needs to identify the correct set of recipients for a given event, and deliver the events to the recipients using a proper representation format and medium. Many event-based systems are based on MOM (message oriented middleware), and use asynchronous messaging to deliver notifications. The notification component may need to select the best message transport protocol and physical medium in order to provide fast and reliable operation.

There are two frequently employed mechanisms that are used in realizing event monitoring and event delivery functionality:

1. *Source-initiated communication.* In this model, clients register their interest directly on a component that is capable of publishing the required event. The component provides the necessary interfaces for registering filters and recipients. This model is also called the observer design pattern, and it is, for example, used by the Java event system. This model requires that the components are located using some mechanism and it may be insufficient if there are many subscribers and complex filters. Introducing an abstraction layer between the consumers and publishers of events can solve these issues.
2. *Decoupling of event source and client.* This model, also called the event notifier pattern, is related to the observer pattern, and provides a logically centralized component that decouples the consumers and publishers of events. The CORBA Event Service and Notification Service [78] are examples of notification models, where the event channel is a centralized object that decouples consumers and producers. These approaches support anonymous one-to-many and many-to-many communication [79].

For heterogeneous and dynamic environments we require decoupling of producers and consumers. Event-based communication is well suited for supporting adaptive software, because components that use events are loosely coupled, and asynchronous messaging tolerates disconnections. In order to support adaptive components on mobile terminals, at least partial support for event monitoring and notification needs to be provided on the terminals. This local event bus would allow run-time subscriptions to both local and external components, and support the reconfiguration of these components. This system would support both distributed services and applications, and also parts of applications that are local to the terminal, in order to be aware of when changes that meet their filters occur, and allow them to adapt and reconfigure accordingly. Event notification in peer-to-peer and ad hoc environments requires that each peer is capable of deciding whether or not an incoming event is destined for any local component, and also determining the set of recipients to whom to forward the event.

3.5.4.3 Distributed Application Framework

One of the main requirements for the distributed application framework is to allow the creation of new services from existing ones. There are two types of service combination that can be created.

1. *Horizontal combinations (federations).* These allow the provision of a unified view on services of the same type but a limited domain. If there are, for example, services that allow one to locate a device inside a closed area, e.g. buildings, a corresponding federation service could offer a service to locate the device in a city by horizontally

combining the existing building services. To that end, the federation service has to disseminate a query to the corresponding single services and to integrate the answers into a single one.

2. *Vertical combinations (virtual services).* These allow the provision of services that are not offered as such, but can be provided using existing services of different types and some glue. To that end, the virtual service has to generate queries to the single services out of the query to the virtual service, and to compute an answer to it using the answers of the single services. A virtual copier service, for example, could use a scanner and a printer service to scan pages and to send the images to the printer, so the specified number of copies is produced.

One aspect of combination services is that they can be executed on every device that is able to provide some computational power, as they rely not on special resources available only on certain devices. If they can use the combined services remotely, such services can be created and placed on demand.

There are some approaches in this area, but these need to be modified so they fit into the overall WG2 scenario and work even in the absence of a central organizational authority and a third party service infrastructure-provisioning environment.

As, in 4th generation networks, communication-wise the user will be able to use a mix of telecom provider networks, hot spot area networks and privately owned LANs, the user should also be able to use services provided by network operators and any third parties. This idea can even be developed further to a system where even third parties' service infrastructure components can provide not only communication, but services. Such components could include means to discover services, to provide services, etc.

Apart from the economic advantages for the user, a system allowing such a third party service and infrastructure provisioning would have the following advantages:

- A need for a central administrative party can be avoided. In a system where there is no central party that earns money by providing an overall system, there is only weak motivation to operate central resources in a scalable and reliable manner. In rare cases, a distributed, hierarchical administration works along organizational structures (e.g. in the DNS case). But sometimes, especially in cases where manual configuration of the corresponding components cannot be done (e.g. in isolated ad hoc networks), even this option is not possible.
- A system can develop incrementally when any party can provide partial functionality. One problem of systems requiring a certain level of infrastructure is the necessary critical mass. Only if this critical mass of infrastructure and services can be provided, can enough users use the system, so it becomes, in return, economically relevant for more parties to provide services. This critical mass can be reduced significantly if any party can provide services or infrastructure that spans only a limited domain and still can be used in this domain.
- In isolated ad hoc networks, the infrastructure needs to be provided by third parties. In ad hoc created networks without a connection to a larger network, the necessary infrastructure and services need to be provided by the participating devices, as no dedicated infrastructure can be assumed by definition.

The adaptation into the run-time changes in the system structure requires configuration of the service infrastructure. In addition, a user centric approach requires that such an adaptive configuration needs to be automatic. For example, when a node is plugged into the system, the services of it need to be available and capable of being discovered by the other system users. Therefore, the execution framework must support autoconfiguration and service discovery.

3.5.4.4 Perception Service

The perception service is responsible for collecting values of perceptions within some time intervals and storing the values. By perception we mean anything that can be observed and be of use for learning agents. This may be, for example, information about system events or the status of some continuous or discrete quantity, such as the battery level of a laptop computer or communication bandwidth. This centralized handling of perceptions offers many advantages. Decisions on which parts of the collected data should be flushed, and when, are much easier to do with centralized data than if all agents had to do this by themselves. Further, previously unknown correlations are more likely to be found with this approach. Also, refining data (e.g. clustering) and avoiding overlapping data is easier.

3.5.4.5 Modelling Services

The primary mission of modelling services is to learn, but they do not use the results of learning by themselves. They just collect data and build models of the phenomenon they are trying to learn and are used as assistants to intelligent applications. They use a learning service to solve a specific learning task defined by the client. During start-up, the modeller is told the type of learning task, which learning algorithm to use and how the data can be obtained. After this, the modeller starts collecting data and building the model, and is ready to answer questions made by the client.

The modeller does not need to understand the semantics of the symbols and values it handles. The learning is purely technical – just following the steps of the learning algorithm. However, the client for which the modeller is working must understand the meaning of the input and output values of the learning task. So, learning does not affect the behaviour of the modeller directly, but it should have influence on the decisions made by the client. We can say that the client is the 'intelligent' one, and the modeller is just its dumb slave that is good at only one thing: learning. The client may act as a teacher, giving feedback about the correctness of the modeller's output. For example, if the modeller makes an incorrect classification, the client may 'penalize' it and possibly also tell the correct answer.

3.5.4.6 Mobile Distributed Information Base

The current trend to use XML as the presentation format in many applications and services requires that a mobile distributed information base should especially be suited to storing XML documents. Its essential characteristics include high availability, consistency and support for weakly connected and disconnected operation that, in turn, calls for efficient and intelligent data synchronization. In an intelligent synchronization process, the following issues need to be regarded:

- policies regarding when, what and how to synchronize;
- conflict resolution mechanism, including conflict resolution policies;
- low-level network transport methods;
- caching and maintaining cache coherency;
- consistency guarantees for synchronized data;
- locking and session semantics; and
- methods for gathering and transporting updates.

Particular question of interest in synchronization include:

- Are unexpected disconnections handled well?
- Do the protocols used save bandwidth?
- Is the architecture suitable for peer-to-peer operation?
- Is the protocol or architecture too complex for mobile devices?
- Is the architecture scalable?
- Is the architecture secure?

Of the current distributed storage systems, Coda [80], Intermezzo [81], OceanStore [82] and Bayou [83] take into account some of the requirements of the wireless world. Coda and Intermezzo are distributed file systems, whereas OceanStore is a highly available and fault tolerant data storage facility designed to be deployed on a global scale. Bayou is a distributed database, designed from the ground up with disconnections in mind.

The SyncML [84] initiative has specified a synchronization framework. The essential parts of the framework are the SyncML synchronization protocol, the SyncML representation protocol, and bindings for various transport protocols, such as HTTP and OBEX. The synchronization protocol defines the high-level interaction between a device and its peer. This entails connection set-up, authentication, synchronization (in several modes) and object ID mapping procedures. The representation protocol presents the protocol messages in detail in terms of syntax, parameters and result codes. The representation protocol also introduces the ability to filter and search the database, as well as execute commands on the peer.

In order to achieve intelligent synchronization, the following issues must be addressed:

1. *Policies.*
   — when and where to synchronize (depending on connectivity, pricing, demand etc.);
   — what to synchronize (automatically fetch files that are important in the current user context);
   — how to synchronize (different methods yield different trade-offs between bandwidth and CPU cycles).
2. *Synchronization across 'similar' data.*
   — synchronization across different data formats;
   — synchronization in the form of three-way merging.

3.5.4.7 Ontology Service

One common way to define the concept of an ontology, especially in the context of knowledge sharing, is to call it a specification of a conceptualization [85]. In the artificial intelligence (AI) community, ontology refers to a construct that has a specific vocabulary and a set of rules that bind meaning to the words of the vocabulary. The ontology may be simply a hierarchy of words, or the words and meaning may be governed by more complex axioms, which are often expressed in first order logic. In contrast with knowledge bases, ontologies do not contain state-dependent information, but are independent of the particular affairs of the world [86].

Standards such as the Resource Definition Framework (RDF), RDF schema and OWL (Ontology-based Web Language) from W3C aim to bring systematic and formal tools for expressing semantics [87]. RDF and RDF-S are mainly used for defining taxonomical knowledge and relations between knowledge, and OWL is an ontology language with well-defined semantics. However, we need to have additional formal logical languages to describe more complex behaviour.

Ontologies allow intelligent software components, software agents, to agree on different concepts. By gaining a consistent and formal knowledge of a resource or a service, the components may use this information to execute that resource and to request actions that affect the state of affairs in the world. Ontological and semantic information may also be shared, extended by other more domain-specific services, publicized and referred by various entities enabling the interoperability of heterogeneous actors [88].

The ontology service needs to support the representation, manipulation and storage of ontologies of varying detail. It is envisaged that top-level ontologies will be specialized into domain-specific and application-specific ontologies. Both domain-specific and high-level abstract ontologies may be extended to incorporate new subconcepts and concepts. The service needs to support reasoning over ontological information, and sharing it with third parties. In a heterogeneous environment, parts of the ontology may become incompatible with other parts of the ontology, and agents or entities may only partially share the ontology. The ontology service needs to be able to cope with the problems of fragmented knowledge by supporting partial mappings and other techniques between ontologies.

3.5.4.8 Semantic Matching Engine

The semantic matching engine is responsible for using the ontological information provided by the ontology service to match and reason over instances of ontological knowledge, such as profiles. The semantic matching engine needs to be able to determine whether a given document of semantic information conforms to ontology. The engine also needs to support the manipulation of semantic information, for example combining two or more overlapping semantic information sets, and set-theoretic expressions over classes.

In a typical scenario, an agent downloads semantic information from a service that contains instructions on how to use the service. The agent has an internal model and representation of how to model and plan actions. The agent may instantiate some model-based procedural programming language code in the context of the downloaded semantic information and the local knowledge base. This allows the agent to apply the requirements and restrictions of the service with respect to the current state of affairs in the world, as perceived by the agent.

It is envisaged that the semantic matching engine may interoperate with an event pattern matcher (complex event processor) and the event filtering language may support the use of an application or domain-specific ontology in the creation of filter instances. Here, a filter may require that a certain property, or properties, is present in order for the filter to match an event that is based on that ontology. In addition, an environment monitoring component could use the semantic matching engine to combine an event modelled after the hardware ontology with the current hardware and software profile, and determine the intersection of the two information sets.

3.5.5 The Open Mobile Alliance

The Open Mobile Alliance (OMA) is one of the various standardization bodies and forums, whose work is most closely related to the standardization of generic service elements, although in the OMA they speak about service enablers. In this section, we give a brief overview of the current status of the OMA [89].

The Open Mobile Alliance was formed in June 2002, and by late November 2002 included almost 300 companies. The member companies represent the world's leading mobile operators, device and network suppliers, information technology companies, application developers and content providers.

The OMA is designed to be a focal point of mobile services standardization work, to assist the creation of interoperable mobile services across countries, operators and mobile terminals. Through a user-centric approach, the OMA ensures fast adoption and proliferation of mobile services. The alliance drives the implementation of end-to-end mobile services, including an architectural framework, open standard interfaces and enablers.

The foundation of the Open Mobile Alliance was created by consolidating the Open Mobile Architecture initiative and the WAP Forum. The Location Interoperability Forum (LIF), SyncML, MMS Interoperability Group (MMS-IOP) and Wireless Village, each focusing on mobile service enabler standardization work, have each signed a Memorandum of Understanding to express their intent to consolidate with the OMA. Additionally, the OMA will be liaising with various other organizations from the mobile and internet industries to leverage existing approved standards and specifications. Recently, the Mobile Gaming Interoperability Forum (MGIF) and the Mobile Wireless Internet Forum (MWIF) have decided to join the Open Mobile Alliance.

The OMA Architecture Framework (see Figure 3.20) identifies three domains (terminal, network and server) and three layers (network infrastructure, platform and applications). The primary concern of the OMA is the platform layer, that consists of a set of service enablers.

The OMA framework is a logical architecture that does not propose any specific topology or physical location of servers. Furthermore, it does not suggest any hierarchy of protocol stacks to be used between domains.

Network infrastructure consists of the mobile network infrastructure that provides basic connectivity, transport and mobility support. Terminal and server application platforms are the middleware solutions that are not application specific. A platform contains

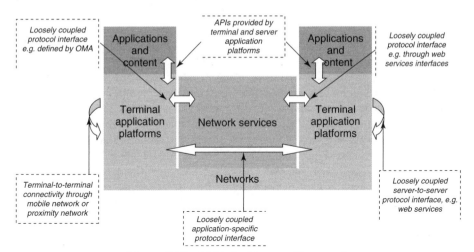

Figure 3.20 OMA Architecture Framework

functionality that can be shared by multiple applications, running either in the terminal or on the server.

Network services are coupled with the network and are hosted by the operator who owns the network. Network services comprise services provided to the server applications platform, for example content push/pull, device management, messaging and location. Network services must have the necessary interfaces that are compatible with the underlying network in order to provide the services.

Applications and content consists of terminal and server applications. Applications use open interfaces (APIs) to interface with the servers or terminals. Applications can be developed either by the terminal or server platform provider, or any third party.

In February 2003, the following specifications were outlined in OMA Phase 1 (Candidate Enabler Release):

- OMA Billing Framework version 1.1;
- OMA Browsing version 2.1;
- OMA Client Provisioning version 1.1;
- OMA DNS version 1.0;
- OMA Digital Rights Management version 1.0;
- OMA Download version 1.0;
- OMA Email Notification version 1.0;
- OMA Multimedia Messaging Service version 1.1;
- OMA User Agent Profile version 1.1.

The OMA IMPS version 1.1 was in OMA Phase 2 (Approved Enabler Releases). In addition, two specifications – WIM: Wireless Identity Module Version 1.1 and WSP: Wireless Session Protocol Version 1.0 – were in a public commenting period.

3.5.6 Conclusions

We have identified and described functionality needed in the upper middleware layer, called generic service elements in the Wireless World Research Forum. We have identified eight generic service elements supporting adaptive applications. The generic service elements described include environment monitoring, event notification, a distributed application framework, a perception service, modelling services, a mobile distributed information base, an ontology service and a semantic matching engine.

When compared to concepts of nomadicity, pervasive computing and ubiquitous computing, we can see similar functionality in the fundamental enablers. When our generic service elements are compared to the service enablers identified in the Open Mobile Alliance, we see an evolutionary path through which a 3G service platform turns into a service platform for beyond-3G systems.

3.6 References

[1] Wellner, P., Mackay, W. and Gold, R., 'Computer Augmented Environments: Back to the Real World,' *Communications of the ACM*, **36**(7), 24–26, 1993.

[2] Eckardt, T., Magedanz, T., Ulbricht, C. and Popescu-Zeletin, R., 'Generic Personal Communications Support for Open Service Environments,' in *Proceedings of IFIP World Conference on Mobile Communications*, Canberra, Australia, September 1996.

[3] Martin, D., 'The Open Agent Architecture: A Framework for Building Distributed Software Systems,' *Applied Artificial Intelligence*, **13** (1–2), 1999.

[4] Breugst, M. and Magedanz, M., '*On the Usage of Standard Mobile Agent Platforms in Telecommunication Environments*,' 5th ACTS IS&N Conference, Antwerp, Belgium, 25–28 May, 1998.

[5] Raymond, K., '*Reference Model of Open Distributed Processing (RM-ODP): Introduction*,' Center for Information Technology Research, University of Queensland, Australia, 1995.

[6] ITU-T Recommendation X.901, '*Information Technology – Open Distributed Processing – Reference Model: Overview*,' ISO/IEC IS 10746-1, Geneva, 1997.

[7] ITU-T Recommendation X.902, '*Information Technology – Open Distributed Processing – Reference Model: Foundations*,' ISO/IEC IS 10746-2, Geneva, 1995.

[8] ITU-T Recommendation X.903, '*Information Technology – Open Distributed Processing – Reference Model: Architecture*,' ISO/IEC IS 10746-3, Geneva, 1995.

[9] ITU-T Recommendation X.904, '*Information Technology – Open Distributed Processing – Reference Model: Architectural Semantics*,' ISO/IEC IS 10746-4, Geneva, 1998.

[10] IST Programme Advisory Group, '*Scenarios for Ambient Intelligence in 2010*,' final report, IPTS Seville, February 2001. Available at: www.cordis.lu/ist/istag.htm.

[11] Pfeifer, T., '*Automatic Conversion of Communication Media*,' dissertation, Technical University of Berlin, Institute for Open Communications Systems (OKS), 1999.

[12] FAIN, '*Policy-Based Management Approaches for Active and Programmable Networks*,' available at: www.ist-fain.org/deliverables/FAIN.paper.Vf.pdf.

[13] Hawkins, R., 'Looking beyond the .com bubble: exploring the form and function of business models in the electronic marketplace,' in H. Bouwman, B. Preissl and C. Steinfield (Eds), '*E-Life after the dot-com Bust,*' Springer/Physica, Hanburg, 2003.

[14] Timmers, P., 'Business Models for Electronic Markets', *Electronic Markets*, **8**(2), 1998.

[15] Weill, P. and Vitale, M., '*Place to Space: Migrating to eBusiness Models,*' Harvard Business School Press, Boston, 2001.

[16] Porter, M., '*Competitive Advantage: Creating and Sustaining Superior Performance,*' Free Press, New York, 1985.

[17] Methlie, L. and Pedersen, P., '*Understanding business models in mobile commerce*,' paper presented at WWRF 3, Stockholm, September 2001.

[18] Bouwman, H. (Ed.), '*Business Models for Innovative Telematics Applications: Case Study Protocol*,' Telematica Institut Report, 2002.

[19] Kallio, P., 'Business Models in Wireless Internet Service Engineering,' WISE project paper, 2002.

[20] Parolini, C., 'The Value Net. A Tool for Competitive Strategy,' John Wiley & Sons Inc., New York, 1999.

[21] Stabell, C. and Fjeldstad, O., 'Configuring value for competitive advantage: on chains, shops, and networks', Strategic Management Journal, 19(5), 1998.

[22] Håkansson, H. and Johansson, J., 'A model of industrial networks', in B. Axelsson and G. Easton (Eds) Industrial Networks: A New View of Reality.' Routledge, 1989.

[23] Normann, R. and Ramirez, R., 'From value chain to value constellation: Designing interactive strategy,' Harvard Business Review, 71(4), 1993.

[24] Welling, I. (Ed.), 'Description of Selected Business Cases,' IST-2000-25172 TONIC Deliverable number 1, 2001.

[25] Bauer, J., Westerveld, R. and Maitland, C., 'Advanced Wireless Communications Infrastructure: Technical, Economic and Regulatory Conditions of UMTS Network Deployment,' Delft University of Technology Report, October 30, 2001.

[26] Essler, U. and Whitaker, R., 'Re-thinking E-commerce Business Modelling in Terms of Interactivity,' Electronic Markets (anniversary edition): Business Models, 11(1), 2001.

[27] Alt, R. and Zimmermann, H.-D., 'Guest Editors' Note', Electronic Markets (anniversary edition): Business Models, 11(1), 2001.

[28] WWRF, The Book of Visions: visions of the Wireless World, Version 1. December 2001.

[29] Afuah, A. and Tucci, C., 'Internet Business Models and Strategies,' McGraw Hill, New York, 2001.

[30] Bohlin, E., Björkdahl, J., Lindmark, S. and Burgelman, J.-C., 'Strategies for Making Mobile Communications Work for Europe: Implications from a Comparative Study,' paper presented at the EuroCPR 2003, Barcelona, March 23–25, 2003.

[31] The Yankee Group, 'The Wireless Access Report,' 2000.

[32] Li, F. and Whalley, J., 'Deconstruction of the telecommunications industry: from value chains to value networks,' Telecommunications Policy, 26(9/10), 2002.

[33] Ballon, P., Helmus, S. et al., 'Business Models for Next-Generation Wireless Services,' Trends in Communications, 9, 2002.

[34] Wehn de Montalvo, U., van de Kar, E., Maitland, C. and Ballon, P., 'Business Models for Location-based Services,' Paper presented at the AGILE Conference in Lyon, 24–26 April, 2003.

[35] Pau, L.-F., 'The business challenges in communicating, mobile or otherwise,' inaugural address, Erasmus Research Institute of Management, November 29, 2002.

[36] Tee, R., 'Contextualizing the Mobile Internet,' Masters thesis, University of Amsterdam, Department of Information Science, May 2003.

[37] Heitmann, M. and Stanoevska-Slabeva, K., 'Impact of Mobile Ad Hoc Networks on the Mobile Value System,' paper presented at the Second International Conference on Mobile Business, Vienna, June 2003.

[38] Berne, M., 'Business Models and Mobile Communications in France, some examples,' paper presented at the Symposium on Business Models for Innovative Mobile Services, Delft, November 2002.

[39] Rakotonirainy, A., Loke, S.W. and Fitzpatrick, G. 'Context-Awareness for the Mobile Environment,' in CHI2000 Workshop 11, The Hague, The Netherlands, April 2000.

[40] Dey, A. and Abowd, G. 'Towards a Better Understanding of Context and Context-Awareness,' in Proceedings of Computer–Human Interaction 2000 (CHI 2000), Workshop on The What, Who, Where, When, and How of Context-Awareness, The Hague, The Netherlands, April 2000.

[41] Pascoe, J. 'Adding generic contextual capabilities to wearable computers,' in The Second International Symposium on Wearable Computers, Pittsburgh, pp. 92–99, 1998.

[42] Dey, A.K. and Abowd, G.D. 'Towards a better understanding of context and context-awareness,' GVU Technical Report ITGVU- 99-22, College of Computing, Georgia Institute of Technology, (ftp://ftp.cc.gatech.edu/pub/gvu/tr/1999/99-22.pdf), 1999.

[43] Arbanowski, S., Waterstrat, H., van der Meer, S. and Popescu-Zeletin, R. 'Open Profiling for Ubiquitous Computing,' in Proceedings of the 1st Workshop on Ubiquitous Computing (PACT 2000), Philadelphia, PA, 2000.

[44] Arbanowski, S. and van der Meer, S. 'Service Personalisation for Unified Messaging Systems,' in The 4th IEEE Symposium on Computers and Communications (ISCC'99), Red Sea, Egypt, pp. 156–163, 1999.

[45] Ceri, S. Fraternali, P. and Paraboschi, S. 'Data-Driven, One-to-One Web-Site Generation for Data-Intensive Applications,' in Proceedings of 25th International Conference On Very Large Data Bases (VLDB'99), Edinburgh, UK, 615–626, 1999.

[46] Schmidt, A., Beigl, M. and Gellersen, H.-W, 'There is more to Context than Location,' *Computer & Graphics*, **23**(6), 893–901, 1999.

[47] Heery, F. and Patel, M., 'Application Profiles: mixing and matching metadata schemas,' University of Bath, September 2000, *Ariadne*, **25**, 2001.

[48] Schweizer, K. 'Portal Total,' *IBM eNews Magazine*, http://www-5.ibm.com/de/software/enews/essay/2001-02-02-ess-1.html, 2001.

[49] Buchberger, R. 'Wenn es persönlich wird. . .' *Webpersonalisierung contentmanager.de*, http://www.contentmanager.de/magazin, June 2001.

[50] Arbanowski, S., van der Meer, S. and Popescu-Zeletin, R. 'I-centric Services in the Area of Telecommunications: "The I-Talk Service" ', in *Proceedings of the 6th IFIP TC6/WG6.7 Conference on Intelligence in Networks*, SmartNet 2000, Vienna, Austria, pp. 499–508, 2000.

[51] Foundation for Intelligent Physical Agents Home Page; http://www.fipa.org.

[52] DAML Services Coalition, 'DAML-S: Semantic Markup for Web Services,' in *Proceedings of the International Semantic Web Working Symposium* (SWWS), July 2001.

[53] Nassen, T., *'Knowledge Is Capital!,'* The World Bank Group, http://www.worldbank.org/ks/articles/knowledge_capital.html, 2002.

[54] Van Kranenburg, H., Lankhorst, M.M. and van den Eijkel, G.C., *'Privacy Protection Related Issues in the Design of a Personal Service Environment for Personalized Mobile Services,'* Paper for the Second Meeting of the Wireless World Research Forum, Helsinki, Finland, May 2001.

[55] Safe Harbor, US Department of Commerce, *'Safe Harbor Workbook,'* http://www.export.gov/safeharbor/sh_workbook.html, 2003.

[56] Day, M. *et al.*, *'A Model for Content Internetworking (CDI),'* Internet draft, http://www.ietf.org/internet-drafts/draft-day-cdnp-model-09.txt, November 2001.

[57] Tomlinson, G. *et al.*, 'A Model for Open Pluggable Edge Services,' Internet draft, http://www.ietf.org/internet-drafts/draft-tomlinson-opes-model-01.txt, July 2001.

[58] Platform for Privacy Preferences (P3P) Project, http://www.w3.org/P3P.

[59] The World Wide Web Consortium (W3C), http://www.w3.org/.

[60] OMG Agent Platform Special Interest Group, Agent Technology Green Paper, Object Management Group, OMG Document agent/00-09-01.

[61] Pokraev, S., Costa, P.D., Pereira Filho, J.G., Zuidweg, M., Koolwaaij, J.W. and van Setten, M., *'Context-aware Services, State-of-the-art,'* WASP/D2.3 Deliverable, Telematica Institute Report Series, 2003.

[62] Odell, J.J., Van Dyke Parunak, H., Fleischer, M. and Bruechkner, S., *'Modeling Agents and Their Environment,'* AOSE Workshop at AAMAS, 2002.

[63] Bagrodia, R., Chu, W.W. and Kleinrock, L., 'Vision, Issues, and Architecture for Nomadic Computing,' *IEEE Personal Communications*, 14–27, 1995.

[64] Cross-Industry Working Team, *'Nomadicity in the NII,'* available from http://www.lk.cs.ucla.edu/LK/lkxiwt/.

[65] Weiser, M., 'The Computer for the Twenty-First Century,' *Scientific American*, 94–104, September 1991.

[66] Weiser, M., 'Some Computer Science Issues in Ubiquitous Computing,' *Communications of the ACM*, 74–84, July 1993.

[67] Demers, A.J., 'Research Issues in Ubiquitous Computing,' in *Proceedings of ACM PODC'94*, 2–8, August 1994.

[68] *Endeavour Expedition*, http://endeavour.cs.berkely.edu/.

[69] *Oxygen*, http://www.oxygen.lcs.mit.edu/.

[70] *Aura*, http://www.cs.cmu.edu/aura.html.

[71] *Future Computing Environments*, http://www.cc.gatech.edu/fce/.

[72] *Portolano*, http://portolano.cs.washington.edu/.

[73] *2K: A Component-Based, Network-Centric Operating System for the Next Millennium*, http://choices.cs.uiuc.edu/2k.

[74] *MosquitoNet*, http://mosquitonet.stanford.edu/index.html.

[75] Bauer, M., *'Event-Management für mobile Benutzer,'* Diploma Thesis Number 1836, Faculty of Computer Science, University of Stuttgart, Germany, 2000.

[76] Sun Microsystems, *Java Message Service (JMS) specification*.

[77] Luckham, D.C., *'The Power of Events: An Introduction to Complex Event Processing in Distributed Enterprise Systems,'* Addison-Wesley Longman Publishing Co., 2001.

[78] Object Management Group, *CORBA Notification Service Specification v.1.0*, 2000.

[79] Gupta, S., Hartkopf, J. and Ramaswamy, S., 'Event notifier, a pattern for event notification,' *Java Report*, **3** (7), 1998.

[80] *Coda file system*, http://www.coda.cs.cmu.edu/.

[81] *Intermezzo file system*, http://www.inter-mezzo.org/.

[82] *OceanStore storage system*, http://oceanstore.cs.berkeley.edu/.

[83] *Bayou storage system*, http://www2.parc.com/csl/projects/bayou/.

[84] *SyncML Representation Protocol* and *SyncML Sync Protocol*, http://www.syncml.org/docs/.

[85] Gruber, T.R., 'A translation approach to portable ontologies,' *Knowledge Acquisition*, **5**(2), 199–220, 1993.

[86] Guarino, N., 'Formal Ontology and Information Systems,' Amended version in N.Guarino (Ed.), '*Formal Ontology in Information Systems*,' Proceedings of FOIS'98, Trento, Italy, 6–8 June 1998, IOS Press, Amsterdam, pp. 3–15, 1998.

[87] Worldwide Web Consortium (W3C), Semantic Web Working Group, 2002.

[88] McIlraith, S., Son, T. and Zeng, H., 'Semantic Web Services,' *IEEE Intelligent Systems*, 46–53, March/April 2001.

[89] Open Mobile Alliance, http://www.openmobilealliance.org/.

3.7 Credits

The following individuals have contributed to the contents of this chapter: Stefan Arbanowski (Fraunhofer FOKUS, Germany); Stephan Steglich (Technical University Berlin, Germany); Radu Popescu-Zeletin (Fraunhofer FOKUS, Technical University Berlin, Germany); Pieter Ballon (TNO Strategy, Technology and Policy, The Netherlands); Olaf Droegehorn and Klaus David (University of Kassel, Germany); Herma van Kranenburg and Johan de Heer (Telematica Instituut, The Netherlands); Erwin Postmann and Christian Feichtner (Siemens AG, Austria); Axel Busboom (Ericsson Research, Germany); Kimmo Raatikainen (Nokia, Finland); Johan Hjelm (Ericsson, Sweden); Fritz Hohl (Sony, Germany); Stefan Gessler (NEC Europe Ltd., UK); H. Ailisto and A. Tarlano (VTT, Finland); Wolfgang Kellerer (DoCoMo Communications Laboratories Europe GmbH, Germany); Francois Carrez (Alcatel, France); K. Kawano and S. Sameshima (Hitachi SDL, Japan) and Sasu Tarkoma (Helsinki Institute for Information Technology, Finland).

4

Cooperation Between Networks

Edited by Rahim Tafazolli, Klaus Moessner, Christos Politis
(University of Surrey) and Tasos Dagiuklas (University
of Aegean)

Beyond-3G systems (B3G) have been envisaged as an evolution and convergence of mobile communication systems and IP technologies to offer a multitude of services over a variety of access technologies. To fulfil the vision, it is necessary to understand the requirements with respect to the support of heterogeneity in network access, communication services, seamless mobility, user devices, etc. It is equally important to promote the necessary research in networking technology by providing a guiding framework of research areas and technical issues with priority. The new architectures and technologies will have to address the fundamental assumptions and requirements that govern the design. All these are being tackled by the Cooperative Network (CoNet) working group (WG) of the WWRF (Wireless World Research Forum). The group is working on a series of white papers outlining B3G visions and roadmap, architectural principles, research challenges and candidate approaches. In this chapter, the CoNet main objectives are outlined, while the most important elements of CoNet architectural principles are examined. Additionally, the flamboyant research challenges of cooperative networks are addressed, and we try to give a glimpse of what might govern their design and encompass their futuristic form by introducing the main network components and technologies.

This chapter documents those research areas considered as being necessary to facilitate the WWRF vision, the four sections of this chapter include:

1. The cooperative networks vision and roadmap.
2. Cooperative network research challenges.
3. CoNet architectural principles.
4. Network component technologies for cooperative networks.

4.1 CoNet Vision and Roadmap

This section defines CoNet's current vision for the networking aspect of B3G systems. The key technical aspects are highlighted and the roadmap to facilitate this vision is discussed.

Technologies for the Wireless Future. Edited by R. Tafazolli
© 2005 John Wiley & Sons, Ltd ISBN: 0-470-01235-8

4.1.1 Vision

The vision of the CoNet working group in the WWRF is to enable seamless communication on mobile devices that operate in networks composed of heterogeneous technologies, by enhancing secure networking capability of B3G systems in the terminal devices, as well as the infrastructure, thus paving the way for cooperative networks.

In 2010, the networking aspect of B3G systems will:

- impose no restriction on, and allow innovation in, business models;
- support the coexistence and convergence of different legacy (wireless/cellular, wired, broadcast, etc) and new networks, including moving and ad hoc networks;
- support the evolutionary deployment of open, secure, extensible, scalable, reconfigurable and manageable B3G systems;
- enable independent evolution of network technologies and services;
- allow stationary and mobile users and devices to access service anywhere, at any time by different means, and allow them to be reached in a controlled manner;
- support pervasive and seamless mobile multiparty multimedia communications and access to all kinds of services, possibly simultaneously, in order to minimize adverse impacts on user experience in the ever-changing networking environment;
- support value-added interfaces to the applications above to enable and enhance their location, context and QoS awareness;
- exploit the interactions with the access systems to optimally manage mobility and use of scarce radio resources.

The desire to deliver seamlessly applications and services across heterogeneous access networks in a B3G system leads to an obvious conclusion that B3G systems would best be built upon the generic IP networking technologies. However, it is also obvious that existing IP networking technologies do not support all the necessary features to enable a B3G system, and need to be enhanced.

4.1.2 Key Technical Aspects

The following key technical aspects highlight the research concerns that need to be addressed in order to design and develop solutions to achieve the vision of CoNets.

4.1.2.1 Moving Networks and ad hoc Networks

The formation and topology of a network needs to become more flexible to support BANs (body area networks) and PANs (personal area networks), vehicular networks, wireless sensor networks, etc. This kind of network typically moves topologically around some IP backbone. The formation of such a network could be infrastructure-less (ad hoc network) thanks to the use of wireless links.

4.1.2.2 Multiaccess

It is sometimes desirable to allow different application sessions to use different access networks, possibly even simultaneously. The multiaccess capability refers to combining and using a number of radio or fixed accesses concurrently.

4.1.2.3 Mobility

Providing mobility support at the network layer has the advantages of being independent of the link layer used in the access systems. However, information from the link layer is essential for more effective and efficient support of seamless handover. For example, link-layer information could be used to predict the need for handover between access systems, and thus prepare handover at an optimal moment and condition. Furthermore, application-level (personal) mobility is foreseen to be an important feature of cooperative networks. Personal mobility refers to the ability of end users to originate and receive calls and access subscribed network services on any terminal in any location in a transparent manner, and the ability of the network to identify end users as they move across administrative domains.

4.1.2.4 Security

The CoNet architecture should provide a securely protected environment in which network elements are deployed and interact. Besides protecting end users and networks from each other through authentication and authorization, the components of the CoNet concept should also be protected from any intrusion or malicious attack. The architecture should provide a security framework to enable protection of private information and data. Furthermore, the CoNet architecture should provide accounting capabilities and further enhance the AAA paradigm. The architecture should be distributed and should consist of security components able to communicate through well-defined interfaces.

To accomplish these, it would be well advised to rely on already existing security algorithms that are stable, resilient and well established. The creation of new mechanisms and protocols is outside the scope of CoNets. Rather, its objective is to build the trust models and security associations between the various components of the distributed CoNet architecture, with scalable access control to distributed components and resources.

The possibly very large number of network operators and content, service and home providers makes the establishment of static trust relations among all of these providers tedious and noneconomical. Therefore, appropriate mechanisms and solutions must be employed that will allow users to roam from one provider to an other without having to explicitly subscribe to the services of each provider, or assuming some pre-established trust relation between the different providers.

4.1.2.5 Multihoming

A host may require multihoming networking at all its network interfaces. This introduces challenging issues with respect to ubiquitous access, identification, mobility security, redundancy/fault recovery, fault tolerance, load sharing and preference settings.

4.1.2.6 QoS

QoS is typically based on intserv, diffserv, or a hybrid architecture of both. While intserv supports per-flow QoS, it suffers from concerns of scalability. On the other hand, diffserv is a more scalable approach but it supports only aggregated QoS flows. The operation of a QoS policy framework needs to be aligned with mobility and AAA operations; otherwise, seamless mobility cannot be achieved for service sessions that require QoS support. Besides this, network operators must be able to dynamically negotiate service layer agreements (SLAs) between themselves, in order to allow data flows of different QoS natures to meet their end-to-end QoS requirements.

4.1.2.7 Multiparty Services and Multicast

IP multicast has the potential to provide efficient and generic multipoint-to-multipoint communications at the network level for use by multiparty applications. IP multicast needs to be enhanced to take into account mobility, security, AAA and QoS, because this way it will provide the optimal support for mobile multiparty multimedia networking.

4.1.2.8 Transport

It is widely known that current transport protocols do not adapt well in wireless communication environments. It is thus necessary to enhance them to optimize their performance. Besides, they need to be enhanced to support multiparty and QoS sessions, which would greatly benefit the development of new multiparty multimedia applications.

4.1.2.9 Network Management

In order to allow efficient and reliable deployment of a B3G system, access routers, access points (or base stations), and ad hoc networks need to be capable of both self-configuration, self-organization and possibly self-healing.

4.1.2.10 Internetworking

In order to ensure connectivity between different networking entities, it is necessary that different naming and addressing schemes converge or interwork. It is essential that user devices are not denied networking connectivity due to naming and addressing incompatibilities.

4.1.3 Roadmap

It is not possible to set a single converged roadmap for technology development to achieve the vision outlined above. This is mainly due to different concerns and priorities regarding technology maturity, technical barriers, market requirements, business models, etc. There will be different potential paths to be pursued, which will be based on learning and leveraging from one another.

The following milestones highlight the research efforts that need to be targeted to accomplish the goals outlined earlier:

- a B3G system in which a user can undertake secure and seamless mobility between heterogeneous wireless (i.e. cellular, ad hoc, WLANs) and wired access technologies;

- B3G systems with interdomain administration, through dynamically established relationships and management support to allow seamless mobility between administrative domains;
- a B3G system capable of supporting QoS sessions between networks and administrative domains;
- a B3G system capable of supporting multiparty sessions between networks and administrative domains;
- a B3G system capable of supporting multiparty multimedia sessions between networks and administrative domains.

These milestones should take into account the following issues:

- fast, seamless and possibly simultaneous handover within and between administrative domains for both terminal and service mobility, and with interface to the link layer to assist more effective handover and better use of radio resources;
- end-to-end application QoS support, addressing QoS issues at the network and transport layers with interfaces to the application and link layers and QoS mapping between layers;
- multiparty communication support in multicast and transport protocols, with interfaces to the application and link layers;
- ad hoc network technologies for infrastructure-less network topology.
- interworking of relevant protocols and protocol versions;
- support of security, scalability, reconfigurability and manageability in all protocols.

4.2 Cooperative Network Research Challenges

4.2.1 Introduction

4.2.1.1 Vision of CoNet

The vision of the CoNet ad hoc working group in the WWRF is to enable seamless communication activities on mobile devices operating in networks composed of heterogeneous technologies, by enhancing the networking capabilities of the B3G systems in the terminal devices, as well as the infrastructure, thus paving the way for cooperative networks [1].

4.2.2 Objectives and Scope

The objective of this section is to provide a framework of the research challenges by identifying the requirements and individual research problems that should be addressed. The requirements set the priorities and the needed enhancements of the network, guided by the architecture principles and the reference model, and the components of the framework will represent the research areas. Individual research items in the framework were derived by identifying technical issues and taking into account the requirements.

The scope of CoNet is to define new networking technologies that are well suited for new applications used over heterogeneous networks (wireless as well as wired) as efficiently as possible. These networking technologies shall, in a cooperative manner, provide users and devices with ambient networking capabilities, while the available resources are used in the optimal way..

This section describes what research areas must be investigated, in order for CoNets to work effectively. It also describes what requirements are imposed on CoNets, and what requirements CoNets impose on applications and services, as well as on the physical layers of these systems. The section spans from systems views to technology implementation, and from service requirements to limitations of the physical world. In a perspective of a network topology, it ranges from the user terminal, the access network, the backbone and all the way up to the applications and services platforms.

However, it is outside the scope of this section to describe aspects on applications and services, as well as physical layer technologies. On the other hand, it is within its scope to describe how applications and services will be supported to fulfil users' requirements as efficiently as possible, by providing them with information through well-defined APIs. It is also within the scope of this section to enable as efficient resource utilization as possible given the currently existing access technologies.

4.2.2.1 Objectives of Research

A CoNet will be considered as a converged network of heterogeneous networking technologies that provides users with seamless and secure multimedia communications and high-speed internet services with rich features and QoS, and also achieves unprecedented levels of flexibility and manageability in network design, operation and service provision. It therefore requires development of innovative networking concepts and technologies, as well as enhancement of the existing technologies (for example, network protocols such as IP), thus it demands new research challenges for this purpose.

The objectives of the research towards the CoNet vision are to develop a new system architecture and design a set of new key networking technologies which are best suited to realizing the network vision. As expected research outcomes, it is desirable to acquire the following:

- a reference model [2];
- solutions for individual key technical components (mobility management, session control, security, etc.), and their integration into cooperative networks; and
- a cooperative networking infrastructure for competing connectivity services, as well as support for applications and services (location services, location discovery, etc.).

The research also fills the important role as a prerequisite for preparing a technical basis for future international standardization and networking products of the beyond-3G mobile network.

4.2.3 Requirements for CoNet

The CoNet vision gives the grand objectives and top-level requirements for the B3G networking systems. In order to discover new features and characteristics that the envisioned networking environment shall have, it is necessary to identify the significant requirements for the target network, in terms of service provisioning and system capabilities.

Identification of those requirements is important as a base by which to investigate and establish design philosophies, principles and guidelines for reference models, network functional architecture and components. In principle, the architectural principles are based on the grand objectives and requirements for the target network. This section presents the areas for which requirements should be elucidated, while analysing the high-level key requirements for each area. The requirement areas include important categories of network capability.

It should be noted that there are interdependencies among the requirement areas. The interdependencies, may be presented in the description of the respective requirement areas.

4.2.4 New Capability Requirements to Support Services

There are a number of general principles governing the CoNet visions, these principles are paramount in fulfilling the requirements for service support.

4.2.4.1 General Principle for Fulfilling Service Requirements by CoNet

In order for CoNet to support the services in fulfilling their requirements, specific reservations must be propagated to different parts of the network.

- Applications and services should be able to request what resources they will need from the network in order to meet the end user's requirements.
- The network should be able to fulfil these authorized requests by attempting to allocate adequate resources where needed. These allocations will be performed on different links, and different links will require different allocations. Hence, the mechanisms for transporting the allocation requests to these links must be link layer independent.
- The network should be able to give feedback to the applications and services about its current capabilities and what resources are currently allocated for the application or service. This feedback should be given independently of the access technology, both as an initial response to the application or service request and also during the session in order to allow the application or service to adapt to current circumstances.
- The communication, both between the applications or services and the network, as well as between the different layers within the network, should be able to use standardized APIs and service primitives, respectively. This should allow existing, as well as future, applications, services and protocols at different layers to communicate and cooperate in a predefined and efficient way.

4.2.4.2 General Requirements for Service Support

It is required that future networks should provide service support capabilities so that new value-added services can be efficiently developed and offered to end users. They include capabilities to support efficient service creation environments. From this standpoint, amongst others, the following major requirements have to be met in future networks.

Support for Value-added Services

Generic services – used by multiple applications – become part of the network. Operators offer them to third party service providers. The network should:

- Provide functionality, not only for the connectivity service, but also for content distribution. For example, the network may be requested to provide intermediary services, such as caching or replicating contents, in order to allow user requests of content access to be optimally routed with less access delay and/or the most suitable quality of service level for content transport.
- Measure and log usage of services in order to provide statistics for SLA, accounting use etc.
- Provide support for service adaptation to the context pertaining to the service usage, so that new value-added services, such as localized or personalized content, fast and secure access to content, etc., could be offered. The context herein will consist of the following categories:
  - *User-related category*, typically consisting of user preferences, user history, user interest, user role, user priorities, etc.
  - *Mobility and location-related category*, typically embedding physical coordinates, velocity, direction of movement, ambient conditions (indoors, outdoors, temperature, humidity, etc.).
- Network and terminal characteristics, like bandwidth, memory and graphic capabilities, screen size of the terminals, etc.
- Non user-related information that, for instance, contains content-related preferences (like presentation format, encoding, etc.), but also business rules that apply.

Service Creation Support

In order to allow for rapid and simplified service creation processes of service providers, the network should provide common functionalities needed to enable the new services for end users. Furthermore, the network should provide third party service providers with a framework that allows the rapid creation and deployment of new services. Finally, the network should allow the end users to have a single point of access for a wide variety of services that they request.

All of the above can be accomplished through the use of open APIs. An effort to standardize such an API has been carried out by initiatives such as Parlay and JAIN. A key barrier to the development of value-added services into a truly open service environment is the lack of standardized interfaces for service creation, management and deployment that can be used by nonspecialists to develop value-added services using commercial software development methods and tools. In B3G systems, services will be created by either customers or providers. Services created by end users are more likely to be customized and configured from existing services, rather than fundamentally new call logic. It appears more realistic to have services created based on rules, i.e. certain conditions triggering certain actions, similar to how email user agents, for example, support message filtering, forwarding and filing.

4.2.5 General Network and Access Requirements

There are a number of general requirements for future networks, in order to achieve the aims of CoNets.

4.2.5.1 General Aspects

- Networks should support different kinds of heterogeneous access technologies to provide users with an optimal access method, based on requirements and preferences of the users. Networks should be able to allow different application sessions to use different access networks, possibly simultaneously.
- Networks should allow for optimal use of available networks and services based on users', applications' and operators' needs.
- Networks should support services with long lifetimes (the session duration between terminal and application server could last for days).
- Networks should support resource control functions within each access network.
- Each access network may support mobility- and security-specific functions.

4.2.5.2 Separation of Control, Transport and Services

- Control and bearer planes should be separated for scalability and architectural flexibility. User data and control planes should be separated to allow independent upgrading and evolution of the planes.
- Control, management and signalling should be separated to provide the distributed intelligence required for user control of the connections/sessions.
- The user environment should be separated from the transport network. The network specifics/characteristics should be transparent to the user so that there is no need to change user parameters (e.g., phone number, address book settings, etc.) when the user changes access networks.

4.2.5.3 Advanced and Efficient Mobility Management

- Different types of mobility (terminal, session, personal) must be supported.
- Mobility management should support mobile users in various scenarios. As an example, mobility management methods must be applied for users who move at both high and low speeds.
- There should be Adaptive mobility management for the change of movement characteristics. Movement characteristics of a user are ever changing. For example, the moving speed of a user changes rapidly. Mobility management applied to the user should also adapt according to the change of movement characteristics.

4.2.5.4 Handover

- Networks should support seamless handover(s) between different kinds of access technologies to provide users with service continuity, while offering an optimal access method for the users.
- Networks must support fast handover(s), handover(s) with minimum data loss, seamless handover(s) with minimum service disruption, as well as handover(s) with minimum resource consumption, which satisfy the desired QoS contracts.

- Networks must support diversified handover mechanisms, for example, local handover minimizing the re-established handover path in terms of length, load and/or cost, network overall handover minimizing end-to-end routing path.
- The mobility management must support both real-time and non real-time multimedia services (e.g. video streaming and web access respectively) and mobile data services.
- After the handover, the mobility management needs to interact effectively with the AAA schemes of the end-to-end network to verify the user's identities and privileges, as the user roams across network domains.

4.2.5.5 Support of Service Adaptation

Networks should be capable of instantly adapting and maintaining services when the available network and terminal resources may change. Three basic cases should be considered:

1. An ongoing service restarts from the point at which it was terminated due to resource exhaustion.
2. An ongoing service changes association from one access system to another.
3. An ongoing service is transferred from one terminal to another.

In scenarios 2 and 3, it may be necessary to adapt the service attributes (e.g., presentation format, resolution, etc.) to the changes in the resource allocation and the capabilities of terminals.

4.2.5.6 Separation of Mobility Management and Session Management

To modularize network operation (mobility management and session management) and facilitate user interaction with the network, the network should support session management independently of the mobility management. This means that the session management will be an independent operation, and it should be able to be operated over multiple access network technologies independently of administrative domains (e.g., roaming). It may provide customers with flexible and comprehensible quality metrics (e.g. QoS) for session management.

4.2.5.7 Transport Technologies

- Network architecture should be able to support several transport technologies (e.g., IP, MPLS).
- Transport technologies used in the network should be common and transparent to access technologies.

4.2.5.8 Open Interfaces

- Open interfaces will ensure a multivendor environment and interoperability among diverse service and network platforms.

- An interface should be provided to allow the network operator to configure the network by using the network nodes/entities provided by the different vendors.
- Open interfaces should support economical creation of various types of new service to the end users and stimulate competitive development of service and network components.
- Interface specifications should be kept simple, and the number of options should be kept to a minimum in order to reduce the cost of the equipment and to improve interoperability between different network elements.
- There should be open and standardized application interfaces for rapid service creation and delivery.
- These open interfaces may even be tailored by the end users wishing to create, customize and configure services to meet their own needs.

4.2.5.9 Provision of Flexible Environments for Applications and Services

- Networks should support applications that can utilize network and transport information to adapt to network conditions in a user-friendly manner.
- Networks should support provision of home-subscribed services and real-time provisioning of services (e.g., provision of subscription to local services advertised to inbound roamers).
- Networks should support common representation of user service profiles, that is, standardized data attributes for facilitating new service creation and improvement of convenience for subscribers.
- Networks should support capability for the users to modify their service profiles within the limits of their subscriptions. Modification of the profiles should be possible through any access systems.
- Service provision by central servers and the terminals should be possible.
- Service architecture should be scalable.
- Customized service solutions that are operator-determined should be allowed, in order to leave the selection scope for the operators to customize their service architecture according to their specific needs.

4.2.5.10 Use of Internet Protocol (IP)

- Extension of standard internet protocols to meet emerging mobile requirements should be used when applicable.
- A single IP network standard, not only for mobile services, but also for fixed services, should be considered.
- At the initial phase, B3G systems will be based on the coexistence of hybrid IPv4 (including firewalls and NATs) and IPv6 networks and their interconnection.

4.2.5.11 Interworking

Backward compatibility with existing networks and mobile terminals should be maintained (e.g., support of the current 2G/3G cellular access systems and other heterogeneous access systems e.g., PSTN, xDSL, WLL WLANs, PANs, ad hoc networks, etc.).

4.2.5.12 Integrated OA&M (Operation, Administration and Management)

- *Unified network management.* Common management should be supported for all nodes in the network with a standardized protocol.
- *Easy, low-cost and scalable OA&M.* Easy and low-cost network management and operation should be facilitated for all service networks. Integrated OA&M should enable easy deployment of new services or expansion of current services.
- *Distributed architecture.* Network management functions should be decentralized in logic modules that are easily upgradeable and located separately through the network.

4.2.5.13 Network Scalability and Performance

The network should be tolerant to heavy load and provide higher throughput, lower transport delay, jitter and packet loss rate than existing mobile networks (e.g., 2G/3G). The network should be provided in a differentiated manner in accordance with the various levels of quality of the services supported.

4.2.5.14 Migration

- In principle, phased and functional blockwise migration at a pace and in a manner that matches the operator requirements, should be possible in order to meet the operational and economic goals.
- It should be possible for a user to carry forward the current numbering scheme (e.g., E.164 assigned to a legacy mobile terminal) to the future mobile terminal.

4.2.5.15 General Terminal Requirements

There are a number of general requirements for terminals. Since terminals interact with networks in terms of functionalities and capabilities, the general requirements for networks and access systems have to be reflected, when applicable, for the terminals as a part of the entire system as well.

- The terminal should be able to be configured by the network and set up by the end user.
- Terminals should support simultaneous reception and transmission of data via different kinds of access technologies.
- Terminals should support extensible programmability of service-specific downloadable components.
- The terminal's user interface and applications should be easily managed by the network operator/service provider/company administrators.
- Functions should reside both in the terminal and in the network to support end-to-end functionality, e.g. end-to-end QoS, end-to-end security, end-to-end service provisioning support, end-to-end mobility management, etc.
- Terminals should support various fixed and wireless access technologies, such as 3G, 4G, WLANs (IEEE 802.11 and HiperLAN/2), WLL (Wireless Local Loop) and PANs (Bluetooth). Furthermore, terminals may have the capability to connect to more than one access system.

- Terminals should be able to efficiently discover the available access systems at the user's current location, before actually initiating an application service (e.g. VoIP, VoD, etc.). Depending on the preferences of the user and the capabilities of the terminal, as well as the network, the architecture should support the selection of optimal configuration of access and transport network resources.
- Terminals should support various QoS classes, QoS negotiations and asymmetrical access QoS (end-to-end and locally applicable).
- Terminals should use the information available on the network, transport and application layers to provide the best possible subjective QoS.
- Terminals should support authentication, authorization and encryption functionalities. Furthermore, dynamic access authorization (for real-time service subscription) should be supported. This type of access authorization feature enables network operators and service providers to dynamically authorize a user to receive services for which the user has not previously subscribed.
- Terminals should support various types of IP address, e.g., static, dynamic, public and private IP addresses.
- Terminals should support IP connectivity from a mobile terminal to any other host. IP connectivity should be established, for example, in an ad hoc network among mobile terminals and other equipment, such as PCs, PDAs and other mobile terminals.
- An open and standard subscriber application environment should be supported, i.e. a software creation environment that is standardized and available to all interested parties for rapid creation and delivery of services.
- Terminals may provide services to more than one user simultaneously.
- Terminals should support several naming contexts, e.g., user identity module (UIM).
- Service usage should be easier for users, for example, terminals and end user services should be able to negotiate through a log-in procedure (user and password) without user intervention.
- Regional/national differences in key interfaces should be eliminated, i.e., no country or regional differences should be recognized in setting up calls and provisioning services.

4.2.5.16 Naming and Addressing

- Networks should have the capability to separate the network address from the subscriber name or number.
- Addresses could support any addressing schemes.
- Networks should allow for an identification that can be assigned to a user/subscriber as long as he/she agrees, while being independent of the addressing scheme given by the network technologies.
- Networks should support static and dynamic addresses for mobile terminal interface(s).
- Networks should support associating public and/or private addresses for mobile terminal interface(s).
- Networks should support mapping between subscriber numbers and addresses or URLs currently used within the network.
- The number or identifier of the destination and/or source should not change, at least during the session, even when the node moves to another access point or different access technology network.

4.2.5.17 Mobility

The requirements for mobility support include the following:

- The mobility management architecture should provide means to support any type of mobility, i.e.:
  - *Terminal mobility*. This refers to an end user's ability to use her/his own terminal in any location, and the ability of the network to maintain the user's ongoing communication as she/he roams across radio cells within the same subnet, subnets within the same administrative or different administrative domains.
  - *Service mobility*. This refers to the end user's ability to maintain ongoing sessions and obtain services in a transparent manner. The service mobility includes the ability of the home service provider to either maintain control of the services it provides to the user in the visited network, or transfer their control to the visited network.
  - *Personal mobility*. This refers to the ability of end users to originate and receive calls and access subscribed network services on any terminal in any location in a transparent manner, and the ability of the network to identify end users as they move across administrative domains.
- Hierarchical fashion. The mobility management must differentiate local movements within an area (e.g. WLAN) and global mobility (e.g. move from enterprise indoors to a public outdoor environment).
- The mobile user can be connected to an access network, but still remain outside global connectivity (e.g. connected anonymously).
- Different types of handover. The handover process can be either hard or soft. In the hard handover, the mobile host can only receive data and control traffic from one base station. As the mobile host attaches to a new base station, the control/data path is abruptly handed over from its current base station/antenna to the new one in a few seconds. With soft handover, the mobile host continues to receive and accept data and control traffic from the previous, as well as its new, base station for a limited period of time. As its name indicates, soft handover smoothly transfers the mobile host's session from its old base station to the new one. In order to deploy soft handover mechanisms in mobile IP packet networks, the packets that are destined for a mobile host shall be routed to both its previous and current locations, during a soft handover period. This routing of packets to the current and previous locations of a mobile host constitutes a logical/virtual soft handover at the IP layer.
- Support of mobility among heterogeneous access technologies. A mobile terminal should be able to address many access technologies – one at a time or several simultaneously. A mobile terminal can also be part of a mobile network. Examples of access technologies are W-CDMA, cdma2000 and IEEE 802.11a/b/g, ETSI HiperLAN/2.
- Seamless handover between networks. The mobility management should provide seamless and fast handover between networks. This may include negotiation between the networks, where crucial information (e.g. QoS or AAA information) is exchanged between the networks before the handover, and where mobility-related signalling may be delegated to another interface or node.
- Support of mobility between different administrative domains. Mobility management should support mobility between different administrative domains. The change of domain can be triggered by, e.g., the user changing access technology and, hence, operator, the

user having multiple subscriptions with different operators or the user moving from one domain area (e.g. country) to another domain area (e.g. country) with the same operator.

- Support of evolution. Mobility management architecture must be flexible enough to support possible evolution, either with possible enhancements of existing protocols, or with the introduction of new ones.
- Efficiency and load balancing. Mobility management requires resources for processing, signalling and storage. Examples of resources are CPU time, bandwidth of wired/wireless line, memory capacity and battery consumption. Resources used for mobility management should be minimized as much as possible. In addition, the load for processing, signalling and storing should be optimally distributed and balanced.

Ideally, all of this should be achieved without sacrificing functionality.

4.2.5.18 Scalability

It is required that one, or several, logical architecture covers from small-scale networks to large-scale networks.

4.2.5.19 Robustness

Robustness against failures of any management-specific component(s) enables the constant provision of mobility management.

4.2.5.20 Easy to Operate and Administrate

OA&M (operation, administration and maintenance) should be easy for operators and providers. Efficient operation is required for OA&M. As the complexity, heterogeneity and dynamicity in future networks increases, more and more of the OA&M should be automatic and adaptive.

4.2.5.21 Access Security

Security must be implemented for preventing unauthorized access, as well as to ensure privacy. Access restrictions and limitations in the access rights of a user must be possible. A mobile terminal should be authenticated when it attempts registration and access. Also, the terminal should be able to authenticate and authorize the network(s) offered to the terminal. To ensure the security of the system, the user's access to the networks and services must be secured and must allow all involved entities to identify the user in a secure and assured manner. This requires, on the one hand, securing any signalling protocols the users might be using for utilizing the offered services, and preventing malicious usage by deploying protective measures (i.e. use of firewalls and NATs and policy control).

4.2.5.22 Location Confidentiality

The mobility management should hide the user's actual location if the user wishes this. Trusted entities (e.g., user, network), however, should be able to get the location information of users for route optimization and/or location information services. But, the location information of particular users should be concealed from nontrusted entities.

4.2.5.23 Route Optimization

The path for users' data flow should be optimized in terms of length, load and/or cost. Optimum paths should minimize latency and should allow for better resource management.

4.2.5.24 Resource Optimization

Resources are to be used as efficiently as possible in order for as many users as possible to simultaneously be connected and communicate using the shared resources.

4.2.5.25 Support of a Moving Network which Consists of Several Nodes

A moving network consists of several nodes, including mobile terminals. By controlling all nodes in a moving network as one group, a network operator can manage efficiently the mobility of them; since this can reduce the network resource consumption for signalling compared with the case of controlling them individually. For example, PANs and mobile terminals in a moving vehicle can be controlled as a moving network. Also, user data traffic sent to and from nodes within the moving network can be concatenated and thus save network resources.

4.2.5.26 Support of Roaming between Heterogeneous Network Providers

Most solutions for enabling roaming of users between different network providers assume the existence of pre-established trust relations between the involved partners. To avoid the establishment of a trust relation between each possible provider, a more scalable approach, based on trust brokers, is often used. However, this scenario assumes that any two providers that would like to enable user roaming must be registered at this broker. With the advent of next generation networks, the concept of a network and service provider is taking a much more general scope. Currently, the number of network providers is relatively small, with those providers having a large scale in terms of the number of supported users and covered geographical area. In next generation networks, any coffee shop offering wireless LAN access would qualify as a provider. Further, the definition of a home provider might change as well, allowing, for example, a banking, entity to act as the home provider of a user and authenticate his identity. Registering all possible providers at a single trust broker would surely not scale. This would then lead to having different brokers, each maintaining trust relations to only a small number of providers. To allow for roaming between any two entities, the foreign network would need to be able to discover the home provider of the user and which trust brokers could be used for establishing a trust relation with the home provider. The experience gained for private key distribution architectures, shows that for such schemes to work, they need to be distributed, i.e. with no central components, and lightweight.

4.2.5.27 Modularity

To meet different requirements for mobility, such as regional differences of mobile node density and required QoS differences for each session, mobility management should be modularized. For example, one module for urban area management, one for a rural area, one for real-time services and one for best-effort services should be prepared. This modularization allows operators to select and adopt the modules they require.

4.2.5.28 Keeping Required QoS

Depending on the user's preferences, the QoS of all connections should be kept before and after completion of handover, to at least the same level as the QoS negotiated previous to the handover, whenever possible.

4.2.5.29 Taking the Effects of Mobility into Account

Future entities, i.e. applications, as well as communication protocols at different layers (i.e. the link layer, the network layer, the transport layer, etc.), should be able to differentiate between different situations generating packet losses and errors, and these entities should also be able to adapt to these different situations.

4.2.5.30 Transport

Requirements for transport of data in future networks include the following:

• Transportation of data should be performed as efficiently as possible with regard to application and service requirements on one hand, and with regard to network capability on the other hand. With the introduction of new access technologies, applications and services, as well as new terminal capabilities and user preferences, the dynamicity increases. In order to support as efficient transportation of data as possible, the different networks, terminals and applications must be able to adapt to new environments.
• All users' data traffic and signalling should be on IP transport.
• Interfaces between the next generation of IP network and legacy transport networks are necessary.
• Transport technology agnostic, i.e. architecture and operation independent of the utilized transport technology.
• Transport networks should support broadcast and multicast.
• Separation of access technologies from transport technologies. The access technologies utilized should be transparent to the common and standardized transport infrastructure.

4.2.5.31 QoS

This section presents the requirements for QoS support in future networks. These are as follows:

• Networks should support QoS mechanisms that are not tied to any specific link technologies, but which instead provide a common basis for QoS coordination across multiple access technologies, and standard-based interoperability across multiple network domains to provide an end-to-end QoS solution.

- Sufficient quality of service has to be provided to applications and services in the future. New applications will be developed with new requirements on service and network support. At the same time, new access technologies will be developed that have different characteristics. New terminals with advanced capabilities and requirements will be developed, and users will have diverse requirements and needs to fulfil, depending on where and what they are doing. Applications should request resources from the network at start-up time and continuously receive network capability feedback from the network. These negotiations should use standardized APIs to support simplicity and interaccess support. The specific link layer reservations are hidden from the application.
- Networks should support various QoS schemes supporting different classes of services and applications. That is, the network should support simple and scalable solutions, such as differentiated services for flexible, widely used and less demanding applications. For applications with strict QoS requirements, a less scalable QoS scheme that could require special or explicit QoS signalling should also be supported.
- To allow efficient usage of QoS resources, the provisioning of QoS resources needs to be coupled with application signalling. Initiating and controlling the coordination between the QoS provisioning and the application signalling should be done by either the network or the user.
- There should be support for adaptive QoS. One should be able to specify bandwidth ranges or multiple bit rates that an application can operate, and associated control, or prioritization, values that will ensure an agreed level of performance to enable resource allocation decisions.

4.2.5.32 Capabilities for Supporting Interoperability with Other Internet Services

- Networks should have the capability to support end-to-end QoS across multiple operators, and a framework for QoS negotiation between operators.
- Networks should support different QoS parameters in uplink and downlink for wireless links.
- Network resource (e.g., CPU processing, memory usage, bandwidth) consumption for QoS control should be minimized.
- QoS mechanisms should scale to large numbers of flows and large networks.
- Networks should be able to control the QoS in the units of packet, flow and session.
- The status of QoS parameters or classes should be allowed to be reported by the network function(s) to the application server and/or end users involved in a session.
- The capability to change the QoS of an application instance at any time should be available in case of environmental or network traffic condition change through negotiation or prearrangement between the user and the operator.
- Negotiation of QoS and roaming service capabilities can be performed either manually or automatically.
- The QoS required by each individual media component can be negotiated prior to the execution of the media component (establishment) or during its execution.
- There should be support for the control and negotiation of QoS end-to-end when operating with a third party provided service platform.
- There should be support for multiple levels of static QoS (negotiation of parameters before the session set-up) as well as dynamic QoS (negotiation of parameters while the session is in progress).

- There should be support for enforcing QoS across administrative boundaries by both network techniques and business policy enforcement between companies and networks.

4.2.5.33 Connectivity Session Management

Connectivity session management is about setting up, maintaining and terminating connectivity as requested by service session management. The notion of a *session* represents the evolution and outgrowth of the call control in present telecommunication systems. Such a session is an abstract view given by the service domain of connection-related resources necessary to deliver a particular service. Present use of a session reflects the notion of voice and videoconferencing services. Session servers in NGN will be able to provide interactive, multiparty, multiconnection multimedia communication services. Multiple qualities or grades of these services will need to be supported. Connectivity session management is the capability (or a set of functions) to determine the specifications of the connectivity sessions, based on the requirements of the service sessions, to distribute the information of the connectivity sessions to the relevant network entities and terminals for setting up, maintaining and terminating the connectivity, if required, and to feedback the capability information of the network entities and terminals towards the service sessions. The specification parameters of a connectivity session include network addresses and QoS parameters:

- Suitable interfacing between service and connectivity session management is required, each at an appropriate abstraction level and independent of technology, to support unbundling of the innovation cycles of applications and networks.
- The network should be able to respond to connectivity requirements of service sessions, independently of the content type.
- The network should provide notifications and statistics to service sessions, so that a service session can react to network conditions, and, in turn, change connectivity if required, or to account for the use of resources, based on information originated in the network, etc.
- Connectivity session management should support different connectivity and flow models, for example peer-to-peer, multiparty, etc., and this should be easily expressible in the service session management.
- Connectivity session management should be amenable to many different types of end link, whether they are wired or wireless, including mixtures, so that, for instance, mobility support is typically available, i.e., not the exception but the regular case.
- Connectivity session management should be a scalable and distributable control of each service session.
- Session restart. An intelligent mechanism must be employed such that session is restarted/resumed from the point where it was terminated.
- Session renegotiation. While the user hands over to an access network with different requirements and characteristics, sessions must exhibit adaptation in terms of bit rate and QoS requirements, depending on dynamic resource availability.

4.2.5.34 Security

User terminals must be fully identified, independently of the access technology. Unauthorized terminals should be prevented from gaining access to the networks. However, in some cases, anonymous access to networks should be allowed (e.g. tourist information, push advertising, pay-per-use). In these cases, anonymous users can only have access to limited services, and an internet connection may be restricted in terms of security and bandwidth.

The terminals must be able to authenticate (and authorize) different networks, independently of access technology, as long as the terminal can use the access technology.

4.2.6 Research Framework

This section presents the framework components that require prioritized research challenges for the CoNet vision. We intend to describe overall problem areas and generic technical issues to be addressed, so as to provide guidance for defining the scope and specific targets of respective ongoing and future research activities/projects in the field of networking technologies for the CoNet vision.

4.2.6.1 Multiaccess Capability

Description
Future communication systems will include a variety of access network technologies which will be used by different types of terminal and application, each with different characteristics and requirements on network resources, delay, security, etc. Users will have access to different devices with different capabilities and will use different access network technologies. The employment of different access networks also depends on the user's subscriptions and profiles. Some of the devices will be able to access multiple access technologies, possibly simultaneously, while some other devices may only be able to use a certain access network technology. The various access network technologies will differ in offered bandwidth, coverage, cost, capabilities, etc. Some applications will be able to adapt to changes in network characteristics, while others will not.

The goal is to provide users with the capability to select an optimal access method. What is optimal may vary over time, depending on the type of device, the availability of different access network technologies, what type of applications the user is running, or simply what the user prefers at the moment.

Technical Issues and Research Priorities
In order to enable a system for multiple access network technologies, a number of different issues need to be investigated. The following lists some of these issues.

- The device needs to be provided with access technology independent capabilities (API and protocol) to detect what different access networks are currently available, and to learn the characteristics of these access networks. These characteristics may, for example, be the type of access network technology, available bandwidth, QoS, cost, authentication technology and operator. Furthermore, these characteristics should be provided in a structured way that allows the introduction of new access network types to the system.
- Mobile terminals should be able to scan in a specific environment to discover candidate available access networks and register some policy issues (e.g., provided services,

provided QoS, billing, etc), as well as current status of the access networks (e.g., available bandwidth, congestion, etc). Alternatively, the network may provide the terminal enough information about the currently available access networks, so the terminal will not have to scan.

- There is a need for mechanisms to decide which access network technology to use. Either the user could decide at each instant what access network he or she prefers, or the terminal could make the decision, or there could be more advanced mechanisms for access selection, based on parameters such as device capabilities, operator policies, offered services and application requirements.
- The decision leading to handover per data flow between network technologies, i.e. vertical handover, may depend on many issues, such as supported services, available and required bit rate, round-trip delay, battery charge, cost of connection, user location, as well as user direction and speed, long-term connectivity, roaming agreements and prediction of network congestion, in addition to signal strength.
- In any roaming scenario there are always two main trust relations that need to be established, one between the involved providers, and one between the user and the providers. To establish a user–provider trust relation, the user needs to authenticate himself to the provider and vice versa. While service and network access providers are often the same legal entity, one usually differentiates between the access to network services, i.e. receiving an IP address and being able to send and receive IP packets, and access to multimedia services such as IP-telephony or streaming.
- Preferably, the user should not have to close down and/or restart the applications/sessions when moving between different access networks. To enable this, we might need a solution for mobility management that is tied to the log-on procedures for the different access networks. The solution for mobility management needs to fulfil the requirements of the user, and be compatible with existing solutions in the different access networks.
- An API towards the applications should be designed so that the application can communicate requirements for access selection to the network or device, and also so that the application can obtain information about the access network characteristics currently used. The benefit for the application to be able to obtain that kind of information would be to adapt to current conditions, in order to provide an optimized performance experience to the user.
- There is a need for a multiaccess paging mechanism. It should be possible to page a terminal with multiple accesses, even if some of the interfaces are switched off (to reduce power consumption).

Mobility Management
The architectural changes in the networking infrastructure are dictated first by constantly increasing levels of heterogeneity in modern networks, what they connect and what they transport.

Single access will shift towards multiple, simultaneously used access networks – multihoming. A single address space will shift to multiple, coexistent address spaces – IPv4, IPv6. We are entering the era of merging networking technologies – telecom, datacom, as well as embracing other autonomous systems such as broadcasting, ad hoc networks, etc. These trends further increase heterogeneity in the communicating environments.

Single device usage will shift towards multiple device usage: PDAs, smart phones, laptops, digital cameras, etc., used interchangeably or at the same time for different purposes. Secondly, all networked entities will become mobile: users changing devices, devices changing access networks, access networks allocating network (IP) addresses dynamically and from different address spaces. PANs will become a reality in the near future and are inherently mobile. The use of networking in vehicular and transport areas will also emerge, with completely new applications, e.g. in the telemetric area, which put new requirements on the capabilities for mobility support. In essence, devices and gadgets without networking capabilities today are moving towards being networked, as they serve as sinks and sources of digitally formatted information and content, as well as running applications with networking needs.

Proposals for interplanetary internet are already being designed within the Internet Research Task Force. This research work will set new requirements for mobility because extraterrestrial routing equipment located on space vessels, such as satellites, cannot be placed in fixed points of space.

Considering the above background, this section presents the research issues in the area of mobility management for CoNet. Mobility management primarily consists of two major functional components that are: mobility routing management, including handover; and location management that provides functionality of location updating and paging.

4.2.6.2 Mobility Routing Management

Description

Given the above-mentioned trends, architectural principles used to build networks relying on static mappings between users, devices, addresses, access networks, etc., do not work well in such a future heterogeneous, dynamic and highly mobile environment. Thus it will require additional effort in revising and rethinking communication architectures so that they absorb all of the above trends and function reliably under the above mentioned conditions.

Mobile terminals having multiple access network interfaces must be able to use, at any given time, the access technology that best suits the needs of the actual situation. The selection shall be based on requirements and preferences of the terminal (user) and application, as well as policies of the network operator. The network shall have means to support a change of access technology during active communication. In addition, mobile users having multiple terminals should be able to switch terminals and resume communication from the point where it was terminated (personal mobility). Both of these mobility types should be supported with the smallest disturbance possible.

For the user, the change of network should be transparent, causing only a change in quality level, e.g. the lower resolution in video picture when the device hands over to a network supporting a lower transmission rate. It also depends on the user's needs–which of the many simultaneously available networks should be chosen for the transmission? For example, the video stream may be handled via another network, while the data transmission is concurrently delivered over some other network. If the new network(s) cannot support the QoS requirements from the active application(s), these applications either have to adapt to the new network's capabilities (e.g. by switching from a video conference to an audio conference) or the application has to be terminated, depending on the user's preferences.

One of the research challenges in maintaining connection to the network (e.g., internet) and providing real-time services for mobile users, is to achieve low latency handovers and minimize packet losses. The need to communicate efficiently on the move and to minimize service disruption caused by handovers is becoming increasingly important for delay sensitive applications such as internet telephony (VoIP) and video streaming. Fast handover and minimum (ideally none) data loss may be required to ensure that any disruption is unnoticeable by the user, and hence to achieve seamless handover. There have been many related studies in this area. An extension to this research would be to investigate whether, and how, to formulate handover algorithms by utilizing additional information, such as the mobility profile of the terminal (e.g., velocity and direction).

The issue of wireless communication is twofold; the traditional means of wireless communication is to have fixed base stations providing network connectivity in their range (WLANs), but more and more interest has been expressed in the concept of ad hoc networking, where mobile users autonomously create multihop networks among themselves, without any management infrastructure or controlling of the communication from a central point (base station/access point). An interesting issue is also to merge together these two concepts, meaning the ad hoc network could connect to the core network via base stations. The issue of wireless sensor networks has its own research issues. All of these scenarios should be considered when designing mobile routing.

A lot of research has been conducted on ad hoc networks, and specifically on ad hoc network routing. Outdoor ad hoc wireless networks may include a 'large' number of communication nodes that may be arranged in sparse network topology. It is foreseen that these nodes will have a great degree of mobility and in the case of large-scale networks, a quantitative measure of mobility is directly related to the path selection route. This necessitates the adoption of a strategy that is capable of evaluating the probability of path availability over time and minimizing the reaction of routing tables due to topological changes. As an effect, within an ad hoc infrastructure, hierarchical routing schemes must be introduced. In such schemes, the amount of information required to keep each node up-to-date with respect to changes in the network topology is proportional to the number of nodes and the rate of topology change. A synergy between DLC routing and IP schemes will enable efficient usage of radio resources and reduction of network-layer routing, and organize the nodes in clusters based on geographical and physical relationships among the nodes. The IP-layer clustering will organize the nodes in logical groups according to the logical and functional relations among the different nodes.

One extension to this research would be to investigate whether and how, to support new types of application, such as applications with QoS requirements, and how to support different types of overlay network, in the mobile ad hoc network environment, where dynamicity such as network topology change is remarkably large.

Technical Issues and Research Priorities
A large number of various technical issues needs to be addressed in future research in the area of mobility management in order to achieve the required mobility routing capability in the future networks. The following lists some of the issues, but this is not an exhaustive list.

- *Mobility architecture.* This architecture should support mobility on a number of levels. This includes: support for such mobile entities as off-system users (e.g. humans), applications/agents, services being mobile in the network, hosts and devices, groups of mobile entities (nesting), as well as individual data flows and sessions (e.g. between devices);
  - — devising a coherent layering of mobility support, including its impact on the whole resulting protocol stack;
  - — support for dynamic use of intermediaries operating on various layers;
  - — optimization of data flow paths from various perspectives, including aspects of quality of service, security, privacy and business relations of providers;
  - — mobility support as end-to-end schemes versus via mobility intermediaries.
- *Group mobility.*
  - — formation and mobility of networks. This includes routing issues, service/capability discovery, etc.;
  - — aggregation of data flows and their mobility;
  - — solutions to group mobility by nesting in several levels;
  - — support for mobility in multicast scenarios.
- *Session continuity.*
  - — a relatively seamless support with efficient, timely and predictable behaviour and harmonized with transport protocols. Relative here means that there is a broad spectrum of what an application and user sees as 'seamless';
  - — support for context and/or state transfer;
  - — efficient and smooth support for both local and global mobility, including interdomain handovers;
  - — access technology and network independence.
- *Naming and addressing.*
  - — a common name space for all networked entities. In a networking environment which comprises a large diversity of access technologies, where a multitude of different terminals should be reachable, a common name space will provide the transparency necessary;
  - — uniqueness of names and addresses. To unambiguously address a certain user without unintentionally reaching another, or causing conflicts due to the existence of several instances of the same address, a scheme that ensures uniqueness of network names and addresses is of eminent importance;
  - — the nature of names and addresses, e.g. cryptographic, random and assigned by authorities. The authority for assigning network names and addresses should operate so that misuse is avoided and the uniqueness of names is ensured. A proper assignment will enable name resolution of network entities;
  - — necessary name resolution systems. Name resolution systems are required in a network when different networking devices with a uniform name space are all enabled to reach each other.
- *Mobility triggering.*
  - — interaction with layers above and below to ensure fast and efficient mobility support;
  - — network conditions' impact on mobility.

- *Proactive mobility support.*
  - capabilities to proactively track, predict and influence mobility patterns of the networked mobile entities.
- *Context transfer schemes.* These should be integrated within the mobility management. In many situations, the change of communications path includes a change in communications between the host and access networking. During the handover, it is essential to preserve the context of IP flows. The IP flow context that might be useful to transfer could include security context. QoS and AAA information.
  - secure security context transfer in handovers between heterogeneous access technologies or network types. When a handover occurs, timing constraints may forbid performing a full new access procedure, including authentication and key agreement. Instead, the security context may have to be transferred between points of attachment in the network that trust each other. The precise nature of the transferred security context needs to be specified, and the security for the discovery of points of attachment and of the transferred context, need to be studied. While security context transfer in horizontal handovers (access technology remains the same) has been resolved in certain cases, security context transfer in vertical handovers (the access technology changes) is largely unsolved;
  - secure security context adaptation in handovers between heterogeneous access networks. It may not be sufficient to merely transfer the security context in a handover, but the security context may need to be adapted according to the new environment. For instance, the IP address in an IPsec security association may change, or different cryptographic mechanisms or schemes to protect communication traffic could be used.
- *Mobile ad hoc networks.* Enhancement for efficient support of peer-to-peer applications over mobile ad hoc networks, where some of the issues are:
  - route discovery for multihop ad hoc networks;
  - terminal and neighbourhood discovery mechanism;
  - new metrics should be considered for the ad hoc routing infrastructure (i.e. longevity of a route, availability of information associated with network load in a certain route, network load in the intermediate relaying nodes, overhead expressed as the average number of routing control messages determined by taking into account the number of connections and the mobility between the end hosts);
  - to investigate ways for the best interactions between peer-to-peer applications and lower layers to enable efficient maintenance of peer-to-peer connections, like an overlay network, on top of the mobile ad hoc network.
  - to investigate mobile ad hoc network routing mechanisms and interoperability between inside and outside the mobile ad hoc network, that are optimized for the requirements of peer-to-peer applications while considering the environmental constraints;
  - interaction between IP mobility and ad hoc mobility. Each ad hoc node will support two mobility protocols. The intra-based mobility management will enable a mobile node that participates in an ad hoc cluster to move either within the same cluster, or between neighbouring clusters. One important issues for macromobility concerns the employment of efficient location management schemes;

- Appropriate extensions of current micromobility protocols to interwork in an ad hoc environment;
- network security and mobility routing. Wireless mobile ad hoc networks bring new security challenges to the network design. Generally speaking, mobile wireless networks are more vulnerable to information and physical security threats than wireline networks. The introduction of security mechanisms in an ad hoc network is a challenging task due to the broadcasting of wireless messaging across multiple destinations, the vulnerability of channels and nodes, the dynamic variation of network topology and malicious corruption of information associated with modifying routing information. Therefore, the absence of infrastructure makes the introduction of security mechanisms, such as the use of certification authorities and online servers inapplicable in ad hoc networks. Communication operation in free space exposes ad hoc networks to attack, as anyone can join the network and either eavesdrop or flood messages. Ad hoc network attacks can be classified as passive or active. In passive attacks the attacker just listens to the channel and attempts to discover nodes of information. On the other hand in active attacks, packets are flooded within the networks. Active attacks can be grouped into impersonation. DoS and disclosure. One other challenging issue regards the requirement for mechanisms that encourage/enforce users to behave as 'good citizens', allowing hosts to relay packets for the benefit of other hosts, making their data available;
- power awareness;
- system throughput;
- clustering/logical grouping.
- *Sensor network.*
  - extreme power efficiency with adequate connectivity;
  - fault tolerance;
  - failure detection;
  - addressing and naming;
  - data aggregation and fusion from many sources.
- IP mobility management architecture and schemes need to be developed. These may utilize salient features and capabilities of existing protocols (e.g., Mobile IP, SIP, IP micromobility protocols). Mobility management schemes that are hierarchy-oriented (separating the mobility of the core from this at the access), and/or multiprotocol in nature, should be investigated.
- Two challenging problems should be addressed in the context of handover:
  - how quickly the mobile node can send and receive IP packets subsequent to handover; and
  - how to ensure that the disruption caused by the handover is minimized, once the mobile node has established IP connectivity.
  The former refers to the fast handover problem, whereas the latter strives to make the handover process seamless by utilizing efficient mechanisms, such as context transfer between last-hop routers.
- Integration of mobility management architecture with QoS and AAA infrastructures needs to be investigated from the viewpoint of the requirement that active communication sessions should not be degraded at handover, depending on the user's preferences.

4.2.6.3 Location Management

Description

Location management includes two major processes that enable the network to discover the current attachment point of the mobile node for call or packet delivery. The first process is location updating, in which the mobile node periodically notifies the network of its new access point, allowing the network to revise the user's location profile. The second process is paging, in which the network is queried for the user location profile and the current position of the mobile node is found.

Location management can be used within two areas. The first area is how to keep track of inactive nodes that are in sleep mode or similar, pseudoactive, modes. Location updating entails a mechanism that supports tracking of both inactive and active mobile nodes. The other area is to provide support for optimized routing and location-aware services.

Paging is about locating a node in sleep mode, where the network does not exactly know where the inactive node currently is. Thus, a paging mechanism enables savings in terms of battery power consumption at the mobile host, by supplying approximate location information to the network. If another node tries to establish a communication request to a node in sleep mode, the network has to find this node, wake it up, and subsequently establish the connection. If the incoming communication request is for a time-critical connection (e.g. voice), the network should be able to establish the connection within a limited time.

In order to reduce power consumption and to limit the amount of signalling in the network, inactive mobile terminals should not need to update their respective location information at each move. The network should have support to track terminal mobility at a lower granularity. For example, wireless cells may be grouped into location areas, such that mobility of idle terminals within these location areas triggers no signalling traffic. In this case, when a terminal has an incoming call, it must be searched for (i.e., paged) within a location area.

Besides predefined location areas, networks may support custom or dynamic location areas, as well as custom paging strategies. These specialized mechanisms are adapted to the typical mobility pattern, or special characteristics, of a particular mobile terminal.

Technical Issues and Research Priorities

There are several research issues that have to be investigated in order to enable efficient location management within the networks. Some of these are:

- What architectures and entities are required to support location updating and paging?
- How can paging be supported in the IP layer?
- How large or small should be the location areas in which the inactive mobile nodes can move, without updating the network about their position?
- Determine the trade-off between the power saving for the mobile nodes and the response time to locate the mobile nodes within reasonable time.
- Support of group location updating, e.g., for efficient utilization of network resources.
- How will a node that is in sleep mode detect that it is being paged?
- When the network pages a mobile node in sleep mode, what paging strategy should be used?

- Should there be one paging strategy suitable for all mobile nodes, or should the paging strategy be based on node characteristics, such as the speed of the node, the load of the network, etc.? Support of custom paging strategies that utilize user's context information is an important research issue.
- Traffic and mobility patterns affect the performance of location area/paging constructs. The old well established relations found for cellular speech use might be put aside with extensive packet based application use. For example, support of custom and/or dynamic location areas that utilize context information needs research.
- Support of paging through heterogeneous access systems are of interest for research.

4.2.6.4 User Management

Description
During the last decade, we have seen a significant number of paradigm shifts in communication. It is necessary to emphasize the following:

- The user focus shift from 'getting connected' to the network, towards 'getting services' mediated by networks.
- Communication is becoming more and more personalized.
- There is a growing demand for support of 'peer-to-peer' communication between users that enables them to exchange services with each other.
- Collaborative work and collaborative spaces facilitating group interaction between users are already popular forms of communication on 'fixed' internet, but are expected to include an increasing number of 'mobile' users.

Future mobile terminals will be equipped with several network interfaces, which may be of different access technologies, both wired and wireless. Different requirements of different applications can result in a preference for the interface that should be used. Network connections should be placed in the best possible interface, based on these requirements. During communication, changes in the availability or characteristics of an access network behind an interface may result in a situation where already established connections would need to be moved from one interface to another. For this purpose, a variety of mobility management protocols supporting handovers between interfaces have been proposed. Some of these protocols move all traffic from one interface to another at once, while some protocols allow simultaneous communication over different interfaces. However, the current solutions do not propose any means for the user or application to be able to dynamically influence the interface selection during the operation of a mobile device. An interface selection mechanism needs to be developed for multihomed mobile hosts. The mechanism should allow for dynamic decision making during the operation of a mobile device. The access selection should be controlled by rules or profiles defining which access/network/provider to be used for a certain traffic flow. These can be defined by the users, application service providers or network service providers. The actual decision should be based on the adaptation of these rules into availability and characteristics of the interfaces and access networks at any given time.

Users may have different access rights, preferences, settings, traffic and mobility characteristics or other properties. A set of mechanisms to store, manage and utilize user-related information must be part of the architecture. This information base may be contacted when a user demands access to the network or to a particular service. In addition, the information stored about a particular user may influence the handling (e.g., mobility management) of the particular user. The user management functions may potentially be combined into a new type of network entity.

Mobile users, application service providers and network service providers may define preferences and requirements regarding the use of interfaces and access networks. The interface selection decisions are based on these preferences and requirements when they are evaluated against gathered information about transport characteristics. The operation of the interface selection mechanism must be continuous, as the information may change at any point of time, e.g., a network interface may become unavailable. Therefore, there has to be a policy database and mechanism by which to hold and maintain rules for interface selection. Policies provide different network entities a possibility to control the placement of a mobile host's traffic flows into different network interfaces. Interface selection policies describe the preference of different network interfaces in various situations. On account of the policies, the local routing mechanism routes outgoing IP packets into available interfaces. The preferred available interface is always used, and if it becomes unavailable, e.g., when a user moves out of a wireless network coverage area, the connections will be moved to the next preferred interface.

User management related information described above should be stored in policies. These should be general enough to cover all needs related to user management. There are several reasons that motivate the need for policy-based user management, the main ones being:

- The cost saving potential of policy management techniques and tools. Network service providers and application service providers can manage users using policy management techniques to their advantage. Using these, network operating costs can be reduced.
- The need for an open, generic, standard and relatively simple user management system that can be used by network service providers and application service providers to manage and develop services and applications. This is especially important in future multiaccess, multiprovider networks.

A principal goal of this research is to specify the policy management system and related APIs. These specifications should be independent of network architectures and protocols. A related, and equally important, goal is to specify an information model for policy management.

An information model is a scheme that represents policy information in a flexible but unambiguous manner. One can also speak, informally, of a policy domain. A policy domain encapsulates all policy information about a given domain, e.g., a policy domain for authentication services, or a policy domain for QoS services. An information model will help to ensure that policy domain definitions are well defined, unambiguous, extensible and reusable. More specifically, the user management research area should specify:

- A policy information model. The information model must be general enough to allow for the definition of policy domains across the various areas, e.g., access selection, network management, charging and billing, etc. It must also be general enough to allow for the definition of third party application-specific domains. For example, the Policy Core Information Model (PCIM), as specified by the IETF policy framework working group, could be used as the basis to develop a policy information model.

- Definitions of policy domains for applications and network-based services.
- Definitions of policy APIs and related operations for use by service providers and network providers.

Technical Issues and Research Priorities

1. *Profile management.*
   - To investigate, specify and test approaches for the user for specifying and setting their preferences and profiles.
   - To investigate, specify and test approaches for the provider to execute and realize the users' preferences.
   - To investigate, specify and test approaches for the provider to set their own profiles and policies.
2. *Profile exchange.* When roaming between different networks owned by different network providers, the user profiles need to be exchanged. This is needed to enable the foreign provider to provide the user with services in accordance with the user's preferences and profile privileges. This issue involves achieving a balance between information revealed to the foreign provider and the required amount of exchanged messages between the providers.
   - Policy management architecture. Where are the various policies stored – in the network or in the terminal (or both)? Where are they enforced?
   - Policy structure. The internal structure of the policies. Hierarchical structure for policies (user/application/operator).
   - Policy APIs. What are the interfaces and protocols in the system?
   - Security. Regarding policy storage, transportation, etc.
   - Local user policies may be used by users and applications to define their preferences and requirements for the network services and application services. How are these taken into consideration with regards to service and network-originated policies?
   - User customized communication services and personalization. How can these be implemented using policies?
   - User interfaces. Usage of the system must be easy and low-cost.
   - Policy types. Long-term (subscription) and short-term (pay-per-use), etc.
   - Grouping of users. Gold, silver, bronze, etc.
   - Performance issues in a policy-based communication system.
   - Definition of information related to user management.

4.2.6.5 Naming and Addressing

Description
Identities and addresses are of key importance in the construction of any communications system or network. The way in which identities and addresses are used may have implications for the system as a whole. These are the key entities around which many functionalities are built. A prime usage is to identify the other party intended for communication and establish a session.

The possibility to find a name, and hence a node, will depend on how and where the name is published. Names will be taken out of name spaces where they are unique. Different name spaces will exist that make up private communities for different purposes and usages. Cooperation, dependencies and even reuse of name spaces can be of great importance. Certain applications may, e.g., have their own name spaces, corporations may have their own name spaces, etc.

A host in the internet has an alias name and an IP address. The IP address is an identifier for the host. In general, there exist two types of identifier: identifiers with, and without, a naming trust relationship. The naming trust relationship means that an identifier can be tied unconditionally to an identity. The naming trust can be gained with cryptographic methods, thus a public key is an example of an identifier with naming trust, as it can be proved using the public key pair. Currently, the host in the internet has an identifier without naming trust: the IP address. There is no way the host can unconditionally tell to other hosts that the address actually belongs to it.

The usage of addresses in the current architecture leads to a name space problem: identifiers without naming trust are used to identify hosts. Identifiers without naming trust can, however, be used for addressing purposes, and these identifiers further can be bound to identifiers with naming trust. As an example, the host identity in the host identity protocol (HIP) can be used to identify a host and IP address of the host used to route packets (addressing).

Currently, the IP address semantics is overloaded and it is used for identity, identification and addressing purposes. Finally, in the internet, the name space problem is related to security issues. Therefore, an IP address is insufficient from the security point of view to identify a host. An identifier that fulfils the security requirements must be defined. An identifier must enable multihoming and mobility requirements.

When TCP/IP was introduced, hosts were static and IP addresses were naturally used, both for host identifying and routing purposes. Since then, the characters of hosts have changed; mobility and multihoming have set new requirements for identification. Existing solutions suffer from the historical payload. Mobile IP and several other existing solutions actually use the separation of identity and routing information. However, the separation has been done using multiple IP addresses in different roles. This is not optimal. The existing solution suffers from unsolved security problems which can be worked out using cryptographic identifiers. Those identifiers are separated from the routing information.

Technical Issues and Research Priorities
The following gives some of the issues to be investigated, but this is not an exhaustive list.

- What kind of identifier – common to all interfaces – should be used for hosts?
- Where should this identifier be located and how should it be bound to addresses, etc.?
- How to generate the identifier.
- How to receive and where to store such an identifier.
- How to bind an identity to an identifier.
- How to prove an identity.
- How to transfer and exchange the identifiers.
- How the new identifier could be used in mobility, multihoming and security.
- How to solve the migration issues.
- Effects on application layer.

4.2.6.6 Quality of Service Differentiation

Description

The next generation of wireless communication networks will consist of multiple access technologies and communication systems. In addition to this heterogeneity in communication models, new terminals and applications with new communication patterns and different QoS requirements are evolving. Future terminals will not only consist of single entities, but will consist of BANs. PANs and other groups of network devices, all with different capabilities and requirements.

Currently, IP is seen as the foreseeable integration technology for the next generation of systems, often called all-IP multiaccess systems. As mentioned above, different applications and terminal types with different requirements will communicate using different access technologies, but they will all use IP as network layer technology. This implies that the network/transport layer must support and enable QoS negotiations and renegotiations between terminal and application requirements and network capabilities, and these negotiations should use standardized APIs to support simplicity and interaccess technology functionality.

The network/transport layer is responsible for propagating, reserving and enforcing the negotiated QoS requirements to the different links and networks on the end-to-end connection requested by the application. The specific reservations are, however, hidden from the application.

Simplicity is a key issue in making the next generation wireless systems become a success, in particular, simplicity in how applications' QoS requirements are propagated to the network layer and vice versa. The link layer characteristics will be hidden from the application, but the QoS capabilities of the link layer will be propagated up to the application layer. This also includes application and session adaptation when a session is transferred from one terminal to another with different capabilities and requirements. For example, a video conference on a laptop connected to a WLAN is transferred to a cell phone on a 3G mobile network, and the session is adapted to become an audio conference for the participant.

Also, a policy server is a key component for achieving QoS. With respect to practical application of QoS, policy is a combination of enforcement and decision making that permits calls to be initiated, established and torn down between networks over the internet, based on certain rules.

Technical Issues and Research Priorities

Some of the prerequisites are:

- The QoS control mechanism that is available for the applications should support the control of QoS over multiple access technologies and multiple wireless hops.
- Applications should be able to control QoS without being aware of the specifics of the actual access network they are running over.
- Support of both quasi-real-time (e.g. streaming) and real-time applications (e.g. video-conferencing).
- The QoS mechanism should provide application developers with appropriate granularity of QoS control through open APIs.

- It should make it possible for resource-scarce air interface technologies (e.g. cellular) to dynamically optimize air interface resources (spectrum allocation, power control, etc.).
- The QoS control mechanism should be flexible so that it is applicable to non access technology-specific terminals.
- The QoS mechanism should support the development of adaptive applications and services that are capable of changing their QoS (e.g. bit rate) requirements while switching among heterogeneous access networks.
- End-to-end QoS should be supported over a single operator domain, as well as over multiple operator domains.

The following summarizes some of the arising research issues. This is not an exhaustive list.

- What API parameters should applications use to specify their QoS requirements, including applications that can adapt their needs to diverse access technologies?
- What mechanisms and protocols should applications use to transfer these QoS requirements to the various access networks assuming access agnostic applications?
- How should such protocols interwork with QoS mechanisms in an IP-based core network?
- How can this access agnostic QoS requirement set be supported over the various access networks, and how can such mechanisms work seamlessly in intra- and interaccess handover cases?
- How can such protocols support adaptive applications?
- How could the protocols be optimized for efficient transfer over narrowband interfaces?
- For intercontinental communications with distances longer than the circuit switched Long Haul Hypothetical Reference Model (5000 km), how can we manage QoS in all its shapes and aspects for B3G networks?
- Which mechanisms would offer both QoS guarantee and QoS adjustable granularity, independently of whether it is for the access or the backbone?
- Can all traffic types (bursty, nonbursty) coexist in the same classes, in fully loaded networks? If not, how should they be treated?
- QoS control at middleware level. Much work has to be done on the metering, interpretation and processing of QoS information, such that stable and scalable solutions for adaptation will be spawned. Full QoS management will not always be feasible. Important research subjects, in our view, have to do with the question of which cases need end-to-end QoS management (including handover situations), and when will adaptive applications suffice? It will become important to develop and design policy frameworks, programming languages and enforcement mechanisms that reinforce such application and end user dependent behaviour, and are able to map them efficiently onto network-level concepts.

4.2.6.7 Resource Management

Description
Resources in the access network, particularly the radio interface, will continue to represent a scarce resource and potentially a bottleneck in communication flows. The network should incorporate mechanisms to manage the utilization of these resources. Resource

management will be related to mobility control, selection of access technology, power consumption and QoS provision.

1. *Access resource management.* In current networks, the resource handling is supported between different access types. However, the handling of radio resources is different (e.g. W-CDMA and GSM). Naturally, there are even larger differences between these technologies and, e.g., WLAN technologies. One of the main ideas of an ambient network is to make efficient use of a large number of different access types, each of them being the best choice for a certain type of service requirement, environment, cost, etc. This will be even more important when new access technologies are added to the already existing ones. In order to make this happen, a number of problems need to be resolved.

   Currently, there is little support for comparable access descriptors among technologies. As a result, it is difficult to make an intelligent choice between different accesses, and a fast change of access when needed. Seamless mobility for real-time services across different access networks is not possible today (with some exceptions, e.g., voice between GSM and W-CDMA). Currently, it is not really possible to coordinate the resource usage between accesses, and it is thereby difficult to use the resources efficiently. The conclusion is that a joint handling of access resources among different access types needs to be developed and a suitable architecture needs to be defined.

2. *Adaptive resource management.* One of the important components in resource management is QoS control, which is a mechanism to control resource allocation to different applications based on their requested QoS levels and negotiation results. There have been a great number of research works in the area of QoS control. One extension to this research would be to investigate adaptive QoS control technologies, taking into account various factors in resource allocation to applications. The factors could include pricing policy on resource usage, while prioritizing the chosen user.

Technical Issues and Research Priorities
The following lists some of the research issues, but this is again not an exhaustive list.

1. *Access resource management.*
   - Access resource management architecture:
     — functional distribution and interfaces;
     — interaction between common and access-specific parts;
     — planning, deployment and service mapping for multiaccess;
     — impact of different architectures.
   - Management of access resource consumption in a multiaccess environment:
     — QoS provisioning per flow;
     — resource optimization and balancing;
     — access discovery and coordination;
     — coordination of generic and access-specific resource management functions;
     — provision of access descriptors for access selection.
2. *Efficient and adaptive resource management.*
   - Definition of adaptive protocols for user data and control message communication.

- Provision of efficient congestion control schemes for multimedia applications that integrate knowledge of functionalities and capabilities of the lower layers with the requirements of the applications.
- Investigation of adaptive QoS control, considering various factors relating to resource allocation in the network. There are various research items in this area. The items regarding adaptive QoS control considering pricing/revenue management for resource usage include development of models and mechanisms for optimal adaptive control.
- Definition of adaptation mechanisms for applications, combining knowledge of available resources, forward error correction and loss conditions.

4.2.6.8 Service Support

Description

On a broad scale, today's communications environment is characterized by being composed of a multitude of autonomous systems, such as radio and TV broadcasts, cellular phone networks and the internet. In addition, the media environment is separate, with their own autonomous systems for media delivery, such as streaming servers, digital cameras, MP3 players and so on. These are controlled and managed separately, as seen by both end users and providers, and are not viewed as being components of a 'connected' media and communications system.

The future vision is that these systems are all connected to each other in various ways, so that user-relevant information and media can be delivered and adapted with no, or at least diminishing, system boundaries. An example is the emergence of PANs, where different end user devices get connected to share services and media. This convergence of technologies will provide new communication channels of a vastly heterogeneous nature, where mobility and dynamicity on different levels of the systems are main characteristics.

The merge will also present means for new delivery channels of applications and services, but will also stimulate new usage patterns, leading to even more opportunities for services and applications to be provided.

Heterogeneity will take place on various levels. Typically, this includes a large range of different devices operating as sinks and sources of information, content and media with different capabilities, access networks with different characteristics, as described above, varying user preferences and contexts, different formats of media where the media format and type (e.g. visual, audio) may have to be preserved, or may be allowed to change, depending on the context.

Dynamicity and mobility will also occur on different levels of the system and for various reasons: change of access, such as vertical and horizontal handovers with different characteristics; change of device for ongoing communications and sessions or parts thereof; change of media format to match different device capabilities; and change of media type due to change of user context. These are just a few mobility scenarios.

In summary, there are basically two main motivations for service support:

1. Bridging heterogeneity, mobility and dynamicity to deliver an intuitive and smooth communications experience to the end user regardless of the used device, access, media or context. In many respects, this will also be characterized by simplicity, convenience and a certain degree of automation.

2. Stimulating and enabling rapid development and deployment of applications and services. This includes means for providing toolkits to third party developers and providers of applications.

A requirement for feasible support of a large degree of heterogeneity is that functionality that is common across applications can be reused. These functions shall hence be extracted and identified as autonomous and generic functions for use by any application.

In order to allow an application-stimulating environment, a toolkit for application development is needed that gives access to information and capability in the network. The above support functionality may, to a large extent, be viewed as autonomous support services to the actual applications. These services may hence be provided openly for any party to use as they are, or aggregate to higher levels of specialization for further open provisioning. This will require means for service publication, discovery naming and description techniques.

The set of service support functions shall be open ended, with new functions added as demand arises. In order to facilitate application development, service support functions shall be made available through open application programming interfaces (APIs) or via networking protocols, depending on how they are to be accessed and used, and at what system level.

A service provisioning environment, which forms an adaptivity layer between the network and the applications, should provide the application designers with a higher level of abstraction, hiding the underlying network heterogeneity. This environment should handle the underlying network resources as a single, integrating computing facility, enabling seamless service provisioning in heterogeneous, dynamically varying computing and communication environments.

4.2.6.9 Security and Privacy

Description
Networks, as well as terminals, should include mechanisms for mutual authentication and authorization of each other, in order to authorize or deny access to networks, as well as to specific services. Control signal protection is needed, both from users towards networks and between network entities.

It must be possible to protect control messages so that the identity and moves of a mobile user cannot be revealed to eavesdroppers or by active attacks. The actual location of a user should normally be hidden from other mobile terminals, as well as from its correspondents. While in some cases giving location information may be part of a service, this should be dependent upon a user-selectable option. The network should also allow for mechanisms to encrypt user data for privacy. Location privacy should be supported.

New trust models are emerging [3], and the current security solutions need enhancement to be valid in the future. Efficient and secure solutions for media routing, adaptation and the support for building special solutions like secure closed user groups are either missing or too complex. Many access and backbone operators will, in the future, carry peer-to-peer traffic, and this will impose a greater need for end-to-end security, since the trust relations become more diffuse and harder to manage. Multicast and broadcast of prime content necessitate security mechanisms for access control. The functions to support location-based and context-based services should be under user control and have adequate privacy protection. There will be a requirement for the support of multiple subscriptions. Functions for multiaccess, QoS, session management, network access, etc. will require a wide range of security mechanisms.

Security between different entities that do not know or trust each other, in ad hoc networks, has to be supported. Security in different types of overlay network has to be guaranteed. Operators and partners sharing network resources have to be granted access to some network resources in other partners' networks, while some network resources should be strictly private. Payment systems and charging have to be user friendly, trusted and allow for different payment methods. Future terminals will contain sensitive and secret data, such as keys and passwords. Security in intertechnology roaming and handover is little understood and has to be further researched. Reconfigurable terminals and network components have to be protected from abuse. Methods for supporting secure, smooth handovers and access point roaming have to be developed. Anonymity has to be supported at all layers in order not to reveal the user's identity.

Technical Issues and Research Priorities
The following lists some of the research items regarding security and privacy:

- single sign-on (SSO);
- privacy control management;
- digital rights management (DRM);
- security in architectures for public access;
- call control security;
- end-to-end security;
- multicast and broadcast;
- context management and user control;
- network control;
- mobility management;
- ad hoc networks;
- PAN, PN, HAN, CAN, VAN security;
- security for shared network resources;
- payments and charging;
- terminal and configuration management;
- security modules;
- secure infrastructure;
- heterogeneous networks;
- secure reconfigurability.

4.2.6.10 Moving Networks

Description
A moving network (NEMO) consists of one or more mobile routers (MRs), with a number of devices connected to them. These networks can change their point of attachment to other networks because the NEMO physically moves or changes in topology. When the NEMO changes the point of attachment, the entire NEMO changes the point of attachment as one entity.

NEMOs appear in different sizes, from small groups with a few devices attached to the MR, up to NEMOs on ships and trains that might have several hundred or more mobile nodes (MNs) attached to them. The administrative characteristics will differ between NEMOs as well, the MRs and the MNs may be administered and owned by the same or different entities. In small NEMOs, the MRs and the MNs are typically owned by the same entity, while in large NEMOs, the MRs and MNs are typically owned by different entities. This affects, amongst other things, the level of trust between the different nodes. Different types of MN have different requirements on location updates and connections to external networks (home operators) when the MNs are not actively participating in a session. A laptop might not require instantaneous location detection to a peer in another network, while a cell phone has to be reached 'immediately' when someone is calling it. This puts different requirements on what, and how much, update traffic will be sent through the MRs.

Technical Issues and Research Priorities
Here are some research areas that could be addressed:

- Addressing.
- Scalability.
- Speed, some NEMOs may be moving at high speeds, which puts requirements on handovers and location updates, etc., as well as frequency of these handovers and location updates.
- Nested mobility.
- Access control and security.
- Multihoming, a mobile node can be attached to multiple ISPs/operators, and there can be multiple MRs in a NEMO, and/or a NEMO can be also multihomed (MR having multiple accesses).
- Mobility management.
- Aggregation of signalling and user data traffic between MRs and external networks.
- Should mobile nodes be aware of being attached to a NEMO, or should this be hidden from them? Should it be hidden from their operators? From their peers?
- How to join and leave a NEMO. What mechanisms are required?
- What type of traffic will be sent through the NEMO (control traffic, conversational multimedia traffic, best-effort traffic, etc.)?
- How to implement AAA mechanisms in NEMO environments.
- How to treat different types of traffic. How to prioritize, distribute and terminate different traffic flows.
- How to support traffic between NEMO and external networks, and within a NEMO. Is special infrastructure needed to support a NEMO (e.g. for trains)? Should satellites, existing cellular infrastructure or specially developed network models be used? How to bundle traffic between NEMO and external access points.
- How to deal with location information of users in a NEMO.

4.2.6.11 Personal Area Networks

Description
A personal area network (PAN) consists of different personal devices, such as a PDA, a laptop, a cell phone or a camera, which are able to communicate directly with each other.

These devices may also be able to communicate with devices outside the PAN, either by connecting directly to those external devices, or by using one of the PAN devices as a mobile router (ad hoc operation).

If a PAN device can access different access networks, or if different devices in the PAN can access different access networks, a similar situation as the one described in Section 4.1.2 occurs. The aspects of multiaccess systems, such as the access network technology that can offer optimal connectivity to the internet, are valid for these scenarios as well. However, certain PAN aspects need to be identified and investigated in order to provide multiaccess capabilities to a device in a PAN. Some of these issues are device detection and PAN formation, access detection with PAN aspects, presentation of access capabilities and access selection with PAN aspects.

Technical Issues and Research Priorities
The following lists a few topics that need to be investigated in order to allow a PAN device to find and use the optimal access method for its communication.

- The user device, that is, the device that the user is currently running his or her applications on, needs to detect what other devices are present in the neighbourhood, and how these devices together can form a PAN. The different PAN devices also need to be authenticated and authorized to each other.
- If the devices in the PAN are owned by different users, the question of authentication, authorization, security and legal aspects in general becomes even more complex.
- The user device needs to be provided with capabilities to detect which access-capable devices in the PAN can provide access for it. These access capabilities need to be formulated, communicated and presented to the other devices in the PAN. These access capabilities may include performance characteristics of the device and networks that this device is currently connected to, and need to be structured in a generic way to ease the introduction of new device and access network types.
- There may be multiple user devices in a PAN sharing the same access device and access network. This may affect the performance, and the traffic flow from each device needs to be secured from eavesdropping and interference from the other devices.
- The PAN needs to be able to handle the case where different user devices in the PAN use different access-capable devices and access networks, and get assigned IP addresses belonging to different address spaces.
- How will the PAN be able to interact with the external network? Is there a certain hierarchy amongst the devices that ensures that external communications are handled by a PAN gateway? Will this hierarchy also assist in distributing PAN-wide personal preferences?

4.2.6.12 Transport Networks

Description
A transport network is responsible for transmitting data packets between users, as well as forwarding data packets to other transport networks. Different transport networks interconnected to each other enable end user devices to communicate with each other, but the end user devices are usually not seen as parts of the transport network. Transport networks are based on network infrastructure, such as routers and servers.

Currently, the heterogeneity between different transport networks is enormous, both from the type of operator that provides the transport network functionality, and from the scale of the transport network and the operator's network type. Examples of operator types are access network provider, backbone provider, service provider and omniprovider. Examples of the network scale are intranet, regional networks and international networks, while examples of network types are LANs/MANs/WANs, access networks, backbone networks, fixed networks and mobile networks.

This heterogeneity in transport network models affects how different transport networks are built, while the functionality and infrastructure of a wide-area mobile phone system differs greatly from that of a fixed LAN. The diversity in transport network models will increase in the future, with new types of service, new network technologies and new requirements from users, as well as new communication models and communication patterns. The goal of this research is to provide new solutions, as well as enhancements to already existing solutions, in order to build and support different future transport networks, from a per-network or per-service or per-operator view, as well as the way that these different access networks can interconnect and intercommunicate in order to provide end-to-end connectivity.

Technical Issues and Research Priorities

- Transport architecture and functionality.
- Enhancement of Routing Protocol (OSPF, IS-IS, BGP) regarding resilience and QoS.
- Protection switching (L1, L2, L3) vs. rerouting.
- Robustness aspects, e.g. traffic conditioning and policing.
- Redundancy (link, node, network).
- Availability in a carrier-class network.
- Service requirements for protection.
- Transport interfaces (ForCes, GSMP, H.248).
- VPN technologies and solutions.
- Interoperator interfaces (SLAs, etc.).
- Router elements: access routers, edge routers, core routers, border GW routers.

4.2.6.13 Internetworking

Description

Mobile terminals attached to a network must be able to access services residing inside the network, as well as to communicate with terminals located in other networks. Communication with legacy IP nodes should also be possible. There are two main areas within the internetworking area that have to be investigated. The first one is how to enable devices located on different types of network to communicate; the other is how different network operators will cooperate and interoperate in order to enable worldwide internetworking.

Related to the first area, are several issues like naming and addressing, IPv6/IPv4-public/IPv4-private addressing, DHCP, network capabilities/application requirements, NAT/firewall/proxy traversal, infrastructure to support application layer translators, etc. A terminal must be able to switch from an Ipv4 network to an Ipv6 network without breaking any active sessions.

Different naming schemes, e.g., DNS and E.164, have to interwork to allow users and terminals with different backgrounds to interact with each other. A terminal accessing a network has to be uniquely identified in order to enable communication between itself and other terminals. The terminal should be identified as the same terminal, independent of which and how many accesses the terminal has at any time. This terminal identification should also allow the terminal to be reached when its physical access point is located behind a NAT, proxy, firewall or similar device.

Terminals with different applications and capabilities should be able to share information, even if they do not understand each other. This can be accomplished by sending the data through intermediate translator nodes that change coding to enable communication.

In order to allow different network operators to share infrastructure, there must be methods for the operators to negotiate QoS levels for different traffic flows belonging to different operators. These SLAs should be formed in such a way that they can be negotiated and renegotiated dynamically and automatically. Another interoperator area is how to support VPNs that extend over several network operators' networks.

1. *Roaming.* Mobile terminals should have a means to access networks other than their home operator's network. In such roaming cases, user authentication and authorization may be based on appropriate protocols between the home and visited networks, such as AAA. Calls addressed to roaming mobile terminals should reach them in the visited networks.
2. *Service mobility.* To serve a human user in the world of cooperative networks, the technology must provide a maximum of personalized services seamlessly across networks, with a maximum of ease of use and a minimum of cost. This necessitates an interworking structure, in which an unbundling set of services, domains and networks is established. Furthermore, open systems conforming to an appropriate interworking architecture are needed to unleash as many services and applications as possible. For this purpose, the federation of domain-specific services needs to be investigated, and the relevant issues include, for example, designing proper interfaces, enabling service mobility (across different access networks, multiple domains and devices), providing appropriate session control mechanisms and solutions for accounting, billing and revenue sharing. Service mobility support for a user moving between administrative domains could be achieved by direct interaction between both administrative domains, or by a trusted third party when direct interaction is not possible because of technical or other requirements from either domain.

4.2.6.14 Network Management

Description
There is great variety amongst current networks, and the variation will probably increase in the future. A network can be global, regional or local. The network type can be a backbone, LAN, cellular wireless or fixed. An operator can be an access provider, backbone provider or service provider. For instance, an IP-based radio access network is large (several thousand or more routers), geographically widespread, centrally managed (from at most a few management centres) and management is often in-band (no dedicated management network) to reduce cost. Because of the lack of homogeneity, the current network management is complex and tedious.

Future network management architecture will include mechanisms and entities that allow network operators to configure, monitor and manage the operation of individual entities, as well as the entire system. Some of these functions will be automated and potentially distributed (e.g., autoconfiguration, path restoration), others may be made available from appropriate management tools and interfaces.

It is probably impossible to have one network management solution that fits every network and every operator. But there is a great need to reduce the complexity and to get more standards into the area. Much of the management that is done manually today should be automated in the future.

Technical Issues and Research Priorities

- *Management architecture.* Since networks are growing and becoming more complex, it is of great interest to network operators to control their operating expenses and, in particular, their costs for network management, while retaining flexibility to adapt as networks are evolving. Instrumental in this is the architecture of the network management system:
  — distributed vs. centralized management;
  — to what extent should the management system be distributed over the network nodes?
- *Protocol-related aspects.* The design of powerful and effective/low-overhead element management protocols is also an open issue. Although SNMP is widely accepted today as an element management standard, there are still a number of issues that mandate further research into the development of such element management protocols.
- *Tools.* Powerful but still user-friendly tools for network management are needed. Tools include capacity planning tools, traffic engineering tools to measure link load, latency, jitter and packet loss.
- *Configuration management:*
  — *Multivendor networks.* In a network with equipment from different vendors, e.g. in a LAN or Metro network, all configuration management is vendor specific. This means that there is one management system or set of methods from each vendor. To reduce complexity, standardized tools or protocols for configuring the network are needed.
- *Complex networks.* One issue is how to configure complex networks. The goal is to treat the network as a distributed system capable of providing services, instead of a collection of individual network elements. This greatly simplifies the handling of complex configurations (e.g. add a new service class to a differentiated services network versus configure queuing and scheduling parameters consistently in 1000 routers). The research topics in this area are: effective modelling of logical functions and services (e.g. to avoid redundancy), algorithms to map the logical configuration to element settings, to compare the current state of the network with the desired configuration, and detection and resolution of inconsistencies.
- *Configuration change analysis.* In the case of central management, it is crucial to keep the connection between the management centres and the managed devices alive. Any failure to do so results in a costly site visit by an engineer. Therefore, all configuration changes must be carefully analysed in order to reveal and handle the possible side effects. Any configuration change that can have some effect on the connectivity must

be the subject of such analysis. This includes interface configurations, IP addressing, routing configurations (static routes as well as dynamic routing protocols).

- *Self-configuration*. In order to ease the management of a network, as much as possible of the configuration should be automatic. One way of doing this is to have plug-and-play enabled nodes. This functionality requires new algorithms and protocols.
- *Performance management.*
  - *Traffic monitoring*. To optimize performance, we should be able to continuously monitor the traffic in the network. How is this best carried out?
  - *Network monitoring and fault management*. How do we best detect and report abnormal network behaviour? Faults should rapidly be detected, isolated and corrected. How should alarms be managed? Could self-healing networks be a reality?
- *Capacity management*. For smooth network operation providing the required throughput, investigation is needed to estimate network capacity in environments with a very large number of users, and also potential local traffic bottlenecks created by them.

4.2.6.15 General Networking Issues

Robustness

As new and more time-critical applications and services are being introduced, new requirements and constraints are introduced on the networks. Users of voice services are used to high availability and reliability of the telecommunication world. As a result, these users will expect at least the same level of availability and reliability from new time-critical services from the internet. IP routing protocols, e.g. OSPF, BGP, etc., are designed to be robust and to ensure that connectivity is automatically restored if a connection is broken. However, these protocols require some time, in the order of seconds or more, to recover from a broken connection. During this time all packets are dropped.

Network robustness is about how to minimize the end user perception of topology changes and network (e.g. link or node) failures. Some of these methods are called *protection switching methods* and *path protection methods*. These methods work at different protocol layers, and they are designed to bypass network errors in different ways. However, they all try to minimize the time it takes for the network to set up a new path or connection to circumvent an erroneous link or node.

Another factor that may affect the robustness of a network environment, such as the cooperative network envisioned in this text, may be the use of policy management for making decisions, for example, on which access network should be selected, which device is the preferred one for a specific application/service, etc. This 'late-binding' functionality may cause the network environment to be vulnerable for policies that are conflicting, which may lead to unspecified behaviour.

Reconfigurability

Network reconfigurability should be researched as a means to provide different applications and services with ubiquitous and sufficient network resources. With a huge number of applications and services, as well as huge numbers of different devices and user requirements, network reconfigurability is a way for the networks to adapt to current requirements and environments. For example, A video conversation requires different resources than other services do.

To allow new and different applications and services to reside on the same terminals, the access and transport network resources could adapt to the currently used traffic pattern, i.e. the access and transport network should be able to be dynamically reconfigured. This will allow new applications and services, as well as new network technologies, to be launched as they are being developed. Networks and devices will automatically download and use new network technologies while these technologies are being launched.

This will impose several requirements on the applications as well as on the networks:

- The application must be able to reserve resources in the network.
- The network should be able to identify which network configuration (including which access technology) will provide sufficient resources for the current application or service in the most efficient way.
- What is considered the most efficient use of network resources might change during a session, and the network should be able to dynamically reconfigure to provide sufficient resources in the most efficient way possible.
- Not only the access technology, but also the entire network architecture, should be reconfigurable.

Investigation is also needed on making the wireless environment aware of, and capable of optimally adapting to, changing environments and network load. Efficient estimation of network load and resource availability might lead to more efficient resource utilization by multiplexing data transmissions over time, frequency and space.

4.2.7 Conclusions

With the vision toward secure seamless networking technology beyond 3G networks, the CoNet ad hoc working group in the WWRF has presented in this section the research challenges and technical issues to be addressed and fulfilled. As one of the primary aspects for capturing significant research scopes and directions, the requirements for future networks in terms of convergence, functionality and capability are also presented. The framework of research components and requirements will constitute fundamental guidance that encourages the emergence of new concepts and ideas for achieving the vision for future networking technologies. We realize that the research framework consists of a great variety of issues and requires a huge amount of effort and collaboration worldwide.

4.3 CoNet Architectural Principles

4.3.1 Introduction

Any complex system design project needs a set of broad architectural guidelines that provide generic guidance when specific design choices have to be made. Designing cooperative networks (CoNets) for the wireless world of the future is indeed a complex system design project involving a huge number of researchers and independent activities. Thus, it is essential that the researchers working on CoNets share common ground (i.e. architectural guidelines) to guide their work. This section documents the guidelines discussed and agreed in WWRF WG3.

The guidelines in this text have been formed according to the best understanding of the contributors and reviewers during the time of writing. However, since a significant amount of CoNet research still lies ahead, the guidance given now may turn out to be obsolete. Thus, the research community should be encouraged to challenge the guidelines given here and propose new ones, if new research results so demand.

4.3.1.1 Objectives

The objective of this text is to document the architectural guidelines and reference model discussed and commonly accepted in WWRF WG3. The guidelines documented in this text shall provide guidance for research projects studying cooperative networks. The reference model is needed to structure the research work.

4.3.1.2 Scope

This text focuses on guidelines and principles concerning cooperative networks. This text aims not to define any network architecture for CoNets; the research projects studying CoNets may define several independent architectures, each of which can comply with the guidelines depicted in this text. However, this text depicts a high-level reference model that can be used to structure the research work and to map functionalities of different architectures.

4.3.2 Architectural Principles/Guidelines

This section describes architectural principles which are commonly recognized concepts and/or basic requirements for studying related technologies and designing future IP network systems. The following architectural principles should be applied when designing cooperative networks.

4.3.2.1 Layered Approach

A layered approach means that the functionality of the system is grouped into distinct layers, which each form logically separate subsystems. Any layered model has at least three layers: an application layer, a connectivity layer and an access layer. The application layer can be further divided into a service application sublayer and a service support sublayer. The connectivity layer can be further divided into a network control sublayer and an (IP) transport sublayer. Furthermore, the access layer may contain several independent access networks, which can also function simultaneously. Therefore, there are five sublayers, the service application sublayer, the service support sublayer, the network control sublayer, the (IP) transport sublayer and the access layer, as depicted in Figure 4.1.

The layers should have well-defined interfaces and they should be functionally independent of each other. The layered approach is required to ensure easy adaptation of heterogeneous access technologies, related technology changes and flexible support for rapid service innovation.

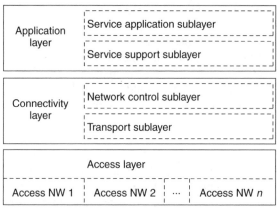

Figure 4.1   Layered approach

4.3.2.2 Independent Functional Blocks, Modularization and Reuse

The CoNet architecture should define each functionality as an independent functional block. This promotes a modular architecture, where different building blocks can be combined as needed to realize complex functionalities. For instance, functionality required to process and route user data, to handle control signalling, and to deal with network management, should be separated into different functional blocks, which can be called user, control and management planes, respectively. It should be noted that each layer of the architecture has its own, separate set of functionality for each plane; hence, nine different functional blocks can be identified, as Figure 4.2 illustrates.

Further, more functional blocks can be identified in the control plane, e.g., mobility management and session management functionalities should be separated. Similarly, the user plane can be divided into smaller blocks, e.g., QoS and security.

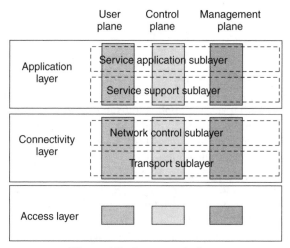

Figure 4.2   Layers and planes

The realization of the functionalities as independent building blocks allows the introduction of advanced functionality, when needed, without changing the whole architecture. For instance, modularized mobility management allows operators to adopt the mobility components they need, e.g., advanced mobility management to support, say, mobile networks and concatenated mobility, could be built, using modular building blocks when and where needed.

The building blocks should be reused, when applicable, in conjunction with different access technologies and realization of different services. The architecture should avoid implementing the same or very similar functionality multiple times. For instance, the access layer should contain clearly identified building blocks that are inherently access technology specific in nature, and building blocks that can be utilized with several access technologies. In a similar manner, the application layer should have common building blocks that can be reused by any service and service-specific building blocks.

4.3.2.3 Multiple Services and Service Creation Support

The CoNet architecture should support a multitude of services, including services that are not yet invented. While the efficient support of high-quality voice is also important for future networks, the network should be designed to be able to handle various types of application and/or media types. The application layer of the CoNet architecture should include application creation environments that meet the requirements of the big and small application developers, and facilitate different business models. It is assumed that the services of the future are based on IP technologies.

4.3.2.4 Cooperative Connectivity

The CoNet architecture should ensure connectivity between all the entities of the network. The connectivity layer of the architecture should provide consistent connectivity across access technologies for any service; this requires consistent device mobility, QoS support, AAA support, etc.

The connectivity layer provides cooperation across various realizations of the network; hence, it is referred to as *cooperative connectivity*. The connectivity layer shall be independent of the various transport technologies used to link the nodes of the network together.

In the CoNet architecture, access technologies should be separated from transport technologies, in order to deploy easily new access or transport technologies. Separation of access technologies from transport technologies makes the access technologies utilized transparent to the common and standardized transport infrastructure.

Currently, the feeling is that the connectivity layer will be based on IPv6, future developments, however, may change this.

4.3.2.5 Heterogeneous Access Accommodation

The CoNet architecture should accommodate heterogeneous, existing and forthcoming, access technologies in a consistent system that hides the technology from the end user and facilitates the most efficient usage of the spectrum resources. The CoNet architecture should both support session continuation (i.e. handover) when the used access technology is changed, and possibly simultaneous usage of access networks. The user should be able to roam globally across different access technologies and administrative domains without any disruption to services or a need to manually change any user parameters.

4.3.2.6 End-to-End Approach

The CoNet architecture should include the endpoints of the communications (i.e. terminals) as part of the communication system. The CoNet architecture should support end-to-end negotiation of QoS parameters, security settings, etc. Furthermore, the CoNet architecture should be able to provide the negotiated QoS, security level, etc., from end-to-end.

The end-to-end approach does not mandate that all the functionality should be located in the endpoints. On the contrary, the functionality can also be provided in a hop-by-hop, and/or edge-to-edge, manner. For instance, QoS service can be implemented differently in different network segments; CoNet architecture must be able to ensure that the QoS provided by each segment of the communication path provides sufficient QoS to satisfy the end-to-end QoS requirements.

The CoNet architecture should ensure that the interoperability of heterogeneous endpoints is maintained in such a way that end-to-end communication is enabled between all endpoint types.

4.3.2.7 Security

Besides protecting end users and networks from each other through authentication and authorization, the components of the CoNet architecture should be protected as well from any intrusive action. Assuming that the CoNet architecture should be as open as possible, to encompass multiple business models where application and service development could be achieved by any party, the architecture should also provide a security framework to enable protection of private information and data. The CoNet architecture should provide accounting capabilities, and probably follow an architectural framework similar to the AAA paradigm. More generally, the architecture will be distributed and will consist of security components able to communicate through well-defined interfaces. Secure mechanisms such as Diameter [4] and Radius would provide the necessary secure communication framework where multiple networks cooperate. Establishing security should fundamentally consist of composing the security functions from the basic components. The thrust of the architectural framework definition reduces to defining the base components and the communications protocols and mechanisms to achieve distributed security. The list of components should include the elements required to achieve intra and inter-domain security.

It would be advisable to rely on already existing security algorithms that are stable, resilient and well established. The creation of new mechanisms and protocols is not in the scope of the architecture definition. Often these new creations have led to very inefficient security mechanisms, such as WEP in wireless LANs for instance. The objective of CoNet is to define the architectural framework, not the mechanisms. Indeed, a component-based architecture can evolve easily through upgrading, even replacement, of components, if open and modular in their original/primary conception.

The architecture itself would contain an independent common framework in the systems part of the CoNet architecture, and a number of private security systems for the networks themselves. The network-dependent part should ideally use the same components and necessarily use the specified interfaces to achieve interdomain security. This may require the design of interworking functions by the networks. The CoNet architecture should not be concerned by the interworking function that remains the exclusive responsibility of the networks. However, the definition and identification of the basic components that would ease (at least reduce the complexity of) the interfacing with legacy and current security architectures and networks is an important goal. These goals should drive the creation of the CoNet security architecture.

Following this reasoning and the fact that segments in the end-to-end path of wireless networks require well-adapted security mechanisms, different algorithms should be used for the air interface, Layer 2 and Layer 3 security. In each layer, only appropriate techniques for the said layer should be used. For the air interface, the methods should be simple and efficient and definitely not as complex as in Layer 3 security.

In the CoNet architectural framework, communications between networks are achieved through reference points and interfaces. The communications, and the exchange of information via these points, have to be secure.

The security architectural framework shall not differ from that of the global CoNet architecture in terms of functionality and building blocks. To this avail, the technology-dependent security building blocks should be clearly identified for legacy and current security technology. The CoNet security architecture should take into account these blocks and ensure the interworking with its own common building blocks that could be potentially reused by any security service and any specific building block.

Security in the CoNet architecture is actually viewed as an ensemble of building blocks that resides in any of the user, control or management planes. This also holds if the architecture is viewed from the application, connectivity and access layer view angle. Security could, hence, be achieved by assembling a number of building blocks to provide the required security functionality. This is the architectural framework that needs to be created and specified.

One objective is to build the trust models and security associations between the various components of the distributed CoNet architecture. This includes the authentication of the components (building blocks) of the architecture itself, and the authentication of the data and software sources when data and software is retrieved for installation, adaptation and reconfiguration. Access control to distributed components and resources, and the communication between the components of the architecture, must be secure. These aspects relate mostly to the architecture itself and the interworking that will be in place within the CoNet architecture. Nonetheless, issues related to the end-to-end connections where the terminal is part of the network and the user is included in the global security process, must be addressed.

Consequently, a number of aspects must be investigated to propose a number of reference architectures for the security infrastructure. Namely, a dynamic security framework should be provided by the global security infrastructure to enable: privacy/confidentiality;

fulfilment of authentication; authorization and access control roles and integrity and nonrepudiation. This requires a focus on: the security framework scalability; usability; reusability and interoperability. This will require the adaptation of existing standards and proposals of new security mechanisms.

Security management must also be accounted for. Currently, security requirements are merely fulfilled in one single domain. A multidomain environment linking, for instance, session establishment, access control and handover control, would lead to multidomain security, thus leading the CoNet architecture to a cooperative security framework (which will be multilayered and multifunctional, just as the CoNet architecture).

The requirement for the security framework and architecture is to map clearly to the cooperative network architecture as a whole. In particular, a direct mapping of the security architecture to the interconnectivity architecture should be achievable. Thus, interconnectivity (interfaces and data exchange via the interfaces) must be secured via the security architecture. This means that security is also provided for all objects or components (also building blocks) of the CoNet architecture.

4.3.2.8 Interworking with Legacy Systems

The CoNet architecture should interconnect with existing networks (e.g. 2G/3G cellular systems and noncellular systems), and support connection of subscribers to private IP networks.

4.3.2.9 NW Control, Operability and Maintenance

The purpose of the NW control and maintenance system is to enable communication for all users and over all network capabilities, but at the same time, to deal with illegitimate usage of the common resources.

The scope of the NW control and maintenance system is internetworking over multiple network types, multiple network operators, multiple access types, and multiple user types. The system also bridges the control over any resource attached.

The motivation for the architectural principle is that, as there are many users and operators who are cooperating over the CoNet, the network control, operability and maintenance needs to be cooperative as well. Borders of responsibility will be overlapping and the need for status information will be unanimous. At the same time, the security aspects have to be regarded and protection against malicious attacks needs to be handled. The organization of the control and maintenance will, as a consequence, become multilayered, multioperational and multifunctional.

Efficient control, which needs to be added to the many heterogeneous networks, is a signalled, automated, secure, real-time management of (re-)configuration, policies, addresses, traffic engineering, accounting and security. Already today there are parts of this NW control available, like AAA with Diameter [4], MPLS with IGMP, or policies for BGP, but there is a need to organize these into a system of systems and to complement them with real-time signalling control. Adaptive services and adaptive network resources will require interactive signalling to enable reconfiguration, and possibly retrieving new functionality over the network.

Privacy protection has to be executed in a cost-sensitive manner. For this purpose, privacy of virtual networks should promote, to a certain extent, sharing of resources from networks:

- It is clear that wasted resources is a concept of the past. Therefore, optimization of resources is required.
- Network control and maintenance should include new efficient routing and survivability algorithms, which can be used in the context of dynamically configurable networks, by taking into account location management schemes, congestion resulting from potential nodes or link failures, quality of service and also security requirements.
- The control layer has an understanding of global networking needs or goals. It manages the forwarding layer so that the goals are reached. In this way it is possible for the control layer to reconfigure the forwarding layer.
- Self-organizing networks may be possible.

The CoNet architecture is expected to comprise a number of networks exhibiting different capabilities in terms of coverage, capacity, transmission rate, transmission delay and transmission cost. Networks might be in a cooperative or competitive relation with one another.

Part of the access domain in the wireless world may not be centrally organized in the future, and may even provide infrastructure-less connectivity. Nodes may come and go, and may be loosely associated with each other, forming alliances whenever and wherever needed. Such nodes, that could be part of the personal, the local, or even the global, sphere would temporarily cooperate to provide connectivity in an ad hoc manner, or would share application resources, such as peer-to-peer networks. Devices incorporated could reach from mobile phones or simple sensors to PDAs and portable PCs. While some results have been achieved on dynamic routing within such loose aggregates, topology forming issues would have to be solved as well: what criteria/metrics should be used to form the most suitable connectivity topologies? How much centralized control should be exercised, and how much self-organization may be allowed? How persistent and powerful need such self-clustering aggregates be, in order to support global connectivity and provide access to general web services, to support multimedia communications or less demanding voice services? Could QoS be supported by self-organizing and infrastructure-less networks? Availability of central control and management, as has traditionally been used for QoS support, would necessarily be relatively low. It is a tempting vision to think of personal area and local area devices that intermingle and participate in external communications only temporarily, while forming a closed world network most of the time.

A network being a member of the CoNet has to build up relations with other networks in order to provide the expected global connectivity, and support the demanded access versatility. The relations between certain networks are supposed not to be statically configured, but dynamically established to meet the requirements of individual traffic flows. The creation of these dynamic relationships is a first aspect of self-organization. Functionalities to consider are appropriate routing mechanisms, QoS provisioning across network boundaries, as well as service and access discovery techniques.

Having drawn a vision of cooperating networks, the same vision holds for the cooperation between network elements to provide multihop connections spanning a number of autonomous network nodes. Both traditional ad hoc networks, as well as sensor networks, are supposed to be integrated in the CoNet architecture by building on the same procedures to organize cooperation between them.

The dynamic organization of relations between networks and individual network elements is supposed to result in a CoNet structure that adapts its topology to meet the demands of varying traffic patterns and transmission. This kind of adaptation should also be adopted within the networks participating in the CoNet structure, to ease its administration and operation, i.e. network nodes should be, to the largest extent, self-configuring, and resources should be distributed among network nodes dynamically to cope with varying traffic volumes and traffic characteristics.

4.3.3 Architectural Framework

An architectural framework supporting the research work on CoNets should aid the structuring of research tasks and identify relevant fields of research. Parallel development within identified fields of research should be possible, without losing the connection to the overall objective of achieving a complete, functioning system implementing the vision of CoNets. The IST project WSI, the initiator of the WWRF, has focused on developing a reference model to ensure that research towards 4G is carried out in the envisaged harmonized and open manner. This reference model is proposed to the WWRF to prove its usefulness in structuring the research work of the research community gathered within the WWRF.

4.3.3.1 The WSI Reference Model

The following sections describe the WSI reference model as of November 2002.

Basic Structure of the WSI Reference Model

The reference model developed covers all aspects of the wireless world from business models and user issues down to radio interfaces. The reference model describes the grand building blocks of the wireless world and how they interact at reference points. The reference model captures user scenarios and different views. The combined definition of business models and reference points enables the early definition of roles and business relationships, as well as assumptions on business topology and market value chains and value networks.

Following the sphere model, which was developed by the WSI 'Think Tank' during 2000, the reference model identifies the building blocks of the wireless future. The main achievement is the definition of a 'communication element' that acts as a communication entity in the different spheres. The wireless world will consist of a huge number of these communication elements. Therefore, the communication elements will have a generalized structure. Depending on their current communication context, they can be logically placed in different spheres (e.g., global, local and personal sphere). The different spheres reflect the vicinity of building blocks with respect to a user communicating in the wireless world. So, the local sphere can be seen as some kind of body area network (BAN), the local sphere serves local networking infrastructure, and the global sphere is responsible for global connectivity.

A communication element can be understood as the representation of a certain device or node in the network, as visualized in Figure 9.3. The functionalities integrated in the communication element are provided by different building blocks. The assumption is that the reference model should separate content processing, control, and management functions into their own end-to-end planes and subsystems. The architecture should not allow a mixture of these three functions in specification. Therefore, a subdivision common to all building blocks of the communication element has been defined.

A communication element will consist of four different building blocks (see Figure 9.2):

1. *Cyberworld.* hosts all types of applications, ranging from machine-to-machine, man-to-machine, peer-to-peer, etc. It does not host, however, application-specific functionalities. It relies on a generic service infrastructure provided by the 'Open service platform' and exploits it to implement applications and services. It will have to dispose of means to generically describe and explain their characteristics and demands to ensure that the underlying infrastructure is being used efficiently and to the user's satisfaction.
2. *Open service platform.* is responsible for providing a flexible and generic service infrastructure to the Cyberworld, in order to facilitate the creation of new services according to users' and operator's needs. The restrictions imposed on the creator of services have to be reduced to a minimum by providing reusable generic service elements.
3. *Interconnectivity.* can also be referred to as the networking part of the wireless world reference model. The functions located there take care of logically linking communication elements from different spheres together, and maintain and manage these links, even when they are subject to change of network topologies or access networks.
4. *Access.* implements all aspects of the physical connection(s) between different communication elements. These may be either radio or other types of connection. Due to the hierarchical structure of the reference model, a connection in higher spheres could use multiple connections in underlying spheres, relying on services provided by the 'interconnectivity' block.

Reference Points
The building blocks that form communication elements are connected by *reference points*. The early identification and specification of these reference points will enable more flexible communication systems than we will have with 3G systems. There are *vertical* and *horizontal* reference points. The vertical reference points are defined interfaces between the building blocks of the communication elements. A connection can also take place between communication elements that reside in different spheres using horizontal reference points.

The reference points (see Figure 9.4) represent well-specified points of contact between the building blocks. This specification will cover so-called generic vertical functions that have to be provided by all reference points. Vertical functions provide certain functionality through all the building blocks by addressing the dedicated problems and technologies of each building block. At the moment, the WSI project has identified nine different functions that have to be provided (see Figure 9.4).

The reference points between the building blocks are crucial elements for the precise technological description of the model. The functionalities that the different blocks have to provide at these reference points will have to be well-defined, complete and generic, in order to assure the proper functioning of the model and to allow treating the building blocks as 'black boxes' from the viewpoint of the adjacent blocks. The reference points can be divided into two categories:

- those that provide an interface between different building blocks in one communication element; and
- those that 'virtually' link, equal building blocks of different CEs, thereby possibly spanning several spheres.

4.3.4 Summary

The WSI Reference Model describes the kind of elements that can be used to set-up future wireless communication systems. It identifies how these elements can interact, what functionality they have to provide and how the elements can be assembled.

4.4 Network Component Technologies for Cooperative Networks

4.4.1 Introduction

4.4.1.1 Objectives and Scope

The objective of this section is to provide several possible solutions for research challenges, such as mobility management and quality of service differentiation, which were identified in Section 4.2 of this chapter. The possible solutions should take into consideration the requirements identified in Section 4.2 and the architectural guidelines identified in Section 4.3 (and [5]). Multiple candidate solutions can be included, related to all the white papers in CoNet. This section also provides the evaluations of the possible solutions, to show how the possible solutions can be effectively quantitatively based on the requirements and architectural guidelines.

It is expected that participants in the WWRF, and/or regional research projects, may refer to this text for further studies, and standardization bodies may consider its contents and the results of further studies for selection of candidate technical solutions.

The scope of this section is the research area identified and documented in previous sections of this chapter. However, it should be noted that selection of a solution is outside the scope of this section.

4.4.2 Solution and Evaluation of Component Technologies

4.4.2.1 Mobility Management

The prioritized requirements for mobility management for an IP-based IMT network platform are:

- mobility;
- resource optimization;
- route optimization;
- location confidentiality.

The technical issues and research priorities are

- to minimize service disruption caused by handovers;
- to achieve low latency handovers and minimum packet losses;
- to reduce power consumption of the mobile terminal;
- to implement architectural guidelines;
- a layered approach.

In this solution there are routing managers and location managers in the network control sublayer, and the managers manage routing information and location information separately. The managers are triggered from the access layer for access initiation of the mobile terminal, handover, location update and the like. The managers can not be directly accessed from mobile terminals. This enables the managers to be protected by the access layer from malicious end users, and to evolve independently of the access layer. In other words, this solution relies on the access layer in access security. The reference [10] proposes a solution for an access layer with multiaccess capability, which triggers the managers.

The routing manager creates, updates and deletes the routing cache in the transport sublayer, i.e. in the routers, for the mobile terminal managed. The location manager manages the MT's location information for paging to the MT. The routers forward packets and perform address exchange procedures for packet routing and rerouting. This enables mobile terminals to communicate at any time without recognition of their locations or movements by the peer terminal. In addition, this mechanism can also realize an efficient utilization of radio frequency, mobile terminal battery power and network resources. Figure 4.3 shows the general architecture of the proposed solution.

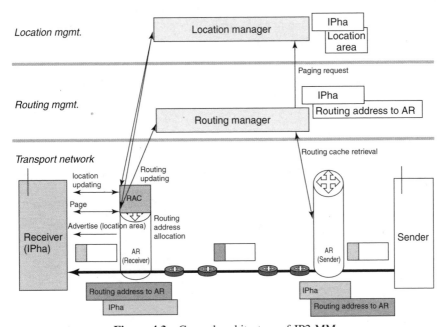

Figure 4.3   General architecture of IP2 MM

The radio access coordinator (RAC) is an entity of the access layer and triggers routing updates at handover, and location updates.

4.4.2.2 Separation of IPha and IPra

The current internet is based on a fixed network technology in which the IP address is used as a terminal identifier, as well as a routing identifier. CoNet and IP2 target the 'all internet mobile environment', in which mobile networks and the internet are converged into one world. In this environment, the IP host address (IPha) of the MT (mobile terminal) should not have to be changed in order to maintain communication, even when the IP routing address (IPra) changes due to its movement.

The IPha is used only to identify an MT, while the routing address is used only to transport packets within networks. Basically, the IPra includes location information and is used only in networks. On the other hand, the MT cannot know the IPra by exchanging the IPha and IPra at the edge router that is deployed between the MT and network, and packets sent to an MT are not encapsulated at any node in the network. Therefore, the proposed routing mechanism can prevent the location information of the MT from being disclosed to the peer terminal. In addition, it can reduce the packet header overhead caused by encapsulation.

Figure 4.4 shows the separation of IPha and IPra. When the access router (AR) of the peer terminal receives the packet for the MT, that is, the destination address of the packet is IPha, it exchanges the destination addresses from IPha to IPra, according to the routing cache that is a mapping of IPha and IPra and is managed by the routing manager. Then the AR forwards the packet to the mobile terminal, according to its routing table. The packet can be routed to the AR of the MT since the AR advertises the route to the network prefix of the IPra. The AR of the MT receives the packet and exchanges the destination addresses according to the routing cache from IPra to IPha. Then the AR forwards the

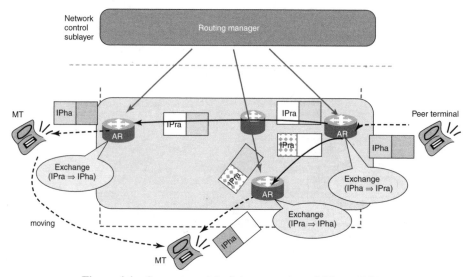

Figure 4.4   Concept model of the separation of IPha and IPra

packet to the MT with the IPha as the destination address. If the AR of the peer terminal does not have the routing cache for the MT's IPha, the AR requests the routing manager of the MT to set the routing cache for address resolution. When the MT moves from an AR to another AR, the routing manager is triggered from the RAC in the access layer, as already mentioned. Then it updates the routing cache in the AR of the peer terminal and creates that in the new AR of the MT. Before the update and creation, the new AR allocates the MT a new IPra and notifies it to the routing manager. The access layer or routing manager triggers the new AR to allocate the IPra.

4.4.2.3 Combination of Routing Management and Location Management

Current cellular systems, such as IMT-2000, have adopted location management for saving the battery power consumption of mobile terminals. This requirement remains in the CoNet and IP2 environment.

Location management consists of location updating and paging. In the proposed solution, both capabilities are implemented by the location manager. As shown in Figure 4.4, the location manager tracks the MT's location information for paging, and pages the MT according to the location information, in order to activate routing management. Paging is requested by the routing manager when it is requested to set the routing cache from a router, and stores no routing information of the MT, i.e. no IPra is stored in the routing manager. The paging is done through the access layer. The RAC in the access layer receives the paging request indicated from the location manager. When the MT receives the paging signal, the paged MT establishes the access link and it causes the trigger from the access layer to the routing manager. The routing manager manages the routing cache in the routers, as mentioned in the previous section.

4.4.2.4 Local Entities and Global Entities

As shown in Figure 4.5, we propose the separation of local and global entities in mobility management. Generally, LRM and LLM control the transport network nodes located in the receiver's local side, and HRM and HLM control the transport network nodes of all other networks.

4.4.2.5 Local and Global Routing Management

A change in the receiver's AR causes a change of routing address and, if the routing cache in the sender's AR contains the routing address allocated by the receiver's AR, the routing cache must be updated. To reduce the number of routing cache updates on the remote side, and to reduce the time for updating of the routing cache in the network, an intermediate router should be an anchor router (ANR) of the receiver.

The ANR is a common intermediate router while the receiver moves within a group of ARs. While the receiver moves within the group, routing cache updating can be localized within the group and the ANR. LRM controls the localized updating without involving the HRM. ANR allocates a routing address for the receiver. Routing cache in the sender's AR does not contain the routing address allocated by the receiver's AR, but that allocated

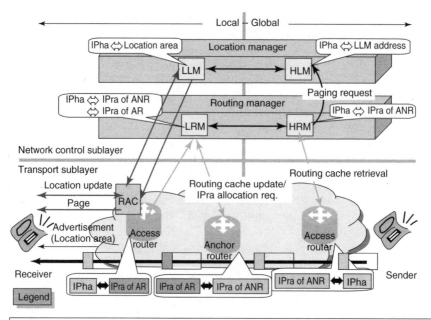

Figure 4.5   Architecture of local and global mobility management

by the ANR. ANR stores a routing cache which consists of a routing address allocated by itself and that allocated by the receiver's AR. It changes the destination address field according to the routing cache.

The AR of the receiver allocates a routing address for the receiver and sends the newly-allocated routing address to the LRM. LRM manages the routing addresses for the receiver, allocated by the ANR and AR. LRM sets the routing cache in the AR and selects an ANR for the receiver. If the selected ANR is the same one previously selected, it updates the routing cache in the ANR with the newly-allocated routing address. If not, LRM requests the selected ANR to allocate a routing address for the receiver and sets the routing cache in the ANR. LRM then sends the routing address newly allocated by the ANR to the HRM. HRM also manages the routing address for the receiver, which is allocated by the ANR, and sets the routing cache in a transport network node on the sender side, i.e. the sender's AR.

Therefore, HRM and the sender's AR are involved in routing cache updating only when the ANR changes. When the ANR does not change, the amount of signalling and time for completion of routing cache updating can be reduced. These conclude the improvement of network performance and quality of communication services, respectively.

4.4.2.6 Local and Global Location Management

Paging requires a multicast mechanism, since an agent for location management requests multiple ARs to page the receiver. The request should not be multiple unicast messages, but one multicast message. If a location management agent is only in the home, multicast routing information for paging must be spread all over the network in order to convey the paging request message from any home agent to any access routers. It is very difficult to set the multicast routing information in all routers in the world. Consequently, it is better that a local agent send the paging request to the access routers in the location area, and routing information for multicast should be localized for easy maintenance. In addition, the amount of signalling can be reduced as well.

HLM manages the LLM address, which manages the location area where the receiver roams. HLM requests the LLM to page the receiver. The trigger of the paging request comes from HRM when it doesn't store the routing address for the receiver. LLM manages the location area where the receiver roams, LLM receives an update notification of location area from the receiver via an AR when the receiver detects the change of location area. LLM sends a paging request to the ARs in the location area.

The request message is a multicast message within a local network. ARs that receive the paging request page the receiver via the access points in the location area. An AR may receive the response and the AR allocates a routing address for the receiver and sends it to the LRM.

When the LLM that manages the receiver's location area changes, i.e., the LLM receives update notification from the receiver that it has not managed, the LLM sends an updating message to the HLM with its address.

4.4.2.7 Routing Management Procedures

The following sections show the proposed procedures for routing management and packet delivery.

Location Management Procedures
The proposed procedures for location management are as follows:

Relationship with Other Solutions
ANR relocation with buffering is a solution for local routing management. It can be combined with this solution.

This solution is for mobility management of a mobile terminal. The moving network in IP2 in section 4.4.2.9 is in line with the general concept of this solution and provides mobility management of a moving network.

Proactive Anchor Router Relocation with Buffering Handover
Prioritized requirements:

- mobility;
- resource optimization;
- route optimization.

Technical issues and research priorities:

- achieving low latency handovers and minimum packet losses;
- implementing architectural guidelines;

- a layered approach;
- independent functional blocks, modularization and reuse.

Proposed Solution
We propose proactive anchor router relocation with buffering handover to reduce handover disruption time and to eliminate the packet loss and the redundant routing path [5, 6]. In this method ANR is located in the network. It terms a region and has packet rerouting and buffering functions. When an MT moves to a new AR within the same region, ANR changes the destination for the MT from a previous AR to a new AR. ANR buffers the packet sent from the CN during handover. ANR is always located on a router, which lies at the junction of the routing path from a CN to the previously connected AR, and that to the new AR. When the MT moves to a new AR, ANR is relocated, if necessary. More detail can be seen in Figures 4.4 and 4.5 based on mobility management for an IP-based IMT network platform.

Anchor Router Relocation with Buffering Handover on IP2 Mobility Management
First, we describe the ANR relocation using Figure 4.6. LRM in the network control sublayer performs routing control in the local network.

We assume that MT#C is connecting to AR2 and communicating with MT#M. MT#C is expected to move to either AR1 or AR3, which are geographically adjacent ARs of AR2, for the next AR. RM selects a router as ANR that lies at the junction of the routes from MT#M to AR1, AR2 and AR3. In the case of Figure 4.6, ANR1 is selected as ANR. In this example, ANR1 is selected while MT#C moves from AR1 to AR4.

Next, we assume that MT#C connects to AR4. In this case, MT#C will move to AR3 or AR5. The ANR1 region is configured with AR1, AR2, AR3 and AR4. AR5 is located on the outside of the region. If ANR1 is used as ANR after MT#C moves to AR5, the

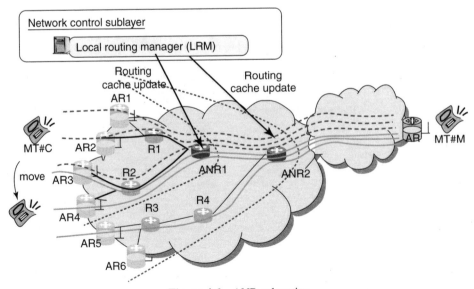

Figure 4.6   ANR relocation

routing path from ANR1 to AR5 becomes redundant. Therefore, RM needs to change the ANR to eliminate the redundant routing path, even if MT#M moves from AR4 to AR3. Before the next handover occurs, RM proactively relocates ANR1 to ANR2, which lies at the junction of the routing path from MT#M to AR3, AR4 and AR5. In our proposal, the packet sent from MT#M to MT#C is always forwarded on the optimal routing path, because ANR is located on the optimal path. As a result, packet misordering is prevented, and the ANR relocation process is independent of the handover process. Therefore, the relocation process does not affect handover disruption time.

Next, we describe the buffering procedure using Figure 4.7. We assume that MT#C is connected to AR2 and communicates with MT#M. Then it will move to AR3. The following gives the outline of the buffering handover procedure.

1. When the signal strength degrades at AR2, the radio access coordinator (RAC) at AR2 informs the RM that MT#M will move to a new AR.
2. When the RM receives this notification, it sends a buffer indication (BuI) to ANR1 to request buffering.
3. If ANR1 receives BuI, it starts to buffer the packet destined for MT#C.
4. When MT moves to the coverage area of AR3, the RAC at AR3 informs the RM that MT#C connects to itself.
5. When the RM receives this notification, it sends a forwarding request (FR) to ANR1.
6. ANR1 forwards the buffered packets to AR3.

This method prevents packet loss during handover. Moreover, this handover method can be expanded as follows. When ANR1 receives BuI, it starts to buffer the packet, but continues to forward to the previously connected AR simultaneously. At this time, ANR assigns the sequential number to the buffered packet. Then, MT#C moves to a new AR, it memorizes the sequence number of the last packet received while it is still connected to AR2. When MT#C connects to the new AR, it notifies ANR1 of the next sequential number of the last received packet. Then, ANR1 forwards the following packet. In this

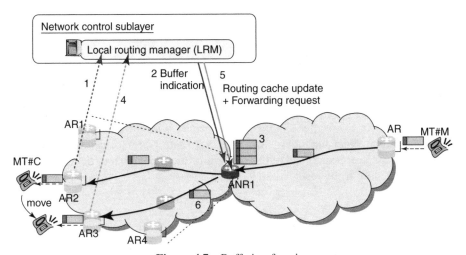

Figure 4.7 Buffering function

expanded method, handover disruption time in the networks can be minimized, because MT#C can receive the packet at the previous AR until just before handover.

Anchor Router Relocation with Buffering Handover on Another Mobility Management System

We described that our procedure works based on the IP2 architecture in this text. But our proposal can work on another architecture, such as Mobile IP. In this architecture, when MT is about to move to the new AR, it sends BuI to ANR. The ANR is notified by the connecting AR in advance. If the ANR receives BuI, it starts to buffer packets destined to the MT. After the MT moves to the new AR, it sends FR to ANR. Then, the ANR forwards the buffered packets to the MT. As described above, our proposal does not depend on the architecture.

Relationship with Other Solutions

Our proposal is a solution to local routing management in IP2 mobility management (see section 4.2). It can also apply to moving networks to reduce packet loss during handover.

Evaluations

Mobility Management for an IP-based IMT Network Platform

- *Efficient use of radio resources:* No overhead of encapsulation over radio interface.
- *Efficient use of wired link resource:* No overhead of encapsulation on links over the network. Route optimization.
- *Location confidentiality:* The peer node cannot know the location of the mobile terminal, since it cannot know the routing address of the mobile terminal. In addition, location confidentiality can be guaranteed with route optimization.
- *Minimization of service disruption caused by handovers:* Local routing management reduces the time for rerouting and loss of packets.
- *Reduction of power consumption of mobile terminal:* Location management makes the mobile terminal dormant when it does not need to be active.

Some Quantitative Evaluation

In IP2, the IP address for identifying the mobile terminal (MT) differs from the address used for routing the packets [60]. They are called the IPha (IP host address) and the IPra (IP routing address). In order to respond dynamically and efficiently to terminal moving and rerouting, IPha and IPra should be separated, the IPha being utilized only to identify the terminal.

The AR has to manage the cache table for address conversion from the IPha to the IPra and vice versa. Also, it executes a signalling procedure with the MT and MM entities, in addition to existing router functions, such as packet routing and forwarding. Therefore, it is assumed that the processing load of the AR will be more than that of the existing router.

The issues needing to be resolved are:

1. Size of the cache table for address conversion.
2. Increase in signalling procedure overload.
3. Buffer overflow caused by packet buffering.
4. Delay in packet transmission.

In this contribution, we evaluate points 1, and 4, to study the feasibility of the AR by using computer simulation and Motorola's network processor (C-5 NP).

Cache Table Size at the AR

The AR maintains two cache tables for address conversion. One is the table to send packet (TSP), which is used to convert the destination address from the IPha to the IPra, and the other is the table to receive packet (TRP), which is used to convert the IPra to the IPha for the packets to the MT.

The amount of cache memory is evaluated by using computer simulation [6]. The MT under the AR is assumed to be in one of the three states dormant, idle or active, with certain probability. When an MT becomes active, it communicates with ten fixed CNs sequentially, and each communication length is 20 seconds on average, obtained by exponential distribution. Also, the recurrence interval of calls is 2.5 seconds on average, obtained by exponential distribution. 100 000 000 CNs are assumed to exist in the network and are randomly selected by MTs. The entry in the TRP of the AR is deleted 240 seconds after the MT stops communicating.

Table 4.1 shows the simulated maximum number of required cache memories for TRP and TSP when the number of MTs under one AR is varied. Here, we assume that one entry of the TRP or the TSP is 32 bytes. In the case of using the Motorola C-5 network processor (C-5 NP) [7] as the AR, an AR could support more than one million MTs, because the C-5 NP can have 64 Mbytes table look-up memory. Therefore, the issue on the amount of memory is irrelevant to the bottleneck of the AR.

Packet Forwarding Capability and Delay of the AR

The AR of the IP2 needs to convert the IPha to the IPra and vice versa. It also sends queries to the mobility management (MM) if the AR does not have the information on the cache table for address conversion. In addition to the address translation function, the routing cache management feature is implemented. In this simulation, it is assumed that the AR already has the mapping information for IPha and IPra for all IP2 packets. That is, the AR does not need to query the IPra.

The pps value result for packets based on IP2 architecture is 1 209 677 pps, only 7% smaller than that for IPv6 packets, that is, 1 304 348 pps. This difference is relatively small, and the result proves that the address conversion process does not impact the performance of packet forwarding in the AR. The packet forwarding delay of the AR is also simulated for both IPv6 packets and the packets based on IP2 architecture, and is shown in Figure 4.8. In summary, the delay for the IP2 packets is a little larger than

Table 4.1 Variation of cache size with number of terminals covered by one AR

	Number of terminals			
	10 000	100 000	1 million	10 million
TRP	17.6 KB	176 KB	1.76 M	17.6 M
TSP	225 KB	2.25 M	22.5 M	225 M

* 1 entry = 32 bytes

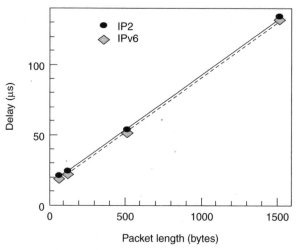

Figure 4.8   Forwarding delay at AR

that for IPv6, but the difference is very small. As for the packet forwarding capability, one could say that the impact of IP2 address translation is relatively small for the packet forwarding delay of the AR.

Proactive Anchor Router Relocation with Buffering Handover
Proactive anchor relocation with buffering handover was compared with HMIP with smooth handover in our experimental system. Figure 4.9 shows the simulated network model. In the experiment, we used UDP traffic to confirm packet loss and misordering.

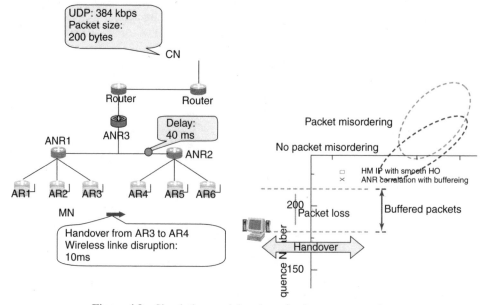

Figure 4.9   Simulation model and received sequence number

The figure shows the sequence number of UDP packets received by the MT. In HMIP with smooth handover, when the MT moves to a new region, it needs to register a new MAP to the CN. Therefore, we can confirm that the packet sent from the CN during handover is lost. And there are two routing paths during handover. One is from the CN to the MT by way of the old ANR, and the other is from the CN to the MT on the optimal route. We can confirm that there is packet misordering after handover.

In our proposal, ANR is always located on the optimal position, before and after handover. The packet sent from the CN to the MT is buffered at ANR during handover and they are forwarded on the optimal path. This confirms that packet loss and packet misordering do not occur in our proposal.

4.4.2.8 Quality of Service Differentiation

Solutions:

- Multiaccess IP QoS.

Prioritized requirements:

- QoS;
- QoS mechanisms should not be tied to any specific link technology;
- open to new applications and new types of QoS requirements;
- support interoperability with other internet services;
- support end-to-end QoS across multiple operators;
- QoS mechanisms should scale to large numbers of links and to large networks;
- implementation of architectural guidelines;
- a layered approach.

Proposed Solution
It is assumed that the end-to-end packet delivery service is provided at the IP layer. The quality of this delivery service determines the end-to-end QoS that is available for the applications, while the end-to-end path may consist of several domains that can be quite different in terms of QoS capabilities. Clearly, in order to provide some form of end-to-end QoS, these heterogeneous QoS mechanisms need to be coordinated. Therefore, from the application's perspective, the key elements of the end-to-end QoS control are related to the following aspects (see also Figure 4.10):

1. The mechanisms and interfaces that allow the application (and application level entities) to specify the QoS requirements. For instance, the application may be associated with a configuration or management tool that allows the specification of default (qualitative or quantitative) QoS parameters assigned to the application. Alternatively, a configuration tool may provide default QoS settings for a group of applications.
2. The QoS requirements need to be transported to each domain along the end-to-end path. Some domains may use this information to configure domain-specific functions and resources, such that the required QoS can be provided within the domain. When a new domain becomes part of the end-to-end connection (due to mobility or, for

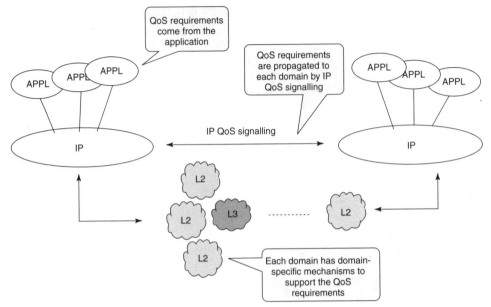

Figure 4.10   A high-level view of a multiaccess multidomain scenario

instance, access selection), QoS-related information may have to be transported to this
new domain.
3. Each domain may have specific mechanisms to support QoS requirements. Examples
   of such mechanisms include per-flow admission control and resource reservation (e.g.
   in a cellular wireless network), scheduling and prioritization mechanisms (e.g. in a
   differentiated services IP domain) or resource overdimensioning (e.g. in a best-effort
   IP network).

The key elements of end-to-end QoS control include the specification of the QoS
requirements, the distribution of these requirements to domains and the domain-specific
QoS mechanisms.

The end-to-end QoS architecture of Figure 4.11 builds on the high level view of
Figure 4.10 and takes into account the above three aspects of end-to-end QoS provisioning. A communication session typically begins with exchanging application and session
layer information. Such information exchange allows the involved parties to negotiate
service capabilities and application-specific attributes, such as encoding formats, version
numbers, etc. The communicating parties may proceed by requesting an appropriate bearer
service from the network. This phase requires that QoS information be distributed to all
domains of the end-to-end path. Specifically, the requirements on the end-to-end QoS
information distribution mechanism include:

- It must be applicable over various QoS technologies.
- It must be clearly separated from the control information being transported.
- It must be independent from the application.

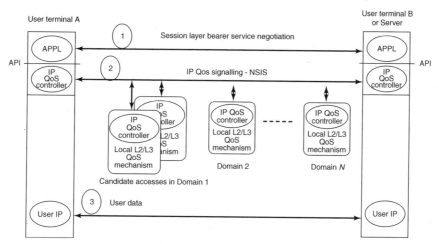

Figure 4.11 Setting up an end-to-end IP QoS bearer by means of a QoS bearer specification mechanism

- It must be addressed independently of the flow identification. That is, the identification of a state associated with a user flow must be independent of the flow identifier itself.
- It must support both uni- and bidirectional reservations.
- It must allow efficient service re-establishment after handover. In particular the distribution mechanism must be able to interwork with seamless mobility and handover protocols.
- It must allow any domain along the end-to-end path to exercise admission control and resource reservation. It must enable a domain to communicate the result of an admission decision to involved parties.

Once the QoS information is distributed along the end-to-end path, each domain may configure its respective mechanisms (if any) to provide the necessary QoS for the user data flows.

4.4.2.9 Moving Networks

Technical issues and research priorities:

- Group mobility
  — moving network;
  — mobility.
- Support of a moving network which consists of several nodes
  — personal area networks (PANs).
- Mobility.
- Resource optimization.
- Route optimization.
- Location confidentiality.
- Minimization of service disruption caused by handovers.
- Achieving low latency handovers and minimum packet losses.

The above five things are the same requirements as for host mobility research. In other words, we should meet them, even if not for moving networks.

The proposed solution follows the architectural guidelines, a layered approach and independent functional blocks, indicated in a previous section of this chapter.

Proposed Solution

To support moving network mobility, some techniques are proposed in the IETF NEMO WG [8]. However, all these proposals are based on Mobile IP, therefore there are several open issues, matching those of Mobile IP from the mobile carrier's point of view, regarding triangulation and route optimization techniques [9]. For instance because of the packet overhead due to encapsulation, we cannot achieve route optimization and location confidentiality at the same time. In addition, these proposals assume that the prefix inside the moving network is stable, even if handover occurs. Therefore, it makes it possible to hide the mobility of a moving network from MTs; however, it causes more packet overheads due to several encapsulations, or by using the routing header option to support route optimization when compared with host mobility.

As an alternative solution four concepts are proposed, i.e. *care of prefix (CoP), aggregate router (AGR), concatenated routing management*, and *hierarchical address management*, to achieve the same quality as that of host mobility. However, location confidentiality and route optimization cannot be kept simultaneously, which is an open issue in Mobile IP. In addition, mobility management for an IP-based IMT network platform (IP2) is proposed as one of the possible approaches for the next generation mobile communications network beyond-3G systems [5] to resolve this open issue.

Care of prefix, according to NEMO's assumption is that the IP packet prefix inside the moving network is constant and stable, and the packet header size gets bigger compared to the host. To resolve this problem, it is necessary to identify the care of address of a mobile terminal in a moving network as well as the MT which can be anywhere in the core network. The care of prefix enables this. More concisely, the CoP changes the prefix inside the moving network every time the moving network moves to another access router, AR, and gets the correct topological address (CoA) of the MT in the moving network. When the moving network connects to AR, the AR assigns a new prefix that which shows that the prefix inside the moving network has changed. Each MT in the moving network uses this prefix to generate the CoA. In this way, the CoA of MTs in the moving network can be identified from anywhere in the core network. In addition, the allocated prefix is original for each moving network, in order to avoid generating duplicate CoAs even when handover occurs. The prefix in the conventional router advertisement signal to MTs not in the moving network and CoPs are different from each other. However, in the CoPs all CoAs of all MTs in the moving network are changed when handover occurs, therefore a number of binding update signals to the home agents (of all MTs and all CNs), in order to realize route optimization, are sent according to the characteristics of the moving network, as shown in Figure 4.12. Aggregate router and concatenated routing management can solve this problem, and are described in the following sections.

The *aggregate router* (AGR) handles the mobility management of a moving network. This is a router like the modified mobile anchor point (MAP) in the hierarchical mobile IPv6 (HMIPv6) [11]. In addition, the AGR aggregates the update signals (U-Plane) to the ARs of all CNs. This reduces the number of binding update signals toward CNs that the

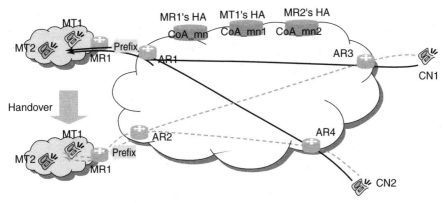

Figure 4.12   Care of prefix

moving network is communicating to. Furthermore, the packets destined for the moving network are always sent via the AGR. Therefore, AGR should be optimally located, considering route optimization for each communication path. Moreover, if the moving network moves too far, we should relocate to a different AGR. In addition, there are trade-off relationships between route optimization, handover delay and AGR relocation frequency. For instance, if the AGR is located near the moving network, it may achieve route optimization and reduced handover delay, but AGR relocation may occur frequently. On the other hand, if located far from the core network, the AGR cannot achieve route optimization and reduced handover delay, but in this case. AGR relocation rarely occurs. Therefore, it is important to study AGR location selection algorithms (Figure 4.13).

In *concatenated routing management* (concatenated HAs) each home agent (HA) only holds information of an MT's whereabouts/attachments (e.g., in mobile router-1), while the

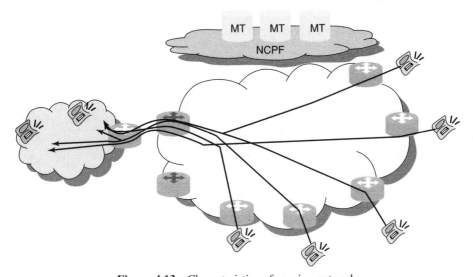

Figure 4.13   Characteristics of moving networks

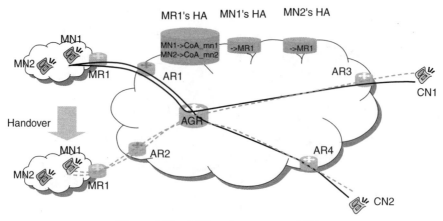

Figure 4.14   AGR and concatenated HAs

MRs mobile router-1 (MR$_1$) HA maintains all the binding information for home address (HoA) and CoA. In this way, we can realize the handover for the moving network only by updating the MR$_1$'s HA (Figure 4.14).

When the moving network moves, it is necessary to inform the AGR and MR$_1$'s HA of the updated CoA, since these have changed. Therefore, the data volume of binding update signals is very large. The proposed *hierarchical address management* concept resolves this issue. Concretely, in this concept, the CoA of each MT in the moving network consists of *shared* information and *individual* information. Shared information indicates the location of the moving network, which changes whenever a handover occurs. On the, other hand, individual information indicates the location of each MT in the moving network of each MT and it does not change, even if handover occurs. The care of prefix enables this address management, because the access router delegates the individual prefix to the moving network, in order to avoid generating duplicate CoAs. In more detail, MR$_1$, AGR and MR$_1$'s HA manage the binding information as mentioned above, thus we can realize a handover of the moving network by updating only the shared (common) information (refer to Figure 4.15).

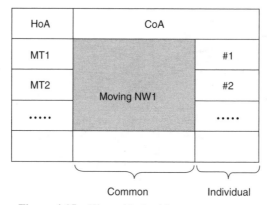

Figure 4.15   Hierarchical address management

Summing up, hierarchical address management, including an aggregate router and concatenated routing management, makes it possible to support frequent handovers.

IP2 Routing for Moving Networks

IP2 resolves open issues in Mobile IP as mentioned above. Therefore, by applying four key technologies for IP2 routing management, as shown in Figure 4.16, we can get a better solution.

The IP2 architecture is as shown in Figure 4.16 and introduces an AGR and concatenated routing management in the network control sublayer [5]. In this proposed architecture, we can categorize two routing mechanisms distinguished by location confidentiality (see figure 4.17). That is, to where we keep location privacy. One case is that the gateway of IP host address (IPha) [5] and IP routing address (IPra) [5] is MR. In this case, MR deals with IPra, which indicates the location of MTs, so we cannot achieve location confidentiality at MR. Therefore it may be better that MR is owned by carriers. Another case is that the gateway of IPha and IPra is AR. In this case, MR does not deal

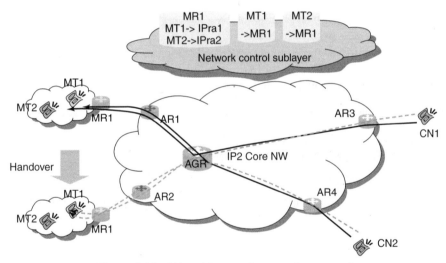

Figure 4.16   IP2 architecture for a moving network

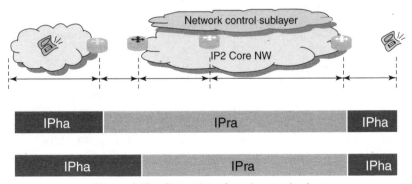

Figure 4.17   Categories of routing mechanism

with IPra so we can achieve location confidentiality at MR and at MTs. Therefore, users can own MR.

Basic Routing Mechanism
This routing mechanism uses three key technologies, i.e. an aggregate router, concatenated routing management and hierarchical address management. The procedure for transmitting a packet from MT#A to MT#B in moving networks is as follows:

1. MT#A sends the packet whose destination IP address is IPha#B to AR#1.
2. When AR#1 receives the packet from MT#A, it asks the network control sublayer about IPra of MT#B.
3. AR#1 exchanges IP address from IPha#B to IPra#1 after it knows the IPra of MT#B.
4. The packet is sent from AR#1 to AR#2 via AGR#1. In this case, AGR#1 exchanges IP address from IPra#1 to IPra#2, as well as the anchor router (ANR) [12].
5. After AR#2 exchanges IP address from IPra#2 to IPha#B, it sends the packet to MT#B via MR#1, by means of IPha host routing, e.g. MANET [13] routing protocols.

When handover occurs,

6. AR#3 assigns IPras to all MTs in the moving network and informs the network control sublayer of the new IPras.
7. The network control sublayer updates MR#1's routing management table and instructs AGR#1 to change the IPras of all MTs in the moving network.
8. After the seventh procedure, we realize rerouting.

To summarize, the important operations in the basic routing mechanism are:

- The address exchange points are AR and AGR.
- The routing from AR to MR is IPha host routing.

Care of Prefix Routing Mechanism
This routing mechanism uses four key technologies, i.e. care of prefix, an aggregate router, concatenated routing management and hierarchical address management.
    First of all,

1. AR#2 delegates prefix to MR#1.
2. MR#1 assigns this prefix to MTs in the moving network as IPras. In this way, we can get a topologically correct network until MR#1. (This is care of prefix).

The procedure for transmitting a packet from MT#A to MT#B in moving networks is as follows:

3. MT#A sends the packet whose destination IP address is IPha#B to AR#1.
4. When AR#1 receives the packet from MT#A, it asks NCPF about the IPra of MT#B.
5. AR#1 exchanges IP address from IPha#B to IPra#1 after it knows the IPra of MT#B.
6. The packet is sent from AR#1 to MR#1 via AGR#1 and AR#2. In this case, AGR#1 exchanges IP address from IPra#1 to IPra#2, as well as ANR. AR#2 resends the packet to MR#1 by means of prefix routing, because it knows that it delegated the prefix of IPra#2 to MR#1 at first.
7. After MR#1 exchanges IP address from IPra#2 to IPha#B, it sends the packet to MT#B.

When handover occurs,

8. AR#3 delegates prefix to MR#1.
9. MR#1 assigns IPras to all MTs in the moving network and informs the network control sublayer of the new IPras.
10. The network control sublayer updates MR#1's routing management table and instructs AGR#1 to change the IPras of all MTs in the moving network.
11. After the tenth procedure, we realize rerouting.

To summarize, the important operations in the CoP routing mechanism are:

- The address exchange points are MR and AGR.
- The routing from AR to MR is IPha host routing.
- Care of prefix gets a topologically correct network until MR, by delegating prefix from AR to MR.

Related Work
The buffering scheme of ANR relocation with buffering is applied for this solution to improve handover performance. IP2 routing seems to be the ideal solution to apply these four key technologies, as shown above.

Evaluations
The key technologies, i.e. CoP, AGR, concatenated routing management and hierarchical address management, realize support for the same quality of host mobility, even in moving networks. In addition for routing mechanisms in IP2, the two routing mechanisms outlined above meet the requirements, that is, resource optimization, route optimization, location confidentiality, and frequent handover support. In more detail, route optimization depends on the location of AGR, because all packets destined for moving networks are via AGR. As to location confidentiality, the basic routing mechanism supports location privacy for MR and MTs, the care of prefix routing mechanism, on the other hand, keeps location privacy only for MTs. This depends on the carriers' policy of location confidentiality. In addition, without considering location confidentiality, the care of prefix routing mechanism is better, because the routing from AR to MR is prefix routing, which is more efficient than the host routing in the basic routing mechanism. Furthermore, when handover occurs, in care of prefix, the number of signals between MR and AR is one round. On the other hand, in the basic routing mechanism, it is the same as the number of MTs in the moving network, because a new AR has to know all the IPhas of all MTs in the moving network.

4.5 References

[1] WWRF CoNet Ad hoc Group, '*CoNet Vision and Roadmap*, Version 1.0,' November 2002.
[2] WWRF CoNet Ad hoc Group, '*CoNet Architectural Principles, Draft Version 0.0.7*,' November 2002.
[3] Ericsson Research, Katholieke Universiteit Leuven Research & Development, Nokia Research, Royal Holloway and Bedford New College, University of London, Siemens AG, Stichting Telematica Instituut, Vodafone Group Services Ltd., '*Security in WWI, Security Research Topics*,' presented at the WWI meeting in Dusseldorf, Germany, September 2002.
[4] DIAMETER Base Protocol, IETF DRAFT, http://search.ietf.org/internet-drafts/draft-ietf-aaa-diameter-11.txt.

[5] Sawada, M., Okagawa, T. and Yabusaki, M., '*Mobility Management for IP-based IMT Network Platform*', WWRF WG3 CoNet interim meeting, June 2003.

[6] Nishida, K., Okagawa, T. and Shinagawa, N., '*Study on Processing Load of AR for IPha/IPra Address Conversion in IP2*,' IEICE General Conference, B-6-226, March, 2003.

[7] http://e-www.motorola.com/webapp/sps/site/taxonomy.jsp?nodeId= 01M994862703126.

[8] NEMO WG, http://www.nal.motlabs.com/nemo/.

[9] Perkins, C. and Johnson, D., '*Route Optimisation in Mobile IP*,' draft-ietf-mobileip-optim-11.txt, work in progress, September 2001.

[10] Yumiba, H., Imai, K. and Yabusaki M., 'IP-based IMT Network Platform,' *IEEE Personal Communication Magazine*, **8**(5), 18–23, 2001.

[11] Soliman, H., Castelluccia, C., El-Malki, K. and Bellier, L., '*Hierarchical Mobile IPv6 Mobility Management (HMIPv6)*,' draft-ietf-mobileip-hmipv6-07.txt, October 2002.

[12] Isobe, S., Iwasaki, A. and Okagawa, T., '*Proactive Anchor Router Relocation with Buffering Handover*,' WWRF CoNet Interim meeting, June 2003.

[13] MANET WG, http://protean.itd.nrl.navy.mil/manet/manet_home.html.

4.6 Credits

The following individuals have contributed to the contents of this chapter: Hong-Yon Lach (Motorola, France); Armardeo Sarma (NEC, Germany); Raimo Vuopionperä (Ericsson, Finland); Toshikane Oda (Ericsson, Japan); Christoph Lindemann (University of Dortmund, Germany); Christos Politis (University of Surrey, UK); Dorgham Sisalem (Fhg-FOKUS, Germany); Gerard Hoekstra and Harold Teunissen (Lucent, The Netherlands); Johan Nielsen (Ericsson, Sweden); Josef Urban (Siemens, Germany); Marko Palola (VTT Electronics, Finland); Masahiro Sawada (NTT DoCoMo, Japan); Osvaldo A. Gonzalez (Motorola, Spain); Peter Schoo (DoCoMo Eurolabs, Germany); Tasos Dagiuklas (University of Aegean, Greece); Wouter Teeuw (Telematica Instituut, The Netherlands); Takashi Koshimizu (NTT DoCoMo, Japan); Sami Uskela (Nokia, Finland); Dionysios Gatzounas (Intracom, Greece); Henning Sanneck (Siemens, Germany); Carl-Gunnar Perntz and Raimo Vuopionperä (Ericsson, Sweden); Djamal Zeghlache and Hossam Afifi (INT, France) and Andreas Schieder (Ericsson, Germany).

5

End-to-End Reconfigurability

Edited by Klaus Moessner (University of Surrey), Didier Bourse (Motorola) and Rahim Tafazolli (University of Surrey)

This chapter aims to define the WWRF SDR reference models, it analyses the overall problem of SDR architectures and identifies the main SDR research themes to be investigated in the next decade. The document follows a top-down approach and the reference model targeted encompasses systems and networks (including core and access network/base stations), the hardware issues in both RF and BB sides, and the data and control/management interfaces between the various building blocks of the reconfigurable environment.

In order to support and complement SDR terminal and network reconfigurability, additional intelligence for network reconfigurability functionality will be required. To this end, since the reconfigurability aspects in, and of, all OSI layers should be adequately addressed, the provision of holistic solutions towards the support of reconfigurable mobile environments implies the introduction of advanced reconfigurability management and control. Furthermore, terminal and network reconfiguration is closely coupled with optimal service and RRM provision. The multitude of available services, with highly diverse requirements from the network and terminal, creates the need for a dynamic and intelligent way of adapting the network and terminal to enable optimal service provision by using adequate RATs. This adaptation encompasses the abstraction of the complexity related to the terminal capabilities.

5.1 Reconfigurable Systems and Reference Models

5.1.1 End-to-End Reconfigurability

In 2G networks, services provided to mobile users are either rigidly integrated in network equipment, or developed with proprietary tools by mobile operators or equipment manufacturers. However, in 3G and beyond, an open marketplace is expected to emerge, where a huge number of diverse services will be developed by independent software vendors (ISVs), including radio software and application and service software, which typically will not be targeted solely to mobile networks. This situation creates the need for more

Technologies for the Wireless Future. Edited by R. Tafazolli
© 2005 John Wiley & Sons, Ltd ISBN: 0-470-01235-8

flexible networks that can be adapted dynamically to the requirements of the multitude of services and RATs software that are provided to them. Thus, network and terminal reconfigurability becomes an issue critical to the successful development of the 3G and beyond market, according to the expectations of the market players that have invested in this technology, as well as the end users. Requirements and architectures for the support of reconfigurability management on the network side are addressed in subsequent sections of this chapter, as well as in [1–3].

Network and terminal reconfiguration to best accommodate different service profiles, may be required at any time during the execution of a service, thus, network reconfigurability is required in order to:

- complement and support terminal reconfiguration;
- support protocol and RAT adaptivity;
- support flexible service discovery and provision;
- support protocol/software download;
- support policy provision, based on terminal, service, user and network profiles;
- support the provision of self-planning networks.

Some examples of the types of reconfiguration action that would be necessary in a mobile network include:

- Quality of Service (QoS) provisioning;
- charging and billing;
- dynamic and secure software download.

Although the capability of a network to be dynamically reconfigured is, by itself, a very powerful tool for terminal reconfiguration support, service adaptation and delivery, the involved parties (users, operators, service providers) can only realize the full potential of reconfigurable systems if such features are accessible. If reconfigurability were only usable by a restricted number of operators or equipment vendor-provided services, the potential impact of the technology would be greatly limited. Opening up the reconfigurability capabilities to service provision platforms and applications is expected to create a dynamic environment, where flexible, personalized services will be accessible by the user at any time and within any environment. To this end, there is a severe need for network reconfiguration functionality to be a part of open, standardized interfaces that provide access to mobile networks.

5.1.1.1 Reconfigurability, Adaptability and Flexible Service Provision

Reconfigurability management and control is essential for the support of network-wide reconfiguration actions (including support for terminal reconfigurability). The basic requirements for reconfigurability management for 3G and beyond include: support for flexible business models with novel dynamic services; dynamic reconfiguration based on profile; service provision requirements; dynamic reconfiguration based on MT/access capabilities; and dynamic reconfiguration based on policy provision (QoS, charging, etc.).

The mechanisms and functionality for reconfigurability management and control depend on various aspects. Some basic aspects that affect reconfiguration management span across all system levels. Based on some of these reconfigurable features, reconfigurability management can be split into: service provision dependent reconfiguration management, mobile terminal capability, profile, policy, and location dependent reconfiguration management.

Reconfiguration management also encompasses triggering and control of reconfiguration actions in various layers on mobile systems and networks. It incorporates management of protocol and software downloading, as well as policy provision, in order to enhance or change mobile terminal and network component characteristics and capabilities [4]. The main processes for implementation of reconfiguration procedures that are directly related to reconfigurability management and control, include the identification of the reconfiguration context, this may be done by spatial scooping of the technological surroundings of the (reconfiguration) requesting entity, and by identifying affected elements in the communication and computing infrastructure. The identification of viable alternative solutions through capability exchange and negotiation procedures under specific policies (e.g. maximization of system features), leads to the decision on a solution and respective implementation by taking into account (e.g. user) preferences, alongside strictly technical considerations (type-checking). Finally, the physical deployment of the identified solution will follow by reserving necessary resources, then downloading, installing and activating the identified solution. This solution may be protocol stack-related components, service-related tasks, etc.

Service and protocol adaptability is an important feature for the implementation of reconfigurable and adaptable environments. Management of adaptability coincides with reconfigurability management. It encompasses protocol, service and system adaptation, based on identified information handling related to change of status in various functional elements (resources, mobile environment, etc.). Based on the context change and identified policy, specific protocol elements and service tasks may be triggered, in order to activate or accomplish the overall desired behaviour. These actions may be supported by wider reconfiguration actions and protocol and software downloading on the network and mobile system. It entails protocol adaptivity in various layers. This can be achieved by the support of new generic approaches in protocol design and implementation, where a basic, or generic, protocol part will be supplemented by downloadable protocol extensions that further specialize the protocol functionality and behaviour.

Protocol download is required to achieve and manage reconfigurability also on the higher layers of a system. It requires mechanisms for the identification of protocol elements to be downloaded, the management of protocol libraries, the versioning and conformance. Finally, caching mechanisms of protocol elements for performance purposes should also be investigated.

An important aspect related to the reconfigurability and adaptability support and management, is the introduction of intelligent service and RAT management platforms (its generic architecture is depicted later in Figure 5.1) that act as mediators between SPs,

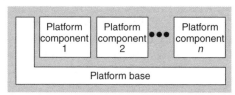

Figure 5.1   An open platform with its base and components

end users and mobile network operators, thus simplifying the extremely complex task of service management and provisioning, as well as reconfiguration actions in various layers in the protocol stack, terminal and network, employing various policies [5, 6]. The platform functionality would include:

- A service and RAT provision and reconfigurability management middleware enabling reconfigurability and adaptability actions in various layers in mobile networks, systems and terminals. These actions could include service-level operations like service deployment and modification of service tariffing policies, as well as lower layer actions like monitoring of traffic flows,
- Generic, reusable mechanisms for adapting applications, as well as service provision procedures to the current context (identified, among others, by terminal capabilities, network characteristics, user locations and status).
- A one-stop shop offering mobile users highly personalized, location and context-aware service discovery and access through a unified, customizable interface.

The generic framework introduced, incorporates aspects related to the combination of flexible service provision and best access and terminal connection selection, terminal and network reconfiguration in various layers, in order to support better overall service provision, and terminal connectivity (taking into consideration the various user/terminal/network and service profiles). Important features are the introduction of open APIs in order to support service registration, discovery and profile provision, and additionally to trigger reconfiguration actions towards the network and terminal in various layers.

5.1.1.2 Reconfigurability Applications

The whole approach towards SDR technology revolves, in principle, around three axes:

1. Increasing software implementations of radio applications on generic hardware platforms.
2. Device reconfiguration.
3. Network reconfiguration management.

SDR functionality can already be found in existing products, mainly BTS equipment. The use of programmable DSPs permits one to change (reconfigure) the equipment functionality implemented in software. However, in the future, the focus will be on how to design the equipment to facilitate this change, make it part of the system operation (dynamic) and, finally, allow the core network to exploit it in meaningful ways. In this spirit, a number of technical issues need to be addressed in order to make radio equipment (terminals, base stations) reconfigurable, and to enable networks to exploit these reconfiguration capabilities and to optimize the networks' performance.

SDR literature survey and market studies identify two popular use-case scenarios. First, the scenario for over-the-air software upgrades, and secondly mode and standard switching [7]. These two scenarios are closely related in terms of both technical requirements and deployment times. Over-the-air software upgrades may occur either when the device is not in use (offline), or when it is (online). Depending on the case, different R/T constraints need to be satisfied. A specific case of software upgrades is the device adaptation to changing environments or requirements by means of algorithm diversity [8]; different algorithms for the same functions may be better adapted to different operational situations. The multimode/multistandard operation scenario also has different manifestations. Switching may be offline or online, and may or may not necessitate prior downloading. It can be user, terminal or network initiated. This depends on the reasons (e.g. coverage, QoS, spectrum availability, etc.) for a mode/standard switching. Finally, instead of switching from one mode to another, it may also be necessary to make the device operate in two different modes at the same time, one for the downlink and one for the uplink.

5.1.2 SDR System and Supporting Network Reference Models and Architectures

5.1.2.1 Reconfigurability in Future Mobile Communication Systems

In this section, the importance of reconfigurability or adaptability for mobile communication systems is highlighted. A very high-level system architecture, focusing on reconfiguration, is presented. Reconfiguration in middleware is also discussed and other technologies that support flexibility and reconfiguration are mentioned, e.g. active networks. This section ends with some requirements derived from this overall view.

5.1.2.2 An Abstract Model for Open Platforms and Reconfigurability

Reconfigurabilty is the main aim of the approaches described in this chapter, adaptability is a function within the reconfigurability concept, it applies to all layers within a system. The abstract architecture model ranges from radio transmission mechanisms, to applications visible to the end user. It is structured in abstraction layers, which encompass transmission, networking, middleware (including service support), as well as applications and services.

An abstraction layer may consist of several open platforms providing the actual computational behaviour. Each such platform consists of a stable and minimal platform base, plus several platform components, which can be added or removed to address different requirements at different times (Figure 5.1).

This approach can be seen as a central concept in future mobile networks, because flexibility is a key requirement for future mobile systems. Our notion of adaptability is broad, in order to cover all system aspects, from reconfigurable radio parameters to applications. Due to this broad scope, different forms of adaptability are considered. The concept of an open platform is only an abstract idealization of reconfigurable entities, either hardware or software.

5.1.2.3 Adaptable Network Element Model

In this section, the overall architecture of network elements is described, with the focus on the networking and transmission layers. More detailed descriptions, including the middleware layers, can be found in [9].

In this architecture, four platforms are considered, each programmable with configurable components. For terminals, typically a more compact subset is selected, due to resource constraints.

- The middleware platform can be structured in the local, virtual execution platform, the distributed processing platform, and the service support platform.

- The computing platform serves as a general-purpose platform for processing state protocols, e.g. routing, QoS signalling or connection management.
- The forwarding engine is in the data path of a network node and it connects network interface platforms, e.g., by a switch matrix. This engine can be implemented as dedicated hardware or as a kernel module of common operating systems. The forwarding engine is programmable for performance-critical tasks, which are performed on a per-packet basis.
- The network interface platforms are medium-specific for different wireless or wired standards. They can be configured or programmed to adapt to new physical layer protocols, or for triggering events in higher layers.

The key technologies are the following:

- Software-defined radio technologies, as discussed in the remainder of this chapter.
- Programmable, active network nodes [9] will enable the fast deployment of new services by using open interfaces to the network resources. Active network technology typically provides an execution environment that is tailored for networking protocols. Depending on the requirements, this can be on the middleware layer or on the networking layer. In both cases, the forwarding engine platform typically provides filtering of packets to be handled in the active network environment. With this technology, new protocols can be installed on the network elements, which use the lower layer resources for creating new services.
- Adaptable and open middleware platforms.

From the operation point of view, the main requirements for the platforms are:

- Reliability for uninterrupted services, in particular when updating services.
- Remote management for the central configuration in large networks.
- Security with respect to attacks from outside. Since we assume that the network is owned by an operator, the main security risks arise from external interfaces.

5.1.2.4 Terminal Reconfiguration Control

Since reconfiguration is ubiquitous in network elements, controlling is a major challenge for the operator. This holds, in particular, for the terminal, where smooth reconfiguration is essential for a positive user experience.

Over-the-air download of software will enable the dynamic provisioning of mobile application services to reconfigurable terminals. Depending on the characteristics of the software download, which includes radio parameters, protocols and software bug fixing, time constraints, traffic management and security considerations have to be addressed.

The motivation of network operators for SDR is to provide the best services to the user, but also to ensure a positive user experience. For this, the software update has to be controlled from the network side, in order to optimize application requirements and network resources. The key requirements from a network operator point of view are:

- Secure and reliable distribution of software.
- Integration with user profiles, applications and billing.

- Scalability, requiring a distributed execution of software updates.
- Support for heterogeneous network infrastructure and terminal diversity.

In order to fulfil the above requirements, a flexible and reliable architecture for SDR control is needed.

5.1.2.5 Network Adaptability/Active Services

Since networks are composed of numerous and various elements, supervision and management is complex. Active networking technology offers a transverse approach to adapt the network from the server to the terminal, and dialogue occurs directly through the active nodes. An active network is a complementary approach which is added to existing networks.

Active networks provide smartness to nodes, which means that providers and users can be differentiated by establishing policy on request at each node: this constitutes a way to ensure service delivery. Some advantages provided by this technology are: reliable multicast protocol (redundancy avoidance), dynamic routing and QoS booking for each node. The advantage for the provider is network adaptability, and the advantage for the user is a content-aware service. For example, configurability can be delegated to the last node to adapt the service, depending a link quality and terminal characteristics and to unload the network and the server.

5.1.2.6 Software Defined Radio Architecture Model

Defining and designing SDR is a challenge and needs knowledge of various skills that cannot be covered by a single person or entity. To maximize the efficiency of the design work, a common background and understanding has to be defined, using an unambiguous language like UML.

In Section 5.3, some design patterns are described, using a formal language well known and used by electricians. At the system level, the same procedure has to be undertaken: definition of concepts, architectural patterns using a formal language to avoid misunderstandings or ambiguity. Even if not written, SDR is based on object-oriented concepts: Radio is an object that can be instantiated in various ways from FM receiver to multichannel, multiband transmitter–receiver.

The various 'models' proposed by the SDR Forum, TRUST, etc., have commonalities and discrepancies. They all refer to the same concept or object, but have been instantiated in different ways. The OMG (Object Management Group, www.omg.org), creator of UML, has developed a method to support the design of systems named Model Driven Architecture, MDA (OMG is also the creator and promoter of CORBA, but there is no direct relationship between CORBA and MDA).

The MDA initiative allows one to define and design a system at a high level, without explaining the partitioning between hardware and software, nor the detailed design. Such a platform independent model (PIM), can be shared by various companies, each company possessing one or multiple platform-dependent models, corresponding to the instantiation of the PIM for a dedicated platform. The OMG leads an initiative to define PIMs and PSMs for SDR components.

5.1.3 High-level Reconfiguration System Model

Being an enabling technology for the communication systems beyond 3G, reconfigurability benefits both network operators and users. Whereas the former can maximize radio network planning, as well as rapidly incorporating new enhancements and improvements, such as personalized and customized services, the latter's flexibility to select services 'on demand', with equitable QoS and perceived value for money, is also enhanced, alongside the provision of worldwide roaming capabilities.

End-to-end reconfigurability allows for a plethora of heterogeneous environments integration, consequently coping with the diverse availability of access schemes, converging in an all-IP backbone. Subsequently, provision of means for interworking on a common platform, leading to horizontal and vertical handovers, service negotiations and global roaming constitutes a pivotal factor, should the reference model capture the aforementioned requirements.

Accordingly, the definition of a 'reference model for end-to-end reconfigurable mobile communication systems' should be based on both legacy and future proof infrastructure, modelling radio access, transport and service technologies, as they relate to reconfiguration, and separating them from each other. Hereafter, the presented proposal aims to compile various existing architectural approaches, so as to design the basic structure required to support end-to-end reconfigurable systems. The focus is centred on identifying necessary (re)configuration controlling management instances, in both terminal and infrastructure (network), from structural and logical points of view.

First, we denote 3G cellular systems as a reference model for the complexity they involve in contrast to other plainer environments, such as satellite and broadcast systems (i.e. DVB, DAB), wireless LAN (i.e. HiperLAN2, IEEE 802.11) and wireless personal area networks (i.e. Bluetooth); even though the latter must also be considered to develop a seamless mobile network like the internet, the necessary evolution and enhancements to bring reconfigurability to cellular systems, constitute the grounds applicable to the rest. This chapter shows, at a basic level, physical and functional separations to allow a network supporting reconfiguration capabilities to fit within the context of the IMT-2000 family of networks. It does not attempt to develop aspects of 3G networks that are highly specific to that implementation.

Conceptualization of the system depends on dividing the network architecture into subsystems, based on the nature of traffic (i.e. packet and circuit switched domains), protocols (i.e. access and nonaccess stratums, control and data) and physical elements (i.e. access and core network structure). Accordingly, Figure 5.2 builds on these bases to present a perspective of (re)-configuration management (CM) instances, essential protocols and structural elements.

On the bottom of the figure, the structural approach illustrates a relationship of 3G network elements of diverse domains involved in an end-to-end reconfiguration process. The functionalities implemented on these elements present a framework whose capabilities would certainly be enhanced by reconfiguration services. Whether this should constitute a different network element (i.e. server, proxy) or, on the other hand, be embedded into the same elements, is beyond the scope of this chapter. Hence, instead of identifying the physical elements that incorporate CM instances, these are only depicted at the system level they operate. One way or another, functionalities and protocols should be likewise standardized, leaving an open choice of implementation.

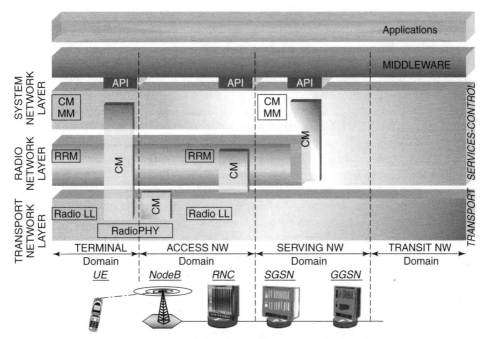

Figure 5.2 High-level reconfiguration system model

The top of the figure presents the logical architecture by separating the protocol structure into three layered subsystems, namely transport, radio network and system network layers. The protocols within each of the layers operate across multiple interfaces and belong together in the sense of interworking. This refers to protocols of the same layer which extend across multiple network elements. Thereby, they comprise a set of protocols that contribute to distributed execution of a common systemwide function, as discussed in more detail in [7].

CM functionality extends across all layers of communication systems. The range of CM varies depending on the protocols and functionalities implemented in each element. Furthermore, CM instances are located in various domains for the functionalities required during the reconfiguration process, and expand across several network elements. The terminal and the base station, or node B, comprise the elements performing the most dramatic processes, such as hardware and physical layer reconfiguration. Accordingly, these two elements need CM instances at those levels. Note that CM reaches partially to the radio network layer, for a minimal set of radio resource management processes are located over there, and therefore, are subject to reconfiguration too.

It is also worth stressing the importance of having a CM element at the radio network domain. The reason is twofold. First, for the advantages it brings [10] shows that using inter-radio network signalling, easily achievable in an all-IP environment, offers implementation and capacity benefits compared to conventional dynamic channel assignment (DCA) algorithms. Thus, a CM entity supporting the system at this level can considerably improve the overall performance by triggering terminal and base station reconfigurations, as well as performing a crucial role during radio access bearer establishment during vertical handovers. The bottom line is a better switching process for the user, with enhanced throughput and minimized delay which, in turn, means optimization of the reconfiguration QoS.

Secondly, CM in the radio network domain proves valuable for network operators by making it feasible to better administrate the extremely expensive and scarce spectrum resource, leading to radio network planning optimization. The proposals for dynamic spectrum allocation between differing systems are facilitated by flexible system selection schemes, as proved in [11]. Concisely, CM instances in the access network domain could interact with elements that perform dynamic spectrum allocation, reporting current and target load conditions, depending on the implemented reconfiguration capabilities.

Finally, CM has to interwork with system network entities, such as communication (connection, session) and mobility management. High-layer interactions are envisaged during procedures for authentication, global roaming and software download. The latter constitutes one of the key research areas; for specific cases (i.e mass terminal upgrades), it tends to rely on multicasting technologies managed in the top subsystem, and thereby within the scope of the presented framework.

Reconfigurable networks will use object oriented software technologies for distributed systems. The scalability and modularity advantages, which enable efficient methods for software provisioning, as well as fast deployment and high flexibility, have been highly disseminated by TINA [12]. Middleware is an open, broadly applicable and platform-neutral technology that uses object-oriented techniques to hide most of the underlying complexity introduced by differing systems.

The benefits of incorporating a glue-layer are straightforward, considering the vast amount of different environments and access schemes and the subsequent proprietary protocols. Provided that adequate APIs are available, middleware copes with migration of objects from one platform to a remote platform, conforming a whole set of interworking functions.

In the following, some architectural approaches will be presented, highlighting the correlation to the high-level reconfiguration system model.

5.1.3.1 Heterogeneous Network Architectures and SDR Network Functions

SDR terminals must be supported by new network entities or functions, and a fast download and reconfiguration management must be provided by networks. Before we introduce these new functions, we will define the interworking between different networks and the relationship to reconfigurable terminals.

In future systems, a certain deployment is required to support the interworking between different radio subnetworks, in particular between 2G, 3G and 4G systems. Generally, we consider subnetworks capable of running different radio access technologies (RATs). Currently, different scenarios towards the interworking between different radio subnetworks exist, or are under investigation.

If the interworking between different access technologies is nowadays controlled by the core network (loose and tight coupling cases), information is spread over several network elements, i.e. increased delay for intersystem handover due to the high number of involved network elements. Therefore, a need for evolutionary solutions for interworking exists, in order to maximize the benefit of the coexistence of different radio access networks.

5.1.3.2 Management and Location of Profile Data for Mode Switching

In order to support fast reconfiguration or intersystem handover, and the related mode negotiation procedure among heterogeneous networks, the cooperation between terminal and network should be highlighted. One important key of the cooperation between the network and the terminal is the information (user/terminal/service/network profile) needed by the mode negotiation and radio resource management for intersystem handover, or any other adaptations. This information or profile data describe the capability of terminals, service, network and user preferences, and present an important decisive element in adaptation (intersystem handover) decisions.

Two key features should be considered for mode switching (intersystem handover):

1. A database for system deployment supporting intersystem handover between coupled RAT subnetworks. The database can be organized as a table for storing network topology.
2. Management and storing of terminal (user/service) profile data in RANs and CNs, depending on network interworking (e.g. tight vs. very tight coupling).

5.1.4 SDR System Architectures

A suitable network architecture based on IP paradigms needs to be further developed, which supports reconfigurable terminals and access networks, manages software download traffic, provides reconfiguration support and facilitates joint radio resource management. The possible architecture could take as a basis the TRUST network-centric architecture [2, 7], presented in Figure 5.3. The network-provided supporting entity serves as a proxy instance for negotiations with other network entities, in particular the serving reconfiguration manager (S-RM) in the access network and the home reconfiguration manager (H-RM) in the core network.

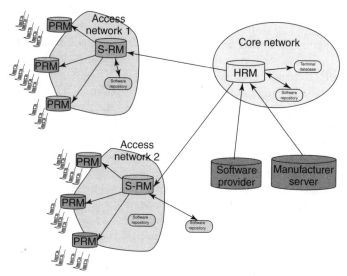

Figure 5.3 Network-centric architecture. Reproduced with permission from [2]

The S-RM is used in each access network and is dimensioned according to the number of users that an access network can support, and what the expected services for the access network are. The H-RM is used to minimize delays, use information on location and status of terminal, and negotiate between terminal and S-RM a strategy for reconfiguration. Both S-RM and H-RM are key elements for upgrade of a large number of terminals, and they relieve currently deployed network entities from several management and security issues that have a critical impact on network performance.

The SDR functional architecture developed in TRUST [2] reflects elements that are needed in the terminal architecture and in the network. The terminal architecture consists of reconfigurable components and terminal resource system related parts, and on the other hand, reconfiguration process control and configuration parts, which are the core of this architecture. Additionally, network implications during the reconfiguration process are also addressed [2]. This functional architecture may be mapped on terminals' and base stations' physical architectures.

The objectives of the reconfiguration architecture are to support a fast and reliable reconfiguration and to use the available resources in an efficient manner. Furthermore, a common functionality is required to avoid a continuous adaptation of the backbone infrastructure in case of the possible heterogeneity of future radio access networks. Therefore, the main responsible components and functions are located in the particular radio access network and additionally on the terminal.

The reconfiguration of a terminal may be initiated by the terminal, but the process can also be triggered by an external entity. In the case of a new hardware driver version, it is inefficient to inform each terminal separately. The use of multicast would help to optimize the content delivery.

We define a terminal reconfiguration serving area (TRSA) as the area of the PRMs served by an SRM. The TRSA encloses all PRMs connected to one SRM. The TRSA is not restricted to one single radio access network or technology, and can therefore be larger and cover several radio access networks (e.g. if the total cover of an area is achieved by different WLANs), or smaller than a single radio access network and cover only a part of the access network (e.g. if a single radio access network covers the whole of a continent).

All the reconfiguration signalling from and to the TRSA leads either across the external connection of the SRM, or across the inter-PRM interface, if neighbouring PRMs with overlapping cells exchange information.

Figure 5.4 shows an example of a terminal reconfiguration serving area. In this area, different radio access technologies are located. There are three hot spot areas with two access points (e.g. IEEE 802.11 or Hiplerlan2) in each area, with lower range but high maximum available bandwidth, and one cellular access point belonging to the same TRSA. The neighbouring PRMs with overlapping cell coverage are coupled with each other by the inter-PRM interface, and every PRM has a connection to the local SRM.

An SDR terminal could perform many different types of reconfiguration. Among others, the firmware associated to different hardware parts might be updated, for example by new DSP algorithms that might improve connection to existing or new access networks. New applications might be downloaded, e.g. updates to new versions of WAP. Also, new codecs for existing applications might appear, such as new voice or image coding, allowing better compression. Malfunctioning terminals might be repaired by means of new software if the problem was provoked by some bug in the old one, without the need of expensive and

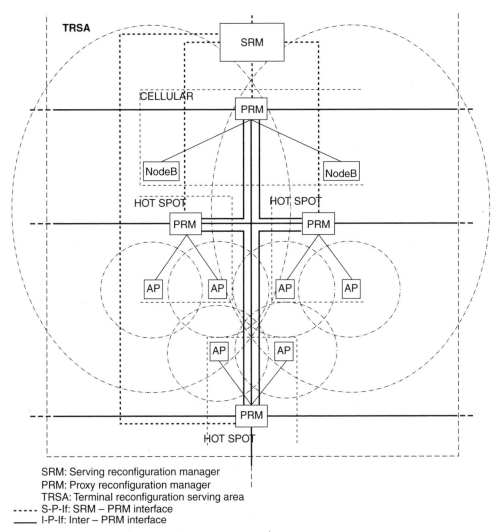

SRM: Serving reconfiguration manager
PRM: Proxy reconfiguration manager
TRSA: Terminal reconfiguration serving area
----- S-P-If: SRM – PRM interface
——— I-P-If: Inter – PRM interface

Figure 5.4 Example of entities and interfaces of a terminal reconfiguration serving area (TRSA) with different radio access technologies

inconvenient shipments. Even new transport protocols, or improvements in existing ones, could be adapted by an SDR terminal after downloading the required files. A software terminal allows the reconfiguration of every layer.

A special mention should be made for massive software downloads. This could be the case, for instance, of fixing an existing bug by means of a software update. Apart from appropriate planning (scheduling it for low traffic times within the network, using hot spots offering capacity for a high number of users), there is the need to minimize the impact in the radio link. Multicast or broadcast can solve, or at least minimize, this problem but they also introduce extra requirements in network functionalities. Support for multicast/broadcast, or adding reliability, are crucial aspects for software downloads, especially for the case of mass upgrades, also, mobility is an important aspect in case the user performs a handover while receiving a download from a multicast/broadcast session.

There are also some reconfigurations specific to vertical handovers, changing to a different radio access technology might imply software download if the RAT has not been used, or the terminal storage capabilities are limited (and the required software needs to be deleted). In addition, the whole, or part of, the algorithms required in order to negotiate with the available networks prior to taking the decision to reconfigure, might be offered for download. For example, the steps required in the algorithm, the measurements the terminal requires or the mapping functions in order to compare parameters of different networks, might be downloaded.

The state of the terminal influences greatly the requirements of the reconfiguration process. For example, the possibility to scan the radio frequencies changes radically, depending on whether the terminal is idle or in connected state. A terminal in idle state does not have strong time constraints for monitoring of alternative radio access technologies or performing software download, but, on the other hand, the necessity of reconfiguration might be quite limited.

A terminal in connected state has more reconfiguration triggers, but also has higher constraints. Since the terminal has a connection running there is not much time left in the terminal to perform monitoring of different available networks, as it can only use transmission interruptions for the frequency testing.

A vertical handover to another access technology might have different time restrictions imposed, among others, by the state of the terminal, or by the cause triggering the handover. The higher the constraints are, the more important it is to have assistance from the network side retrieving as much information as possible. For example, for a faster identification of possible modes, the network could inform the terminal about available radio technologies in the vicinity of the terminal's position. The terminal can now limit the scanning to the corresponding frequencies. Furthermore, the network could provide the terminal with additional helpful information, for example, different measurements from the neighbouring networks, services provided and their general QoS. Also, short-term information (currently occupied resources) could be provided.

If the PRM (proxy reconfiguration manager) is aware of the requirements placed by the current session(s) running at the terminal, and also of some user requirements, preferences and terminal capabilities, it may provide a further selection of RATs fulfilling these requirements, avoiding unnecessary negotiations.

If the required software modules for reconfiguration are not available, the terminal has to request them at the PRM. This might also be a decision parameter if the reconfiguration process is urgent. A mode for which the required software is available at the terminal, or at the PRM, might be preferred above another one for which the software must be downloaded from a server, since this introduces further delay. If the PRM is aware of the neighbouring cells, it could obtain and store the software required for reconfiguration, the terminal could download these modules in periods when no connection is established, or might perform the download in the background with quite a low priority, without impacting the active sessions.

If the download is crucial for reconfiguration, this should take place with high priority, the active connections might get their priority reduced, or even be stopped/cancelled, to allow completing the download in time. This scenario highlights the importance of having the modules cached somewhere close to the terminal, the PRM is seen as the most appropriate entity for storing this kind of software. The SRM (serving reconfiguration manager) might collect less frequently requested downloads as well.

Another point concerning reconfigurability is the general download of application updates or new driver versions where time constraints are, in general, not so strict. The network could schedule the download and execute the process in a moment of low load and therefore use the resources more efficiently. For this, the terminal should provide its reconfiguration requests with a priority indication for the download.

In the case of an externally initiated reconfiguration, e.g. by the manufacturer's server or HRM, the terminal is notified about the available module. If the terminal is in a busy state and currently not able to receive it, the terminal should inform the PRM. Otherwise, if the transmission succeeded completely, the terminal should acknowledge the reception. If errors or packet loss occur during the download process, the terminal could request some retransmissions at the PRM.

5.1.4.1 Proxy Reconfiguration Manager

The PRM is in charge of negotiating and obtaining all kinds of information from the network, in order to minimize interactions on the wireless link, and also to avoid wasting terminal resources in these negotiations and information obtainment processes. The PRM acts on behalf of managed reconfigurable terminals. Here is a list of the functionality required in a PRM:

- information broker for the terminal;
- autonomous service discovery and mode negotiation (required during mode negotiation);
- download management (interworking with other RRM functions);
- terminal classmark awareness;
- records terminal classmark and capability information;
- caches measurements of terminals operating in specific modes (required during mode monitoring);
- caches negotiations of terminals requesting the same bearer services (required during mode negotiation).

5.1.4.2 SRM and HRM Role and Location

The main idea is to have a hierarchical distributed architecture, which minimizes the network load and speeds up the software download. The home reconfiguration manager (HRM) is located in the home network of the terminal and is informed by providers about new software upgrades. In this case, the HRM notifies the availability of new software to the SRM in the radio access networks, and forwards the software to them in case of a mass upgrade. If a request for a software download arrives at the HRM, it is also responsible for the authorization of the terminal in case of a request to download licensed software. Another point considers accounting of software download. For this, the HRM uses a charging repository, which is updated if appropriate software is downloaded.

The SRMs are located between the PRMs and the HRM. One SRM is connected to several PRMs and is responsible for the provision of reconfiguration software to the attached PRMs. Thus, the SRM manages, on the one hand, a large database of software modules for the reconfiguration process and is, on the other hand, able to get non-available software, e.g. from external servers or the HRM. Unlike the PRMs, which aim to reduce the delay and necessary memory, and are caching only a small amount of files, SRMs have access to large software repositories and store a larger amount of files. Furthermore, the SRM could be involved in the control of the mobility, allocation of resources and security of moving terminals. This includes procedures required for vertical handovers, location update and interworking between different radio access technologies, in order to provide the desired QoS.

If, in the case of a mode switch, the necessary mode software for another radio access technology is not already stored on the terminal, the required software modules must be delivered by the PRM. Because of the large number of different radio access schemes and different terminals, the PRM does not store every possible software module. Therefore, this reconfiguration architecture includes a hierarchical reconfiguration management architecture. To speed up the reconfiguration process, every PRM caches the most frequently used modules within its access network. One difficulty in the sped up reconfiguration process occurs if the necessary software modules are neither available at the current PRM, nor at the new PRM. For this, the PRM contacts its serving reconfiguration manager (SRM) and informs it about the appropriate software. Now, the SRM is responsible for the provision of the software and forwards it to the requesting PRM.

5.1.4.3 Reconfigurable Modules and Protocol Stacks

The reconfigurable SDR terminal architectures have been subject to consideration since the middle of the 1990s. The typical implementation of SDR Forum SW and HW implementation for an SDR military platform (SPEAKeasy) is presented in [13]. The relationships between HW and SW in SDR, ensuring open platform characteristics, are presented in [14]. Both concepts advocate and promote the introduction of application programming interfaces (APIs), as well as the inclusion of a virtual machine (VM).

In order to have an interoperable environment for different SDR components, JTRS software communication architecture (SCA) based on CORBA, which is also adopted by the SDR Forum, has been defined [15]. The SCA aims to define a middleware that allows baseband modules, modulation modules and protocols working together over different SDR platforms.

The SDR Forum supports the idea that application portability is an essential concern for software radio, which is why the forum promotes the software radio architecture (SRA). One of the SRA goals is to be able to port a waveform application from one SDR set to another.

Portability of SRA applications is mainly based on two technologies:

1. Standardized layers.
2. A component approach.

The processing environment and the functions performed in the architecture impose differing constraints on the architecture. An application environment profile (AEP) has been defined to support portability of waveforms, scalability of the architecture, and commercial viability:

- An operating system (OS) based on POSIX specifications shall provide, as a minimum, the functions and options designated as mandatory by the AEP.
- A communications middleware layer based on CORBA.
- The core framework (CF) layer of the SRA which specifies the configuration management framework in order to install, deploy, run, stop and dismount applications.

To be SRA conformant, applications are supposed to settle above these three layers. Furthermore, this approach means that portability relies on the component approach (http://www.omg.org/), SRA conformant applications are supposed to be split into separate components whose ports are connected together at installation time.

To achieve better performance and real-time requirements for wireless communication systems, non CORBA-based implementation of the SDR framework is also considered as another solution. Different representations of the reconfigurable SDR terminal architectures coexist [7, 16], and one of the main challenges in SDR equipment design resides in the proper description of HW/SW architecture and related interfaces.

For flexible usage of a description of algorithms, and more flexibility of implementation of the terminal, the functionality description and the hardware realization should be as independent as possible. A complete independence probably cannot be achieved, but this characteristic has to be pushed as far as possible. Each communication standard requires a complex algorithm implementation. With the introduction of flexible and reconfigurable realization, a further dimension of complexity is added. All these complexities have to be handled, which can be by modularization and abstraction. This also implies the detailed definition of module interfaces. All the algorithm descriptions for a reconfigurable SDR terminal can be downloaded and, clearly, the acceptance for terminal reconfiguration will be strongly influenced by the time needed for this.

5.1.4.4 Reconfiguration Security

A communication network is exposed to the following basic threats: messages can be intercepted, modified, delayed, replayed, or new messages can be inserted; a network, and provided resources, can be accessed without authorization, and they can be made unavailable by denial-of-service attacks.

Generic security services defeating these threats are:

- *Confidentiality:* Data is only revealed to the intended audience.
- *Integrity:* Data cannot be modified without being detected.
- *Authentication:* An entity has, in fact, the identity it claims to have.
- *Access control:* Ensures that only authorized actions can be performed.
- *Nonrepudiation:* Protects against entities participating in a communication exchange and later falsely denying that the exchange occurred.
- *Availability:* Ensures that authorized actions can, in fact, take place.

Further security services are anonymity (user identity, tracking), traffic flow confidentiality, or even protection of the fact that communication is taking place.

Reconfiguration will allow one to change, extend and upgrade functionality that has, until now, been fixed. Without suitable protection mechanisms, the flexibility and increased functionality of reconfigurable equipment could be misused in severe ways. It is therefore essential to ensure that reconfigurable equipment can be configured only in safe, authorized ways, where the preferences and expectations from the perspective of all involved stakeholders are respected. A reconfiguration architecture comprises security services and mechanisms that implement the reconfiguration security objectives.

The following objectives specify the reconfiguration security services and mechanisms required:

- Secure download and execution of software is required to prevent malicious activity from intentionally misbehaving, or nonfunctional, software. Both restricted execution environments (sandbox) that control access to system resources, as well as restrictions on the source of software (either the provider or the download server) can be used.
- Reconfiguration control ensures that only legitimate reconfigurations can take place, i.e. that only authorized reconfiguration managers that are trusted to obey the user and network preferences, can initiate and perform a reconfiguration, and only of permitted parts (reconfiguration classes). The signalling traffic has to be protected, and information used for the reconfiguration has to be reliable.
- Access to a user's private information, such as his preferences and the current status of his reconfigurable terminal, and also himself, has to be controlled. It should also be possible for a network operator to keep internal details of his network confidential.
- Reconfiguration allows one to flexibly adapt the radio interface during operation. Regulatory bodies pose requirements on radio equipment concerning user safety, electromagnetic compatibility (EMC immunity) and radio spectrum use (EMC emission). It has to be ensured that a terminal can be reconfigured only in such ways that regulatory requirements are met.

Although security and configuration management mechanisms should prevent the activation of an nonfunctional, or malicious, configuration, fault management procedures could be introduced to detect misbehaving or nonfunctional terminals and perform corrective actions.

5.1.4.5 Secure Software Download and Execution

Dynamic software download is a key technology for reconfiguration. Malicious software could invalidate properties required for type approval or assured in a statement of conformance, but it could also lead to other types of harm. For example, it could circumvent other security mechanisms required for secure network access to a cellular network or a company's intranet, or it could send a user's private data to unauthorized parties, or make the device simply unusable. The device could also be manipulated to operate against the user's interest, for example by calling premium rate services in the background, or by implementing a surveillance function (bug). To prevent harm from potentially malicious software, two basic approaches can be taken:

- *Sandbox method:* In this method, downloaded software runs in a restricted, controlled execution environment (sandbox). Software executing in the sandbox can access only functionality that is considered to be safe; thus, a misbehaving program cannot cause harm. This approach is taken, for example, for Java MIDlets [17].

- *Trust-based method:* In a trust-based approach, only software from trusted providers is accepted. A solution should be based on signed code, where the provider signs a software package using a digital signature (e.g. RSA/PKCS#1 or DSA). Signed code allows the receiver to verify the provider and the integrity of the received software package. The 3GPP 23.057 MExE standard requires that core software may be installed only when it has been authorized by the manufacturer or other authorized party [17]. Especially for open radio platforms, on which third party software can be used, a more flexible solution seems to be required.

These two approaches can be combined. Although the software is executed in a controlled execution environment, it is either accepted only from trusted sources, or it receives increased permission to access restricted functionality when it originates from a trusted provider. For example, the MIDP 2.0 supports trusted applications and privileged domains, so that only trusted applications can access sensitive APIs. Mechanisms are defined to sign and authenticate applications, using an X.509 public-key infrastructure [18].

The download process has to be protected to prevent illegitimate triggering of a software download, removal or modification of required software on the reconfigurable device, or on the download server, and illegitimate access to download servers. Furthermore, the usage, copying and forwarding of software may have to be controlled.

5.1.4.6 Reconfiguration Control

Restrictions on the reconfiguration have to be in place to ensure that only safe configurations are activated, that are not only in-line with regulatory requirements, but that also meet end user and network preferences and expectations. The checks on whether a configuration is safe, can be performed by the reconfigurable terminal itself, or by a trusted support node in the network.

It is necessary to define who is actually allowed to initiate and perform what types of reconfiguration. The manufacturer is expected to be responsible for the implementation of protection mechanisms that ensure the regulatory compliance and protect the terminal from malicious core software. But it is also necessary to control which software is actually downloaded and used, when a reconfiguration takes place and to which mode. These decisions could be under the control of the end user, the network operator, the communication service provider, or an independent service provider. Service providers and network operators want to provide a reliable service that meets the expectations of the end users. The activation of a configuration that either does not work, or that does not provide the services requested by end users, should be prevented, to ensure a high user satisfaction and to minimize the need for expensive customer care.

The signalling traffic between the nodes involved in the reconfiguration has to be protected, for example using IPsec or TLS. Furthermore, policy information is required that defines the correct, trustworthy entities and their permissions. Both read and write access to information used for the reconfiguration as profiles, preferences, and information on the current status and configuration, has to be protected. Such information may be stored and accessed only by trustworthy entities, and illegitimate modifications have to be prevented.

5.1.5 SDR System and Network Research Challenges

To support the development of software defined and reconfigurable radio and networks, the research on SDR system and software architectures will have to address the following main areas:

- system, network and protocol architectures supporting reconfigurable equipment;
- network-centric reconfiguration support;
- terminal-centric reconfiguration support;
- an immediate impact is expected on:
  — efficient use of the existing network resources;
  — enhancing the capabilities for adaptable wireless service provision;
  — adaptive and customizable QoS provision by using 'the right network for the required service';
  — global access, roaming and service availability.

The three main research areas include a number of technical, regulatory and business issues to be addressed. On the technical side these are:

1. *System, network and protocol architectures for reconfiguration management:* Based on IP-based RAN and CN architectures and the architectures already defined in TRUST/SCOUT and Mobile VCE Core Project, the requirements for a network infrastructure supporting terminal reconfiguration should be assessed. Proxy reconfiguration functions need to be elaborated in the network and, dependent on the network infrastructure, different approaches need to be developed and evaluated. Distributed architectures for such an infrastructure, as well as particular additional components required in the network, should be developed. The infrastructure has to provide support for several functions, in particular, for enabling distributed profile management, mode negotiation and mode switching decision making, efficient strategies for user-centric mode selection, based on context information, and ensuring QoS of software download. Furthermore, new efficient handover procedures for software download need to be defined. Download strategies for fulfilling QoS demands need to be investigated (e.g. in mass upgrade scenarios). A protocol architecture should be developed, which allows efficient cooperation (coupling) between different RATs. Protocol improvements on the link layer especially are to be investigated. The protocol architecture should allow reconfigurable protocol layers to insert, remove and suspend their services and to interwork with other protocol layers. The need for a framework for reconfigurable protocol stacks which addresses all layers is needed. Supporting security architecture and functions are needed to allow for secure downloads and reconfiguration.

2. *Network-centric support reconfiguration:* The terminal reconfiguration can be at the initiative either of the terminal or the network. A local intelligence on the terminal must interact with the network, other terminals, or new software on another terminal. Procedures and signalling for reconfiguration tasks between communicating entities must be defined. There is also a need for specification of a minimum set of signalling for the purpose of controlling reconfiguration procedures, namely, request, control and management processes. The research should also include the following points:

- analysis of service discovery in support of the reconfiguration process, and implication of wireless middleware and agent technology from the terminal point of view;
- automatic modulation and RAT recognition in dynamic spectrum environments to facilitate spectrum on demand functionalities;
- development of standardized and generic protocols for vertical handover negotiations;
- identification of requirements and constraints on wireless middleware;
- definition of wireless middleware services as transaction management or interaction management.

3. *Terminal-centric support reconfiguration:* The terminal reconfiguration manager, as part of the local intelligence on the terminal, takes care of the successful reconfiguration tasks. The research will have to focus on the following technical issues: management of mapping of software on the hardware platform; management of lock-up, crash, reset and recovery mechanisms; management of run-time insertion, removal, and configuration of software and the modules running on the SDR hardware platform through the operating system and hardware abstraction layer (HAL) decorrelating the HW platform from application software; definition of the open APIs necessary for reconfiguration processes and development of an object-oriented approach for platform design and realization.

Global circulation and issues like standard compliance testing of software reconfigurable terminals are the initial topics that need to be addressed by the standardization bodies, with potential interaction with regulators. Furthermore, the provision of a global connectivity channel for software radios and its implications could be investigated and, finally, research areas such as usage of competitors' network infrastructure, unified billing and global roaming support have to be explored.

Cross-network integration and global software radio connectivity and support will have various beneficiaries, the results of this research, the implementation and application of these technologies will influence:

- global mobility and connectivity to wireless communication systems;
- service-specific choice of networks;
- flexible application execution platform and network adaptation;
- global regulation and standardization;

A transverse challenge is to define a generic model of software-defined radio, the result of this research will:

- create the necessary background to exchange without misunderstanding;
- structure the collaborative work based on a common basis or model;
- allow component intercommunication;
- allow the definition of a framework: configurability, reconfigurability, control and management.

The reconfigurability support and domains are synthesized in Figure 5.5.

It is inevitable for potential research into the system and software architecture for software defined radio systems to include complete access networks into the reconfiguration processes, and also to introduce reconfiguration management techniques that cover

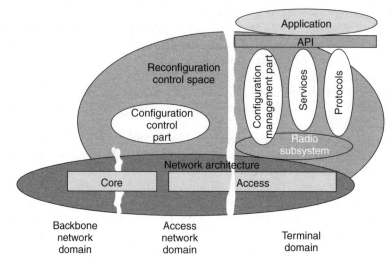

Figure 5.5   Reconfiguration support in reconfigurable access networks and terminals

the complete reconfigurable communication system. Figure 5.5 illustrates the cooperation and interaction of network and terminal reconfiguration management. The principle distinguishes between network and terminal reconfiguration management parts (i.e. the configuration control part and the configuration management part, respectively) but combines them in a 'reconfiguration control space'.

5.2 RF/IF Architectures for Software-defined Radios

5.2.1 SDR RF/IF Reference Models and Architectures

A key issue in the design of RF transceivers will be the interaction between the analogue front-end and the digital baseband part. There are a number of RF/IF related nonidealities that can be tackled by means of correction loops in the digital domain (DC-offsets, IQ-phase and amplitude imbalance, LO-feedthrough, digital predistortion, etc). To get an optimized solution in terms of costs, power consumption and performance, modern transceiver architectures have to support a tight interworking of the RF front-end, the IF-section (analogue and digital), and the digital correction loops in the baseband part. This tendency is strengthened by the advent of RF-CMOS, which may ultimately lead to a 'true' single chip transceiver. Since the development of mobile communication terminals is mainly cost driven, RF-CMOS is a strong contender as the key semiconductor technology for future RF transceivers.

5.2.2 Receiver Architectures

Much of the work done in the field of SDR research focuses on the software part, and is based on an approach that can be called an ideal SDR architecture (see Figure 5.6). Evidently, this architecture lacks a realistic evaluation of the upcoming technological developments related to the analogue front-end part. If one also takes into consideration

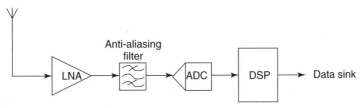

Figure 5.6 Ideal SDR-based receiver

the expected power consumption (one of the key problems of today's mobiles) of such an ideal SDR, it is evident that the direction of research should turn to more 'realistic' front-end architectures. Therefore, feasibility studies of SDR realizations, based on different front-end architectures, should be performed.

The primary distinction between receivers is the number of stages taken to downconvert a signal to baseband. Direct conversion takes one downconversion. Superheterodyne receivers employ two or more. In general, complexity increases with the number of downconversions. There is great interest in direct conversion architecture as a component of an SDR receiver. This interest stems from the fact that direct conversion solves the image signal problem quite neatly. An overview of current state-of-the-art architectures for highly integrated RF-ICs can be found in [19]. Direct conversion architectures are alternatively known as 'zero IF' or 'homodyne'.

5.2.2.1 Homodyne Receiver Architecture

The homodyne receiver structure will bring forward vastly different implementation issues when compared with the heterodyne topology. To understand these issues, suppose that the IF in a heterodyne receiver is reduced to zero. The LO will then translate the centre of the desired channel to 0 Hz, and the channel translated to the negative frequency half-axis becomes the image to the other half of the same channel translated to the positive frequency half-axis. The downconverted signal must be reconstituted by quadrature downconversion (or some other phasing method), otherwise the negative frequency half-channel will fold over and superpose on to the positive frequency half-channel. The simplicity of this structure offers two important advantages over a heterodyne counterpart:

1. The problem of image is circumvented because wIF = 0. As a result, no image filter is required. This may also simplify the LNA design, because there is no need for the LNA to drive a 50 W # load, which is normally necessary when dealing with image rejection filters.
2. The IF SAW filter and subsequent downconversion stages are replaced with low-pass filters and baseband amplifiers that are amenable to monolithic integration. The possibility of changing the bandwidth of the integrated low-pass filters (and thus changing the receiver bandwidth) is a major advantage where multimode and multiband applications are concerned.

On the other hand, the zero-IF receiver topology entails a number of issues that do not exist, or are not as serious, in a heterodyne receiver.

Since, in a homodyne topology, the downconverted band extends to zero frequency, offset voltages can corrupt the signal and, more importantly, saturate the following stages. There are three main modes in which DC-offsets are generated.

1. First, the isolation between the LO port and the inputs of the mixer and the LNA is not infinite. Therefore, a finite amount of feedthrough exists from the LO port to the mixer or the LNA input. This 'LO leakage' arises from capacitive and substrate coupling and, if the LO signal is provided externally, bond wire couplings. This leakage signal is now mixed with the LO signal, thus producing a DC component at the mixer output. This phenomenon is called 'self-mixing'. A similar effect occurs if a large interferer leaks from the LNA or mixer input to the LO port and is multiplied by itself.
2. A time-varying DC offset is also generated if the LO leaks to the antenna and is radiated, and subsequently reflected, from moving objects back to the receiver.
3. Large amplitude modulated signals that are converted to the baseband section via second order distortion of the IQ mixers also lead to time varying DC offset. The spectral shape of this signal contains a significant component at DC, accounting for approximately 50 % of the energy. The rest of the spurious signal extends to two times the signal bandwidth before being downconverted by the second order nonlinearity of the mixers. The cause for the large signal content at DC is that every spectral component of the incident interferer is coherently downconverted with itself to DC. In order to prevent this kind of DC offset, a large second order intercept point (IP2) of the IQ mixer is necessary.

3GPP compliant receiver front-ends need approximately 80 dB gain. The baseband amplifiers contribute most of this gain. That means that even small DC offsets (in the range of several mV) at the mixer outputs may lead to DC levels sufficient to saturate the analogue to digital converters (ADC). In time division multiple access (TDMA), systems' idle time intervals can be used to carry out offset cancellation. This kind of offset cancellation cannot be used for systems with continuous signal reception. Here, the natural solution for DC offset cancellation is high-pass filtering. This approach is especially suited for wideband systems. From a system level point of view, DC-free coding can help to diminish the performance degradation of high-pass filtering.

I/Q mismatches are another critical issue for the zero-IF receiver topology. Fortunately, digital correction loops can tackle the problem of I/Q phase and amplitude mismatches from a data point of view. A reasonable level of I/Q match is, of course, still required.

Research Issues

- DC offset;
- maintenance of I/Q balance;
- maintenance of low second order distortion.

5.2.2.2 Heterodyne Receiver Architecture

Figure 5.7 shows the heterodyne receiver structure. This architecture first translates the signal band down to some intermediate frequency (IF), which is usually much lower than

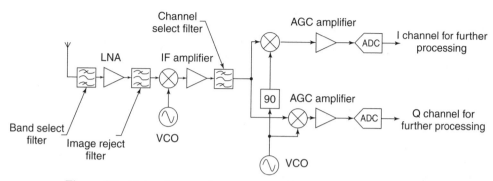

Figure 5.7   Heterodyne receiver structure with two hardware conversions

the initially received frequency band. Channel select filtering is usually done at this IF, which relaxes the requirements of the channel select filter. The choice of the IF is a principal consideration in heterodyne receiver design (see Figure 5.7).

As the first mixer downconverts frequency bands symmetrically located above and below the local oscillator (LO) to the same centre frequency, an image reject filter in front of the mixer is needed. The filter is designed to have a relatively small loss in the desired band and a large attenuation in the image band, two requirements that can be simultaneously met if $2\omega IF$ is sufficiently large. Thus, a large IF relaxes the requirements for the image rejection filter, which is placed in front of the mixer. On the other hand, it complicates the design of the channel selection filter, because of the higher IF. In today's cellular systems, the channel selection filtering is normally done with surface acoustic wave (SAW) filters.

Another interesting situation arises with an interferer at $(\omega wanted + \omega LO)/2$. If this interferer experiences second order distortion, and the LO contains a significant second harmonic, then a component at $|(2.(\omega wanted + \omega LO))/2 - 2\omega LO| = |(\omega wanted - \omega LO)| = \omega IF$ arises. This phenomenon is called the half-IF problem [10]. It is typical of problems that arise if filtering is removed from a radio. Not only do intermodulation problems become exaggerated, but previously unconsidered harmonic distortion problems begin to emerge. Harmonic and intermodulation distortion problems occur when distortion is present in the early stages of RF hardware. Linearity of the LNA and mixer is important with SDR design. Phase noise and LO distortion also contribute to this problem.

A major advantage of the heterodyne receiver structure is its adaptability to many different receiver requirements. That is why it has been the dominant choice in RF systems for many decades. However, the complexity of the structure and the need for a large number of external components (e.g., the IF filter) cause problems if a high level of integration is necessary. This is also the major drawback if costs are concerned. Furthermore, amplification at some high IF can cause high power consumption, particularly if a high degree of linearity is to be maintained.

Due to the fixed receive bandwidth of the heterodyne receiver structure, caused by the external IF filter, the multimode and multiband capability can only be implemented by using a separate IF section for each mode. This would result in high costs and a complex receiver structure.

Research Issues

- image signals – flexible preselect filters – receiver linearity;
- single chip solution;
- channelization filtering – flexible IF bandwidth.

5.2.2.3 Digital IF Receiver Architectures

A multiple conversion receiver is shown in Figure 5.8. Its advantages are: ·

- Good selectivity (due to the presence of preselect and channel filters).
- Gain is distributed over several amplifiers operating in different frequency bands.
- Conversion from a real to a complex signal is done at one fixed frequency, therefore a phase quadrature, amplitude balanced, local oscillator, is only required at a single frequency.

  Its disadvantages are:

- The complexity is high.
- Several local oscillator signals are required.
- Specialized IF filters are required.

Superheterodyne architecture

Although the multiple conversion stage of Figure 5.8 only shows two explicit down-conversions (one in the RF hardware and one in the DSP (digital signal processing)), further conversions can be done in the DSP via the processes of 'subsampling'. If the linearity of the receiver hardware can be maintained to a sufficiently high level, then the channelization can be performed in the DSP. This can mean that the variable bandwidth IF filter is not necessary. One of the significant stumbling blocks to implementing the superheterodyne architecture would then be removed.

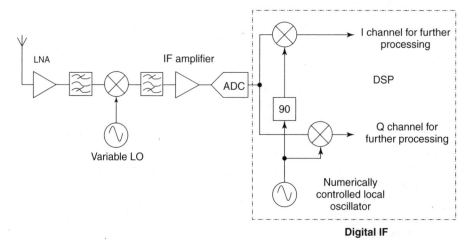

Figure 5.8 Superheterodyne architecture with digital second downconversion

The receiver architecture described above may represent the best choice for an SDR receiver design today, given that the two principal disadvantages of direct conversion are very difficult to get around with current technology. With this architecture, the first conversion may be done in RF hardware, and all of the others in DSP. Two hardware downconversions will probably be required by the radio LAN (HIPERLAN/2 and IEEE 802.11a) which operates in the 5 GHz region. With this architecture, the ADC needs to have an analogue bandwidth equal to the IF frequency, and a sampling speed at least equal to twice the bandwidth of the wanted signal.

If a superheterodyne architecture is chosen, then the problem of image signals must be seriously addressed.

Research Issues

- hardware linearity.

5.2.2.4 Low IF Architecture

Low IF architecture represents an attempt to combine the advantages of a superheterodyne structure, with the advantages of direct conversion architecture (see [21] and [22]). Having a low IF means that the image rejection requirements are not as onerous as with the superheterodyne structure, and the fact that the LO signal is not the same frequency as the wanted signal, minimizes the DC offset problems inherent in the direct conversion architecture.

Its advantages are:

- The DC offset problems associated with the direct conversion architecture can be overcome, whilst retaining most of the benefits of this architecture.
- Lower complexity than the superheterodyne approach (but slightly greater than the direct conversion).

Its disadvantage is:

- Better image rejection is required from a low IF receiver than that required of the direct conversion receiver.

Research Issues

- design of complex image filter – image reject mixer design;
- linearity.

5.2.2.5 Six-Port Technology for Terminal Receivers

A zero-IF receiver based on six-port technology is another possible approach to solving the bandwidth requirements of an SDR implementation. Examples of this technique can be found in [23] and [24]. Compared to conventional Gilbert cell-based mixers, the six-port device has an advantage in power consumption and is inherently broadband. On the other hand, there are problems concerning calibration of six-port devices and commercial products are not yet available.

Research Issues

- calibration of the device;
- commercial production of a six-port mixer.

5.2.3 Transmitter Architectures

In designing the transmitter, like the receiver, there are choices of architecture to be made. Although some of the comments made about certain receiver architectures are directly applicable to the transmitter, there are some differences in emphasis. This section will highlight these differences.

5.2.3.1 Direct Conversion Architecture

If the transmitted carrier frequency is equal to the local oscillator frequency, the architecture is called *direct conversion*. In this case, modulation and upconversion occur in the same circuit. The architecture in Figure 5.9 suffers from an important drawback. Through a mechanism called *injection pulling* or *injection locking* the transmit LO spectrum is corrupted by the power amplifier (PA). The problem worsens if the PA is turned on and off periodically, as is the case for TDD systems.

Problems also arise if the system has to fulfil tight requirements on output power range, which is usually necessary in W-CDMA systems. Most of the gain has to be achieved in the baseband section, leading to high linearity requirements for the baseband filters and the modulator. Furthermore, the LO lies always in the transmit band, which causes high requirements on the LO-RF isolation. I/Q phase mismatches are also an issue when using direct upconversion. Even a small error in the phase shifting network may lead to a severe degradation of the error vector magnitude (EVM). Direct conversion architecture can only be a successful option if its shortcomings (LO feedthrough, I/Q mismatches) are corrected by means of digital correction loops.

Research Issues

- I/Q balance in the local oscillator;
- distortion of the LO signal by the PA;
- baseband linearity.

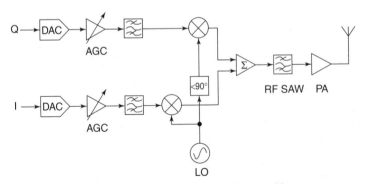

Figure 5.9   Direct conversion transmitter architecture

5.2.3.2 Heterodyne Architecture

The second possibility of signal upconversion, which circumvents the problem of LO pulling in transmitters, is to upconvert the baseband signal in two steps, so that the PA's output spectrum is far from the frequency of the VCO's. An advantage of two-step upconversion over the direct conversion approach is that, since quadrature modulation is performed at lower 'fixed' frequencies, I and Q matching is superior. On the other hand, an IF filter (in most cellular applications, again, a SAW filter) is needed, which can increase costs considerably. If high integration is an important feature, then both heterodyne transmitters and receivers can cause problems. Trying to find intermediate frequencies for the transmit and receive section, that do not lead to spurious frequencies falling, for example, in the receive band, may prove to be impossible. This is especially true if single chip transceivers are to be utilized.

Research Issues

- spurious distortion products;
- IF filter with configurable bandwidth;
- single chip solutions.

5.2.4 The Analogue/Digital Interface

Besides component issues on ADCs and DACs there are two basic architectures of digital transceivers that can be distinguished from their A/D interface.

- *Synchronous A/D interface:* This means that the A/D interface is clocked with a standard-dependent clock rate, i.e., the clock rate is changed for each air interface the transceivers support.
- *Asynchronous A/D interface:* This means that the A/D interface is clocked with a fixed rate.

The sample rate of the digital signal must be converted between chip/symbol/bit rate and the rate of the A/D interface by means of digital sample rate conversion.

Research Issues

- high-quality clock generators for a synchronous A/D interface;
- power-efficient implementation of digital sample rate conversion for an asynchronous A/D interface.

5.2.5 Critical Components

This section will review issues that have arisen out of architectural considerations. Current state-of-the-art solutions to these issues will be examined. Necessary research to extend the usefulness of these solutions to a practical SDR design will be identified.

5.2.5.1 Filter Functions within a Conventional Receiver

In any radio receiver that employs a superheterodyne architecture, filters are required to perform three functions. These are:

1. Band-limit the signal to the frequency of interest. This function is often referred to as 'channelization' and is usually implemented within the digital baseband processing of the receiver.
2. Suppress the image signal with respect to the wanted signal. This function is performed at the first opportunity in the receiver chain.
3. Suppress of out-of-band 'blocker' signals to prevent them generating sufficient 'in-band' power to interfere with the wanted signal.

It should be noted that if the receiver components were perfectly linear, then it would not be possible for out-of-band signals to generate in-band products, and a filter to achieve this function would not be needed.

In practice, some degree of nonlinearity exists in all the amplifiers and mixers that make up the receiver chain. This means that some degree of bandwidth constriction is essential, at an early stage, in the receiver.

5.2.5.2 Device Linearity

A useful measurement of distortion for receiver design purposes is third order intercept (TOI). TOI is the value of input (or output) power at which the third order intermodulation distortion product rises to meet the fundamental output power. The second order intercept point (IP2) has also been shown to be important with zero-IF architecture.

Blocker specifications are an important factor in determining the allowed distortion performance of a receiver. Reference to these specifications for a GSM system shows that if the receiver has a 5 MHz RF bandwidth to accommodate the 5 MHz channel of UMTS, then any blocker within ±2.5 MHz of the centre frequency may introduce distortion products into the wanted GSM channel. Blockers at these frequencies are allowed to be up to −23 dBm. Calculations based on the minimum specified GSM signal and the maximum allowable 'in-band' interference due to the blockers, will show that the minimum input TOI will be +20.5 dBm. This is an onerous linearity specification, and obtaining devices (mixers and amplifiers) with this specification is currently not possible when considering cost and power-consumption issues.

We can infer from this result that the receiver of an SDR should be extremely linear. In general, for best overall distortion performance, the components where the signal is at its highest power levels require the best distortion performance.

A large body of research has been undertaken on linearizing PAs (see, for example, [25]). To enable SDR receiver design, work needs to be undertaken on linearizing receiver components. Warr et al. [26] discuss the linearization of wideband RF receiver amplifiers.

Work on linearizing mixers was undertaken under the IST TRUST programme (see [27] and [28]). A new technique known as *frequency retranslation* has been developed under this project.

5.2.5.3 Mixer Linearization

The inherent nonlinearity of mixers is particularly acute in SDR applications. Here, the broadband receiver front-end 'sees' not only the wanted channel, but also a number of nearby signals. A nonlinear mixer will downconvert all of these received channels, together with the wanted channel, to IF. During this frequency translation process, in-band interference caused by the nearby signals will be added to the wanted channel, making it potentially more difficult, or even impossible, for the receiver to correctly detect the wanted signal. This places demanding filtering requirements on a broadband receiver front-end to reject the out-of-band unwanted channels (blockers) entering the mixer. Filtering out strong interfering nearby channels at high RF frequencies is difficult. In a traditional radio application, the frequency of transmission and reception will be fixed, and the filter parameters will be set only for these known frequencies. This is incompatible with the SDR concept, and filtering out the blockers of multiple standards will be a challenging task for the RF designer, thus making a linear mixer for an SDR front-end an attractive proposition.

5.2.5.4 Image Signals and Variable Preselect Filters

Image signals are a problem that must be dealt with if a superheterodyne receiver is being employed. One possible way of achieving this is to make use of a variable preselect filter. An image reject filter will always operate at RF, therefore, the option for the design of a flexible preselect filter is, at present, limited to realization as either a distributed component design or as an MMIC.

There have been a few variable MMIC filter designs reported in the literature, notably that of Katzin *et al.* [29]. The device described here was produced as a prototype MMIC for the Hittite Corporation. Two versions were produced, both exhibiting ~100 MHz bandwidth. One had a centre frequency that could be swept from 1500 to 2000 MHz, and the other a centre frequency sweepable from 1980 to 2650 MHz. The filter unfortunately never progressed beyond the prototype stage, due to lack of sufficient dynamic range [30].

Nonlinearity is introduced into MMIC designs from two sources. First, it occurs because a varactor is often used to tune the filter. Secondly, a nonlinear active device is often used to compensate losses in MMIC filter components. Distributed component microwave filters might be electronically tuned by the following techniques:

- Varactor diode tuning at some strategic point on the filter structure.
- Constructing the filter on a substrate whose dielectric constant could be electrically varied.
- Switching parts of the transmission line so that the physical characteristics of the filter structure could be changed.

Switching the component parts of a filter, in and out of circuit, using microelectromechanical structures (MEMS) seems to offer an interesting solution [31]. The use of electromechanical switches will mean that the filter is composed entirely of linear components and, therefore, the dynamic range of the filter should not be prejudiced.

5.2.5.5 ADC/DAC Issues

The dynamic range requirement of an A/D converter in a radio receiver is determined by a set of parameters closely related to the given application. These parameters include the required SNR for baseband decoding, peak to average signal ratio of the given modulation scheme, channel noise already present at the ADC input and, most importantly, the residual signal range after the gain and filtering functions preceding the ADC (often the highest residual interfering signal over the weakest desired signal). The bandwidth requirement, on the other hand, depends on the IF frequency, input bandwidth (signal as well as those interferers to be tolerated), as well as a certain amount of oversampling desired by the baseband section of the receiver.

A software-defined radio expects the ADC to cope with the worst-case combination of the above parameters, in order to cope with a wide range of possible applications. Calculations based on 2G and 3G radio standards point to dynamic range requirements in excess of 100 dB, and speed of the order of several hundred mega samples per second.

If converters consuming several watts, or more, power (those perhaps found useful for an SDR radio on a naval vessel or a juggernaut) are to be excluded from commercial portable applications, then suitable ADCs do not exist today for SDR applications. Given that typical radio ADCs have a dynamic range between 60 and 70 dB (10 ∼ 12 bits) today, and the dynamic range improvement in the past has occurred at a rough rate of 4 bits (24 dB) per decade without power constraints, serious and sustained long-term research is absolutely necessary if the SDR dream is to be fulfilled within the next decade.

In the short-run, reconfigurability can be achieved by designing converters that target the most useful combinations of applications. In [32], for example, a very low-power, IF sampled ADC has been described for dual mode GSM/W-CDMA applications. For the combination of radio standards described in the early part of this section, it should also be possible to find a suitable and clever architecture that fulfils all requirements, without excessive power consumption.

The generic question of how to provide the desired bandwidth and resolution on demand, must rely on a substantial advancement of the state of the art. Promising architectures for such improvement include those based on sigma–delta modulation and pipelining, as well as hybrids thereof. Extended use of self-calibration techniques, as well as digital pre/postprocessing techniques, may also be a very useful way forward.

Scaling of semiconductor technology in the next decade will help us improve speed. Downward scaling of allowable supply voltage, however, makes dynamic range issues even more acute than they are today. While signal range is forced down by available headroom in supply voltage, the lower limit of thermal noise, that presents itself in the form of kT/C noise in a sampled-data system, does not scale with the transistor feature size.

Research Issues

- low-power, high-performance ADC architecture;
- trade-off in the definition of RF receiver and A/D converter;
- dynamic reconfiguration of architecture;
- kT/C noise and dynamic range;
- timing jitter in high-IF converters (clock jitter of LO and aperture jitter of the ADC);
- low voltage design;
- device matching;
- intelligent self-calibration schemes;
- digital error correction by means of preprocessing (for the DAC) and postprocessing (for the ADC);
- broadband synthesizer and its impact on phase noise characteristics/system performance.

The LO of any SDR receiver must operate over a wide frequency range. The tuning voltage available at the input terminals to a VCO, on the other hand, will be limited. To cover the wide tuning range would necessitate a large tuning sensitivity for the VCO. The higher sensitivity would result in increased LO spurious components and phase noise:

- Increased LO phase noise and spurious outputs would further increase the attenuation demand on RF filters.
- Increased phase noise would be problematic in broadband systems like 802.11a (OFDM carriers would lose their orthogonality, effectively increasing the noise floor).
- Increased LO phase noise can cause increased baseband noise in zero-IF and low-IF designs. This happens because the decreasing correlation between instantaneous LO phase at LO input, and LO leakage at RF input, will produce this noise. This would also create problems in wideband designs.

5.2.5.6 Digital IF Processing

All digital signal processing that lies in between the baseband processing and the AD interface should be named *digital IF processing*. A particular receiver architecture where the naming fits perfectly is the *digital IF receiver architecture* (see previous section). However, the other receiver architectures, as well as transmitter architectures, require digital signal processing apart from the baseband processing.

Digital IF processing on the receiver side prepares the digitized signal in a way that pure baseband processing can follow. Digital downconversion of the channel of interest to baseband, channel filtering, and sample rate conversion belong to the field of digital IF processing on the receiver side.

Respective tasks can be identified on the transmitter side. The fundamental problem of realizing digital IF processing is the very high sample rates that require very high clock rates of the underlying hardware platform. This results in a high power consumption. Besides this, different air interfaces require different types of digital IF processing that must be performed on a common hardware platform.

Research Issues

- design of a common power-efficient hardware platform for digital IF processing;
- commonality between Tx and Rx schemes.

One important consideration when selecting the architecture for implementing a mobile terminal is to try to exploit common points in receiver and transmitter implementation. This will lead to a reduced complexity, and important cost savings. Receiver and transmitter designs share a number of problems which, if not necessarily the same, are more or less 'duals' of each other. The research in the preprocessing filter in reception, and the postprocessing filter in transmission have common points. Linearization techniques, applied to transmitters and receivers, amplifiers and mixers, will be similar.

Linearization techniques are applied to a receiver to enable the receiver to increase its capacity to handle blocking signals. Linearization techniques are applied to the transmitter to enable the transmitter to reduce the output of spurious emissions. Linearization techniques are also applied to both the transmitter and the receiver to protect the integrity of the modulation.

The architectures for the transmitter and receiver chains of the TRUST Project follow the multiconversion strategies, with one intermediate frequency almost identical. The use of similar IFs will give us the possibility of employing a common design for the IF reconfigurable filters, with all the advantages this will bring.

An important consequence of employing the heterodyne transceiver architecture stems from the fact that the choice of the IF is closely linked to the duplex distance of the system. It is desirable for an SDR to have the choice of independent transmit and receive bands. Based on this requirement, we will also have a variable duplex distance between those bands, which increases the number of spurious responses dramatically. A highly integrated solution relies on a frequency planning concept, which ensures that there are no low-order spurious signals in the receiver band, since isolation between Tx and Rx paths in a highly integrated transceiver is critical. A direct conversion architecture does not suffer from any limitations of the duplex distance between Rx and Tx bands.

5.2.6 SDR Design Choices

5.2.6.1 Active and Shadow Receiver Chains

There is a school of thought that says that if a reconfiguration is to be truly transparent, then there should be two RF chains available in the transceiver. One channel would be active, and the other would be performing housekeeping tasks (such as looking for high-power interfering signals), or reconfiguring prior to transparently taking over the traffic.

Fairly obviously, the alternatives are:

- a single transmit–receive chain;
- an active and shadow transmit–receive chain;
- a hybrid in which some of the elements of a transmit–receive chain are duplicated.

Single Transmit – Receive Chain

This architecture relies on the fact that mobile protocols are designed to cope with random blocking of the information transfer. All current standards can cope with this, as they are designed to operate in a harsh multipath environment. One could make this assumption about future standards.

It is possible, if the user tolerates service breaks of (say) one second within the current generation of services, that a single RF chain architecture could be perceived as being transparent.

Advantages

Complexity and cost could be substantially reduced.

Disadvantages

There is no real possibility of properly monitoring the receiving conditions of other standards. The definition of the target standard would have to be based on some other criteria (maybe it could be defined from the base station).

In real terms the standard availability and functionality information would have to be held at the network level and accessed by the mobile for full transparency with this system.

Active and Shadow Transmit and Receive Chains

The two transmit and receive RF subsystems should be identical in order to guarantee a proper switching of their functions (both should be capable of working as active or shadow units). The two subsystems should also be prepared for implementing the monitoring functions that will define the success of the reconfiguration.

Advantages

- Monitoring the RF environment to:
  — provide cancellation or filtering of signals likely to upset reception of the wanted signal (nearby high power blockers);
  — determine which air interfaces are available to the user.
- Providing 'over the air' downloads whilst the main receiver is providing the link for the user.
- Being set up as a receiver prior to 'mode switching', so that the mode switch is 'transparent' to the user.
- Any failure in the RF chain would not make the terminal useless. The other chain could always act as a reserve unit for those critical cases.

Disadvantages

- The cost and complexity could be prohibitive.
- The power consumption would be high, imposing critical demands on the batteries.

Hybrid Architecture

This alternative resolves some of the complexity overhead of having two RF chains. One part of the receiver and transmitter chain would be designed with the active/shadow philosophy (all the mixers, local oscillators and intermediate frequency stages), whilst other parts would be common (the low noise and power amplifiers). There could be two options for a switching strategy for these single element/dual element combinations. If the receiver were reconfigured in the usual way, we would end up with an architecture whose transparency would be somewhere between that of the single chain and that of a dual chain. The power consumption, complexity and cost would also be in an intermediate position.

The other option, which could certainly imply a different performance, would be based on a wideband permanent operation of the single or common blocks. The linearity of these common LNA and PA blocks would be critical. The low noise amplifier would have to be capable of simultaneously managing two different standards, which would be processed by different receiver units. The power amplifier would also have to do something analogous on transmission.

With this architecture, the terminal would be fully transparent, and the standards could be monitored in order to define the target one. The power consumption, complexity and cost would not be as critical as in the active/shadow pair.

An example of a possible active-shadow chain transceiver is shown in Figure 5.10. The reconfiguration management module (RMM) manages the set-up of the active and shadow chains. This block diagram accommodates the following situations:

- Switching between antennas to optimize the antenna for the frequency band being transmitted/received.

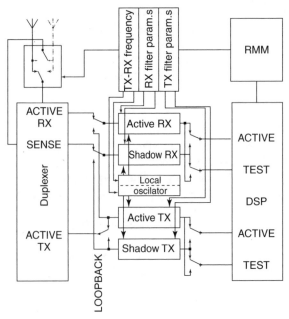

Figure 5.10 Possible block diagram of an active shadow chain transceiver

- Back-to-back testing of the receiver transmitter combination before it is switched live to air. The RMM would need to set up LO frequencies and insert appropriate attenuators to manage this situation.
- The ability to 'test' the ether for high-power interfering signals, or new air interface standards that the user might prefer to use. This facility could utilize the antenna in current use by the active chain, or it could use the second band antenna.

Research Issues

- pragmatic realization of doubling the transmit and receive functionality.

5.2.6.2 Separate Antennas for Individual Systems

The antenna performance is generally optimized for narrowband operation. Increased insertion loss for both transmitter, as well as the receiver, and reduced input filtering are likely to be the price one would have to pay for using a single antenna for multiple systems. Electronic tuning of the antenna is a possibility [33].

Antenna duplexing is complicated for an SDR. This is because the modulation may be either FDD or TDD, so the duplexing must be capable of dealing with signals that are, in the one instance, not separated in time, or in the other, not separated in frequency. Potential duplexing solutions involving cancellation of the unwanted signals are possible.

The required antennas are to be operational over perhaps at least three octaves bandwidth. At present, there are several broadband antennas available to cover these wide bandwidths [33], however, it is not clear if these antennas are suitable for SDR base stations and terminals. For terminals, the required antennas will be small and inexpensive. Although triband antennas [34] are available, they do not necessarily have the required bandwidth, nor are they compatible with the given real estate within a terminal. There is a distinct deficiency in the type of broadband/reconfigurable antennas needed for SDR applications, and further research is required.

Potential Approach

The potential approach should be to consider both the terminal and the base station antennas separately, as, clearly, the sizes of the antennas play a significant role in this area. Initially, a specification for the required antenna will be developed, and possible antenna types will be identified and evaluated against these specifications.

Several types of antenna are envisaged as possible candidates for the SDR applications:

- the conventional type and hybrid antennas [33];
- softboard and semiconductor types [35];
- dielectric resonator antennas [36].

In the first approach, conventional radiators will be appraised to see if they meet the required specifications. These antennas will then be used with some emerging technology, such as MEMS switches, to derive some hybrid form of antenna that offers a potential for meeting the SDR specification for both/either the terminals or the base stations.

The second approach uses semiconductor-compatible material, such as gallium arsenide (GaAs, er approx 10), or softboard (er approx 2), to print various antennas to achieve the required coverage and other specifications.

The third approach uses dielectric resonators as the radiating element. The efficiency of these radiators is extremely high and the size is now much reduced as very high dielectric material is used. These antennas also provide the much-needed bandwidths for the present applications.

It is envisaged that a systematic research approach to the evaluation will lead to antennas which will meet the required SDR specification.

Other novel types of radiating antenna also need to be developed for the SDR applications.

Research Issues

- electronic antenna tuning;
- combined FDD, TDD duplexing;
- research into hybrid antennas;
- dielectric resonator;
- other novel antenna types.

5.2.7 Future Directions

Up until this stage, this section of the text aimed to establish an account of the technological obstacles that impede the development of the RF component of the SDR. Summarizing these obstacles briefly, we can say that they revolve around the issues of how we deal with image signals, and how we deal with high-power interfering signals. The former problem essentially boils down to either development of flexible preselect filters or solving the problems associated with the zero-IF receiver. The latter problem can be resolved by improving RF linearity, or by the development of some form of interference cancellation.

The following is a list of suggested research topics that can help develop the enabling technologies required to solve these problems and to aid the evolution of SDR hardware design. Note that all except one of the topics relates to issues already raised. The diplexerless radio has not been discussed, as the issues it raises are on the borderline of the RF system. It is still an important hardware issue, however, and must be given due consideration. MEMS technology is specifically examined, as it provides the possibility of innovative breakthroughs that established technologies are incapable of providing.

5.2.7.1 MEMS Technology and its Impact on Software Radio Front-end

During the last few years, the sophisticated techniques associated with the manufacture of integrated circuits have begun to be adapted to manufacture mechanical systems on a microscopic scale. Such systems are referred to as MEMS, or microelectromechanical systems. The most obvious electrical application of such a system is the MEMS switch. A typical MEMS switch (see [31]) is envisaged to be usable to electrically alter the line lengths from which a filter is comprised. Because such a switch operates by making physical contact, and not by utilizing a semiconductor device such as a PIN diode or an FET, little or no distortion will be introduced.

There are a number of other MEMS that may be used to provide a filter tuning action, for example, interdigitated capacitors, where the fingers are capable of mechanical movement relative to one another. References [33] and [37] provide an excellent introduction to this exciting development. Other potential MEMS components include low-loss, low-power mixers, and high Q resonators for low phase noise oscillators.

5.2.7.2 Agile Linear Frequency Translation

The aim of this research would be to build what essentially amounts to a linear mixer capable of operating over a wide frequency range.

5.2.7.3 Flexible Linearity Profile (FLP) Amplifiers

The object of this work would be to extend previous work on linear LNAs. Linearity performance peaks can be adjusted to 'follow' high-power interfering signals, where high linearity is most important.

5.2.7.4 Diplexerless Frequency-agile Radio Front-end

An SDR transceiver will be operating in an environment in which signals may be either FDD or TDD. The object of this work would be to implement an appropriate high-isolation porting technique for the physical coupling of the transmitter and receiver amplifiers to the antenna port.

5.2.7.5 Interference Cancellation or Filtering

The object of this research would be to develop a means of detecting a nearby high-power blocking signal. Means would then be devised to remove the influence of this signal from the receiver, either by cancellation or adaptive filtering.

5.2.7.6 Adaptive Preselect Filters

This research would continue work, already mentioned, on electronically tuned preselect filters to eliminate image signals.

5.2.7.7 Frequency-agile Zero-IF Receiver

This work would look at techniques by which the front-end of a zero-IF receiver can be made to operate successfully over a wide frequency range.

5.2.7.8 Novel Upconversion Techniques

The transmit path is responsible for most of the power consumption of an RF front-end. Since low-power RF front-ends are a key issue for mobile terminals, new transmit architectures should be researched. One possible starting point would be a transmitter based on a two-point modulation architecture. Such an architecture should be able to handle arbitrary signal modulation formats. Incorporated into such a transmitter could be an efficient power amplifier (PA) control (e.g. via controlling the supply voltage depending on the necessary output power).

5.2.7.9 Digital IF Processing

This work aims to design a common power-efficient hardware platform for digital IF processing. The work can include investigations on the advantages and disadvantages of a synchronous or asynchronous analogue digital interface.

5.2.7.10 Technological Roadmap for Receiver Architectures

In this section we will summarize the technology necessary to make an SDR transceiver a reality. We will then try to place the development of these 'enabling technologies' into some form of time frame.

In order to try to obtain some feel for the rate at which the hardware for an SDR transceiver may evolve, we have produced Table 5.1. This table lists the technological development required to realize particular transceiver architectures.

We will now try to assess the order of developmental time frame we are talking about for the development of these technologies.

Quadrature Local Oscillator
The realization of an accurate quadrature local oscillator is crucial to the development of direct conversion and low-IF receivers. It is crucial to the development of a direct conversion receiver because all mobile radio systems use some form of complex modulation, and,

Table 5.1 Comparison of technologies required to realize an SDR receiver using various architectures

	Quadrature local oscillator	DC offset compensation	Image reject mixing	Variable image filtering	High linearity receiver amplifiers	High linearity mixers
Direct conversion	√	√	√	×	√	√
Low IF	√	×	√	×	√	√
Superheterodyne	×	×	×	√	√	√

by definition, the direct conversion receiver will only have one downconversion available. This downconversion therefore needs to be complex. In addition, some image cancellation is required as part of this receiver. 40 dB image rejection is typically required for a direct conversion receiver. This requires the local oscillator to exhibit phase quadrature of better than 1 degree accuracy, and amplitude balance of better than 0.1 dB.

A low-IF receiver also requires a complex signal to provide part of the image filtering that is required with this type of receiver. Thus, it also relies on there being a local oscillator that has accurate phase quadrature and amplitude balance over a wide frequency range. Because the image of a low-IF receiver is separated from the wanted signal by twice the (albeit low) IF frequency, the image rejection requirements of a low-IF architecture are significantly greater than the requirements of a direct conversion receiver (typically 70 dB).

Trying to set a time frame over which development of an accurate quadrature oscillator may take place is difficult. The problem has been solved to a limited extent today. Agilent Technologies have developed a chip that is claimed to generate a phase quadrature local oscillator over a frequency range that would be very useful for mobile communications use (see [38]). This chip is a combined high-pass low-pass network, with servo amplitude control. It is used in Agilent instruments, and presumably gives their instruments a commercial advantage. For whatever reason, the chip does not appear to be commercially available.

Analog Devices have a direct conversion quadrature demodulator chip available (the AD8347, see [39]). This chip appears to offer quadrature phase shift with a typical accuracy of about $1°$ at a spot frequency (1.9 GHz). Spot amplitude accuracy is quoted at typically 0.3 dBm at the same frequency. Phase variation over the range of 800 MHz to 2600 MHz is also about $1°$. Amplitude balance variation is not quoted.

This is getting close to the performance that we require. It is difficult to predict where development will go from here, but a five to ten year period for the development of a suitable wideband quadrature oscillator seems a reasonable assumption.

A quadrature oscillator is required for the superheterodyne architecture. This oscillator is only required to operate at a fixed single frequency. It will most likely be implemented in DSP and, as such, will not present a significant problem.

Offset Compensation
The only architecture that requires offset compensation is the direct conversion receiver. There are a number of techniques for compensating the DC offset of a direct conversion

Table 5.2 Estimated time frame to develop technologies required for a fully functioning SDR receiver

Technology	Development time
Quadrature local oscillator	5 to 10 years
Offset compensation	5 to 10 years
Image reject mixing	Available now
Variable image filtering	3 to 5 years
Linearity	3 to 5 years

receiver. Some rely on measuring the DC offset during the idle period of a TDM [40], others rely on using data streams with zero DC component, or placing a high-pass filter of sufficiently low frequency in the data stream [41]. None of these techniques are directly applicable to an SDR receiver, as the type of modulation is not known *a priori*. It is possible that the techniques could be refined for an SDR, and therefore a development time of five to ten years is assigned to this technology.

Image Reject Mixing
As discussed previously, the image rejection requirements of a direct conversion receiver are comparatively low, and can be achieved almost routinely, provided the magnitude and phase balance of the local oscillator is reasonable. Image rejection requirements for a low-IF receiver are more stringent. Filtering, rather than image reject mixing, will be used for image signals when the receiver employs superheterodyne architecture.

Variable Image Filtering
Developmental time for this technology has been assigned three to five years. There are no obvious extreme performance requirements, and the process should represent the adaptation of existing technology.

Linearity
The need for high linearity mixers and amplifiers has been explained earlier, and various methods of linearizing LNAs and mixers have been developed. Extending the usefulness of these techniques to a wide operational bandwidth is a current research interest. A developmental time frame of three to five years has been placed on this technology.

The time taken to develop all of these technologies is summarized in Table 5.2. From this table, it can be deduced that the superheterodyne architecture is seen as being the most likely architecture to be developed as a fully functioning SDR receiver in the short term. As has been emphasized in other parts of this chapter, the zero-IF architecture is seen as providing a better long-term solution, because of the possibility of achieving a single chip implementation and other issues described here.

5.3 SDR Baseband

5.3.1 SDR BB Reference Models and Architectures

The baseband architecture of reconfigurable SDR equipment has to be designed so that a total, and also a partial, reconfiguration of the baseband is possible. In the case of switching from one radio access technology to another, total reconfiguration is required, which usually requires significant changes in the functionality, behaviour, and interfaces of the baseband modules. Partial reconfiguration is performed if only parameters of one, or several, baseband modules have to be changed without changing the air interface standard. Certain modules, for example, may be reconfigured in order to improve QoS, whilst remaining on the current operating standard.

If the baseband should be reconfigured without interrupting the ongoing service, a second so-called 'shadow transceiver' has to be provided in the baseband architecture. However, since cost and complexity of the required hardware platform is significantly increased, alternative, but less flexible, solutions have to be investigated. Duplication of baseband modules, for example, can be avoided by designing generalized modules that can switch their operating parameter during run-time. This suboptimal, but technically feasible, approach requires a reconfiguration table, listing all modules and their parameter requirements for all standards supported by the SDR.

The baseband architecture of reconfigurable SDR equipment has to be designed so that it can execute today's and tomorrow's wireless access schemes on a future hardware platform, and can be integrated into an overall system architecture that allows one to reconfigure the platform by executing a program. These contrary design goals suggest defining two reference baseband architecture models, one applied for emphasizing software-related issues, and the other for hardware-related issues. Note that the two generic architectures complement each other and should not be seen as contradictory.

5.3.2 Algorithms

5.3.2.1 Harmonization and Classification of Transceiver Algorithms

Different air interfaces use different algorithms for baseband processing, e.g., equalization algorithms, filtering algorithms, the RAKE receiver algorithm and DFT-based algorithms. A harmonization based on a generalization of these algorithms might lead to a set of very few common basic algorithms, for which a common software platform, as well as a common hardware platform, can be designed.

The same principle applies to synchronization. Synchronization is crucial for every receiver, however, it depends on the air interface. A harmonization of the synchronization algorithms should lead to a *synchronization engine* that is applicable to virtually all known air interfaces.

Research Issues

- classification of receiver and transmitter algorithms;
- design of a generic synchronization engine;

5.3.2.2 Interworking between the RF/IF Part and the Baseband Part

A tight interworking between the RF part and the baseband processing part of a terminal promises to enable an adaptive calibration, as well as compensation of mismatches, offsets, nonlinearities, etc. of the analogue RF/IF part at run-time, by means of digital signal processing.

5.3.3 Software Architecture of Reconfigurable Baseband

A generic software architecture for a reconfigurable baseband has been defined in the IST project TRUST [42]. This architecture can be well embedded into an overall system and network architecture that can support and exploit software-defined radio (SDR) equipment. It comprises a baseband management module that controls the configuration of the baseband. In the following, the key features and components of the proposed software architecture of the baseband are briefly described.

The TRUST baseband subsystem is adaptive because of its ability to reconfigure itself. The software architecture of the reconfigurable baseband is based on object-oriented methodology. Each module of the baseband transceiver chain is reconfigurable by instantiation of an appropriate class and/or reinitialization with new parameters. It is assumed that the software (class) of each module (modulator, FEC decoder, etc.) is available (downloaded and stored in the baseband library), error free, and compatible.

In order to reconfigure the baseband without interrupting the ongoing services, the baseband architecture must support dynamic creation and binding of new, or modified, modules. As a result, instantiation of downloaded classes must be administered through dynamic binding, whereby the required functionality of a given class is only made available at run-time, whilst the structure of the downloaded class is known *a priori*.

Figure 5.11 illustrates the reconfigurable baseband architecture, which consists of the following components:

- *Reconfigurable baseband management module (R-BBMM):* The overall controller of the R-BB subsystem. It is responsible for negotiating reconfiguration, creating active and shadow transceiver chains and controlling the run-time behaviour of each module.

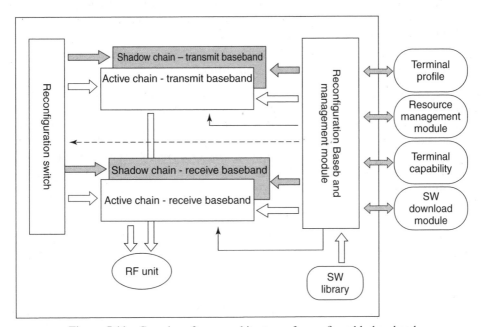

Figure 5.11   Generic software architecture of reconfigurable baseband

- *Active baseband transceiver chain:* The currently operating baseband chain. Each chain consists of several modules, and each module is referred to as a *baseband processing cell* (BPC).
- *Shadow baseband transceiver chain:* The post reconfiguration baseband transceiver chain. It contains references of BPCs that are kept unchanged from the active chain, and one/more new BPCs. The shadow chain does not interfere with the active chain, and constituent processes are kept mutually independent.
- *Baseband software library:* This contains the baseband active and shadow configuration maps, corresponding to the active and shadow transceiver chains respectively. A configuration map is a list of baseband modules (type, functionality, algorithmic identity, etc.), their associated interface definitions and interconnections. The library will also store all the baseband module classes that are currently in use, and those that were used previously.
- A 'read-only' default configuration map, together with all the associated module classes and parameter lists. This would allow the baseband to confidently reconfigure to a known, working standard, which is compliant with the user profile.
- A complete copy of the previous working baseband configuration. This should include the configuration map, associated baseband classes, parameter lists, operating standard, host network registrations, etc. Such a store would allow the terminal to return to a fully working baseband configuration, without requiring a power off reset.
- *Reconfiguration switch:* A typical ON/OFF switch. It implements the ON/OFF signal from the R-BBMM, in order to switch the shadow chain ON and the active chain OFF.

The configuration map defines the baseband chain, in terms of functionality of constituent modules and their interfaces and interconnections. The R-BBMM uses this map to create, and then connect, the different BPC objects. In order to reconfigure the baseband, the R-BBMM creates a shadow baseband chain in compliance with the agreed configuration map for reconfiguration.

5.3.4 Hardware Architecture of Reconfigurable Baseband

Figure 5.12 illustrates a generic baseband hardware architecture for a software-defined radio which can be applied to the design of user terminals and base stations for mobile communication systems beyond 3G. It is well suited for the execution of a variety of current and future wireless access schemes, and provides the flexibility to reconfigure the physical layer transmission scheme under software control. This architecture covers at least some of the competitive hardware platform currently available on the market. It can also be viewed as a generic architecture for a reconfigurable baseband that is implementable as a system-on-chip (SoC), containing several general purpose and digital signal processor cores, memories, and dedicated, as well as reconfigurable, hardware blocks.

The radio signal received at the antenna is first processed with a wideband radio front-end, before it is fed to the digital baseband (BB) signal processing unit. The signal is downconverted with an analogue mixer to an intermediate frequency (IF) band, sampled with a wideband, high-resolution A/D converter, and fed to the digital radio front-end.

This digital, reconfigurable IF/BB module performs channelization. It extracts from the received signal a single, or several, user channel(s) for further processing at the baseband, and thus implements the signal processing tasks 'downconversion', and 'filtering'.

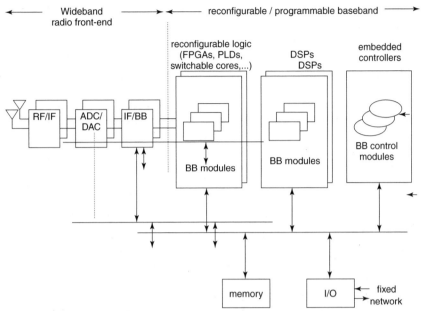

Figure 5.12 Generic hardware architecture of reconfigurable baseband

Depending on the required radio access scheme and selected user channels, different channel frequency characteristics and parameters have to be implemented. These numerically intensive functions have to be executed on a dedicated programmable device with today's available technology; however, we expect that in some years from now, these functions will also be implemented on a DSP.

Most of the baseband signal processing functions of the various wireless access schemes are executed as independent BB modules, either on reconfigurable logic or digital signal processors (DSPs). The required signal processing modules and their configuration parameters are given by the selected wireless access scheme. We have to determine these modules for each candidate scheme, identify the similarities in the different functions, and try to use common reconfigurable or switchable logic devices, or DSP code, to implement them. This is a challenging task considering the widely differing nature of the different signal processing functions of the above radio access schemes, as well as the wide range of involved radio interfaces.

At the same time it would be preferable to make the software modules independent of the hardware platform. This would allow upgrading of the hardware platform without changing software modules. Something like a resident compiler, or java interpretor, is an option. Another requirement would be to minimize the reconfiguration data to be downloaded, while constraining the total memory requirement.

This hardware platform should be partly reconfigurable on the run, while some part is executing some functions. An ideal solution would be a combination of independent hardware modules and hardware-independent software modules linked together by process control module. All signal processing modules should have standard interfaces so that they can be plugged in and out in any order in a system implementation. This would also allow switching off of some hardware modules to save power in good operating environments.

Another issue to be considered would be selection of the system clock. The least common multiple of the basic clocks of the chosen systems is unrealistically large for practical implementation. This would necessitate a new approach for timing and rate/frequency control. It would be advisable to have a single system clock and clock generation schemes, approximating the basic clocks of different systems on average. Other solutions incorporating multiple clock/PLL-based schemes would be bulky.

The BB control modules are implemented as independent processes on a real-time operating system (RTOS) on an embedded controller. One process is responsible for selecting the required BB modules according to the selected standard, configuring the parameters of the modules and determining their execution order. The executable FPGA macros and DSP code implementing the BB modules are stored in memory devices. They can be downloaded to the memory over the fixed network or over the air (OTA) before the baseband transceiver operation is started. The reconfigurable logic devices, DSPs, embedded controller, memory and external I/O devices communicate with each other via a local system bus, which allows the parallel implementation of the baseband signal processing functions for several radio transceivers. Since the hardware architecture shown in Figure 5.12 is a generic one, more advanced mechanisms, other than conventional bus architectures, can also be applied to interconnect the various hardware blocks or cores.

Another requirement of the baseband processing is some standard way of defining its capabilities, like processing power, RAM/ROM memory size, RF capability, ADC/DAC resolution, etc. A protocol to communicate this information to the higher protocol layers and, in turn, to the network is required. This will allow the network to assess the terminal capability before entertaining any upgrading in terminal capabilities.

In the next few sections, we identify the BB modules of several wireless access schemes by briefly summarizing their baseband functions. As potential candidate systems, we selected the cellular systems GSM and W-CDMA, and the WLANs 802.11a and 802.11b.

5.3.4.1 Generic Baseband Receiver Chain

A generic baseband processing block diagram of the receiver chain for the four identified systems is given in Figure 5.13. IF processing is not included in the block diagram. In conventional implementations, this is done in the RF front-end. In the case of software radio, it is likely that these functions would also form part of the baseband processing. The included functions would be channelization and decimation. These functions would easily fit to a parameterized block with parameters centre frequency and channel bandwidth.

Even though the blocks are shown in a sequential way, the actual implementations may be parallel, or may even combine some of them. The idea here is to capture the critical blocks involved in the signal processing. Even though an attempt is made to identify critical parameters of some of the blocks in the attached boxes, these are not complete. The complexity of the systems and the different modes/data rates of different systems would not allow a complete parameterized representation of the systems. Some of the common data rates and their critical parameters are shown in the table. In general, it can be said that the different systems have different types of framing and encoding requirements. Hence, identifying common building blocks that can be parameterized to meet their requirements is a really challenging task.

The channelization block is followed by an optional filtering block. In the candidate systems, only W-CDMA specifies a filter at this stage. This is followed by essentially

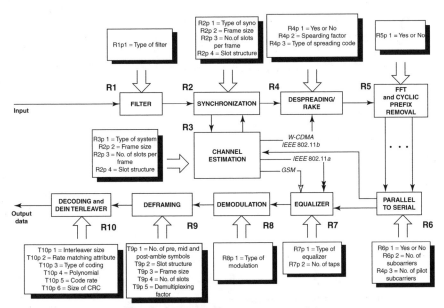

Figure 5.13 Baseband receiver module

two parallel blocks: channel estimation and synchronization. The performance of the receiver is largely decided in these blocks. The channel estimation block is mainly for estimating the critical multipath profile and, as such, would be closely linked to equalization/Rake combining. Hence, their implementation would be highly dependent on the system characteristics.

The synchronization function is expected to estimate the frame and code symbol timings, as well as frequency offset, and do the necessary correction. This would be based on parameters like frame size, number of symbols/frame, number of slots/frame and the synchronization information in the slot. This would estimate the correction function, and again is dependent on the individual systems.

The synchronized information is sent for despreading/Rake combining in W-CDMA and 802.11b systems, and FFT processing in 802.11a. These functions are completely different and these blocks are missing in GSM. This is followed by equalization in non spread-spectrum systems. Finally demodulation, deframing, decoding and descrambling of the received data is carried out. The order of these functions is dependent on the individual system and the particular service chosen. It would be possible to develop common blocks for these functions with the parameters and frame format for a particular service downloaded for activating that service.

5.3.4.2 Generic Baseband Transmitter Chain

A generic baseband processing block diagram of the transmitter chain for the four identified systems is given in Figure 5.14. The user data is fed to the framing block. The encryption of the data is carried out in higher layer protocols, or is optional in W-CDMA and 802.11a and 802.11b. Only in GSM is the encryption carried out in the baseband Layer 1 processing.

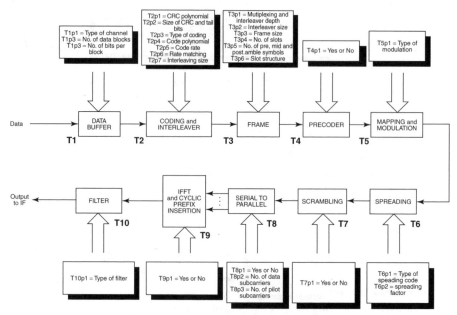

Figure 5.14 Baseband transmitter module

The framing information is different for different services in the same system and, hence, is difficult to parameterize. The details of the framing for different services have to be stored separately and fed to the framing module. This is followed by coding, interleaving and framing blocks. In 802.11a, scrambling is done before channel coding. Here also, different services use different formats and would have to be stored separately.

A common block for all the systems is possible with sufficient memory depth. This is followed by a precoder in GSM and 802.11b systems. These blocks are followed by modulation, spreading, scrambling-like functions and, finally, the spectrum shaping filters. Some of these blocks may not be present in all the systems, or the order may be different. These call for a flexible architecture, which can place the processing modules in any order, as defined in the standard.

5.3.4.3 Other Baseband Transceivers (PAN/DAB/DVB)

The baseband architecture of reconfigurable SDR equipment has to be designed so that, not only the transceivers of the cellular systems like GSM and W-CDMA and 802.11 based WLANs can be efficiently implemented, but it should also allow the integration of technology to be used in today's and tomorrow's personal area networks (PANs), as well as satellite and broadcast systems. Since these systems require some additional functional baseband modules not shown in the generic transmitter and receiver chain given in the last two sections, we will now briefly discuss the required transmit and receive modules for the PAN 'Bluetooth' and the digital broadcast systems DAB and DVB.

Personal Area Network 'Bluetooth'
Bluetooth has been accepted by the IEEE 802.15 working group as a standard for enabling wireless ad hoc connectivity between portable and/or fixed electronic consumer products.

It can manage within a small local area up to three synchronous connection-oriented (SCO) links, mainly for speech transmission at a rate of 64 kbit/s, and up to seven asynchronous connectionless (ACL) links supporting symmetric or asymmetric data transfers at a maximum rate of 433.9 and 723.2 kbit/s, respectively.

The radio subsystem is operated, like 802.11b, in the globally available unlicenced ISM frequency band at 2.4 GHz, covers distances of up to 10 metres with a transmission power of less than 1 mW, and applies frequency hopping in conjunction with a TDMA scheme for transmitting data at a symbol rate of 1 Mbit/s over the air.

Figures 5.13 and 5.14 show the baseband modules required for the transmitter and receiver chain of Bluetooth. Figure 5.14 shows the transmitter chain. User synchronous (US), user asynchronous (UA), or user isochronous (UI) data, and link manager (LM) control information are sent via the corresponding logical channels to the transmit (Tx) SCO or ACL buffers. The stored information represents the payload to be transmitted over the radio link. Before its transmission, the payload is protected by appending cyclic redundancy check (CRC) bits, ciphering, whitening and optionally encoding with a rate 1/3 or 2/3 forward error correction (FEC) code. In parallel, a packet header is assembled by the link control (LC) module and stored in the Tx header register. The header is also protected with header error check (HEC) bits, whitened and encoded with a rate 1/3 FEC code.

A radio frame is obtained by first concatenating the filtered and coded header and payload information, and then preceding the resulting bit string with the access code. Finally, the radio frame is forwarded to the analogue radio front-end for its transmission at a frequency $f(n)$. The value of $f(n)$ is provided by the hopping-frequency-select module.

The corresponding receiver chain is shown in Figure 5.13. When the access code correlator detects the arrival of a radio frame at a frequency $f(n)$, a trigger event starts the processing in the receiver chain.

The header information is extracted from the received frame, decoded, dewhitened, and stored in the receive (Rx) header register. When the HEC check is successful, the receiver can start decoding, dewhitening and deciphering the payload information, and store the packet in the Rx SCO or ACL buffer, depending on the packet type received. From the Rx buffers, the payload is carried via the logical channels US, UA, or UI to the synchronous or asynchronous I/O port. The link control module configures, monitors and controls the transmitter and receiver chain. In the case of packet data transmission, the LC module also coordinates the retransmission of erroneously received data packets with an automatic repeat request (ARQ) scheme. It also carries out the connection set-up protocol, authentication and power management control.

Digital Audio Broadcast Systems

Digital broadcast systems have been investigated since the middle of the 1980s. The main reason for going digital is to make the receiving quality of broadcast services comparable to that of CD reproduction. It gives listeners interference-free reception of high-quality sound and, simultaneously, can provide additional service information like dynamic text labels (e.g. programme, titles, names of artists), or switching to traffic reports or other services.

In 1992, the EUREKA-147 DAB system (digital audio broadcasting) was recommended worldwide by the Inter Union Technical Committee of the World Conference of Broadcasting Unions. In December 1994, the system achieved a worldwide standard given in ITU-R recommendations BS.1114 and BO.1130 for terrestrial and satellite sound broadcasting to vehicular, portable and fixed receivers in the VHF/UHF range. It is an accepted European Standard (ETS 300401, March 1997), adopted by the European Telecommunications Standards Institute (ETSI).

The DAB system is a sound and data broadcasting system with high spectrum and power efficiency. It uses digital audio compression techniques (MPEG 1 Audio Layer II and MPEG 2 Audio Layer II) and, by this, achieves a spectrum efficiency equivalent, or higher, than that of conventional FM radio.

The DAB transmitted data consists of a number of audio signals sampled at a rate of 48 kHz with a 22-bit resolution. This audio signal is then compressed at rates ranging from 32 to 384 kbps, depending upon the desired signal quality. The resulting digital data is then divided into frames of 24 ms. DAB uses differential QPSK modulation for the subcarriers.

A null-symbol (or a silence period that is slightly greater than the OFDM symbol length) is used to indicate the start of the frame. A reference OFDM symbol is then sent to serve as a starting point for the differential decoding of the QPSK subcarriers.

Differential modulation avoids the use of complicated phase-recovery schemes. DAB uses a rate 1/4 convolutional code with a constraint length of seven for error correction. The coding rate can also be increased using puncturing. Interleaving is used to separate the coded bits in the frequency domain as much as possible, which avoids large error bursts in the case of deep fades affecting a group of subcarriers. DAB is designed to be a single frequency network, in which the user receives the same signals from several different transmitters. This greatly enhances spectral efficiency. Even though there is a delay in the reception of signals from different transmitters, this situation can be considered as a multipath situation, and can be easily handled by selecting the guard interval properly. Further, this can be considered a form of transmit diversity that the DAB receiver can take advantage of.

Digital Video Broadcasting (DVB)

Digital video broadcasting (DVB) is a standard for broadcasting digital television over satellites, cables and through terrestrial (wireless) transmission. DVB was standardized by the ETSI in 1997. The following are some important parameters of DVB. It has two modes of operation: the 2k mode with 1705 subcarriers, and the 8k mode with 6817 subcarriers. 1/2 DVB uses QPSK, 16-QAM or 64-QAM subcarrier modulation. DVB uses a Reed–Solomon outer code (204,188, $t = 8$) and an inner convolutional code with generator polynomials (177,133 octal), combined with two layers of interleaving for error control. Pilot subcarriers are used to obtain reference amplitudes and phases for coherent demodulation. Two-dimensional channel estimation is performed using the pilot subcarriers, which aids in the reception of the OFDM signal.

5.3.5 Critical Components

5.3.5.1 Application-specific DSPs

Application-specific DSPs have far better performance than general purpose DSPs. In the future it will be more and more important to have a design methodology for application-specific DSPs that enables short design cycles with moderate, or low, costs. Fundamental criteria for the design of application-specific DSPs, and thus for respective research, are:

- *Parameterizable basic architectures (PBAs):* Such basic architectures enable a fast adaptation of the baseband processing platform to a given application as early as in the design phase. One PBA must be found for each application class. Among these are the signal processing machine (SPM) and the control processing machine (CPM).
- *Application classes:* To find a set of application classes it is necessary to investigate different systems, applications and kernal algorithms. Fundamental characteristics are extracted, leading to a classification.
- *Parameterizable instruction set architecture (ISA):* This is the interface between programmers and hardware. Its main characteristic is to map the scalability of the hardware platform to the command complexity. The structure must remain unchanged, independent of the underlying hardware architecture.
- *Parameterizable design and programming tools:* These are tools like assemblers, debuggers, simulators, etc. Similar to the ISA, the tools must be easily adaptable to the different architecture alternatives. It is important that the usage of the tools remains unchanged, independent of the underlying hardware architecture, so that the programmer can develop a kind of familiarity.

Research Issues

- design of a parameterizable basic architecture for application-specific DSPs;
- identification of application classes;
- design of a parameterizable instruction set architecture for application-specific DSPs;
- design of parameterizable design and programming tools for application-specific DSPs.

5.3.5.2 DSP Operating Environment

DSP is used for power efficiency, but DSP software is not only power consuming algorithms. Reconfigurability is requested for DSP software, as for other parts of the baseband, but should take into account the power efficiency.

Modem software can be split into three parts (see Figure 5.15):

- *Waveform applications:* algorithms, 'pure' data processing, waveform sequencer.
- *Modem manager:* control, supervision and management.
- *Platform services:* services awaited for each platform but implementation dependent of the environment.

This approach allows one to separate software modules which are only waveform dependent, from those which are platform dependent. As a consequence, the software structure shows three types of interface:

- *I-SWAPIs:* interface services for waveform applications provide communications channels able to establish links to/from the digital transceiver, BB, etc., for control and data.
- *miRTOS:* modem interface for RTOS defines a reduced set of RTOS services dedicated to the DSP environment.
- *SMAPIs:* services for management applications provide a dedicated set of services to manage the DSP software and/or the waveform.

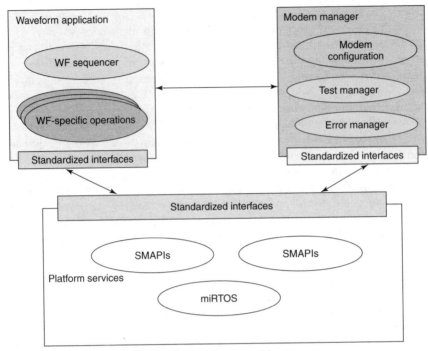

Figure 5.15   Modem architecture

Figure 5.15 is an illustration of the DSP operating environment approach. The advantage of this software design is that it separates waveform from platform, which is the major SDR difficulty, and so facilitates waveform management, update and configuration.

Research Issues

- design and standardization of SWAPI interface definition;
- design and standardization of miRTOS interface definition;
- design and standardization of SMAPI interface definition.

5.3.6 Future Directions/Research Goals

The primary goal of the research in the field of the SDR-BB concept is a generic hardware architecture which is able to cover as many current and upcoming wireless standards as possible. Baseband processing of the future multistandard terminal or base station will set very demanding requirements in computational performance and flexibility.

The baseband hardware architecture should be designed and implemented with reasonable costs and energy efficiency. The current baseband hardware architectures typically consist of the following elements:

- general purpose processors (GPPs);
- digital signal processors (DSPs);

- special purpose hardware blocks (e.g. Viterbi/turbo en-/decoder, (I)FFT, correlator, correlation filter);
- busses and interconnections; and
- memories.

It is very probable that the future baseband processing architectures will also consist of different kinds of processing, storage and interconnection resources. As an addition to the list above, different kinds of reconfigurable technologies will be used as processing elements. The reconfigurable technologies include configurable processing cores, FPGA-based approaches, embedded FPGA-based approaches and approaches based on arrays of processing elements. The special purpose hardware blocks will still be needed for the most demanding algorithms, due to their superior silicon area and energy efficiency.

Research is needed to define an optimum hardware architecture for a given wireless standard set. A crucial issue of the resulting architecture is the resulting implementation complexity, i.e. power consumption, chip area of single modules and the overall resulting architecture. This may also include research on efficient HW structures of reconfigurable logic modules. Design flow for implementation of SW modules for constituent standards on the hardware architecture has also to be considered.

In the recent and current research on reconfigurability in wireless communications, such as EC funded projects TRUST [42] and CAST, the problems of baseband reconfiguration were identified and addressed. With reference to TRUST [42], the top-down design approach was adopted, and it was identified that for the baseband reconfiguration purposes, an additional entity is needed. This entity is termed a *management module*.

In CAST, designers tried to address the problem of reconfigurability using the top-down and bottom-up approaches. Therefore, two additional separate control components were designed for managing the process of the baseband reconfiguration. They are the reconfigurable resource controller [43] and physical layer controller [43]. These two entities can perform the following:

- cover the functionality of the management module as in TRUST;
- allocate physical layer resources;
- optimize physical layer resources in terms of:
  — power consumption;
  — speed of processing; and
  — memory usage;
- implement reconfiguration; and
- accommodate different vendors.

In these projects, the designed baseband software can be characterized as 'passive'. This is because of a need for additional management entities to download, manipulate and control software modules in the process of reconfiguration. In order to eliminate the need for these additional management units, and reduce the requirements for downloading and reconfiguration control signalling, further research is needed in producing new methods for designing the baseband processing software. These methods should enable the baseband software to be 'active'. This means that baseband software should have the ability to change functionality, behaviour and performance, without the need for additional management units. These actions should be completed by baseband software itself. Hence, there will be reduced requirements for downloading and reconfiguration control signalling. This approach should allow the design to be based upon known methods with clearly defined data flow, control mechanisms and interfaces. Procedures for reconfiguration should also be simple and well defined.

In addition, research should be conducted on identifying a specification for enhancing existing real-time operating systems into software-defined radio based real-time operating systems (SDRbRTOS). The SDRbRTOS will be needed to support 'active' baseband processing software, and perform optimization of reconfigurable physical layer hardware.

The overall research process therefore includes:

- thorough analysis of existing wireless standards;
- tracking of trends in upcoming wireless standards;
- analysis and discussion of SDR-BB architecture proposals;
- research and analysis of state-of-the-art and future component technology, including general purpose DSP, memory, reconfigurable logic, semiconductor technology;
- comparison of cost with conventional approaches.

5.4 Reconfiguration Management and Interfaces

5.4.1 Reconfigurability Issues

5.4.1.1 Open APIs for Reconfigurability Support

Value-added services (not necessarily voice-centric) comprising multimedia, location-based, emergency services, multimedia messaging, etc., are a vibrant business area for 3G operators and service providers, and the potential for the provision of adaptable and value-adding services steadily increases. Services based on the mobile subscriber's physical or virtual presence and availability make usability of interactive applications, such as instant messaging and mobile chat, an attractive possibility. Such value-added services will eventually converge with existing location services, as well as with mobile commerce applications. The most important issue, therefore, is that the 'look and feel' of all future services will satisfy the user's expectations towards ease of use for adaptive and user customizable services, as propagated in Chapter 3. The basis for such service adaptability in heterogeneous networks and access technology environments, is the availability of reconfigurable equipment, infrastructure and execution environments.

Architectures for the development and deployment of services by service providers over 3G networks have been introduced by 3GPP (OSA) [44], OSGi [45], Parlay [46], Jain [47], LIF [48], UMTS Forum discussions [49], etc. These platforms and schemes are designed for service support, however, they do not provide open APIs for reconfiguration purposes. Reconfiguration APIs should be introduced in order to enhance the functionality of these service platforms, and to enable network-wide service and service policy-specific reconfiguration.

Reconfiguration actions concern various attributes, the main attributes are 'state' and 'configuration'. Configuration describes the settings of a network device, and includes parameters and classes representing desired or proposed thresholds, bandwidth allocations and traffic classification criteria. State describes the actual state of a device at a particular operating instance like before, during or after a reconfiguration procedure.

A generic service provision and reconfigurability management architecture needs to be based on open APIs for reconfiguration triggering and control, without these, reconfiguration processes would be greatly restricted and limited to a small number of specific platforms. Context awareness features influence the triggers for reconfiguration, hence they need to be taken into account on a generic service provision and reconfigurability management architecture, and need also to be supported through the open APIs (e.g., location/time/environment, etc.-aware policy provision, as described in [50] and [51]).

5.4.1.2 Reconfigurability in IP Networks

On the network side, the foreseeable future will be characterized by the convergence towards an IP-based core network and ubiquitous seamless access (2G, 3G, broadband, broadcast, etc.) in a context of hierarchical and hybrid self-organizing networks (see also Chapter 4). However, mobile users will only fully benefit from such an environment if they are equipped with terminals capable of adapting to the different transmission-service offerings of these networks.

A future generation of reconfigurable SDR terminals will be capable of operating in several, or (ideally) in all, of the different access environments. Such terminals will be able to support the whole range of applications provided in any of the connected networks. To guarantee efficient use of the existing network resources, and to provide to the user adaptive and customizable QoS levels, the terminals need to be able to choose, and adapt to, the right network for the required service. All these capabilities will be offered to the users in a seamless and transparent way, in the form of permanent global access, roaming and service availability.

As propagated by the SDR Forum [52], the SDR concepts impact many layers of a wireless network, and associated benefits will be realized from the physical layer to the user applications plane.

A number of SDR and reconfigurability research initiatives and groups (i.e. European IST Reconfigurability Cluster Projects [53], SDR Forum [52], Mobile VCE [54], etc.) stress the fact that reconfiguration of mobile SDR terminals requires complex interactions between the terminal and network entities involved, this includes the download of new software modules that are to be installed on the terminal, as well as the discovery of reconfiguration support services (e.g. SW store and download, configuration validation, etc.). The successful development of SDR technology for commercial telecommunication applications will directly depend on the early definition of reconfigurable system and network architectures, including the definition of a separation of intelligence between reconfigurable terminals and the network, as well as the definition and design of reconfigurable equipment that will be required on the network side to support reconfigurability, and to provide the aforementioned reconfiguration support services.

5.4.1.3 Data and Control Interfaces

Reconfiguration of terminals requires somewhat complex logistics, terminal and network information has to be gathered, the software necessary to perform a reconfiguration has to be obtained and it has to be ensured that this software becomes securely downloaded and properly installed in exactly that partition of the software-definable radio hardware platform for which this particular software module has been foreseen. A very coarse structure taking all this into account was defined and propagated during the early days of software radio, when the SDR Forum (at that time called MMITS) advocated a subset derived from the SPEAKeasy architecture. This initial architecture identifies not only the various functional blocks and the software structure of an SDR radio, but it also recognizes sets of interfaces between the various modules of a terminal, (i.e. information (I) and control (C) interfaces), between SW and HW platforms, and also between the HW platform and control infrastructure (see also [55]).

Meanwhile, there are a number of projects and communities approaching the definition of software radio architecture(s). Common to all these example architectures/structures (as listed and shown in [43, 56] is the separation between information and control paths within the radios, as well as the gathering of parameters and general system information, as a precondition to pursue controlled and trustworthy standard-compliant reconfiguration processes.

A reconfigurable communication platform must encompass all system levels [57], [58], ranging from (reconfiguration) supporting networks, reconfigurable base stations to reconfigurability of the terminal's RF and BB modules, whereby also other aspects, like the support for higher layers, may require reconfiguration. It is widely understood that, at least currently, there is no single platform, architectural framework and implementation technology capable of providing a homogeneous structure for SDR equipment.

Additionally, the anticipated flexibility of such equipment requires a modularization of the various system elements [59] and mechanisms to manage complete reconfiguration processes (examples are Mobile VCE's Reconfiguration Management Architecture, RMA [60] and TRUST's Configuration Management Module, CMM [59]).

Taking all this into account, and at the same time aiming for open programmability to enable a 'PC-like' growth and variety of software provided, we require an interface architecture for both control and data domains.

Figure 5.16 abstracts the various interfaces between the different elements within a reconfigurable system. On the terminal side, a management unit (RMA, CMM, etc.) controls and steers the reconfiguration process, whereby TRUST's 'mode switching' feature [53] may be applied to monitor, using a switched shadow chain, the radio environment and to support seamless HOs. MVCE's reconfiguration management controllers [60, 61] are a solution to enable the intended reconfiguration of single modules within the transceiver chains.

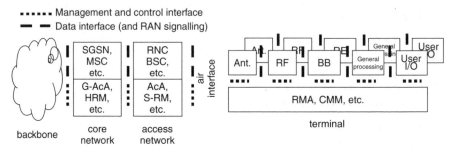

Figure 5.16   Interface architecture for reconfigurable systems and supporting network elements

Based on a logical system and software architecture for reconfigurable equipment, a clear defined and specified set of interfaces for both data and control of reconfigurable equipment needs to be developed. This section aimed to outline and propose an initial approach, applying different available technologies. A focused and detailed breakdown of the management control and data interfaces will follow in the next section.

5.4.2 Management, Control and Data Interfaces

A basis for reconfigurable equipment would be a powerful processing platform and a sensible and structured software architecture capable of implementing the system requirements of the various RATs (radio access technologies). A major task of such an SW architecture is to integrate the different technologies and platforms, and to use the programmability of these platforms to make systems reconfigurable. An architectural framework is required that considers use of open programming interfaces for both the application layer, but also for the lower system levels (i.e. an open programming interface for RAT implementation software).

Assuming availability of the 'ideal' SDR terminal, the definition of the interfaces necessary would be rather simple. However, reality demands the modularization of various functional blocks within the receiver and transmitter chains of SDR terminals, as can be seen in Figure 5.16 and in the sections on reconfigurable RF and BB parts. Reconfigurable terminals have, in contrast to nonreconfigurable terminals, the additional need to provide basic connectivity, in particular if the terminal is not configured to one of the currently available RATs. Such connectivity can be achieved by *polling*, or *scanning* the radio environment, and by reverting the terminal to the required configuration (n.b. this requires that a RAT is identified, and the configuration software necessary is stored within the terminal [59]). Or, alternatively, by specification and establishment of a global connectivity channel, which could implement the required support infrastructure for reconfiguration of SDR terminals [60].

Independent of which approach is being followed, both possibilities require additional features for reconfiguration control and management of the configurations and configurable radio modules/components. The following sections describe the various interfaces of reconfigurable systems, starting with the separation of 'commercial' and 'reconfiguration-related' traffic, followed by a look into the separation within and between the reconfigurable BB and RF parts of SDR terminals. This is followed by the identification of the various control and management interfaces and finally, an overview about possible software structures of SDR radio implementations/configurations.

5.4.2.1 Information Exchange between Network and Terminal

Once available, reconfigurable radios will provide implementations of mobile communication terminals that adhere to given RAT standards (e.g. UMTS, GPRS, H2, IEEE 802.11, etc.). These (software-defined) radio implementations will provide, as a minimum, all services and functions that a 'hardwired' terminal of the chosen standard would offer. However, to undertake reconfigurations during run-time, reconfigurable terminals do require the means to control and manage such reconfiguration processes.

Figure 5.17 depicts some of the principles of this, whereby the data and RAN signalling traffic is handled in the RAT implementation within the SDR terminal (using 'd' (data)

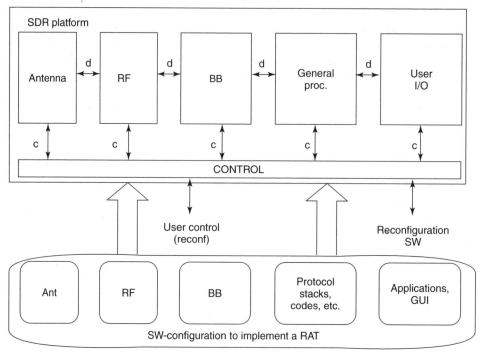

Figure 5.17 SDR architecture – data and control interfaces

interfaces), whilst the reconfiguration-related traffic requires its own (virtual) transport mechanism (conveyed via 'c' (control) interfaces).

5.4.2.2 SDR Platform Module Interfaces

Figure 5.17 illustrates the separation between 'd' and 'c' interfaces, whereby the upper part of the SDR platform contains a number of functional building blocks. This includes: antenna, RF, BB, general processing and user I/O sections, which form the processing platform to execute the software configuration which implements the target RAT. Single building blocks within this structure are vertically separated by 'd' interfaces. The type of information passed through them depends on the actual location within the terminal structure. The 'd' interface between the antenna and RF module provides the various (analogue) RF waveforms, whilst between RF and BB modules, the signals should be digitized already (depending on the scope and capability of the SDR RF platform).

5.4.2.3 Control and Management Interfaces

A reconfigurable radio platform will provide the possibility to develop and apply potentially any (soft-coded) radio implementation. Programmability will enable the execution of system software elements, or even complete SW configurations, that may be obtained from a plethora of different sources, including terminal manufacturer, network operator and also third party software providers. This openness, however, necessitates that reconfiguration and a required reconfiguration control/management cannot be confined or delegated to the terminal only. This implies that, if users are given the possibility to install any available software on their radio terminals, which finally determines the air interface and radio emissions from the terminal, there must be the possibility to control and manage administrative access to the SDR platform and to reconfiguration procedures.

The 'c' interfaces in Figure 5.17 provide the gateway between reconfiguration management and the modules of the implementation platform. Whilst some aspects of configuration management have been researched by TRUST [53] and a reconfiguration management architecture (RMA) has been developed within Mobile VCE [60, 62], the control and management interfaces between the radio implementation platform and the control/management module are yet to be defined.

5.4.3 Software Architecture

5.4.3.1 A Software Architecture for SDR Equipment

SDR equipment will meet many demands; the possibility to implement any air interface standard, together with a suitable support infrastructure, will provide the ability to roam vertically across access networks with one single handset. A change of functionality from one to another RAT is thereby 'just' a matter of changing one software configuration to another. The aforementioned architectures (RMA, TRUST functional architecture; etc.) can support the rearrangement and reconfiguration of the various software modules that implement one configuration. In such a structure, dedicated and open interfaces will allow the addition of new services, without needing modifications of hardware modules.

Apart from the apparent definition of the radio configuration, the software architecture has to provide reconfigurability, i.e. the SW architecture has to be able to modify a radio's configuration software (this is the reconfiguration from one RAT to another). So far, the main focus has been set on the configurability of BB and RF parts, however, not only those parts are required to be reconfigurable 'on-the fly', but also the protocol stacks have to be sufficiently flexible. Since the protocols running at the base station (BS) and mobile station (MS) are basically symmetrical, reconfigurability is applicable to both network elements. Therefore, an operator can build up an infrastructure that easily can be reconfigured to support a new, additional RAT standard when necessary.

In an SW architecture for systems as complex as SDR equipment, reusability of code and software elements is a big concern. New air interface standards typically rely on well-understood protocol stacks of predecessor systems, and SW implementations of RF/BB modules can have similar elements. Cost intensive re-engineering of software can be avoided and software can be reused if designed in a suitable way. Modular or object-oriented software design entails several advantages [63]. Using well-defined interfaces in the complete SW architecture will facilitate the portability and reuse of the same module in different systems. It also facilitates software upgrades, since the respective modules can be exchanged individually.

5.4.3.2 Service Support in Reconfigurable Equipment

To achieve real gain for users (and possible increase in revenues for operators and manufacturers), the issue of unanticipated and undesired formation/appearance of services must be investigated. A reconfigurable communication environment is required to support creation and provision of services; the entire system has to be designed to be as flexible as possible and, therefore, to provide transparent end-to-end reconfigurability and to integrate heterogeneous communication environments. This means that software architectures have to encompass the complete system, ranging from networking, wireless transmission mechanisms to end user services.

A core abstraction, grouping the system into layers, including hardware, network and middleware platform, as well as applications, is shown in Figure 5.18. Instead of strictly isolated layers, cross-layer collaboration will be essential to allow different degrees of adaptability and reconfigurability for the different parts of the system.

The four layers interact with each other using well-defined interfaces (i.e. these interfaces are subject to research and standardization). In addition to their regular tasks and cooperation with neighbouring layers in an operational setting, each layer must also be reconfigurable individually and independently via configuration interfaces. The configuration interfaces are used either to parameterize elements within the layer, or to add, exchange, or remove complete elements. In the following subsections, the middleware and network platform layers are described in detail, whilst the application and hardware platforms are not covered in this section, their interfaces, however (as shown in Figure 5.18), will need to be defined.

Configurable Service and Middleware Platform

Provision and design of a middleware platform for reconfigurable equipment is, from service providers' point of view, the possibility to effortlessly develop and provide new, adapted and customized services. System operators will expect that a middleware platform supports very high flexibility to allow for simple maintenance, whilst being proof for further system evolutions.

The middleware platform abstraction layer can be split into three major logical parts: the *local execution environment*, the *distributed processing platform*, and the *service platform* (Figure 5.19).

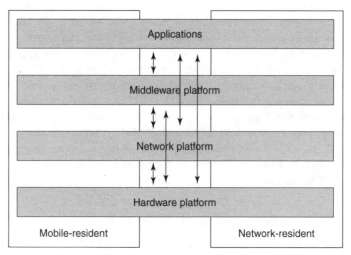

Figure 5.18 4G system abstraction layer groups

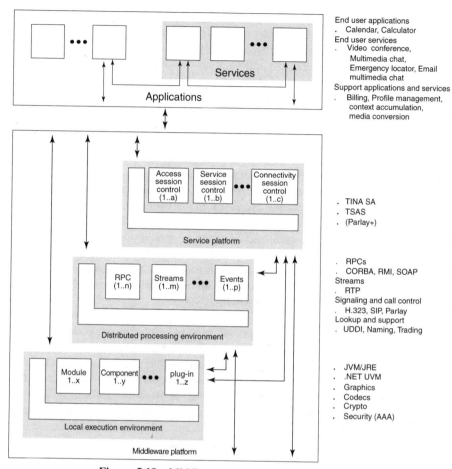

Figure 5.19 Middleware platform and applications

The local execution platform decouples the middleware components from the services/applications, but also from the details and specifics of hardware components and operating systems. The distributed processing platform abstracts the interaction between different system parts, which may reside on different, but also within the same, device(s). Mechanisms supported by this platform may range from simple remote procedure calls to complex streams, calls or resource control.

The service platform offers session semantics to any requesting service. Sessions maintained by the service platform can be separated into access, service and connectivity sessions. Specifications addressing this area include the TINA Service Architecture [64] and OMG's TSAS specification [65]. Extensions to the OSA and Parlay framework APIs [66] provide additional session semantics that may be integrated within the service platform.

The applications abstraction layer is above the middleware platform, it contains mechanisms to interact with layers of both the middleware and network platforms. However, applications not only rely on their interactions with lower layers, but may also act as clients of other applications. Therefore, they are executed in a local execution environment, but rely on the distributed processing platform (in case they are part of a distributed system, spawning more than one process and node within the network). Communication-related services can use this platform and gain access to the interfaces and functionality of standard communication protocols. Applications executed within the application layer can, in general, be grouped into support and end user applications. Services, in contrast, are special applications that require particular session semantics, provided by the service platform.

Configurable Network Platform

Following the open interface approach, all networking components in the system should be based on technologies providing open interfaces and open computing platforms, this includes both radio and networking layers of the architecture model in Figure 5.18. This will simplify the future deployment of new/future radio technologies, protocols and services. Software-defined radio technologies in reconfigurable terminals and programmable, active network nodes will enable the fast deployment of new services and applications. This, however, poses some requirements for the nodes in the network infrastructure, where, similar to the middleware layer, an architecture should include three programmable/reconfigurable components within the layer:

- A computing platform serves as a general purpose platform for processing of state protocols, e.g. routing, QoS signalling or connection management.
- A forwarding engine in the data path of a network node connects the different interface modules (e.g. by use of a switch matrix). This engine can be implemented as dedicated hardware or as a kernel module (as commonly done in most operating systems). The forwarding engine is programmable for performance-critical tasks, which are performed on a per-packet basis.
- The interface modules are media-specific and depend on the different wireless or wired standards. They can be configured or programmed to adapt to new physical layer protocols or to trigger events in higher layers.

Installation of new components and reconfiguration of the functionality within this layer can be performed by reconfiguration management mechanisms. Additionally, to be open programmable, the mobile terminal architecture must include some basic interface components. The main components of a terminal architecture, following the four layer model of Figure 5.18, include:

- a SIM card, e.g. USIM for UMTS, which includes subscriber identities and also a small, but highly secure, execution environment;
- programmable hardware, designed to implement any possible RAT;
- a native operating system which provides real-time support, needed for stacks and certain critical applications, e.g. multimedia codecs.

This again poses some major requirements from operators' and manufacturers' points of view, these include:

- survivability, e.g. robustness to misconfigurations, failures, or misusage;

- security for end user, operator and manufacturer requirements;
- mass-market optimization of hardware and software platforms, balanced with time-to-market and flexibility or upgrade requirements.

5.4.3.3 Services in Reconfigurable Systems

Service deployment and control of terminal and equipment reconfigurations, are complex procedures, this is partly due to the complexity of the technology, but also to the split of responsibilities between operator and manufacturer. Whilst the manufacturer has to provide appropriate and standard compliant low-level code for reconfigurations, the operator actually requires the control of configurations to fit the user needs and the network requirements.

For the widespread adoption of reconfigurable terminals, open interfaces are essential, and may be technically discussed and standardized in bodies like the SDR Forum [52] or by ETSI, or a '4G' standardization organization.

5.4.4 Procedures and Protocols

5.4.4.1 Installing a Radio Configuration

The descriptions for the various algorithms in a radio configuration can, and should be, structured in a hierarchical manner, which generates a highly transparent system. The principle of such a hierarchically organized SW structure for radio implementations is documented in [67], whereby the terminal can store a library of basic algorithms, or even complete radio configurations, and knowledge about how they are to be installed on the HW platform to implement the required RAT.

The principle of such a modular reconfiguration approach is based on the assumption that a terminal consists of a number of 'processing cells' (e.g. baseband processing cell (BPC) as mentioned earlier in this chapter), each of which implements a (high-level) module (e.g. BB/RF) within the terminal (see Figure 5.16). A reconfiguration process could thereby engulf the mere exchange of a few parameters, requiring a very small SW download, or anything up to complete terminal reconfiguration, requiring download of the complete SW structure to implement the required RAT.

The definition and realization of such a hierarchically ordered architecture, must consider a number of issues:

- What is a suitable hierarchy abstraction level and depth?
- Which information does a scheduler need for automatic scheduling?
- How can low-level parameters be set automatically?
- How do parameters effect the overall scheduling?

The hierarchical description of a radio configuration combines two advantages. The first is the abstraction of complex algorithms and complex SW module structures, which, when inherited, can be described with one module name only. The second is the access to every detail within the (hidden) complex structure. The following section describes generic and adaptive protocols as examples for such a hierarchically organized implementation.

5.4.4.2 Generic and Adaptive Protocols

Software-defined radio enables one to flexibly (re-)configure hardware and software within a mobile terminal for a new radio access technology. The functions required are specified for the target RAT, and the SDR is the enabler for the execution of these functions. This raises the requirement that the hardware and software platforms of the SDR terminal must be specified in such a generic way that they allow execution of all functions that may be required.

An SDR is thus characterized by parameterizing generic functions in the RF and BB parts of a mobile transceiver, such that they become compatible to a specific radio technology. In a similar way, the protocols applied above the physical layer can adapt to this paradigm and follow a similar approach by identifying generic functions at different protocol layers and making them configurable by defined sets of parameters. As already mentioned, this would reduce the amount of data to be sent to the mobile terminal in case a new access network needs to be contacted and the terminal reconfigured. Additionally, parameterization would speed up the download and reconfiguration process and could help to facilitate seamless handover between different access technologies.

Therefore, similar to the efforts to parameterize as much as possible in the BB/RF modules, investigations must also be pursued into the structural composition of the protocol steering software. This can be based on a separation of software into *permanent resident code* and *downloadable add-on modules*. Thereby, the permanent resident software must be as generic as possible (i.e. generic software skeleton or generic protocol stack). Using shared resources, the generic skeleton may support efficiency and flexibility at the same time. Assuming that devices are equipped with a generic kernel, the required reconfiguration software can be downloaded in a smart-update fashion, since only dedicated functionality needs to be included. The combination of generic stack and specific supplement then results in software that is able to support a dedicated air interface standard. The subsequent sections will elaborate on the way in which such a generic protocol stack can be constructed in a general way, followed by the more specific example of a 'generic link layer'.

Development of a Generic Protocol Stack

Using the classical ISO/OSI protocol stack reference model to compare the stacks of different RATs, a high degree of similarity can be found. Many of the features of the control software can be implemented as shared resources. Therefore, a certain software development process should be applied, called *design of generic and adaptive protocol software* (DGAPS). Applying DGAPS results in a generic protocol stack that provides a common basis for a number of different systems. Specialization by introducing standard-specific functions to the generic stack stepwise results in a specific realization towards a specific protocol stack.

In a first step (step 1) different systems, say System I and System II, need to be analysed layer by layer to identify their commonalities. A more detailed description of the analysis process, together with a reference implementation, is described in [67]. The number of different systems to be considered may be two or larger. The result will be a specification of a common subset of the access protocol stacks for the systems. Since this stack provides the common characteristics of the considered air interface standards, it is called a generic protocol stack, or protocol skeleton.

The next step is then to develop specifications dedicated to given air interface standards. These include functions that are specific to respective standards and, thus, represent the individual behaviour of a system. Different approaches can be taken to achieve that goal. In order to make use of the object-oriented properties, together with inheritance, it is suggested to implement these parts as subclasses derived from base classes implemented within the generic stack. This is of special advantage if more than two systems are considered; procedures that are common to most, but not necessarily to all, standards still will be implemented within the generic stack. The standard-specific supplements then will have to redefine/overload the respective procedures, and the behaviour required is then achieved. To result in a dedicated air interface standard, the generic protocol stack and the standard-specific supplement, have to be merged.

Definition of a Generic and Reconfigurable Link Layer
In mobile communications, the radio link is, in general, the bottleneck of the end-to-end path, and it is costly, or even impossible, to increase its capacity. Therefore, it is required to utilize the available radio resources in the most efficient way. This role is performed by sophisticated radio physical layers and radio link layers, which are optimized to the radio access technology in use. In Figure 5.20 (left side) a simplified protocol stack is depicted, in which a correspondent node communicates with a mobile terminal. The end-to-end connection is established with, e.g., the Internet Protocol (IP). The radio link layer (here called GLL) and the radio physical layer (PHY) enable data transmission over the radio link.

Assuming that the radio link is the bottleneck, most data which is currently in the process of being transmitted along the end-to-end path is typically either queued or being processed within the radio link layer. This approach has proven to be an efficient design for communication in today's mobile communication networks, where a specific mobile network with a specific radio access technology is used. However, there are certain limitations in the context of cooperating networks, in which seamless communication is desired via a multitude of mobile networks, which may deploy different radio access technologies.

This scenario is depicted in Figure 5.20, where a mobile terminal dynamically selects one of the available radio access networks (RAN A or RAN B) during a session. Each radio access network uses its specific radio link layer and radio physical layer (left side of Figure 5.20). During an intersystem handover from RAN A to RAN B, the radio link in

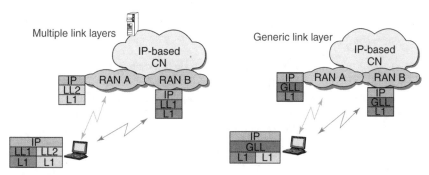

Figure 5.20 Multiple link layer scenario compared to generic link layer scenario (GLL: generic link layer; LL: link layer; L1: Layer 1; RAN: radio access network; CN: core network)

RAN A is torn down and a new radio link is being set-up in RAN B. Such a handover can only be without loss if a further layer of error recovery is applied, e.g. end-to-end on top of IP. But even then, an intersystem handover is neither efficient nor without disruption to the service.

Since different radio link layers have, in general, the same functionality for all radio access technologies, this problem can be solved if the radio link layers are made compatible. The old radio link layer state can then be handed over to the new radio link layer, which continues the transmission in a seamless way. This is achieved by defining a generic link layer, which can be used as radio link layer for all radio links (right side of Figure 5.20).

The generic link layer (GLL) is a specified radio layer protocol, which provides the link layer functions required in every radio link layer [67]. It can be configured in a flexible manner to perform these link layer functions in an optimized way for different radio access technologies with different properties. The generic specification of radio link layer functions enables reconfiguration of the generic link layer, in which the existing communication context at the time of reconfiguration is transformed into a new context within the new configuration. As a result, the communication session can 'survive' the reconfiguration procedure, lossless and without disruption. From a service perspective, it is a seamless reconfiguration.

The GLL concept requires a reconfiguration of the GLL on both sides of the wireless link in the mobile terminal, as well as in the radio access node. This is required in order to seamlessly continue with the old context of the communication. To implement a GLL the following interfaces and reference points have to be defined:

- *The higher layer interface:* Via the interface to the higher protocol layer, data is received for transmission and delivered after reception. This interface further allows configuration of the QoS requirements for the transmission of higher layer datagrams.
- *The physical layer interface:* At the interface to the physical layer, radio blocks are sent to the physical layer for transmission over the radio link.
- *The control interface:* Via the control interface, the generic link layer is configured and reconfigured.
- *The internal interface to embed specific functions:* Via this interface, it will be possible to include a specific function, e.g. a ciphering algorithm, into general functions.

5.4.5 Research Challenges

Complete reconfigurability of SDR equipment will only be achieved if single software modules within the SDR platform can be reconfigured individually. This necessitates clearly defined interfaces between the various modules, and independence between them. To develop a complete framework of interfaces between the building blocks of reconfigurable equipment (terminal/base station, etc.), a number of research issues need to be solved:

- Definition of an overall architecture for future mobile communication systems, based on a set of configurable abstraction layers (programmability is, to a small extent, already

realized in second and third generation systems, but fully flexible component configuration in all layers will be a major breakthrough for the creation and deployment of
fourth generation services).

- The specification of abstraction layer functionality, including the integration of existing
  and emerging heterogeneous systems and networking environments.
- The definition of cross-layer information flows to assist service support on all abstrac-
  tion layers.
- Clear separation of tasks between the different reconfigurable modules (modules are as
  indicated in Figure 5.16).
- Generalization and specification of individual generic interfaces between RF and BB,
  as well as between BB and the GP modules.
- Identification of all reconfiguration control/management related functions.
- Definition of a generic control/management interface between the individual radio modules and the reconfiguration manager.
- The identification of a generic framework comprising policy-based reconfigurable functional entities in various layers and applications, as well as layer-independent reconfiguration management and adaptable service creation, deployment and provision environment.

Research issues that need to be addressed for the introduction of a generic framework
for reconfigurable mobile environments are:

- Identification of open APIs for reconfiguration actions, triggering in various layers
  and components.
- Identification of generic reconfigurability management functionality.
- Identification of advanced network/terminal related applications and functionality for
  flexible service provision, software/protocol downloading/offloading and various access
  systems integration.
- Introduction of adaptability enabling architectures and environments.

There are a number of research issues that need to be addressed in the definition of
generic and adaptable protocols and protocol stacks, these include:

- Definition of elementary commonalities of the various mobile communication systems.
- Resource sharing within telecommunication software.
- Modular software design and interfaces.
- Reusability of software.
- Structure of a generic protocol stack.
- Nature of software extensions to an existent system.
- Composition of an adaptive protocol stack architecture.
- Identification, description and specification of GLL functions.
- Description and specification of GLL interfaces.
- Configuration management and reconfiguration of the GLL.
- A protocol framework for the GLL.
- Integration of the GLL within SDR.

As described in this chapter, various research efforts have been identified and initiated
in different ways. Each result exhibits advantages and discrepancies, parallel implementations and evaluations should be conducted to define the most appropriate one. These

research efforts exhibit a model of SDR based on components or layers interconnected with interfaces, and supported by an operating environment. It is obvious that the next step will be to merge those efforts in the definition of a common and shared model, and to define the associated SDR framework enabling the development of baseband services and waveforms designed for reconfigurability. This SDR framework will also enable the development of the supervision and management services at the network level.

5.5 References

[1] http://mobivas.cnl.di.uoa.gr/

[2] Alonistioti, A., Houssos, N., Panagiotakis, S., Koutsopoulou, M. and Gazis, V., 'Intelligent architectures enabling flexible service provision and adaptability,' Wireless Design Conference (WDC 2002), London, UK, 15–17 May 2002.

[3] Panagiotakis, S., Houssos, N. and Alonistioti, A., 'Integrated generic architecture for flexible service provision to mobile users,' IEEE 12th International Symposium on Personal, Indoor and Mobile Radio Communications (PIMRC 2001), San Diego, California, USA, 30/9–3/10 2001, pp. 40–44.

[4] Houssos, N., Gazis, V., Panagiotakis, S., Gessler, S., Schuelke, A. and Quesnel, S., 'Value Added Service Management in 3G Networks', 8th IEEE/IFIP Network Operations and Management Symposium (NOMS 2002), Florence, Italy, 15–19 April 2002, pp. 529–545.

[5] Panagiotakis, S., Houssos, N. and Alonistioti, A., 'Generic architecture and functionality to support downloadable service provision to mobile users,' 3rd Generation Infrastructure and Services Conference (3GIS), Athens, Greece, July 2001.

[6] Houssos, N., Pantazis, S. and Alonistioti, A., 'Towards adaptability in 3G service provision,' IST Mobile Communication Summit 2002, Thessaloniki, Greece, 16–19 June 2002.

[7] Nakajima, N., Kohno, R. and Kubota, S. 'Research and Developments of Software-Defined Radio Technologies in Japan,' IEEE Communications Magazine, 39(8), 146–155, 2001.

[8] Laster, J.D., 'Robust GMSK Demodulation Using Demodulator Diversity and BER Estimation,' PhD Thesis, Virginia Tech, 1997.

[9] SCOUT webpage: http://www.ist-scout.org.

[10] Qiu, X. et al. 'Network Assisted Resource Management for Wireless Data Networks', IEEE Journal on Selected Areas in Communications, 19(7), 1222–1234, 2001.

[11] Yomo, H. and Hara, S., 'Impact of Access Schemes Selectability on Traffic Performance in Wireless Multimedia Communication Systems', IEEE Transactions on Vehicular Technology, 50(5), 1298–1307, 2001.

[12] http://www.tinac.com/about/about.htm.

[13] www.sdrforum.org, 2002.

[14] Buracchini, E., 'The Software Radio Concept,' IEEE Communications Magazine, September 2000.

[15] http://www.sdrforum.org/tech_comm/mobile_wg.html.

[16] www.mobilevce.com, 2002.

[17] 3GPP TS 23.057: 'Mobile Execution Environment (MExE) Functional Description,' Version 5.0.0, March 2002.

[18] Lachlan B.M., Miodrag, J.M., Haruyama, S. and Kohno, R., 'A Framework for Secure Download for Software Defined Radio,' IEEE Communications Magazine, 88–96 July 2002.

[19] Springer, A., Maurer, L. and Weigel, R., 'RF System Concepts for Highly Integrated RFICs for W-CDMA Mobile Radio Terminals,' IEEE Transactions on Microwave Theory and Techniques, 50(1), 254–267, 2002.

[20] Sagers, R.C., 'Intercept Point and Undesired Responses,' IEEE Transactions on Vehicular Technology, 32, 121–133, 1983.

[21] Crols, J. and Steyaert, M.S.J., 'Low IF Topologies for High Performance Analog Front Ends of Fully Integrated Receivers,' IEEE Transactions on Circuits and Systems – II: Analog and Digital Signal Processing, 45(3), 269–282, 1999.

[22] Banu, M., Wang, H., Seidel, M., Tarsia, M., Fischer, W., Glas, J., Dec, A. and Boccuzzi, V., 'A BiCMOS Double-Low-IF receiver for GSM,' IEEE 1997 Custom Integrated Circuit Conference, pp. 521–524.

[23] Abe, M., Sasho, N., Brankovic, V. and Krupezevic, D., 'Direct Conversion Receiver MMIC based on Six-Port Technology,' European Conference on Wireless Technology, 2000.

[24] Hyyryläinen, J., Bogod, L., Kangasmaa, S., Scheck, H.O. and Ylämurto, T., 'Six-port Direct Conversion Receiver,' in *Proceedings of 27th European Microwave Conference*, **1**, 341–346, 1997.

[25] Mann, I., Beach, M.A., Warr, P.A. and McGeehan, J.P., 'Increasing the talk-time of mobile radios with efficient linear transmitter architectures,' *IEE Electronics & Communication Engineering Journal*, **13**(2), 65–76, 2001.

[26] Warr, P.A., Beach, M.A. and McGeehan, J.P., 'Gain-element transfer response control for octave-band feedforward amplifiers,' *IEE Electronics Letters*, **37**(3), 146–147, 2001.

[27] Nesimoglu, T. and Beach, M.A., '*Linearised mixer using frequency retranslation*,' UK Patent Application No. 0117801.1, 20th July 2001.

[28] Nesimoglu, T., Beach, M.A., Warr, P.A. and MacLeod, J.R., 'Linearised Mixer using Frequency Retranslation,' *IEE Electronics Letters*, **37**(25), 1493–1494, 2001.

[29] Katzin, P. and Aparin, V., '*Active, Self Adjusting, L-S Band, MMIC Filter,*' IEEE GaAs Symposium, pp. 41–43, 1994.

[30] Katzin, P. *Personal Correspondence*, January 2001.

[31] Loo, R.Y., Tangonan, G., Sivenpiper, D., Schaffner, J., Hsu, T.Y. and Hsu, H.P., 'Reconfigurable Antenna Elements using RF MEMS Switches,' in *Proceedings of ISAP2000*, Fukuoka, Japan, pp. 887–890, 2000.

[32] Rostbakken, O., Hilton, G.S. and Railton, C.J., 'Adaptive Feedback Frequency Tuning for Microstrip Patch Antennas', in *Proceedings of 9th International Conference on Antennas and Propagation*, Part 1, **1**, 166–170, 1995.

[33] Johnson, R.C. (Ed.), '*Antenna Engineering Handbook*', 3rd edition, McGraw-Hill, New York, 1993.

[34] Ali, M., Hayes, G.J., Hwang, H.S. and Sandler, R.A., '*A Triple-band Integrated Antenna for Mobile Hand-Held Terminals,*' APS 2002, pp. 32–35, 2002.

[35] Fusco, V., 'Integrated Antennas for Wireless Applications', *Applied Microwave and Wireless*, 22–31, 2002.

[36] Long, S.A., Mcallister, M.W. and Shen, L.C., 'The resonant cylindrical dielectric cavity antenna,' *IEEE Transactions on Antennas and Propagation*, **31**, 406–412, 1983.

[37] Richards, R.J. and De Los Santos, H.J., 'MEMS for RF/Microwave Applications The Next Wave – Part II,' *Microwave Journal*, **44**(7), 142–152, 2001.

[38] Teetzel, A., '*Circuit Design: Design of a wideband IQ modulator*,' RF Design Seminar, Hewlett Packard Corporation, 1997.

[39] http://products.analog.com/products/info.asp?product=AD8347.

[40] Razavi, B., 'Design Considerations for Direct-Conversion Receivers,' *IEEE Transactions on Circuits and Systems II: Analog and Digital Signal Processing*, **44**(6), 428–435, 1997.

[41] Parssinen, A., Jussila, J., Ryynanen, J., Sumanen, L. and Halonen, A.I., 'A 2-GHz Wide-Band Direct Conversion Receiver for WCDMA Applications,' *IEEE Journal of Solid – State Circuits*, **34**(12), 1893–1903, 1999.

[42] www.ist-trust.org, 2002.

[43] WSI, '*Book of Visions for the Wireless World*,' www.ist-wsi.org, December 2000.

[44] 3GPP, '*Open Service Access (OSA): Application Programming Interface (API)*,' Part 1–12 (version 4.3.0), TS 29.198.

[45] Open Services Gateway Initiative (OSGi) *Service Platform*, Release 2, available from http://www.osgi.org/resources/docs/spr2book.pdf, October 2001.

[46] Parlay Group, '*Parlay API Spec. 2.1*,' available from http://www.parlay.org/specs/index.asp, July 2000.

[47] Keijzer, J., Tait, D., and Goedman, R., 'JAIN: A new approach to services in communication networks,' *IEEE Communications Magazine*, January 2000.

[48] LIF '*Mobile Location Protocol*' v2.0.0, TS 101.

[49] UMTS Forum, Report #9, http://www.umts-forum.org/.

[50] Panagiotakis, S., Koutsopoulou, M., Alonistioti, N. and Kaloxylos, A., '*Generic Framework for the Provision of Efficient Location-based Charging over Future Mobile Communication Networks,*' PIMRC, Lisbon, Portugal, September 2002.

[51] Panagiotakis, S., Koutsopoulou, M. and Alonistioti, N., '*Advanced Location Information Management Scheme for Supporting Flexible Service Provisioning in Reconfigurable Mobile Networks,*' IST Mobile Communication Summit, Thessaloniki, Greece, June 2002.

[52] www.sdrforum.org, 2002.

[53] www.ist-trust.org, 2002.

[54] www.mobilevce.com, 2002.

[55] Bourse, D., '*WWRF WG3 SDR Architectures*,' a document circulated in the SDR community within the Wireless World Research Forum, May 2002.

[56] http://www.sdrforum.org/tech_comm/mobile_wg.html.

[57] WSI, '*2nd Book of Visions for the Wireless World*,' www.wireless-world-research.org, December 2001.

[58] Alonistioti, N., Houssos, N., Panagiotakis, S., Koutsopoulou, M. and Gazis, V., '*Intelligent architectures enabling flexible service provision and adaptability*,' Wireless Design Conference (WDC 2002), London, UK, 15–17 May 2002.

[59] Metha, M. and Wesseling, M., 'Adaptive Baseband Subsystem for TRUST,' in *Proceedings of PIMRC 2000*, London, UK, 18–21 September 2000, pp. 29–33.

[60] Gultchev, S., *et al.*, '*Management and Control of Reconfiguration Procedures in Software Radio Terminals*,' 2nd Workshop on Software Radio, Karlsruhe, Germany, 20–21 March 2002.

[61] Moessner, K., '*Reconfigurable Mobile Communication Networks*,' Doctoral Thesis, University of Surrey, UK, 2001.

[62] Moessner, K., *et al.*, '*Software Radio Integration and Reconfiguration Management*,' WWRF, Paris, 2001.

[63] Mitola, J., '*Software Radio Architecture Object-Oriented Approaches to Wireless Systems Engineering*,' Wiley-Interscience, 2000.

[64] Kristiansen, L., '*TINA-C Service Architecture*,' Version 5.0, TINA-C, 1997.

[65] OMG Telecommunications Specifications, *Telecom Service Access and Subscription*, http://www.omg.org/cgi-bin/doc?dtc/2000-10-03.

[66] The Parlay Group, http://www.parlay.org, 2002.

[67] Sachs, J. and Schieder, A., '*Generic Link Layer*,' Wireless World Research Forum, Tempe, USA, 7–8, March 2002.

5.6 Credits

The following individuals have contributed to the contents of this chapter: Nancy Alonistioti (University of Athens, Greece); Didier Bourse (Motorola CRM-Paris, France); Soodesh Buljore (Motorola CRM-Paris, France); Antoine Delautre (Thales Communications, France); Markus Dillinger (Siemens AG, Germany); Rainer Falk (Siemens AG, Germany); Dieter Greifendorf (IMS Fraunhofer Gesellschaft, Germany); Mirsad Halimic (Panasonic PMDO, UK); Tim Hentschel (Technische Universität Dresden, Germany); Quiting Huang (ETH ISL, Switzerland); Apostolos Kountouris (Mitsubishi Electric ITE-TCL, France); John MacLeod (University of Bristol, UK); Ashok Marath (ICR, Singapore); Linus Maurer (DICE, Austria); Stefano Micocci (Siemens MC, Italy); Klaus Moessner (CCSR, The University of Surrey, UK); Nikolas Olaziregi (CTR, King's College London, UK); Parbhu D. Patel (Panasonic PMDO, UK); Santhosh Kumar Pilakkat (ICR, Singapore); Christian Prehofer (DoCoMo Communications Labs Europe, Germany); Tapio Rautio (VTT, Finland); Joachim Sachs (Ericsson, Germany); Andreas Schieder (Ericsson, Germany); Wolfgang Schott (IBM, Switzerland); Matthias Siebert (Aachen University, ComNets, Germany); Joerg Stammen (IMS Fraunhofer Gesellschaft, Germany); Shiao-Li Tsao (CCL/ITRI, Taiwan); Paul Warr (University of Bristol, UK); Thomas Wiebke (Panasonic European Laboratories, Germany) and Manfred Zimmermann (Infineon, Germany).

6

Technologies to Improve Spectrum Efficiency

Edited by Bernhard Walke and Ralf Pabst (Aachen University, ComNets)

A naturally existing trade-off between the transmitted power and the improvement of efficiency in the use of the radio spectrum can be exploited through the application of ad hoc networking of wireless capable nodes. Apart from this, it is common belief that the success of the 'everything, everybody always connected' idea will strongly rely on a wide variety of components, systems and solutions that need to be looked at. Examples are:

- wide area networks (WANs);
- local area networks (LANs);
- personal area networks (PANs);
- networking between appliances at home and in the office;

all of these possibly comprising permanently installed base stations or access points, and fixed or mobile relay nodes to assist serving mobile or wireless terminals.

Technical issues for all the protocol layers also need to be considered. A nonexhaustive list is provided below:

- network layer: addressing, routing, mobility and topology management;
- medium access control layer; centralized vs. distributed, fairness, quality of service support, varying node capabilities;
- management of spectrum vs. device resources (transmit power, link adaptation, processing);
- autoconfiguration: service discovery, network dynamics;
- authentication, authorization and accounting (AAA) and information security and user privacy.

Technologies for the Wireless Future. Edited by R. Tafazolli
© 2005 John Wiley & Sons, Ltd ISBN: 0-470-01235-8

This chapter elaborates on the following topics:

Ad Hoc Networks

Mobile ad hoc networks are formed by wireless devices that communicate without necessarily using a pre-existing network infrastructure. Ad hoc networks are self-configuring, i.e. there is no (central) management system with configuration responsibilities. Some, if not all, nodes in an ad hoc network are capable of assuming router functionality when needed. This enables terminals to communicate with each other when they are out of (radio) range, provided they can reach each other via intermediate hosts acting as routers that relay the packets from source to destination. The structure of the network can change constantly because of the movement of the nodes. Ad hoc networks can be viewed as stand-alone groups of mobile terminals, but they may also be connected to a pre-existing network infrastructure and use it to access hosts which are not part of the ad hoc network.

It can be expected that, in the near future, there will be a proliferation of wireless devices. Ad hoc network functionality, such as self-configurability and independence of existing infrastructures, are key issues in this context. Examples are personal area networks (PANs), body area networks (BANs), home networks, networks of sensors and actuators (e.g. at home, cars, or those for ambient intelligence), or vehicle to vehicle networks.

No less important, the multihop communication capabilities of ad hoc networks can be used to extend the coverage of existing wireless access technologies. Not only is this an interesting approach for cellular networks, but particularly in the case of high-frequency wireless local area networks (WLANs), due to opacity problems.

A third interesting aspect of ad hoc networking has to do with the intrinsic characteristics of ad hoc networks, such as self-configurability and neighbour discovery, which imply that these networks will be a key element for enhancing the interoperability among different wireless technologies.

Relay-based Deployment Concepts

In recent years, there has been an upsurge of interest in multihop-augmented infrastructure-based networks in both industry and academia, such as the 'seed' concept in 3GPP, mesh networks in IEEE 802.16, coverage extension of HiperLAN/2 through relays or user-cooperative diversity mesh networks. The WP on relay-based deployment concepts covers a number of different approaches to exploiting the benefits of multihop communications via fixed and/or mobile relays, such as solutions for radio range extension in mobile and wireless broadband cellular networks (trading range for capacity), and solutions to combat shadowing at high radio frequencies. Further, relaying is presented as a means to reduce infrastructure deployment costs. It is shown that multihop relaying can enhance capacity in cellular networks, e.g. through the exploitation of spatial diversity. Cooperative relaying is introduced as a measure to apply macrodiversity to improve the signal quality at a given receiver. Multihop wireless networking traditionally has been studied in the context of ad hoc and peer-to-peer networks. The application of multihop networking in wide area cellular systems, and in mobile broadband systems for densely populated areas, is expected to yield many benefits.

Spectrum Issues

The text on spectrum addresses spectrum for terrestrial services, and focuses on the 3G/IMT-2000 and systems B3G. The B3G systems will cover a wide range of wireless communications, ranging from low bit rates up to 1 Gbps in hot spots, and from low mobility, local coverage to high mobility, wide area coverage. Areas are described where focused research is required before the spectrum related challenges can be met, and spectrum identified and eventually made available. The areas for research are traffic characteristics of future systems and services, technical developments influencing spectrum efficiency and usage, efficient sharing methods, novel and flexible spectrum use and future spectrum requirements.

The identification of frequency bands is a long international process and it is important to take spectrum requirements into account in early research and to communicate the relevant results to the regulators to enable them to get prepared for the possible new spectrum requirements in time. There is a continuous pressure towards higher spectrum efficiency and greater flexibility. This puts high demands on the new technologies, especially on new air interfaces, antenna technologies, network topologies, etc.

It is expected that an evolved wireless world will require new arrangements for utilizing the rare radio spectrum. Hence, the topics of systems coexistence and network interworking need to be duly addressed. A large number of wireless technologies for varied applications exist and several new technologies are being developed. Moreover, multiple existing wireless technologies are being improved for their field of application. An approach based on dynamic spectrum refarming should be followed, so that the use of radio spectrum <1 GHz for cellular and mobile communications is maximized, simultaneously, more efficient use of spectrum through system overlay and temporary allocation can be achieved. Recommendations for spectrum sharing for the coexistence of several technologies and different applications shall be formulated through the comparison of efficiency/complexity trade-offs.

6.1 Ad Hoc Networking

6.1.1 Ad Hoc Network Characteristics

Networks that we know and use today are based on fixed infrastructure and centralized administration. Before accessing a network, users have to obtain all configuration parameters (get a SIM card for a mobile phone, get a CD for internet access, talk to network administrator, etc.) and usually to configure their devices manually. Even when only a simple connection between two users is required, a lot of configuration work is usually needed to get the connection up and running. The level of technical expertise required for a particular set up may range from a very basic knowledge to high-level technical expertise.

In contrast to these infrastructure-based networks, ad hoc networks are organized spontaneously, when and where needed, solely by nodes wishing to communicate, without the need for the infrastructure support. Ad hoc communication can take place in different scenarios and is independent of any specific device, wireless transmission technology, network or protocol. Nodes are free to enter or leave the network at any time. Networks can significantly vary in size (from a few devices comprising a PAN, to hundreds of sensors comprising a wireless sensor network). All network functions and protocols are distributed and executed by all network participants.

Devices in ad hoc networks can be diverse (laptops, PDAs, camcorders, mobile phones, MP3 players, game stations, sensors, etc.) and with various characteristics like throughput, transmission power, energy resources, size or cost. The common features of all ad hoc network devices are limited energy resources and capability to communicate using one or more wireless technologies. Bluetooth, WLAN 802.11 and UWB are the most frequently considered technologies for use in various ad hoc network scenarios. An example of an ad hoc network of eight wireless terminals based on IEEE 802.11a is given in Figure 6.1.

The following are some of the possible ad hoc networking scenarios [1–6]:

- *Personal use:* noncommercial transfer of data between devices or persons; communication in areas without adequate wireless coverage or short range peer-to-peer communications in an ad hoc group, in which it does not make (economic) sense to use an operator network, as illustrated by a group of hikers wishing to communicate.
- *Commercial use:* setting up communication in exhibitions, conferences or sales presentations.
- *Sensor networks:* communication between, or with, intelligent sensors.
- *Search and rescue operations:* communication in areas without adequate wireless coverage, or when the existing communication infrastructure is nonoperational due to a natural disaster or a global war.
- *Vehicle communication networks:* crash avoidance warning system, safety distance for cruise control for cars, trains, airports, etc.

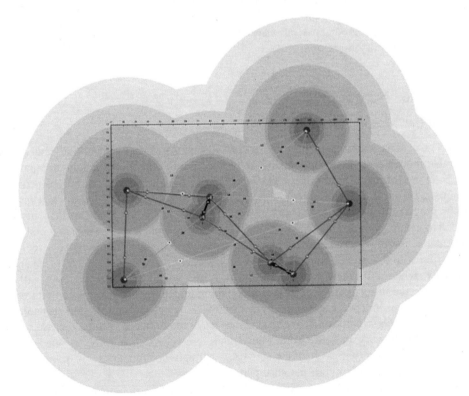

Figure 6.1 ViTAN ad hoc network with eight wireless terminals

- *Military applications:* for fast establishment of communication during the deployment of forces in unknown and hostile terrain.

The main features of ad hoc networks are as follows:

- *Dynamic network topology:* due to the node mobility and radio propagation, network topology is constantly changing. This requires specific network protocol functions for topology construction and maintenance.
- *Distributed nature:* this is an inherent characteristic of ad hoc networks. As there is no permanent central authority, all networking functions have to be distributed across participating nodes.
- *Multihop communications:* due to the limited range of wireless interfaces, usually it is not possible to establish direct communication links with all nodes. As there is no infrastructure to support establishment of multihop routes, the nodes themselves have to run routing algorithms to establish routes in the network, and to forward packets destined for other nodes.
- *Limited bandwidth:* wireless technologies that are envisaged to be suitable for ad hoc networks provide throughputs of a few hundred kilobits per second to a few megabits per second, suitable for many applications. However, the wireless environment is a harsh one and can cause significant error rates, which are aggregated along the multi-hop links.
- *Limited energy resources:* as ad hoc network nodes will usually be battery driven, optimization of energy consumption across all protocol stack layers is extremely important.

6.1.2 Ad Hoc Networking Scenarios, Requirements and Research Issues

In this section, typical ad hoc networking scenarios, their requirements and relevant research issues are presented.

6.1.2.1 Stand-alone Ad Hoc Networks

Stand-alone, isolated, ad hoc networks are the simplest form of ad hoc networks. These networks are established by a certain number of nodes in order to provide, or use, services between themselves. Communication links to other networks do not exist. These are some of the possible scenarios in which stand-alone ad hoc networks are used:

- When people attending a meeting or a conference want to establish a network between themselves to exchange files, presentations or to share applications.
- When people on a bus, train or in a lounge room want to play a network game against each other.
- When firemen are involved in difficult operations inside buildings, towers, etc., or surrounded in forest fires, needing robust communications between each one.
- When emergency and medical services personnel establish an ad hoc network to communicate with each other after a catastrophe has destroyed the regular communication infrastructure.

- When a group of hikers walking through the fog, try to stay close together and avoid losing somebody in a deep canyon or a river.
- When soldiers in a battlefield are exchanging information about their position and the position of enemy forces, or are giving and receiving orders and instructions.

Although it is possible to envisage very large ad hoc networks (soldiers in the battlefield), a more realistic estimation is that networks will usually be small to medium size (up to 100 nodes) with the longest routes not longer than 10 hops. Mobility of nodes participating in the network will be low (people sitting in a bus) to medium (emergency services personnel in a field). Throughput available in the network should be large enough to accommodate various multimedia applications. Since the network is established spontaneously, available services are not known in advance, and appropriate mechanisms have to be provided for service discovery. Ensuring integrity and privacy of information is especially important, bearing in mind that other, not necessarily trusted, nodes relay data. Energy efficiency of all protocols is a must, due to the limited available power resources of devices.

IP and WLAN 802.11 based ad hoc networks have been the most frequent focus of ad hoc networking research so far [7, 8, 9, 10].

The most important research issues in regard to this ad hoc network scenario are described in the following sections. It should be noted that most of the following issues are applicable to other ad hoc network scenarios too.

Network Organization
Ad hoc networks are expected to be capable of self-organization and maintenance. The actual level of the self-organization capability of an ad hoc network will have a profound impact on the user's experience and satisfaction.

Depending on the wireless technology used, various tasks have to be performed in this phase. If Bluetooth is used, devices have to detect each other, establish a communication link and to determine if they have a compatible set of supported services. In WLAN 802.11 networks, the broadcast nature of the communication channel is used to advertise presence of nodes. Once a new node is recognized, address allocation, service discovery, routing protocols, etc. are invoked. Other MAC protocols might need to allocate timeslots, frequencies or codes for transmission and perform network clustering before any network and application level communication can start.

As ad hoc network topology changes frequently and nodes are joining, leaving or changing position in the network, network organization protocols have to ensure a smooth and uninterrupted functioning of the network. Nodes running out of battery should be dynamically freed from routing assignments before they die. If dedicated nodes are used (ad hoc cluster controller), such a node leaving the network can cause problems. The functionality of the leaving node has to be adopted by another node (assuming that there exists a node which supports this functionality). The configuration becomes easier if overlay nodes (no mobility, fixed routes) are used to span an overlay network. But this cannot always be assumed. In WLAN 802.11 based networks, routing protocols take care of some aspects of network maintenance through route maintenance procedures.

Address Assignment
As previously stated, stand-alone ad hoc networks are usually considered to be IP-based networks, and therefore each node is designated with an IP address, assigned to the node in

advance, or dynamically by an addressing allocation server. However, certain applications can use other addressing schemes: wireless sensor networks are usually based on attribute-based naming schemes, where nodes do not have unique addresses, but are designated by their capabilities; in Bluetooth networks, each node has a unique Bluetooth address that can be used on higher communication layers as well.

Here, automatic IP address allocation protocols are considered. Since ad hoc networks do not have a central authority responsible for address allocation coordination, this functionality has to be distributed across all network nodes.

A protocol for addressing autoconfiguration in IPv4 ad hoc networks is proposed in [18]. Addresses are randomly selected from a special part (169.254/16) of the network address space. Duplicated address detection (DAD) is used to eliminate duplicated addresses; this approach uses route discovery messages from a reactive routing protocol like AODV or DSR. DAD is performed only once per node. Hence, the uniqueness of addresses cannot be guaranteed after merging two networks. This approach is not suitable for large ad hoc networks. It should be considered that mobile nodes could have more than one interface to different networks, and therefore may require multiple IP addresses.

Another approach, called Dynamic Registration and Configuration Protocol (DRCP), tries to modify DHCP to an autoconfiguration protocol for wired and wireless networks. Therefore, each node represents a DRCP client and server and owns an IPv4 address pool. The dynamic address allocation protocol (DAAP) is responsible for the distribution of the address pools. Each node requesting a pool gets half of the pool of a neighbouring node. This results in a lot of unassigned addresses in an already scarce IPv4 address space. Network merging is not considered either.

A promising method based on Mobile IP [12], consists of a home address and a care of address that is built by using a distinct prefix for each subnet [13]. The locally assigned address could be used as the care of address, whereas the unique home address could enable the authentication, authorization and, hence, the accounting, similar to the IMSI (international mobile subscriber identity) in GSM.

IPv6 Stateless Address Autoconfiguration (SAA) is another proposed approach. It is a hierarchical solution that works together with the LANMARK routing protocol [14].

Further investigations will show whether private IP addresses could be used as local subnet addresses, and whether the edge router can perform an address translation. Nevertheless, the proposed addressing scheme is basically developed for IPv6, and future approaches have to be developed.

Service Discovery

Ad hoc networks are organized 'on-the-fly', opportunistically, and therefore, available services and service providers are not known in advance. It is the responsibility of service discovery protocols to provide that information. Obviously, as a permanent central service information database does not exist, this protocol has to be distributed across all network nodes.

Service discovery protocols have to enable not only discovery of services available in the 1-hop range, but all services available within a multihop ad hoc network. An efficient protocol should also ensure that services and infrastructure are not underutilized. It should first identify the existence of a service, and then decide if the existent technology can bind to it, and finally establish a session successfully. In order to do that, it should be capable of giving the ability to the devices to announce their presence to the network and describe their capabilities. It should also be independent of the transmission protocol.

Several service discovery protocols have been developed and proposed, primarily for wired networks. In centralized pull protocols, clients pull the services whenever needed from a central component (called the central registry) where all the existent services in the network are registered. Distributed pull protocols pull services from the network itself, and in distributed push protocols, service providers push information concerning the services to the network.

'Universal Plug 'n Play' uses the distributed pull method and relies on HTTP and TCP/IP. Its main drawbacks are that it supports only known devices, and that it does not support many network configurations.

'Jini' is a Java based service discovery protocol that uses the centralized pull method. Its main disadvantages are that it does not support many network configurations, and that its centralized service discovery is not suitable for ad hoc networks.

'Salutation' supports both centralized and distributed service discovery, transport-independent addressing, and device capability exchange. It is designed to function in pervasive and heterogeneous networks, and above most of the network protocols. Its main problems are lack of leasing functionality and complex addressing.

IBM DEAPspace supports the distributed push method and the service description is fulfilled with strings and XML. Its main drawbacks are its word view of services and its emphasis on devices.

The above-mentioned wired network proposals have to be adapted for the wireless world and, especially, for the highly dynamic ad hoc networks. Appropriate strategies have to be investigated, which may include hierarchies for service distribution and announcement.

Routing and Relaying
In ad hoc networks, a direct communication between any two nodes is possible, subject to adequate radio propagation conditions and transmission power limitations of the nodes. However, only in rare cases will direct communication with all nodes in a network exist. Usually, multihop communication paths will have to be used. Care of communication route establishment and maintenance is taken by all network nodes using adequate routing protocols.

Ad hoc network routing is a very challenging task for several reasons: high mobility of nodes which can join and leave the network at any time, thus causing network topology changes and making routing tables obsolete; the bandwidth of the wireless channels is limited and has to be used carefully, thus requiring the routing overhead to be kept at a minimum; the wireless channel is susceptible to various interferences, low throughput and other problems; the limited energy resources of network nodes impose severe constraints.

A number of ad hoc routing protocols have been proposed, primarily as part of the IETF's MANET (mobile ad hoc networks) working group activities [15, 16, 17, 18]. These protocols are designed for IP-based, homogenous, mobile ad hoc networks, and focus on fast route establishment, re-establishment and maintenance with a minimum overhead. Each node in the network is assumed to have identical capabilities (wireless communication interface and ability to perform functions from the common set of services) and a unique IP address. The number of hops is used as the only route selection criterion. Other parameters, like route delay, energy usage, fair distribution of power usage among terminals, load balancing or quality of service are not considered.

The two main groups of the proposed protocols are *proactive* and *reactive* protocols. Proactive protocols continuously update the topological view of the network by exchanging appropriate information among the network nodes and, thus, immediately have a route to a destination when required. A typical example of the proactive group of protocols is OLSR (optimized link state routing) [18]. It is an optimization of the classical link state algorithm, tailored to the requirements of a mobile wireless LAN.

The main problem of the proactive approach in ad hoc networks stems from the fact that topology of ad hoc networks is changing continuously. Hence, a frequent dissemination of topology information is required, which causes a large routing overhead. Also, depending on the traffic pattern in the ad hoc network, it is possible that only a small fraction of routes is used, which leads to a waste of already constrained wireless and computing resources.

AODV and DSR are examples of reactive, or 'on demand', routing protocols [16, 17]. These protocols do not maintain the overall network topology, but instead maintain only those routes that are in use. When a route is not used anymore it is removed from routing tables. When a new route is required the network is flooded with 'route request' messages. When the destination or a node, which has a route to the destination, receives a 'route request' message, a 'route reply' message is generated and sent back to the source node.

One of the critical issues in routing protocol design is neighbourhood detection, i.e. how a node can discover its neighbours and the quality of communication links to them. 'Hello' messages, that each device broadcast periodically, are usually used for this purpose. However, this mechanism is not always reliable, and in some scenarios can be a cause of frequent route breakages and poor network performance.

Wireless transmission consumes a lot of energy and significant savings would be possible if transmit power control was introduced. Instead of always using maximum transmit power, it could be reduced, based on the actual distance between the nodes (depending on the distance, multihop routes could be more energy effective than a 1-hop route, even when direct communication is possible). Another energy saving can be achieved by introducing a 'sleeping' mode [19]. However, before a node take an active part in routing, it has to awake, and that introduces additional latency in the network and signalling overhead.

Routing protocol implementations and performance tests in small (up to ten nodes), WLAN 802.11-based networks were presented in [7, 8, 9]. Power-aware routing protocols are described in [20, 21, 22].

Bluetooth networks differ from WLAN networks; nodes are not peers any more and the communication medium is no longer broadcast. Piconets are the main form of Bluetooth networks and do not require routing, because a master node controls the complete traffic and has direct communication with other piconet members (slaves). Routing is required in scatternets, which consist of several interconnected piconets. The fact that piconets are organized around a master node can help routing protocol design.

UWB (ultra wide band) is a relatively new player in the ad hoc networking field [23]. This technology provides good position location of communicating nodes and this information could be used to support efficient routing in ad hoc networks, as proposed in [24, 25].

Air Interface

The air interface requirements for ad hoc networking are many and varied [2]. They can range from very low-power, low-data rate telemetry and sensor requirements, to very high data rates for high-quality multimedia distribution in the home. Common requirements include coexistence between multiple instances of the same air interface (in the same, or collocated ad hoc networks) and any other air interfaces (ad hoc or deployed).

To reach the requirements, techniques for dynamic frequency selection (DFS), link adaptation and power control have to be included, as in the recent standardized Hiper-LAN/2 or IEEE 802.11 systems. Furthermore, techniques to support QoS have to be added. Current efforts for QoS support in pure ad hoc networks lead to establishment of a central controller (e.g. IEEE 802.11e). Techniques to address this issue include dynamic resource allocation, spectrum sharing and spectrum overlay.

MAC (Medium Access Control) Layer

The MAC layer has to provide efficient and fair access to the wireless medium for all devices, and to ensure reliable data transmission. Current MAC protocols for ad hoc networks could be classified in to three groups, depending on their channel access strategy: *contention protocols, allocation protocols*, and a combination of both–the *hybrid protocols*.

Contention protocols, like ALOHA or CSMA, are based on asynchronous communication models. Collision avoidance is an important feature of these protocols that is realized through some form of control signalling. It has been shown that contention protocols are simple and tend to degrade as the traffic load increases, whereby the number of collisions rises. In overload situations, a contention protocol can become unstable as the channel utilization drops. This can result in an exponential packet delay increase and network service breakdown, since few, if any, packets can be successfully exchanged.

The multiple access with collision avoidance (MACA) [26] protocol uses a handshaking dialogue to reduce the hidden node interference and minimize the number of exposed nodes. Further enhancements are introduced by the MACAW [27] protocol, which includes positive acknowledgements and carrier sensing to avoid collisions. Improvements are also made to the collision resolution algorithm to ensure a more equitable sharing of the channel.

A very similar approach to MACAW is used in the distributed coordination function, DCF, of the IEEE 802.11 standard, with improved collision avoidance [28, 29]. Nodes deliver data packets of arbitrary lengths (up to 2304 bytes), after detecting that there is no other transmission in progress. However, if two nodes detect the channel as free at the same time, a collision occurs. IEEE 802.11 defines a collision avoidance (CA) mechanism to reduce the probability of such collisions.

Allocation protocols employ a synchronous communication model and use a scheduling algorithm that generates a mapping of timeslots to nodes. The mapping results in a transmission schedule that determines in which particular slots a node is allowed to access the channel. This effectively leads to a collision-free transmission schedule. It turns out that the allocation protocols tend to perform well at moderate to heavy traffic load, but these protocols are disadvantaged at low traffic, due to the artificial delay induced by the slotted channel.

Hybrid protocols can be loosely described as any combination of two or more protocols. IEEE 802.15.3 MAC draft standard [30] is one such protocol. It is defined for narrowband 2.4 GHz WPAN applications. The emerging UWB physical layer draft standard IEEE 802.15.3a, adopted to be compatible with the IEEE 802.15.3 MAC standard, will possibly have a few adaptations due to the inherent specificities of the UWB physical layer.

The IEEE 802.15.3 MAC protocol is centrally coordinated, with a piconet coordinator (PNC) that synchronizes the devices (DEVs) and allocates the resources. Even the MAC protocol is a centralized one, the topology is ad hoc and communication is established in a peer-to-peer mode. The PNC can be chosen dynamically, i.e. it is autoclaimed each time a new piconet is created. The main part of the processing power is concentrated in the PNC's hands, but if the PNC disappears, another station can take on its role, which is an advantage over static centralized management.

Radio Resource Management (RRM)

The development of efficient algorithms for RRM is critical from a network point of view, since such functionalities have a significant impact on the fulfillment of QoS requirements, and on attaining higher degrees of spectral efficiency. Radio resource management activities encompass a number of functions:

- Admission control will ultimately decide whether a new flow can be granted, while preserving overall QoS requirements. The admission control would be invoked at each node to make a local accept/reject decision in the framework of cluster-oriented architectures.
- Congestion control mechanisms are invoked whenever network overload leads to unfulfilled QoS requirements for the admitted users (for a fraction of time). When in congestion, some users could experience a reduced QoS margin, not beyond, though, an agreed percentage of time.
- Packet scheduling schemes determine how different flows are forwarded in a specific network node (mechanisms such as priority queues, timers, etc. are used). Priorities can be service-dependent and, for a specific service, transient QoS needs can also be considered. Despite the existence of some degree of flexibility in the choice of the scheduling policy, any sensible approach should target optimizing the overall network performance.

Clearly, in the context of self-configurable radio networks, RRM functionality can no longer be centralized in a specific node. Conversely, a new distributed RRM architecture has to be envisaged, where RRM functionalities are implemented in every single network node or mobile station.

Cross-layer Strategies

Traditionally, in MAC protocol design, little or no attention has been paid to the underlying physical layer features. Thus, most MAC protocol enhancements were proposed with the common idea to suitably manage and avoid collisions. However, the advent of sophisticated signal processing techniques (array processing, multiuser detection, channel coding strategies, etc.) that are able to extract useful signal(s) from noise, interference and unwanted signal replicas, could change many of the underlying assumptions in the conventional MAC schemes. For example, the assumption that more than one simultaneous transmission over the same radio resource (e.g. identical frequency, time and spreading code assignment) inevitably leads to a collision, should be revisited. In other words, making MAC, RRM and upper-layer functionalities aware of the physical layer state information (for example, in terms of diversity-based component status, channel response or interference indicators), could boost system efficiency in terms of resource reutilization, by allowing each mobile terminal to transmit so that an optimal usage of the spectrum available was attained. This strategy departs from those in conventional MAC schemes, where packet collisions should always be avoided and, hence, more than one user is not allowed to share the same radio resource. Given the time-varying nature of those parameters, the envisaged MAC schemes must be adaptive. Accordingly, other inter-layer dialogues can be established in the OSI stack. For example, investigations have shown drawbacks of 802.11 MAC protocol in multihop communication. In particular, the optimization in terms of routing could be improved, by providing some of the information available (SNR information, packet acknowledgments, etc.) at the MAC layer to the ad hoc routing protocol on the network layer [31].

Security

High-level security requirements for ad hoc networks are basically identical to security requirements for any other communications system, and include: authentication, confidentiality, integrity, nonrepudiation, access control and availability. However, similar to wireless communication systems creating additional challenges for implementation of the above-mentioned services when compared to fixed networks, ad hoc networks represent an even more extreme case, requiring even more sophisticated, efficient and well designed security mechanisms [32–39]. These additional challenges are caused by two basic assumptions of an ad hoc system: a complete lack of infrastructure, and a very dynamic and ephemeral character of the relationships between the network nodes.

The lack of infrastructure implies that there is no central authority that can be referenced when it comes to making trust decisions about other parties in the network, and that accountability cannot be easily implemented. The transient relationships do not help in building trust based on direct reciprocity, and give an additional incentive to nodes to cheat.

Ad hoc networks rely on cooperation of involved nodes in order to build and maintain the network. Current versions of mature ad hoc routing algorithms only detect if the receiver's network interface is accepting packets, but they otherwise assume that routing nodes do not misbehave. Whereas such an assumption may be justified when single domain networks are concerned, it is not easy to transpose it on a network consisting of nodes unknown to, and not trusted by, each other. Since ad hoc networks use multihop routing protocols, where each of the nodes, in addition to its own packets, has to forward packets belonging to other nodes, selfish behaviour may represent a significant advantage for a node, saving its battery power and reserving more bandwidth for its own traffic. However, if a large number of nodes start to behave non-cooperatively, the network may break down completely, depriving all users of the services. Non-cooperative behaviour in multihop routing protocols may also result in a denial of service attacks on the network, where malicious nodes join the network for the sole reason of misbehaving and depriving all other nodes of legitimate services. Such denial-of-service focused misbehaviour may consist of dropping (not forwarding) the packets, injecting incorrect routing information, replaying expired routing information or distorting routing information in order to partition the network. Also, bogus nodes may try to attract as much traffic as possible to them in order to be able to analyse it. In general, attacks on a routing protocol can be classified as dropping of data packets, route modifications, dropping of error messages and frequent route updates.

Another challenge is metadata protection, including confidentiality of identity (pseudonym and anonymity), confidentiality of location (traceability) and traffic analysis. The confidentiality of this metadata will gain in importance in the ubiquitous computing environment, where the ubiquitous computing infrastructure could potentially become a tool for a powerful surveillance, making us involuntary participants in a worldwide 'Big Brother' show.

A dangerous attack in civil applications, typically using an open ad hoc environment, may consist of so-called 'sleep deprivation torture' [38]. In this type of denial-of-service attack, the attacker is trying to deprive a device of battery power by keeping it awake and engaging in communication all the time. Strong authentication of communication peers, or some kind of accountability, based on either expensive pseudonyms and reputation mechanisms or micropayments, could be used to prevent, to some extent, this kind of attack, [33, 34].

6.1.2.2 Personal Area Networks (PANs)

Personal area networks consist of networked devices that are located in a confined area (usually a 10 m radius) [40]. These networks are frequently represented as a 'bubble' that surrounds and moves with a person. All devices located inside the bubble are potential PAN members. These devices usually belong to a person, are small in size, battery driven, and the PAN owner is wearing or carrying them. However, any other device that enters the 'bubble' can become a PAN member as well (the person can detect a printer in a room and can use it to print a document, or can 'run into' an access point and use it for internet access). One device usually acts as the PAN controller.

Communication and interactions between devices are opportunistic, depend on user needs and, from the user point of view, should be seamless and automatic.

Various wireless technologies like Bluetooth, IrDA, UWB, HomeRF, etc., can be used. Inherent support for ad hoc connectivity, low power consumption, small footprint and consumption, and wide availability, are making Bluetooth the main candidate for PAN communication technology today.

The main PAN scenarios (see Figure 6.2) are as follows:

- *Cable replacement:* This is a simple scenario that replaces a cable between two devices (or between two parts of the same device) with a wireless connection.
- *Interaction between devices inside the personal 'bubble':* Mobile phone, PDA, laptop, digital camera, headset and other similar devices will communicate in order to enable users to perform certain tasks more easily and efficiently (for example automatic synchronization of daily schedule and contacts database between the PDA and mobile phone, or access to wearable sensors for health care data).
- *Interaction with non-PAN devices:* PAN devices will frequently interact with non-PAN devices to use services these devices provide (to pay a parking fee at the parking meter, or goods at the vending machine).
- *Interaction with other networks:* PANs will interact with other PANs, local and wide area networks to gain access to certain services provided by these networks (access internet, company network, etc.). In this case, one of the PAN devices acts as the PAN gateway (for example, mobile phone that has both Bluetooth and GPRS interface) and controls communication with other networks.

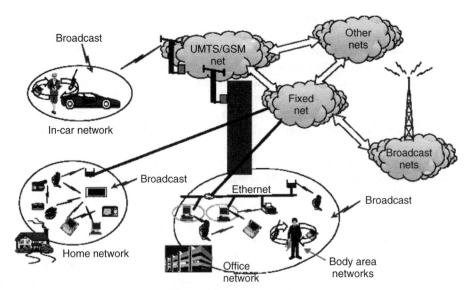

Figure 6.2   PAN scenarios

The future WPAN platform should provide integrated solutions that will enable inter-operability with the existing networks, scalability according to data rates, cost, power and device functionalities, multimode functionality (access to multiple networks), distributed resource allocation (with QoS provisioning) and service discovery.

Research issues mentioned in previous sections, such as network organization, service discovery and security, are applicable for PANs as well. Certain specifics of PAN solutions stem from the requirements for 'low power–low cost' solutions, choice of the wireless transmission technology and the fact that most of the PAN devices are owned by one person. Some of the specific PAN-related issues are as follows:

- *Integration with other networks:* Communication inside the PAN does not necessarily have to be IP-based, and the task of the gateway node will be not only to connect two networks, but also to adapt different protocols and to provide information about the PAN services to the outside world.
- *Network mobility:* When the PAN dynamically changes its point of attachment to the wide-area network, it will cause the reachability of the entire network to be changed in the topology. Without appropriate mechanisms, sessions established between nodes in the PAN and the global network cannot be maintained while the mobile router changes its point of attachment.

6.1.2.3 Interoperability with Fixed/Overlay Networks

Although stand-alone ad hoc networks provide support for many interesting applications, in many scenarios a connection to fixed/overlay networks will be required. This can be primarily achieved by connecting one or more ad hoc network nodes, wirelessly or using a wired link, to a fixed network. These nodes then act as fixed network access points. An example of such combined ad hoc and fixed networks is given in Figure 6.3.

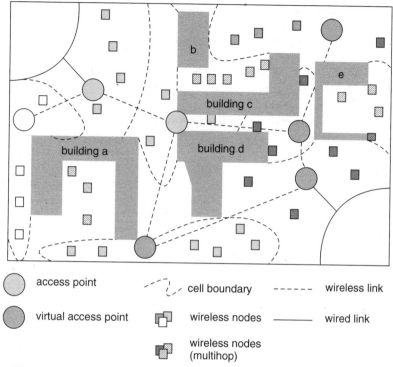

Figure 6.3   An example of a combined ad hoc and fixed overlay network

Wireless nodes are distributed over a given area. Some of the nodes connect directly to a wired access point. Because of the missing infrastructure, not every access point can be hard-wire connected. Therefore, virtual access points are introduced. Virtual access points are connected directly, or over multihop with the wired access points. Wireless nodes can connect to any of the access points, depending on their location and/or signal strength. If a node cannot connect directly to any access point (like the nodes between buildings (b) and (c)) connection can be established using other nodes as routers for multihopping-based access.

Research issues applicable to stand-alone ad hoc networks are also valid in this scenario. There are several issues specific for this scenario:

- *Gateway node role:* Gateway node can act as a bridge or a router device. Depending on the role, different addressing and protocol translation mechanisms have to be proposed.
- *Authentication, authorization and accounting:* When accessing a fixed network, ad hoc network nodes have to be authenticated and granted appropriate access rights. The gateway node can either take responsibility for all underlying ad hoc nodes, i.e. can be the only node seen and authenticated by the overlay network, or can just tunnel the ad hoc network nodes' traffic to an authentication server in the overlay network. If the gateway node is not a dedicated gateway node, but is just occasionally providing access to other networks, who is charged, and how the gateway node or nodes that are actually using services?

- *Addressing:* As in the stand-alone ad hoc networks scenario, the assigned node address has to be unique. Now, however, at least the gateway node has to have a public address along with the private one. If other nodes are to be accessible from outside of the network, then they need public addresses too.
- *Node mobility:* How is the mobility of ad hoc networks supported, i.e. is it possible for an ad hoc network to change its point of access to the fixed network without interrupting current communication, or for an individual node to transfer between ad hoc networks without losing the connection to the fixed network?
- *Gateway service:* How does a mobile node find a gateway – are the gateways advertising themselves, or only responding to requests from nodes? If the network is stable, gateway advertisements produce unnecessary load; but if it is unstable and the nodes have primary internet traffic, the nodes can profit from the gateway advertisements.

6.1.2.4 Integration of Ad hoc Networks into Cellular Networks

Unlike systems providing an ad hoc mode, cellular systems rely on an infrastructure of base stations (BS) and require network planning and operation in licensed radio spectrums. UMTS provides cumulative data rates of up to 2 Mbit/s, which might still not be enough for hot spot areas, where the number of mobile nodes (MNs) per area is very high. To increase the individual data rate of users, WLAN systems are introduced at these places, which can provide transmission rates of about 54 Mbit/s. These systems operate in unlicensed radio spectrums and, generally, offer mobile data communications with very low cost. Nevertheless, transmission power in such communication systems is limited, hence the coverage is limited as well, and interference between such systems is difficult to predict and to control.

Taking into account the advantages, potential and drawbacks of cellular networks, WLAN, and self-organizing network architectures with respect to, e.g., coverage, capacity, mobility, cost of infrastructure and flexibility, it becomes obvious that a combination of them is the logical consequence for the next generation network concepts.

In situations where cellular networking capability and ad hoc networking capability coexist in the same devices (MN), it is possible to utilize a cellular network to assist ad hoc networking. This kind of hybrid network could include centralized servers in a fixed/cellular network to handle ad hoc network topology to assist routing and authentication, but can also be considered as an extension of cellular networks.

While the defining goal of the ad hoc networks is the ability to function without any infrastructure, the goal of these multihop-augmented infrastructure-based networks is the almost ubiquitous provision of very high data rate coverage and throughput.

An evolutionary approach towards the architecture of a hierarchical multihop cellular network (HMCN) can be seen in Figure 6.4. An overlying cellular mobile radio system, e.g. a 3G system, forms the basis of the proposed network architecture. The possible connection of each MN to a BS guarantees full coverage, and this connection can always be taken as a fall back solution in the case where an MN loses connection to any other kind of network it might be connected to. This requires interoperability of the existing and the future networks and the support of vertical handover, i.e. handover between different wireless access networks (intersystem handover). The BS provides access to the backbone, which is most likely to be based on the TCP/IP protocol suite.

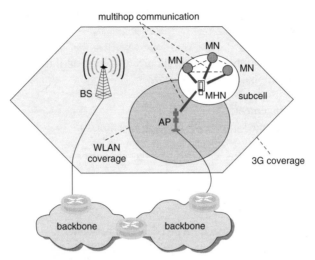

Figure 6.4   A cellular multihop architecture

The next evolutionary step towards a hierarchical multihop network structure is to introduce multihop-capable nodes (MHN), which can be fixed or even mobile (Figure 6.4). With fixed MHNs, the coverage of the APs can be extended. At the same time, a fixed MHN can be connected to a power supply to offer more potent services. Subcells can be established in a self-organizing manner. This means the MHN recognizes its AP and takes over control functionality within the subcell. A typical control function comprises the management of the medium access within the subcell. Furthermore, it provides connections between MNs in the subcell, which can directly communicate with each other by means of a direct mode. Moreover, the AP can provide to the MHN and MNs useful signalling information, like routing information for example. Besides the AP, signalling information can be provided by the overlying cellular 3G systems, too. Of great interest too, is the case where the routing in the subcells is assisted by the overlying 3G systems.

Due to the proposed hierarchical structure, an optimum control of resource allocation can be organized.

Besides fixed MHNs, mobile MHNs are also considered in a further evolutionary step. In the case that the required data throughput cannot be provided any more by the AP or established fixed MHN, due to the increasing penetration and/or to satisfy the future demands for packet data services, an MN can become an MHN and can establish a subcell on demand. These cells can use the same, or different, frequencies. In this case, subcells can be adaptively established. The spectral efficiency of the system can be enhanced when reusing the same frequency in different subcells. In the case of using different frequencies in different subcells, the available transmission rate within the considered cell can be increased.

There are many issues to be investigated on the path towards a successful integration of the multihop capability in conventional wireless networks:

- the advantages and disadvantages of having fixed versus mobile repeaters (routers);
- the advantages and disadvantages of relaying in analogue (amplify-and-forward) versus digital (decode-and-forward) form;

- the load balancing capability by diverting the traffic with repeaters as necessary;
- the signalling overhead;
- relaying interference;
- a possible cap on the number of hops, incurred latency and its impact on QoS;
- complexity and functionality of relay devices;
- scheduling;
- radio resource management;
- novel diversity techniques (macrodiversity for example).

A detailed description of this scenario and the above requirements is given in [1].

6.1.2.5 Wireless Sensor Networks

Wireless sensor networks are a special type of ad hoc network, consisting of numerous tiny devices that integrate radio communication, digital computing and sensing and actuation components [6, 41–44]. After deployment, by dropping from a plane for example, these networks are able to self-organize and work autonomously based on programs stored in each node. There is usually a redundant number of nodes sensing the same phenomena in one area. These nodes use their computational capabilities for event detection and collaborative signal processing, so that higher bandwidth raw sensor data does not need to be sent to the users. Sensor data is collected at a gateway node (static or mobile), responsible for forwarding that data to remote users (over WLAN or UMTS networks, for example).

Users are rarely interested in the readings of one or two specific nodes. Instead, they are interested in specific parameters of the observed physical phenomena. This is the main reason why sensor networks adopt data-centric addressing that uses attributes of information provided by the node as its address. In other words, a user will send a query looking for 'temperature in the XYZ area', not looking for the readings of specific sensors. It is then the responsibility of the routing protocols to distribute the query to all relevant sensors, that will, as mentioned before, execute an appropriate distributed algorithm on aggregate gathered information and formulate a reply for the user.

Possible applications of these networks are numerous, and range from environmental monitoring (water pollution monitoring, for example), military and security applications, to monitoring the structure of buildings, bridges, etc.

Research areas and issues relevant for ad hoc networks in general are applicable for wireless sensor networks as well, but with a significant emphasis on power efficiency of all proposed protocols and solutions. Specific requirements are:

- *Routing:* Routing in wireless sensor networks has different requirements than routing in MANET-like networks. The nodes in a wireless sensor network are usually not mobile, but changes in topology are possible due to battery depletion or node malfunction. Sensor nodes do not have a unique address, but are described using attributes. The energy limitations are even more strict than in other networks. Unlike other networks that act as a medium bringing two parties together (one node accessing the services of another node), sensor networks are more a distributed system than a network that loosely connect independent entities. The communication pattern is also different than in other networks; since sensor data is usually forwarded towards only one node in the network (gateway node), routing protocols in this environment are mainly concerned with establishment and maintenance of routes between sensors and the gateway node. Users' queries are not addressed to a specific sensor but are disseminated through the network based on the required data (a user will ask for the temperature in the south-west part of the forest, not the temperature measured by a specific sensor). Several routing protocols specifically developed for sensor networks have been proposed [45–47].

- *Air interface:* It has to be energy efficient, providing low to medium throughput on a short communication range (a few metres).
- *Network architecture design:* Flat or hierarchical, both architectures have advantages and disadvantages. The former is more survivable, since it does not have a single point of failure; it also allows multiple routes between nodes. The latter provides simpler network management, and can help to further reduce the number of transmissions.
- *Design of localized and distributed algorithms for data processing:* Data processing consumes less power than wireless communication and is extensively used in sensor networks to reduce the amount of transmitted data. Redundant nodes communicate, gather data, process and fuse it and only then forward the commonly agreed reply to users.
- *Interoperability and interworking with fixed infrastructure:* (UMTS, GPRS, GSM, WLAN): Sensor networks will usually interact with other networks via a gateway node. The gateway will have to provide necessary information about capabilities of all sensor nodes in the network to remote users, to distribute user queries through the network and to collect replies and forward them to users [44].

6.1.2.6 Broadcast Ad Hoc Networks

Many times when ad hoc networking applications are identified, a conclusion is made that there is a great variety of different applications and services where ad hoc networks could be used. This variety of applications leads easily to hard requirements on capacity, quality of service and security. This leads to a decision to concentrate efforts on solving these severe limitations of ad hoc networking. However, it could be beneficial to take an alternative view, whereby it is discussed whether such applications exist that can tolerate the severe drawbacks of ad hoc networking.

The network topology management is perhaps the key challenge in mobile ad hoc networks, but it is not needed in all applications, like, for example, in broadcast services. It is enough when the information is broadcast to neighbouring nodes. When the broadcasting operates in the multihop mode, the receiving nodes repeat the broadcast to their neighbours. So, in principle, it is fairly easy to make a network that operates in the multihop broadcasting mode. The question remains, are there applications that could utilize this kind of networking? It is obvious that broadcasting networks could be utilized for different information sharing services in the ad hoc network community.

On top of finding applications for ad hoc broadcasting, there are also some specific research areas and issues that are relevant for broadcast ad hoc networks:

- Broadcast protocols: to find efficient protocols for broadcasting in ad hoc networks, e.g. eliminating useless multiple broadcasting in different network situations, and finding methods that will ensure that the broadcast information will reach all nodes in the network.

• Researching and finding suitable capacity sharing and management techniques in different broadcast traffic loading situations.

6.1.3 Conclusions

Ad hoc networking is a broad research area that covers various technologies, networks and protocols, and appears in different scenarios. Due to the versatility of scenarios and technologies and, hence, a lot of different, sometimes conflicting, requirements, it is not possible to develop one common solution. Instead, different solutions have to be developed for different usage scenarios. Research problems exist on all levels, from physical to application layer issues.

In this text we have defined several characteristic scenarios and explained the most important research issues for each scenario. Some common problems, like routing, have been investigated thoroughly in the last few years and first solutions, which should be used as the basis for further work and improvements, exist. A step away from the traditional protocol layering, and adoption of a cross-layering strategy will probably be required in order to capture all necessary information. Future work should be organized around defined scenarios and should include multidisciplinary research teams working together on all relevant issues.

6.2 Relay-based Deployment Concepts for Wireless and Mobile Broadband Cellular Radio

In recent years, there has been an upsurge of interest in multihop-augmented infrastructure-based networks in both industry and academia, such as the 'seed' concept in 3GPP, mesh networks in IEEE 802.16, coverage extension of HiperLAN/2 through relays or user-cooperative diversity mesh networks. This text, a synopsis of numerous contributions to Working Group 4 of the Wireless World Research Forum (WWRF) and of other research work, presents an overview of important topics and applications in the context of relaying. It covers different approaches of exploiting the benefits of multihop communications via fixed and/or mobile relays, such as solutions for radio range extension in mobile and wireless broadband cellular networks (trading range for capacity), and solutions to combat shadowing at high radio frequencies. Further, relaying is presented as a means to reduce infrastructure deployment costs. It is shown that multihop relaying can enhance capacity in cellular networks, e.g. through the exploitation of spatial diversity.

6.2.1 Introduction

The very high data rates envisioned for the 4th generation (4G) wireless systems in reasonably large areas do not appear to be feasible with the conventional cellular architecture for two basic reasons. First, the transmission rates envisioned for 4G systems are as high as two orders of magnitude more than those of 3G systems, and it is well known that for a given transmit power level, the symbol (and thus bit) energy decreases linearly with the increasing transmission rate. Secondly, the spectrum that will be released for 4G systems will almost certainly be located well above the 2 GHz band used by the 3G systems. The radio propagation in these bands is significantly more vulnerable to non-line-of-sight conditions, which is the typical mode of operation in today's urban cellular communications.

The brute force solution to this problem is to significantly increase the density of the base stations, resulting in considerably higher deployment costs, which would only be feasible if the number of subscribers also increased at the same rate. This seems unlikely, with the penetration of cellular phones already being high in developed countries..On the other hand, the same number of subscribers will have a much higher demand in transmission rates, making the aggregate throughput rate the bottleneck in future wireless systems. Under the working assumption that subscribers would not be willing to pay the same amount per data bit as for voice bits, a drastic increase in the number of base stations does not seem economically justifiable.

It is obvious from the above discussion that more fundamental enhancements are nec-essary for the very ambitious throughput and coverage requirements of future systems. Towards this end, in addition to advanced transmission techniques and colocated antenna technologies, some major modifications in the wireless network architecture itself, which will enable effective distribution and collection of signals to and from wireless users, are required. The integration of multihop capability into conventional wireless networks is, perhaps, the most promising architectural upgrade, and has been proposed for a long time (see [48–51]). In the following, the terms 'multihop' and 'relaying' will be used to refer to the same concept; see also [52].

Multihop wireless networking traditionally has been studied in the context of ad hoc and peer-to-peer networks. The application of multihop networking in wide area cellular systems yields the following benefits:

- Networks applying relaying via fixed infrastructure compared to ad hoc networks do not need complicated distributed routing algorithms, without losing the advantage of being able to move the relays as the traffic patterns change over time.
- Relays are low-cost low transmit power (compared to base stations) network elements, not connected to a wired backhaul dedicated to storing and forwarding data received wirelessly from the base station to the user terminals, and vice versa. They can reduce the propagation losses between the relay transmitters and the user terminals, resulting in larger link data rates, and potentially solving the coverage problem for high data rates in larger cells.
- Since it is possible to have simultaneous transmission by both the base station and the relays, capacity gains may also be achieved by either exploiting reuse efficiency or spatial diversity.
- The relay to user links could use a different (unlicensed) spectrum (e.g., IEEE 802.11x) than the base to user links (the licensed spectrum), yielding significant gains from load balancing through the relays [53].

It is worth emphasizing the basic difference in the fundamental goal of the conventional ad hoc networks and the described multihop-augmented infrastructure-based networks. While the defining goal of ad hoc networks is the ability to function without any infras-tructure, the goal in the latter types is the almost ubiquitous provision of very high data rate coverage and throughput.

In this section, the term base station (BS) will be used for traditional cellular systems, while the term access point (AP) is used for WLAN-type systems. Both have in common that they are directly connected to their backbone network, which distinguishes them from the relay stations (RS). However, it should be noted that with the arrival of new air interfaces and deployment concepts, they might at some point become indistinguishable.

This section is organized as follows: first, an overview of the state of the art in current research on relaying topics is given. The next sections deal with performance benefits gained from relaying when (i) combating heavy path loss and (ii) exploiting spatial diversity. Implications of relaying on routing and radio resource management are discussed next. Section 6.2.6 introduces WMS, an example for a new radio access network architecture based on fixed relays, and the conclusion gives a summary of the key research topics identified in the various contributions to this text.

6.2.2 State of the Art

No smart relaying concept has been adopted in existing cellular systems so far. Solely bidirectional amplifiers have been used in 2G systems and will be introduced for 3G systems. Yet, these analogue repeaters increase the noise level and they suffer from the danger of instability due to their fixed gain, which has limited their application to specific scenarios.

Most existing and standardized systems were designed for bidirectional communication between a central base station and mobile stations directly linked to it. The additional communication traffic between a mobile terminal and a relay, intermediately inserted into a link between mobile terminal and base station, requires additional radio resources to be allocated – one of the reasons that hitherto have hindered the deployment of smart relay concepts.

TDMA-based systems are especially well-suited to introducing relaying, as this scheme allows for an easy allocation of resources to the mobile-to-relay and relay-to-BS links. The first system based on TDM and relaying to connect fixed and mobile was proposed in 1985 [50]. Another method proposed for F/TDMA systems is to reuse a frequency channel from neighbouring cells [54]. The ETSI/DECT standard in 1998 was the first specifying fixed relays (called wireless base stations) for cordless systems, using TDM channels for voice and data communications. The ETSI/TETRA standard specifies a dual-watch function allowing the aggregate traffic of a number of mobile terminals, connected in direct mode, to be relayed by TDM channel switching to a dispatcher panel connected to a BS. Relaying in cellular CDMA-based systems has been investigated by Zadeh *et al.* [55]. Uplink and downlink are separated using frequency division duplex, as is done in IS-95 and UTRA FDD. All these concepts can easily be extended to packet-based systems [48, 49].

A completely different approach is considered in [50], incorporating an additional ad hoc mode into the GSM protocol stack to enable relaying. Similarly, [53] employs relaying stations to divert traffic from possibly congested areas of a cellular system to cells with a lower traffic load. These relaying stations utilize a different air interface for communication among themselves and with mobile stations, which could, for example, be provided by a wireless LAN standard.

ETSI-BRAN/HiperLAN/1 and IEEE 802.11x contain the elements to operate ad hoc networks. ETSI/HiperLAN/2 in the home extension contains an ad hoc mode of operation that allows the nodes to agree on a central controller (CC) to take the role of an AP in a cluster of nodes, but no multihop functions are specified so far in any WLAN standard. Multihop operation, based on wireless relays that operate by alternating on different frequency channels to connect neighbouring clusters, has been proven workable in [57]. In the HiperLAN/2 basic mode (using an AP) it has been shown that multihop operation via forwarding mobile terminals is easy to perform within the framework of the standard [48, 49].

6.2.3 Multihop Operation

This section presents examples of relay implementations and aims to point out the performance benefits that multihop relaying can provide in broadband networks when applied in certain scenarios. Figure 6.5 (left) shows a city scenario with one AP (providing radio coverage to the areas marked white) and four fixed-mounted RS to provide radio coverage to areas 'around the corner' shadowed from the AP.

While the intersection can be covered well by the AP, the close-by streets can only be served if a line-of-sight (LOS) connectivity is available between mobile terminal and its serving station, owing to the difficult radio propagation conditions known for, e.g., the 5–6 GHz frequency band. The RS allow an extension of radio coverage to these streets.

A schematic of this scenario–a multihop network with access point (AP)–is illustrated in Figure 6.5 (middle), where the transmit/receive radius R is shown to be the parameter determining the connectivity of the nodes shown. A fixed relay (S1) would have to route the traffic of the wireless terminals (not shown) it is serving via the intermediate fixed relay (S8) using a low PHY-mode, or via S2 and S8 using a higher PHY-mode and, thus, a higher link capacity, and so forth from S8 either via S9 or directly to the AP. The interpretation of Figure 6.5 (middle) multihop network is that all the nodes shown are either fixed relays or APs, and the mobile user terminals roaming in the area (not shown) are served by the nodes shown. The basic element of Figure 6.5 (left) can be repeated to cover a wide area, as shown in Figure 6.5 (right).

Figure 6.6 shows three examples of concepts for fixed or mobile relaying, see [58]:

(a) Relaying in the time domain with AP and fixed relay operating at the same carrier frequency f1.
(b) Relaying in the frequency domain with AP and fixed relay operating at different frequency carriers.

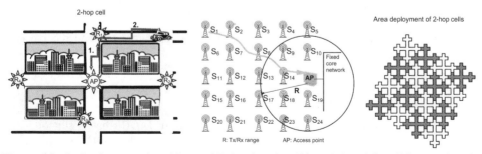

Figure 6.5 Left: city scenario with one AP (serving the white area) and four RSs covering the shadowed areas 'around the corners', shown in grey; middle: schematic of the scenario; right: wide-area coverage using the basic element (left) and two groups of frequencies. (Source: [53])

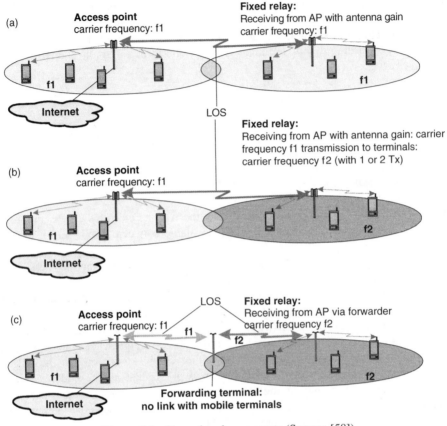

Figure 6.6   Example relay concepts (Source: [58])

(c) Two-stage relaying in the frequency domain, where a fixed mounted RS that is in the range of both AP and relay connects the AP and the second RS by store-and-forward operation, and dynamically switches between frequency carriers f1 and f2. Unlike the second RS, which serves mobile terminals, the first RS's only role is to bridge the distance between AP and the second RS, where this is not possible due to lack of LOS (see also [57]).

In (a) and (b), the radio link is based on LOS radio with transmit and receive gain antennas at AP and relay.

6.2.3.1 Analytical and Simulation Results

An analytical estimation of the bit rate over distance from an AP that is supported by fixed relays to extend the radio range for the approach according to Figure 6.6(a), is shown schematically in Figure 6.7. Without receive antenna gain, Relay 1 would have available only a bit rate equivalent to the value b [Mbit/s], while with receive antenna gain, it achieves a value a [Mbit/s]. A similar consideration applies to Relays 2 and 3.

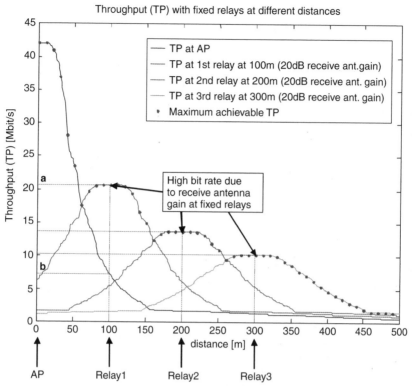

Figure 6.7   Analytical estimation of the extension of the radio range of an AP by relays with receive antenna gain (Source: [58])

The relaying function could be performed according to the ISO/OSI reference model either in Layer 1 (physical) 'repeater', 2 (link) 'bridge' or 3 (network) 'router'. Figure 6.7 and Figure 6.8 show analytical and simulation results according to the solution of a Layer 2 relay, as described in [48, 49] for the HiperLAN/2 standard.

Figure 6.8 (left) shows, for the concept introduced in Figure 6.6(a), the simulated end-to-end throughput between an AP and a remote mobile terminal that is located at a distance d, for different modulation and coding schemes known from the air interface of HiperLAN/2 or IEEE 802.11a [59]. The terminal is shadowed by a building at the street corner and is therefore connected 'around the corner' with the help of a fixed relay. The shaded area under the curves shows the gain in terms of throughput possible from the use of the fixed relay, without which the RMT could not be connected to the AP. It can be seen that the range extension resulting from using the fixed relay is substantial. It is worth noting that smart antenna technology at the AP or mobile terminal cannot provide radio coverage around an obstacle. A relay is currently the only way to achieve this.

It has been stated that the capacity of an AP when using a modem as standardized for HiperLAN/2 and IEEE 802.11a, might be excessive compared to the rate requirements of mobile terminals roaming in its picocell area that is formed by omnidirectional or sectorized antennas and a maximum transmit power of 1 Watt EIRP [60].

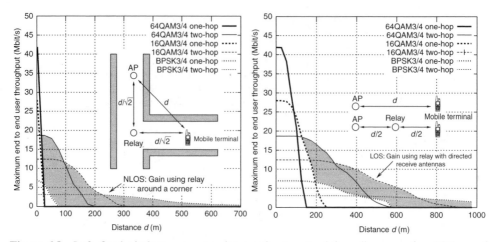

Figure 6.8 Left: fixed wireless router at an intersection to extend the radio range of an AP 'around the corner' into a shadowed area to serve a remote mobile terminal; Right: maximum end-to-end throughput vs. distance for forwarding under LOS conditions with directed receive antennas having a gain of 11 dB (Source: [59])

The fixed relay concept can also be used to extend the range of an AP to non shadowed areas beyond the regulatory EIRP limits, as shown from the simulation results for HiperLAN/2 (Figure 6.8 (right)) according to the scenario in Figure 6.6(a). It can be seen that the radio range can be dramatically increased, especially when using receive antenna gain at both ends. The use of smart antennas and beamforming to reduce the path loss between AP and relays, and to connect multiple relays at the same time, has been known for a long time [50].

As shown in Figure 6.7, the throughput decreases with an increasing number of hops. We now investigate a 2-hop cell with four FRSs, as shown in Figure 6.9. Figure 6.9 shows a 2-hop-cell in the assumed Manhattan scenario [61]. The 2-hop cell is comprised of five

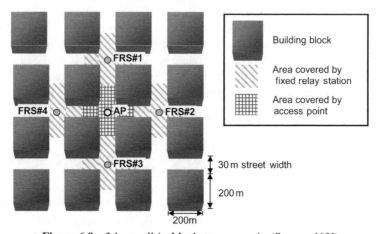

Figure 6.9 2-hop cell in Manhattan scenario (Source: [63])

subcells, one central cell covered by the AP and four relay cells, covered by the FRSs. Each MT will be accessed with a maximum of, two hops. The streets in the assumed scenario are 30 m wide, and the building blocks are sized 200 m × 200 m. Within this text, the MPs are located one in the middle of each intersection, i.e. 230 m apart.

The most simple case of forwarding is in the time domain with equal time share of the capacity [62]. The time frame of a system with four FRSs fed by one AP is shown in Figure 6.10(a). For a better differentiation we will further refer to this case as Case 1. It can be seen that the time slot allocated to feed the MTs in the 'forwarding cell' is split into two parts, one to transmit the data packet from the AP to the FRS (1st hop) and one to transmit the data packet from the FRS to the MT (2nd hop). The time required for the transmission on the first hop, $THop1 = TAP - FRS = DL/TPAP\text{-}FRS(G; dAP\text{-}FRS)$, is mainly dependent on the antenna gain $(G = GR + GE)$ and on the distance between the AP and the FRS, $dAP - FRS$. $TPAP - FRS$ is the throughput available between the AP and an FRS. DL is the length of the transmitted data packet.

Obviously, the solution shown above is very inefficient. Therefore, the overall capacity of the 2-hop cell can be enhanced by exploiting the spatial independency of some of the 'forwarding cells'. Spatial independency in this case means that the cell areas of two or more FRSs are fully shadowed from each other, as shown in Figure. 6.9 for, e.g., FRS#1 and FRS#2, or FRS#2 and FRS#3, etc. In the case of spatial independent 'forwarding cells', neither, e.g., FRS#1 nor any MT in the cell of FRS#1, will cause any interference to the cell of FRS#2 and vice versa.

This allows one to exploit space division for these two 'forwarding cells', resulting in a time frame as shown in Figure 6.10(b). Now, two of the four FRSs can transmit in parallel at a time. This approach, named Case 2, is obviously only possible in a planned infrastructure and has to be adapted individually for each scenario, a fact that is true also for the cases shown below. More sophisticated cases have been introduced and evaluated in [63].

In the analysis, no interference from co-channel APs or FRSs is taken into account. The background noise was assumed −96 dBm. For calculating the relation between received

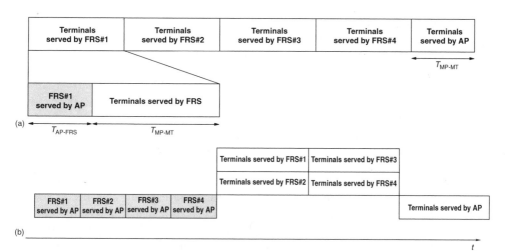

Figure 6.10 (a) Case 1: time frame for pure TDMA relaying; (b) Case 2: time frame with 2 × 2 spatial independent 'forwarding' (Source: [63])

SNR and distance from the transmitter, we use the following simple path loss model [64]:

$$P_{\mathrm{R}} = P_{\mathrm{T}} * g_{\mathrm{T}} * g_{\mathrm{R}} * \left(\frac{\lambda}{4\pi}\right)^2 * \frac{1}{d^{\gamma}} \tag{6.1}$$

P_{R} denotes the received signal power in Watts, P_{T} the transmitting power in Watts, g_{T} and g_{R} are the receiver and transmitter antenna gains, respectively, γ is the wavelength of the transmitted signal, d is the distance between the transmitter and the receiver and is the path loss coefficient. We assumed $\gamma = 3, 5$ for non LOS (NLOS) conditions between the MTs and the FRS or AP, and $\gamma = 3$ for the LOS path in urban areas, as assumed for the connection between the AP and the FRSs. The FRSs are assumed to use omnidirectional antennas to serve their associated mobile terminals. AP and FRS are transmitting on the same frequency, unless otherwise stated. Consequently, it is assumed that only one can transmit at a time. This principle is called time-domain relaying. The investigated multihop system uses the same OFDM modem as used by HiperLAN/2, IEEE 802.11a or the W-CHAMB system [65]. Link level simulation results [66] for the packet error rate (PER) as a function of the signal to noise ratio (SNR) from a 5 GHz HiperLAN/2 system with 20 MHz channel bandwidth, have been used as a basis for the analysis. The throughput of the proposed multihop system has been calculated as physical layer throughput without considering protocol header overhead. Retransmissions from a selective reject ARQ protocol are taken into account. Further, we assume an optimal link adaptation allowing one to operate always on the PHY-mode, which delivers the highest throughput.

To show the gain of the introduced relaying concepts, the capacity of an AP feeding the four FRSs was calculated by means of the following formula, [62]:

$$C = \frac{5 * N_{\mathrm{SC}}}{\displaystyle\sum_{j=1}^{N_{\mathrm{SC}}} \frac{1}{TP_{\mathrm{AP-MT}(d_j)}} + 4 * \sum_{j=1}^{N_{\mathrm{SC}}} \frac{1}{TP_{\mathrm{FRS-MT}(d_j)}}} \tag{6.2}$$

N_{SC} denotes the number of users in each of the five subcells, either forwarding cell or central cell; and d_j are the distances of uniformly distributed users.

In Figure 6.11, the capacities of the different cases of forwarding, as introduced before, are shown over the antenna gain available between the AP and the FRSs. The capacity shown is always the capacity of one AP, feeding the related FRS, i.e. the capacity of the whole 2-hop cell, as the whole traffic is going through the AP. As a reference, the capacity of an AP without FRS is given as a dashed horizontal line. It was further assumed that the users are equally distributed in the cell with a density of 0.1 user per square metre. To calculate the capacity, the covered area was divided into circular areas containing the same number of users.

Naturally, the capacity of the AP increases in all relaying cases with an increasing antenna gain. As for the throughput, the most impressive capacity gain occurs for the cases with three frequency channels, and at best with smart antennas to feed the FRSs in parallel. It is very interesting to see that the capacity of Case 2 is exceeding the capacity of a single AP already, with an antenna gain of around 7 dBi. This is possible due the effect that an increased number of users can be served with high data rates from the AP point of view. Even the capacity of Case 1 comes close to the capacity of the single AP, which

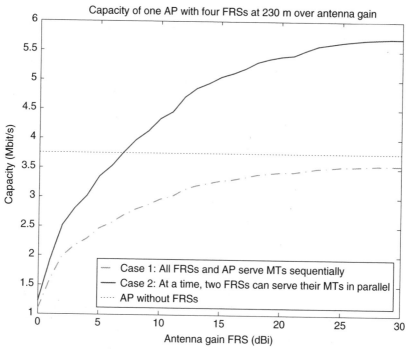

Figure 6.11 Capacity of AP feeding a four FRS 2-hop cell vs. antenna gain for different relaying techniques (Source: [63])

means that the time spent feeding the FRSs is quite low compared to the time required to feed the MTs. This results in a capacity gain for a system with 16 dBi antenna gain and operating, as proposed for Case 2, at 1.3 Mbit/s, which is a gain of 35 % compared to a system without FRSs.

Other benefits of relaying are the possibility of (i) radio resource reuse within the area served by an AP and its related relays, e.g., a frequency reuse within the area served by one AP, and (ii) the cost advantage. The cost to connect the radio access network to the infrastructure (core) network is reduced substantially by reducing the number of points of presence required, e.g., to 20 % for the example shown in Figure 6.5 (left). In a Manhattan grid, according to Figure 6.5, a second tier of relays would enlarge the number of relays to 12, thereby covering 13 intersections. The infrastructure connectivity cost would then be reduced to about 8 % of the cost of a purely AP based system. The number of tiers is bounded by the capacity available from the AP, the capacity per area element required from the terminals roaming in the service area and the delay requirements. It seems advisable not to exceed two tiers of relays in such networks.

6.2.3.2 Performance Limitations of Multihop Communication

So far, it has been shown how forwarding can enhance coverage. But this is not the only advantage of relaying techniques. Multihop communication can also improve the spectral efficiency. In the following, we want to investigate some generic forwarding scenarios with respect to their capacity, based on Shannon's formula. Assuming equally spaced relays between one source and destination, on each link of the multihop connection the data rate has to be n times higher than the data rate of the corresponding one-hop connection to end up in the same end-to-end throughput. The required signal to noise ratio (SNR) for the multihop connection over an additive white Gaussian noise (AWGN) channel for a given bandwidth, W, noise power, N, and received signal power, P_{rx}, becomes [67]

$$\left(\frac{P_{rx}}{N}\right)_{MH} = \left(1 + \left(\frac{P_{rx}}{N}\right)_{one}\right)^n - 1 \tag{6.3}$$

On the one hand, the required transmit power has to be increased to achieve the same data rate, but on the other hand, the distance on the individual links decrease with an increasing number of hops. To incorporate this dependency in the capacity calculation, the receive power is substituted by the transmit power, P_{tx}, and the distance between source and destination, d,

$$\left(\frac{P_{tx}}{N}c_0\right)_{MH} = \left(1 + \left(\frac{P_{tx}}{N}c_0\right)_{one}\left(\frac{1}{d}\right)^\gamma\right)^n \left(\frac{d}{n}\right)^\gamma - \left(\frac{d}{n}\right)^\gamma \tag{6.4}$$

whereas the constant c_0 takes into account the attenuation at 1 m.

The resulting SNR at the transmitter for the multihop connection over the respective SNR value for a one-hop connection is shown in Figure 6.12 (left).

It is evident that the gain for a multihop connection strongly depends on the transmit power, P_{tx}, the distance between source and destination, d, the attenuation coefficient, C_c, and the number of hops, denoted by n. The smaller the transmit power, the larger the distance, and the stronger the attenuation, the more attractive becomes the introduction of intermediate hops. This also matches well with the suggestion to introduce relays in the case of shadowed (non-LOS) links and to enlarge the coverage area. It can further be seen that with an increasing number of hops, the breakeven point is shifted to smaller SNR values and the gradient increases beyond this point.

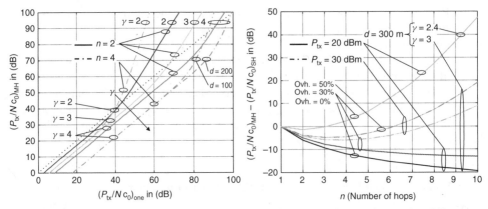

Figure 6.12 Left: required signal to noise ratio at the transmitter for one-hop and multihop connections; Right: relation of SNR for multihop and one-hop connections as a function of the number of hops (Source: [69])

Nevertheless, high gains can be obtained under high attenuation, e.g. approx. 10 dB/20 dB for a $\gamma = 4$ and $n = 2/n = 4$ hops, respectively. Nevertheless, it is noteworthy that with the introduction of intermediate nodes, the attenuation coefficient will most probably decrease, which, in turn, reduces the potential benefit of relaying, not mentioning the additional delay and protocol physical overhead. The latter impact can be assessed by the pessimistic assumption that each transmission requires some fixed overhead, ovh, and n transmissions need n-times that overhead. In this case, the exponent n in equation (6.4) becomes $n \cdot (n \cdot ovh + 1)/(ovh + 1)$. The relation of the SNR for the multihop connection to the SNR for the one-hop connection as function of the introduced number of hops, is shown in Figure 6.12 (right).

Typical transmit powers of $P_{tx} = 20/30$ dBm, a distance of 300 m, attenuation exponents of $\gamma = 2.4/3$, and different values for the overhead between 0 % and 50 % have been chosen. Values below 0 dB indicate a reduced power requirement for the multihop connection compared to the one-hop connection. It can be seen that with an increasing number of hops the gain for relaying is increasing up to a point where the overhead is dominating. For example, for $P_{tx} = 30$ dBm and $\gamma = 3$ the highest gains can be obtained for four and three hops, for overhead values of 30 % and 50 %, respectively. However, the dominant factor is the attenuation exponent, which is expected to decrease to the free-space value with an increasing number of hops. In this case, even for a moderate overhead of 30 %, two hops are the limit to benefit from relaying for small transmit power of $P_{tx} = 20$ dBm and $\gamma = 2.4$. All the aforementioned facts and results lead us to the conclusion that four hops, defined as an oligohop, should not be exceeded to benefit from relaying.

Unlike the concept of multihop communication in ad hoc networks, in hierarchical cellular multihop networks (HCMNs), a reduction of signalling overhead by means of centralized management of the system is envisaged, which, in turn, increases the system capacity [67].

6.2.4 Cooperative Relaying and Virtual Antenna Arrays

6.2.4.1 Concepts

Common to all relaying techniques discussed so far is the 'store-and-forward' operation of the nodes in a relay chain. Each receiver along this chain exploits solely the copy of the information that has been sent by its respective transmitter, while it discards emissions from other transmitters in the chain.

The basic idea of cooperative relaying is to go one step further, by exploiting two fundamental features of the wireless medium: its broadcast nature and its ability to achieve diversity through independent channels. While the broadcast nature is frequently considered to be a drawback, as it leads to mutual interference in a wireless network, the concepts discussed in this subsection exploit the fact that a signal, once transmitted, can be received (and usefully forwarded) by multiple terminals.

In general, cooperative relaying systems have a source node multicasting a message to a number of cooperative, i.e. helping relays, which, in turn, resend a processed version to the intended destination node, see Figure 6.13(a). The destination node combines the signals received from the relays, possibly also taking into account the source's original signal. By performing this combining, the inherent diversity of the relay channel

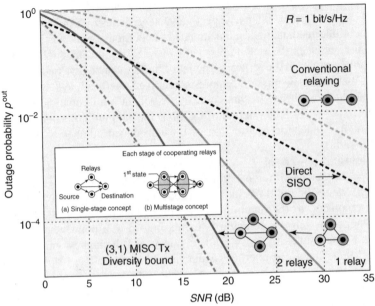

Figure 6.13 Concept of cooperative relaying and performance of cooperative relay networks: outage probability vs. SNR for various scenarios. Cooperative relaying with single antenna nodes achieves diversity gains over direct SISO transmission. The case of conventional relaying with one intermediate relay node is shown for comparison (Source [68])

is usefully exploited. More advanced concepts additionally include successive interference cancellation to maximize throughput. In this context, such cooperative multihop scenarios can be regarded as virtual antenna arrays: each relay becomes part of a larger distributed array. This distributed nature naturally leads to a number of new challenges, among them synchronization, availability of channel state information and appropriate cluster formation.

Like conventional relaying systems, cooperative schemes benefit from path loss savings; moreover, the following gains can be expected:

- the power gain due to the fact that each of the relays may add additional transmit power that is combined in the destination terminal;
- the macrodiversity gain that allows combating shadowing; and
- the diversity gain in the presence of fading.

Finally, it is worth noting that an integration of multiple input multiple output (MIMO) and so-called 'dirty paper coding' techniques may lead to advanced multihop networks with high spectral efficiency.

Cooperative systems can be classified as either decode-and-forward systems or amplify-and-forward systems. In decode-and-forward schemes, the relay nodes fully decode and re-encode the signals prior to retransmission, which poses the danger of error propagation. In amplify-and-forward systems, the relays essentially act as analogue repeaters. This apparently simpler implementation has the drawback of noise amplification. Another classification can be made, based on the nature of the protocol: there are static and adaptive protocols. The latter can be enhanced by allowing feedback and/or signalling between forwarding nodes. While in static protocols, the relay nodes constantly retransmit a processed version of their received signals, in adaptive versions, the relays resend signals only when they believe it to be helpful for the destination node. The adaptation may be done by each relay independently, or jointly for all together, if information is exchanged between the relays. Theoretically, the transmission in up- and downlink directions is fully symmetrical, which is obvious if you reverse the direction of the arrows in Figure 6.13. Practically, the boundary conditions for up-/downlink will often be quite different, as the radio links between AP and relays will exhibit typically good LOS conditions, and may be enhanced by directional antennas, in contrast to the relay-MS links, which are quite unpredictable. As a result, the optimum SFN algorithms for up- and downlink may be different as well, and the benefits may not be entirely symmetrical.

6.2.4.2 Examples and Research Areas

Of the various approaches to cooperative relaying that have been investigated, we focus here on the concepts of single frequency networks and virtual antenna arrays. First, results for amplify-and-forward networks indicate that such systems indeed provide a degree of spatial diversity that is proportional to the number of distributed antenna elements, or, in our context, the number of involved relay nodes [69]. Even under strict power and bandwidth constraints, systems with one or two relays can achieve diversity gains over single input single output (SISO) and conventional relaying transmission for communication over a Rayleigh fading channel [69], see Figure 6.13 as an example [68].

A mature concept of a single frequency network (SFN), as well as of MIMO concepts, is presented in [70]. Fixed relays provide power gains and macrodiversity by forwarding signals from an AP to the destination terminal. Various protocol versions can be implemented, based on the underlying OFDM air interface. To illustrate the potential of single frequency networks, Figure 6.14 shows the achievable performance gain for the simultaneous operation of eight fixed relays and different basic techniques, which can be up to 9 dB in the ideal case. Specifically, the figure compares the cases of random superposition of the individual subcarriers (classical SFN), subcarrier selection based on strongest received power, and ideal phase pre distortion leading to constructive superposition at the terminal. Finally, the maximum ratio predistortion, that additionally optimizes the amplitudes of the signals transmitted from the relays, can be understood as a performance bound, also see [70]. Note that predistortion techniques require channel state information (CSI) at the relays, which leads to comparatively more complex system implementations.

While the former concept is a two-hop concept with one stage of intermediate relays, this can be generalized to a multistage concept, as suggested in the next section. Among others, important research areas are:

- the design of flexible protocols and forwarding strategies that allow integrating cooperative approaches in conventional relaying systems;
- the question of availability of channel state information at the transmitters and relays; and
- the trade-off between the macrodiversity gains and the usage of additional radio resources required by cooperative relaying.

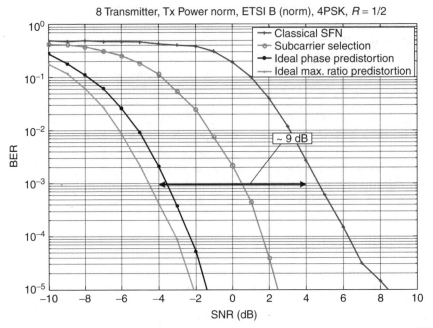

Figure 6.14 Performance gains of single frequency networks with eight transmitters and different algorithms, (for detailed calculation of the gains, see [70])

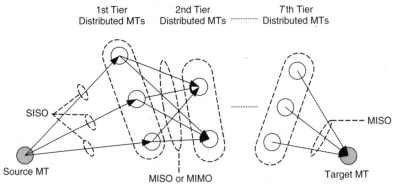

Figure 6.15 Distributed MT network, where a source MT communicates with a target MT via a number of MT tiers, each of which is formed of distributed relaying MTs (Source: [75])

6.2.4.3 Multihop Distributed MIMO Communication Networks

Concept

End-to-end data throughput depends on the capacity of each relaying hop, which is itself dictated by the allocated slot duration, bandwidth and transmission power, as well as the prevailing channel conditions. Telatar could prove that the capacity of an ergodic multiple input multiple output (MIMO) channel exceeds the capacity of a single input single output (SISO) channel by magnitudes [71]. Practical deployment of MIMO channels still remains an open problem, where a tractable approach has been suggested in [72–75]. The concept was termed *virtual antenna array* (VAA) because the t-MT is provided with redundant information by adjacent r-MTs, thus forming a virtual receive array. The concept is equally applicable in the transmission mode, where the adjacent r-MTs cooperate to form a transmit array. This allows the deployment of space–time capacity enhancement techniques, such as space–time block codes (STBCs) or space–time trellis codes (STTCs) to SISO MTs. In [72], the utilization of a generalized form of VAA to mesh networks has been suggested, as depicted in Figure 6.15. Here, the s-MT communicates with the t-MT via various r-MTs in parallel, thus realizing a distributed MIMO channel.

Principle

A source mobile terminal (s-MT) intends to deliver information to a target mobile terminal (t-MT) via a given number of distributed relaying mobile terminals (r-MTs), as depicted in Figure 6.15. The s-MT sends its information to a group of spatially adjacent r-MTs, which form the first tier r-MTs. Knowledge of their mutual existence allows the first tier r-MTs to space–time encode the data stream, i.e. each r-MT transmits only a spatial fraction of the space–time code word, such that the total output from the first tier r-MTs comprises a MISO transmission. The second tier r-MTs receive the data stream, decode it, re-encode and retransmit among each other first (cooperative, i.e. MIMO channels), or directly to the third tier r-MTs (noncooperative, i.e. MISO channels) in the same manner as described above. This process is continued until the t-MT is reached. It is assumed that the tiers of r-MTs are already formed and the terminals know about which spatial fraction of a space–time code word they have to (re) transmit. The distributed encoding process is described in more detail as follows, where an FDMA-based relaying system with T relaying mobile terminal tiers is described, and the example of a space–time block coding (STBC) scheme has been chosen.

1. *Source MT.* The s-MT Gray maps $b0$ source information bits onto symbol x by utilizing an $M0$-PSK (or $M0$-QAM) signal constellation, where $b0 = \log 2(M0)$. The data stream is transmitted on frequency band $W0$ with power $S0$.
2. *First tier relaying MTs.* The first tier r-MTs receive the data on frequency band $W0$, detect it, space–time encode it and transmit it simultaneously on frequency band $W1$ with a total power $S1$. Each r-MT Gray maps $K1b1$ bits onto symbols $x1, x2, \ldots, xK$ by utilizing an $M1$-PSK (or $M1$-QAM) signal constellation, where $b1 = \log 2(M1)$ and $K1$ is the number of symbols per space–time encoding. They are encoded with an orthogonal space–time coding matrix $\mathbf{G}1$ of size $p1 * d1$, where $p1$ is the number of symbol durations required to transmit the space–time code word, and $d1$ is the number of distributed r-MTs (and therefore equivalent to the number of transmit antennas). At each time instant t, the encoded symbol ct, i with $t = 1, \ldots, p1$ and $i = 1, \ldots, d1$ is transmitted simultaneously from the ith distributed r-MT. Clearly, the rate of the first tier space–time block code is $R1 = K1/p1$.
3. *Tth tier relaying MTs.* The Tth tier r-MTs receive data on frequency band WT-1, space–time decode it, space–time re-encode it and retransmit it on frequency band WT with a total power ST. The encoding procedure is the same as described above, where the rate of the STBC is RT.

4. *Target MT.* The t-MT receives the data on frequency band *WT*, space–time decodes it and performs the final detection. If the s-MT deploys a channel code, e.g. a simple trellis code, then the t-S performs the equivalent channel decoding to boost performance.

Each relaying MT tier clearly may use a different signal constellation and STBC. It is only of importance that each consecutive tier has knowledge of the transmission parameters of the previous tier.

Simulation Results

A multistage distributed MIMO network has been simulated with up to five stages (i.e. four MT tiers). The distance between the source and target MTs has been kept constant with $d = 100$ m. This scenario could occur, for instance, in an office building where a remote terminal (s-MT) communicates with other terminals, or a processing central unit (t-MT).

The one-stage scenario corresponds to a direct link communication scenario over 100 m. For the two-stage scenario, the first relaying tier is comprised of various distributed MTs which are randomly located in a squared area of size $d/5$. Only full-rate codes are simulated; therefore, only one or two distributed MTs are uniformly placed into the shaded area. The distance between the s-MT, the centre of the square and the t-MT is $d/2 = 50$ m. For the three, four and five-stage scenarios, the distance between the relaying tiers is chosen to be $d/3 = 33$ m, $d/4 = 25$ m and $d/5 = 20$ m, respectively. In the five-stage scenario, the MT distribution areas touch each other; this corresponds to the case when MTs are uniformly distributed along the path between the s-MT and t-MT.

Furthermore, it is assumed that each terminal transmits at a low average symbol energy of $Es = 1$ nJ measured at 1 m distance. The MT receive noise power spectral density is assumed to be $N0 = -140$ dBm/Hz. This results in a signal to noise ratio of $S/N = Es/N0 = 80$ dB measured at -1 m distance.

The path loss model used is the traditional negative exponential path loss model, where the power loss is inversely proportional to dn, where d is the distance and n the path loss exponent. For the indoor environment, n is approximately between $(3,\ldots,6)$, where $n = 3$ corresponds to a lightly, and $n = 6$ to a densely, cluttered indoor environment.

Ergodic Channels without Shadowing

Figure 6.16 relates to a path loss coefficient of $n = 3$. It depicts the mean capacity in nats/s/Hz[1] versus the number of stages utilized. At each stage, the network performance with and without STBC is compared, where the deployed STBC is the full-rate Alamouti code. Furthermore, the mean, maximum and minimum capacities have been depicted. Note that there is no fair capacity comparison between the communication scenarios with different stages, as the transmission energy was kept constant per node. If a fair comparison was desired, then the total utilized energy to deliver the information from the s-MT to

---

[1] *nat or natural unit*: a unit of information content used in information and communications theory. The nat is similar to the shannon, but uses the natural logarithm instead of the logarithm to base 2. If the probability of receiving a particular message is p, then the information content of the message is $-\ln p$ nats. For example, if a message is a string of five letters or numerals, with all combinations being equally likely, then a particular message has probability 1/365, and the information content of a message is $5(\log_e 36) = 17.9176$ nats. One nat equals $\log_2 e = 1.442\ 695$ shannons, or $\log_{10} e = 0.434\ 294$ hartleys.

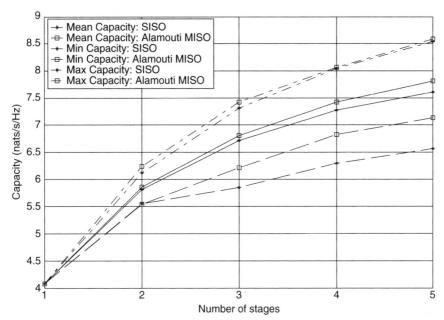

Figure 6.16 Normalized minimum, mean and maximum Shannon capacity in nats/s/Hz versus the number of stages utilized; path loss index $n = 3$ (Source: [74])

the t-MT would have to be equated for all scenarios, as done in [73] and [74]. Such normalization was not performed here, as the primary aim was to compare distributed MIMO with traditional communication networks.

From Figure 6.16 it is clear that the capacity of distributed networks is at least as high as for traditional multistage networks. A five-stage distributed network exhibits a 0.25 nats/s/Hz average capacity advantage over the traditional five-stage network. This leads to an SNR improvement of approximately 1 dB [74]. The average capacity is a useful performance measure if the communication channels are ergodic. Since the positions of the MTs are usually rather static, the channels are not ergodic with respect to the location (mean attenuation); however, the channels are still assumed to be ergodic with respect to the fading statistics.

For the five-stage scenario with uniform MT distribution at each stage, only 10 % of all geometrical MT distributions cannot support a capacity of 6.7 nats/s/Hz if the traditional SISO scenario is implemented; whereas for the distributed MISO case, all geometrical MT distributions can support such a rate. Only 10 % of all five-stage MISO communication scenarios cannot support 7.2 nats/s/Hz. This gives an average capacity advantage of 0.5 nats/s/Hz (= 2 dB [74]) if the respective capacities are to be supported in 90 % of all cases.

Equivalently, while the distributed communication scenario can support 7.2 nats/s/Hz in 90 % of all cases, the traditional network can support such a rate only with close to zero probability (i.e. the outage probability is 100 %).

The achieved gains of 2 dB translate to a transmit power saving of approximately 40 %. This clearly corroborates the advantages of distributed communication over traditional SISO communication scenarios.

Nonergodic Channels with Shadowing
The outage probability of the attainable rate is obtained here for the case when each link is affected by independent shadowing. For the Monte Carlo simulations, only a small shadowing standard deviation of 0 dB was assumed, which is deemed to be realistic for the anticipated communication distances.

Figure 6.17 depicts the outage probability of the attainable rates versus the rate in nats/s/Hz for the three, four and five-stage communication scenarios. Interestingly, for the chosen MT distribution, the three-stage distributed communication scenario does not yield any capacity benefits over the SISO case. Furthermore, the four-stage distributed case yields only small gains. Finally, only when the MTs are densely and uniformly distributed between the s-MT and t-MT, can drastic gains be observed. The latter case corresponds to the anticipated high-density network layouts.

The five-stage distributed MIMO communication network yields a gain of 0.65 nats/s/Hz for an outage probability of 10 %. This a mounts to a power saving of approximately 3 dB, or 50 %.

Future Research
The collection of topics listed below can be seen as a basis for future research:

- The concept of and preliminary research on, virtual antenna arrays.
- The concept of and preliminary research on, distributed MIMO multihop networks.

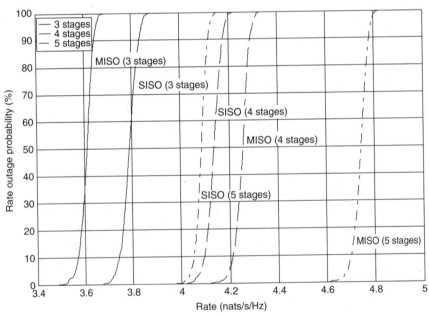

Figure 6.17 Outage probability of a given communication rate in percent versus this rate in nats/s/Hz for the three, four and five-stage single antenna and distributed Alamouti communication scenarios over Rayleigh fading channels with shadowing; path loss index $n = 3$ and shadowing standard deviation of 0 dB (Source: [74])

- A closed capacity expression for the generic ergodic MIMO Rayleigh flat fading channels utilized for resource allocations.
- A closed capacity expression for the orthogonalized ergodic MISO Rayleigh and Nakagami flat fading channels, where the channel coefficients can have an arbitrary attenuation.
- A closed capacity outage probability expression for the orthogonalized ergodic MISO Rayleigh and Nakagami flat fading channels, where the channel coefficients can have an arbitrary attenuation.
- Capacity behaviour of multistage distributed MIMO networks needs to be assessed and simulated; performance should be compared to the nondistributed case and appropriate conclusions drawn.
- Explicit fractional power and bandwidth resource allocation algorithms should be developed, which allow a maximum data throughput.
- Closed error rate performance expressions have been derived for the distributed communication case.

6.2.5 Routing and Radio Resource Management

6.2.5.1 Routing

In multihop networks without a central node in the form of a BS or an AP (i.e., pure ad hoc networks), the main issue is establishing the connectivity; that is, finding a route from each source node to the corresponding destination node. The extensive routing literature within the context of ad hoc networks confirms this observation.

In infrastructure-based networks, multihop communications are facilitated through the use of fixed or mobile relays. When fixed relays are used, the routing problem becomes somehow comparable to that in wired networks, which is an easier problem (more or less predefined). Even when mobile relays are used, routing is still an easier task (in comparison to routing in ad hoc networks), mainly due to the following reasons: first, the BS or AP can assist the mobile terminals in the routing process, and besides, the BS or AP constitutes a common source or sink. Therefore, the issue in such networks is finding the best route (based on some criteria), rather than a route.

In [76], various routing algorithms are proposed for maximizing the network throughput in TDMA/TDD multihop networks which have central nodes to facilitate the scheduling (orthogonal resource partitioning among the hops in a route, in such a way that no additional bandwidth is used due to relaying). It is demonstrated in that study that if the routes are established by taking into account the potential gains due to adaptive modulation and coding, as well as diversity, significant increases in throughput can be achieved.

In [77], routing is considered in a multihop network supported by an infrastructure and communication relations limited to a few hops only. Multiple simultaneous routes become possible, and this makes the choice of the routing algorithm important. An algorithm ensuring that no queue at a relay node explodes for the largest possible set of packet arrival rates is called *throughput-optimal* [77].

Routing becomes more challenging when considering mobile relays. In the MANET subgroup of the IETF, several routing algorithms for mobile ad hoc networks have been investigated. Studies of these algorithms have shown a high routing overhead and low efficiency in network throughput. Based on this observation, it is proposed that routing in the multihop network be supported by an area-wide cellular overlay network [67]. There, a hybrid routing scheme called cellular based multihop (CBM) routing has been studied, where route requests are sent to the base station of the overlying cellular network. This central entity determines the route and responds with a packet comprising a series of mobile nodes willing to relay the data traffic between the source and destination. The service and route discovery is performed in the overlay cellular network, and the packet data transmissions in the microrange multihop network. This exploits both the ability of the macronetwork to communicate with all of its nodes, and the throughput efficiency of multihop transmission in the microrange layer. Results have shown that CBM leads to low packet drops due to wrong route information, and adds little overhead to network traffic [67]. Moreover, it allows fast packet delivery because of quick route establishment and the routing overhead increases only linearly with the number of nodes, which indicates that CBM scales very well with the network size. It is known that the protocol overhead associated with establishing and maintaining the routes in multihop networks may become severe, thus reducing the potential efficiency of multihop techniques substantially. Limiting the number of hops, e.g. employing a single intermediate relay station on any route, would greatly simplify routing complexity.

6.2.5.2 Radio Resource Management

Radio resource management concerns the assignment of base station, channel and transmit power. In view of this, the sensitivity of radio coverage to the selection of relay, relay channel and relay power control are investigated in [54] for a cellular TDMA system where two-hop mobile relaying is employed whenever necessary. Whenever relaying is performed, an additional time or frequency channel is required for the second hop. In [54], an aggressive strategy that does not require any new channels for relaying is adopted: the relay channel is always selected from among the already used channels in the adjacent cells.

Various selection schemes for the relay and the relay channel, from random to smart selection, with and without power control, are considered in [54]. It is observed that, with the proper selection of relay, relay channel and relay power, a significant enhancement in high data rate coverage can be attained through two-hop mobile relaying. The observed trends and the corresponding conclusions are as follows:

- Performance gains due to relaying increase as the number of wireless terminals in the system increases.
- Employing power control in both hops further enhances the performance, especially as the cells get small; the returns due to power control become substantial for interference-limited cells.
- The maximum relay transmit power level is an important factor only in large cells; in small cells most of the benefits are gained with relatively small relay transmit power levels. Therefore, the impact of relaying on wireless terminals' batteries may not be significant in microcellular systems.
- The performance gains are quite sensitive to the relay selection criterion. If the relays are chosen randomly, the performance gets worse in comparison to the no-relaying case (this is analogous to the case where a user is connected to a wrong base station).

Yet, highly suboptimal (with minimal intelligence), but implementationally feasible, relay selection schemes (such as relay selection based solely on proximity through the use of the GPS data available at the base stations) still yield significant coverage enhancements.

- Once a good relay is selected, the performance gains become fairly insensitive to the relay channel selection criterion. Therefore, in systems with limited resources for monitoring and control purposes, the priority should be given to proper relay selection, rather than proper relay channel selection.

Finally, it is worth noting that a relay's energy consumption will increase linearly, regardless of the multiple access scheme used, as more and more terminals' signals are relayed. The increased energy consumption is not critical for fixed relays (since they will be plugged to power lines); however, this increase will constitute a major concern for battery-powered mobile relays. The change in the transmit power of a relay with respect to the load will depend on the multiple access scheme used. In a TDMA system, no additional power will be needed, since a relay will transmit signals to (or from) various terminals in a time division manner. In a CDMA system, on the other hand, a linear increase in the transmit power will be necessary as a result of the simultaneous transmissions. Although the increased transmit power is a concern in both fixed and mobile relaying systems, the undesirable impact (such as increased cost) is more profound in the latter types.

For infrastructure based solutions with fixed relays, the selection of relays is much simpler and more or less predefined. For this case, possible concepts can be based on centrally controlled heuristic methods for relay channel selection within a single multihop cell [78]. Selection criteria involve the mutual interference between relay channels.

6.2.6 WMS: A Prototypical Cellular Relay Network

An example of a new radio access network architecture for a mobile broadband system using fixed relays to provide radio coverage to otherwise shadowed areas, is introduced next. Low-power (1 W) pico base stations using a wireless or mobile broadband air interface at access points (AP) to the core network and at fixed relays to trade the high capacity available at APs against radio coverage range, contribute to ensure the highest spectrum efficiency and the lowest possible transmit power. The so-called *wireless media system* (WMS), shown in Figure 6.18, will provide broadband radio access to moving terminals and is embedded into a cellular radio network to support communications of highly mobile terminals providing a medium transmission rate. The low power requirement for the broadband air interface leads to a picocellular concept relying essentially on multihop communication across fixed relays, and to some extent also on ad hoc networking, using mobile relays at the periphery of the WMS, see [48, 49, 52].

The WMS aims to have a very high multiplexing data rate of between 100–1000 Mbit/s at the air interface for medium velocity terminals, and high deployment flexibility through the use of mass production building blocks for APs and relays. The candidate spectrum bands for operation of the WMS, e.g., beyond 3 GHz or even beyond 5 GHz, will allow very small equipment size for picocellular base stations (access points, AP) and fixed relays, including the antenna, so that the infrastructure can be termed more or less 'invisible'. Fixed relays might use either central or distributed control of the nodes involved,

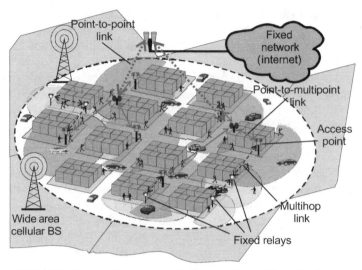

Figure 6.18   Wireless media system: integration with mobile radio (Source: [58])

as described in [50]. The WMS is integrated into a 4 G system shown by means of large hexagonal cells, sharing (i) an IPv6 based fixed core network, and (ii) functions of the cellular system like subscriber identity module (SIM), authentication, authorization and accounting (AAA) and localization to be used in both types of network.

The feeder systems connecting APs to the fixed network might be based either on wire/fibre or wireless, e.g., point-to-multipoint (PMP) LOS radio systems, see Figure 6.18. Both APs and relays appear like base stations to a mobile terminal, and are henceforth referred to as *media points* (MPs).

Figure 6.18 shows, by means of an example, the discontinuous radio coverage available from the WMS in urban areas. The small picocells highlighted around the MPs represent areas where broadband radio coverage is available via preplanned multihop communication.

From Figure 6.18, it is apparent that the WMS architecture scales quite well and is able to provide the traffic capacity of an AP to a small (pico) or much larger (micro) cell built up from many picocells defined by the AP and its related fixed relays, resulting in a very cost efficient and flexible infrastructure.

A mobile terminal must be able to support both mobile cellular radio and the WMS air interfaces, but not at the same time. An operator of a WMS that is integrated into a cellular mobile radio system might apply a mixed cost calculation to be able to trade the low cost for transporting mass data via the WMS, against the high cost of data transport across the cellular systems. This concept of course is applicable also to an indoor service. Further, this operator may end up with a very attractive tariff for all of its services compared to an operator purely relying on 3G technology.

6.2.7 Conclusions

Relaying is broadly considered as a method to ensure capacity and coverage in cellular systems. This section has given an overview on the relaying context, presenting two main concepts. The first is the cooperative use of relays forming virtual antenna arrays to exploit the spatial diversity inherent to multihop communications, leading to substantial increases in the available capacity. The second is the wireless media system, a concept for a mobile broadband system based on fixed relay stations.

Examples for forwarding techniques, as well as a pattern for the wide-area urban coverage using clusters of access points and relays have been proposed. It was shown, for different application scenarios, that multihop communications can provide a substantial increase in link and network capacity when applied in areas suffering from heavy path loss. One benefit for broadband radio systems is that the very high capacity that can be expected from these systems can be traded for radio range that would otherwise be limited due to high attenuation at high radio frequencies. This allows one to meet the expected capacity needs per unit area without the danger of capacity overprovisioning, when relying on single hop concepts only.

To summarize, the especially interesting properties of relay-based systems are that:

- The concept is applicable to cellular mobile radio networks, as well as to wireless broadband access networks, for densely populated areas.
- The radio coverage can be dramatically improved by using RSs to circumvent obstacles ('range extension behind obstacles').
- The large capacity expected from next generation wireless systems, that might be much too large to be used up by the terminals roaming in the small cell area covered by an AP, could be traded against radio range [58] by means of RSs ('range extension in general'). The fixed relay concept especially can be used to extend the radio range of an AP to nonshadowed areas beyond the EIRP limits specified for a given frequency band.
- The RS would allow one to apply higher valued PHY modes of the radio interface for a given link on a hop, thereby increasing the capacity per multihop link, and thereby the end-to-end link capacity.
- Relay-based systems allow radio resource reuse within the area served by an AP and its related relays, e.g., a frequency reuse within the area served by one AP.
- The cost to connect the radio access network to the infrastructure (core) network is reduced substantially by reducing the number of points of presence required, e.g., to 20 % for the example shown in Figure 6.5(left). Much higher reductions appear possible, depending on the capacity available from an AP and the range possible under the power limitation defined for a given frequency band.
- The relaying concept introduced scales extremely well, since the number of tiers formed by relays around an AP only depends on its capacity and the capacity per area element required from the terminals roaming in the service area. In a Manhattan grid, according to Figure 6.5, a second tier of relays would enlarge the number of relays to 12, thereby covering 13 intersections. The infrastructure connectivity cost would then be reduced to about 8 % of the cost of a purely AP-based system.
- Routing for fixed relay based networks does not raise many problems since alternate routes could be calculated once and for ever for a given deployment concept. The dynamic insertion and removal of RSs would not increase the routing related overhead much, since these would be rare events. The additional introduction of mobile relays would be possible too.
- The number of hops in a fixed relay based network will be limited, owing to the capacity limits of the AP and the minimum capacity required at an outmost tier relay. Since the

delay performance is reduced with an increasing number of relays, it is recommended to limit this type of network to two tiers of relays around an AP.
- It can be proven, both conceptually, as well as via mathematical calculations and computer simulations, that the application of relaying in distributed MIMO networks yields significant benefits in terms of power savings or higher data rates. Distributed systems are understood to be in their infancy; thus, major research has to be carried out on various aspects spanned by such communication systems. The emphasis here was put on the physical layer behaviour in terms of capacity; however, equally important are MAC and RRM issues.

A thorough investigation of the relaying concept is an enormously complex task, owing to the very many parameters involved, including propagation and physical layer issues, system issues such as medium access and radio resource management, networking and protocol design, and, finally, implementation-related issues. Moreover, a survey of the literature shows that the analytical understanding of the subject is far from comprehensive yet. This outlines the key importance of future research in the areas of (i) virtual arrays and novel diversity schemes, (ii) multiple access and multiplexing schemes, and (iii) the complex of medium access and radio resource management protocols for multi-hop networks.

6.3 Spectrum for Future Mobile and Wireless Communications

Availability of sufficient spectrum is essential for the deployment of any wireless systems and that will also be true in the future. The usage of wireless services increases, both due to the fact that the number of users increases, and that the usage and traffic by an individual user grows. Current systems and technologies will evolve and new technologies will be developed. There will be new services, which require higher bit rates, and there is a continuing need for higher capacities. As a consequence, new spectrum may be needed.

The identification of frequency bands is a long international process and it is important to take spectrum requirements into account in early research, and to communicate the relevant results to the regulators to enable them to get prepared for the possible new spectrum requirements in time. There is a continuous pressure towards higher spectrum efficiency and greater flexibility. This puts high demands on the new technologies, especially on new air interfaces, antenna technologies, network topologies, etc.

This section addresses spectrum for terrestrial services, and it focuses on the 3G/IMT-2000 and systems beyond IMT-2000. The systems beyond IMT-2000 will cover a wide range of wireless communications, ranging from low bit rates up to 1 Gbit/s in hot spots, and from low mobility, local coverage to high mobility, wide-area coverage. This part describes areas in which focused research is required before the spectrum-related challenges can be met and spectrum identified and eventually made available. The areas for research are traffic characteristics of future systems and services, technical developments influencing spectrum efficiency and usage, efficient sharing methods, novel and flexible spectrum use and future spectrum requirements.

The relevant results from the research need to be communicated to the international regulatory process, which aims to define the future spectrum requirements, and eventually identify the required spectrum.

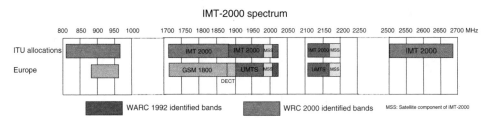

Figure 6.19   Spectrum for IMT-2000, bands identified by WARC'92 and WRC-2000

6.3.1 Introduction

The usage of wireless services continues to increase, both due to the fact that the number of users is increasing, and that the usage and traffic by an individual user is growing. It is predicted that, by the year 2010, there will be 1700 million terrestrial mobile subscribers worldwide [79]. New services are coming, which require higher bit rates, and there is also a need for higher capacities. Therefore, the current systems and technologies will evolve, and at some point, new technologies will be developed. As a consequence, new spectrum may be needed, possibly made available as wider bands than today, because higher bit rates will require wider bandwidths.

6.3.1.1 Progress on Spectrum

The last major global addition to the overall amount of spectrum for mobile/cellular use occurred when the WRC-2000 identified additional spectrum for IMT-2000 (see Figure 6.19) with the aim of having identified a sufficient amount of spectrum for 3G/IMT-2000 until 2010. However, as 3G/IMT-2000 is developed further, and systems beyond IMT-2000 are to be developed, implemented and deployed in the future, more spectrum may be needed globally for further development of IMT-2000 and systems beyond IMT-2000 sometime after 2010.

Today, also, wireless LANs are becoming popular. They are used for deploying local wireless networks both commercially and for private use. A typical application is high-speed internet connection for laptop computers in office buildings, or at some hot spots, such as hotels or airports. Initially, the RLANs have used the 2.4 GHz band, but recently the systems have started to use the 5 GHz bands.

In the 5 GHz range, the bands and the regulations have been different in different countries. For example, there have been significant differences between Europe, USA and Japan, see Figure 6.20. The amount of spectrum has been different, the rules for the usage have been different and, in many cases, the available bands have not proven to be sufficient.

The WRC-2003 identified global spectrum for 5 GHz RLANs. They are the bands that were earlier designated for RLANs by the ECC, see the EU bands in Figure 6.20.

6.3.1.2 ITU Framework for Systems Beyond IMT-2000

At the moment, 3G/IMT-2000 systems are being deployed commercially worldwide. The ITU believes that the capabilities of the systems need to be enhanced in line with user

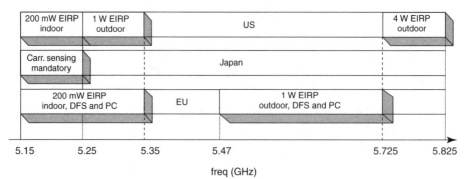

Figure 6.20 5 GHz WLAN bands

demand and expectations, and in line with the technology trends. Work is already under way in various organizations to extend the capabilities of the initial systems. To help meet the ever increasing demands for wireless communication, and the expected higher data rates needed to meet user demands, ITU-R has initiated work on future development of IMT-2000 and systems beyond IMT-2000. The view is that IMT-2000 will evolve, and that 'systems beyond IMT-2000, for which there may be a need for new wireless access technology to be developed around 2010, capable of supporting high data rates with high mobility, could be widely deployed around 2015 in some countries' [79].

The expected new capabilities of systems beyond IMT-2000 are shown in Figure 6.21. There are two parts to the new capabilities: new mobile access capable of peak data rates of 100 Mbit/s, covering high mobility, and new nomadic/local area wireless access capable of peak data rates of 1000 Mbit/s covering low mobility.

6.3.1.3 International Regulatory Process

Since the WRC-2000, the ITU has continued working on systems beyond IMT-2000. The ITU-R recommendation M.1645 states that, to fulfill the framework for systems beyond IMT-2000, it is envisaged that further spectrum may be needed in addition to that identified for IMT-2000 at WARC'92 and WRC-2000. Of course, any need for additional spectrum needs to be well justified before any decisions can be taken.

The ITU process leading to global spectrum identifications is typically taking several years. To initiate the process and be prepared for the future, WRC-2000 already defined an agenda item about the future developments of IMT-2000 and systems beyond IMT-2000 for WRC-2003, and an item to the preliminary agenda for WRC-2007. The WRC-2003 followed those earlier plans, reviewed the progress of the ITU studies and decided that WRC-2007 will have the following agenda item: '1.4 to consider frequency-related matters for the future development of IMT-2000 and systems beyond IMT-2000, taking due account of the results of the ITU studies...' This agenda item allows the WRC-2007 to make the necessary decisions if so required.

Working Party WP8F of the ITU-R has been, and is, responsible for the ITU preparatory work on the IMT-2000 related agenda items. Among the first tasks after WRC-2003 was

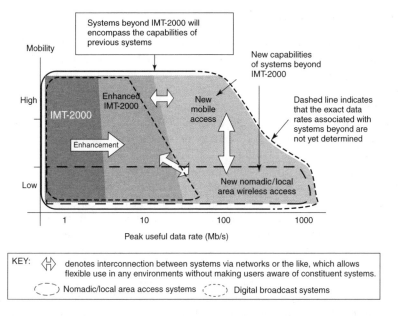

Dark grey colour indicates existing capabilities, medium grey colour indicates enhancements to IMT-2000.
and the lighter grey colour indicates new capabilities of systems beyond IMT-2000

The degree of mobility as used in this figure is described as follows: low mobility covers pedestrain speed,
and high mobility covers high speed on highways or fast trains (60 km/h to ~250 km/h, or more).

Figure 6.21   Capabilities of systems beyond IMT-2000 [79]

to develop and agree a methodology that enables the estimation of the future spectrum
needs. The second step is then to prepare the calculations and to agree on the actual
spectrum requirements. This step also needs relevant and sufficient information on future
traffic, in the form of estimates and forecasts. The WP8F started to work on the spectrum
calculation methodology in its March 2003 meeting in Brazil [80]. So far there have been
some contributions about the methodology and future spectrum requirements, for example
from Japan [81], but the actual work is expected to really start after the WRC-2003. The
WP8F should be ready with the preparations early in 2006 if the conference is held in
mid 2007.

The third step would be the actual identification of the bands by the WRC-2007. The
fourth step would be to agree on the spectrum utilization schemes and, finally, as the fifth
step, the bands should be made available.

During the WRC-2003, a need was identified (COM7/2) for studies to examine the
effectiveness, appropriateness and impact of the radio regulations, with respect to the
evolution of existing, emerging and future applications, systems and technologies, and to
identify options for improvements in the radio regulations. Results of these studies can
be included in reports to WRC-07 for the purposes of considering whether to place this
subject on a future conference agenda. This may have an impact on how spectrum is
made available for future wireless communications.

6.3.1.4 Requirements for Research

One crucial, but yet unknown, element influencing both the methodology and the actual spectrum requirements is the technology for future mobile communications. Any technologies to be employed sometime beyond 2010 are today in the early research phase. The WWRF, as a global research forum, is expected to be in an excellent position to present and to discuss such technical research topics and results that are relevant for the international regulatory process and the work at ITU-R.

As spectrum is becoming a more and more scarce resource, and increased usage, together with higher bit rates, requires higher bandwidths, the demand for more efficient spectrum usage becomes more and more significant. Therefore, any new technologies need to fulfil the requirements for high spectrum efficiency. Spectrum efficiency can be improved in many ways, e.g., by employing new air interface technologies, new antenna technologies, advanced network topologies, etc. Additional ways to improve the overall spectrum efficiency should be sought in the area of efficient sharing, flexible spectrum use and good coexistence, and these should be addressed in research for new system concepts from the beginning.

A realistic view about the capabilities of future technologies is needed in the international regulatory process before spectrum decisions can be taken, and sufficient spectrum be made available. Common ground is needed to reach the necessary level of global harmonization in the spectrum allocations and identifications.

The equipment implementation complexity is expected to increase. It can only be compensated by achieving global markets, best facilitated by making globally common spectrum available. It will also be crucial in the future that mass markets can be achieved to gain benefit from the economies of scale.

6.3.2 Traffic Characteristics of Future Systems and Services

One basic element in dimensioning and designing future communication systems is the forecasted traffic and its characteristics. The UMTS Forum initiated in 1997 a market study about the future 3G mobile services and their usage in 2005 and 2010. Information from that study was used as a basis, when the UMTS Forum calculated the need for additional spectrum for 3G/IMT-2000. The results of the calculations were published in [82], [83] and [84] and they were utilized, together with the developed calculation methodology, in Europe by CEPT, and later by the ITU as a basis when the corresponding decisions were made in WRC-2000.

The UMTS Forum is continuing the studies about the future traffic characteristics of 3G, but with a view of contributing to the decision making about the spectrum arrangements for the additional bands of 3G/IMT-2000 that were identified by the WRC-2000. In Europe, that means the arrangements for the 2.5 GHz band.

There is again a need to predict the future traffic, based on possible services, their characteristics and usage, this time for the years 2010, 2015 and 2020. Obviously, this is a very challenging task as 3G/IMT is only currently about to start. Secondly, the scope is now much wider, as the future developments of IMT-2000 and systems beyond IMT-2000 are to cover a wide range of services and environments, ranging from hot spots to wide areas. Additionally, as part of convergence, the new service delivery mechanisms will allow several parallel technologies to be used for providing the same services. This could reduce the amount of spectrum needed, but it may also require more consistent availability of spectrum.

Japan has proposed, within ITU-R, that new service types, namely ultra high multimedia and very high multimedia, would be used to complement the service types that were used in estimations before the WRC-2000 [81]. This approach has been used also in some other contributions and papers. But in any case, regardless of the terminology, all service types, their characteristics and usage should be predicted, before the traffic and its characteristics can be determined.

For estimating the spectrum requirements, it is important to identify the services and their main characteristics, but for detailed design of the systems there is a need to go deeper and to understand the service characteristics in more detail. There are already some studies about the detailed traffic characteristics of expected future mobile services, such as video services [85]. In many cases, new services will emerge and will first be used through the wireline internet, therefore studying the service characteristics for the internet in detail also helps the dimensioning and detailed design of the mobile protocols and algorithms. Such detailed information is needed for systems research and development.

6.3.3 Technical Developments Influencing Spectrum Efficiency and Usage

The key technical areas expected to improve and influence the overall spectrum efficiency are the air interface technologies, antenna technologies, advanced signal processing, new network topologies and cooperation of networks. All of them are subject to major research, are covered by the WWRF and are dealt with by the various sections of Chapters 4, 5, 6 and 7. This section shall only give a brief overview.

6.3.3.1 Advanced Signal Processing

Two examples of how advanced signal processing could be employed are software-defined radio (SDR) and cognitive radio technology.

SDR
Typically, most radio equipment operates over a fixed range of frequencies, accessing only predefined frequency bands. With future software-defined radio, functions for generation of the transmitted signal and tuning the received signal could be performed by reprogrammable software, meaning that the radio can be reconfigured to allow transmission and reception of the signal over a wide range of frequencies with different transmission formats. This capability would allow much more flexibility over how spectrum is used, e.g., due to national differences.

Cognitive Radio Technology
Current fixed spectrum assignments may result in a significant under-utilization of spectrum resources, depending on the frequency bands, density of active users, time and locations. Recent measurements have shown that the spectrum usage (defined as the active transmission over the frequency band for a given space and time) is very low. This

example indicates the potential for further improvement of spectrum efficiency achievable by dynamic sharing or assignment of spectrum use (see [86]) over a large time scale, and by exploitation the additional RF dimensions such as variable carrier frequency and bandwidth. Cognitive radio technology offers potential to improve the spectrum efficiency. Cognitive radio employs the dynamical time-frequency-power based radio scene measurement and analysis of the RF environment, to make an optimum choice of carrier frequency and channel bandwidth to guide the transceiver in its end-to-end communication with quality-of-service being an important design requirement. The impact of advanced reconfigurable radio transceiver concepts and spectrum usage protocols needs to be considered for the identification of future spectrum requirement.

6.3.3.2 Advanced Network Topologies

It may be possible to improve the overall spectrum efficiency by employing new network topologies. Ad hoc networks, cooperative and multihop networks, especially, are currently under intensive research, and are also covered by the WWRF.

Convergence and Cooperation of Networks

Another important issue is the convergence of services and networks, as the traditional vertically isolated systems and markets are expected to disappear. Communication systems have traditionally been differentiated by services and used spectrum. Today, new services rather begin to complement than replace each other, and this is expected to influence the total demand for spectrum. This is the particular case with new equipment that combines wireless data communications, telecommunications and broadcast reception. For instance, an internet TV could combine some of the functions of radio, TV, PC and phone. This is not only a technology issue, but also an issue of culture and lifestyle.

This will also have an influence on the networks. First, various different systems will offer similar kinds of services to users in a manner where the delivery 'channel' may be chosen by the user, based on availability, price and performance. Later, interworking between the systems is expected to be introduced as part of the evolution of the current systems. This may all lead to some form of cooperation of networks, which may also improve the overall spectrum efficiency and is expected to influence the spectrum requirements. Convergence and the cooperation of networks also needs to be addressed by research, and the results would help in understanding the influence of convergence on the future overall system and spectrum requirements.

6.3.4 Efficient Spectrum Sharing Methods

6.3.4.1 Sharing

Efficient sharing capability is needed as, typically, most bands are allocated for several services in the ITU radio regulations and are, in fact, in use by several systems/technologies. There are no unused bands available, and if new, wide frequency bands are needed, refarming of existing services may not be feasible. Therefore, the new technologies may not get dedicated bands of their own, but may have to be able to use the same bands as other services. This means that the new technologies need to be able to share the bands with existing services.

Quite often, sharing is possible only by applying certain rules or conditions. For example, there may be limitations for spectrum use, output power levels or equipment density (penetrations). Or there may be mandatory technical requirements, for example, on transmission masks, transmitter power control (TPC) or on low-lever protocols.

Sharing may not only be needed between the new technologies to be deployed and the existing services in the same bands, but it may also be needed between several parallel new networks based on the new technologies, so that the networks can efficiently share the same bands, in the same way as RLANs may use the 2.4 GHz and 5 GHz bands in an uncoordinated manner today.

In any case, the aim should be that sharing with other services should be simple to implement, but efficient in use. The requirements for sharing should be taken into consideration, even if the target bands are not yet decided and actual other services are not yet known. Obviously, detailed studies and definitions can only be made when the actual bands are known. However, as time goes on, it is very likely that some actual candidate bands will be defined and announced, and this information can be used in the work. At the time of writing, only Japan has announced its national intentions to make the bands 3600–4200 MHz and 4400–4900 MHz available for systems beyond IMT-2000. Any research before the WRC-07 has to examine those bands, but also other potential spectrum ranges should be considered.

6.3.4.2 Coexistence Issues

In addition to efficient sharing, coexistence is also important for any future technologies.

The deployment of future networks should be spectrum efficient and easy from the spectrum coordination point of view. This means that the coordination needed for transmitters and systems in the same, or adjacent, frequencies should be minimal, or at least straightforward, and that there should be no need for large guard bands between the bands assigned for each system. This may imply requirements for spectrum scenarios and arrangements, radio interface design and network topologies. Good coexistence is important for any future technologies. Knowing the actual bands would be beneficial, but the work can be started based on initial assumptions.

6.3.4.3 Spectrum Efficiency Definitions and Measurement

It is envisaged that future new spectrum arrangements will allow much more efficient and flexible utilization of the frequency resource. However, the classical information theory based spectrum efficiency definition cannot reflect the emerging reality of the spectrum usage e.g. with dynamic spectrum sharing. For example, a more generalized spectrum efficiency definition could be the (maximum useful information bits transmitted)/(the spectrum resource impacted), where the spectrum impacted = (bandwidth impacted) × (geographic space impacted) × (time denied to other users). Hence, spectrum efficiency will be used as a fundamental metric to evaluate the performance for new air interfaces and will impact new spectrum allocations. Further study on appropriate spectrum efficiency definitions and measurement is required.

6.3.5 Novel and Flexible Spectrum Use

The new technologies can open up new possibilities to use the radio resources, frequency spectrum and the deployed infrastructure in a more flexible and efficient way, possibly leading to new network deployment and spectrum utilization schemes. Due to the impact of new technologies on system and network configurations, and the possibly required new spectrum schemes, a close cooperation between the research community and the international regulatory process is needed so that the regulations can evolve in time to adapt to the foreseen developments.

6.3.5.1 Flexibility and Scalability

Identification of new spectrum will be much easier if the new technologies can use spectrum in a flexible manner. It is also expected that the ability to use spectrum in a flexible manner may increase the overall spectrum efficiency. Current examples of flexible spectrum use are dynamic frequency selection (DFS) for 5 GHz WLANs, or multiband operation of cellular systems. So far, the required flexibility has been only relatively simple and built on well defined system internal properties. In the case of DFS for 5 GHz RLANs, the overall definition and implementation process has been complex and time consuming. That clearly demonstrates the practical problems when different systems have to be taken into account. In the future, the methods can probably be enhanced by additional logic and by some interworking between the systems. Possibilities for flexible spectrum use in future mobile communications forms an important research topic, and the potential benefits of the methods to be employed should be investigated carefully.

Scalability is considered as an important characteristic of the future systems to deal with the requirement for various services requiring different bandwidths, together with the possible need to operate in frequency bands of different width.

6.3.5.2 Dynamic Use of Spectrum

It may also be useful to consider alternative methods for spectrum assignment and use, rather than the current 'fixed block' approach normally considered. Air interfaces that could cover and use any available spectrum, without interfering with other systems, could provide much higher capacities, but there would also be many issues to be solved [87].

There is some research covering the so-called 'dynamic spectrum allocation'. The IST project DriVe investigated the possibilities of mobile services and broadcasting services using the same bands under certain conditions. One of the starting points of the work was the observation that maximum usage of the spectrum for each service occurs at different times of the day. This issue has been discussed also at the WWRF [88].

6.3.5.3 Interference Management

The FCC is currently working on a concept based on 'interference temperature', which could allow some services to operate just above the noise floor, but without interfering with other services in the same bands. The public discussion of the concept is ongoing and the concept needs to be studied in detail.

6.3.5.4 Ultra Wide Band Radio Systems

Ultra wide band is an impulse radio technology that uses spectrum in a very different way from the traditional use. It operates across a very wide band using very narrow pulses in such a way that the average power and power spectral density is very small. It seems possible to operate UWB as an overlay system in bands that are already in use by other services, without causing harmful interference to them, which would be a benefit from the spectrum coordination or spectrum management points of view.

UWB seems to have many other benefits, such as being capable of very high bit rate transmissions over short distances, and, therefore, it is under intensive research, especially for future short-range communications.

On the other hand, there are services that seem to be harmfully interfered by UWB, and therefore, the regulations in Europe and by the ITU are yet to be defined as the potential interferences are being studied. However, the FCC has already announced regulation for UWB systems when, in 2002, they released a 'First Report and Order' allowing UWB to operate under Part 15 rules with some restrictions.

6.3.6 Future Spectrum Requirements

There are several issues related to the future spectrum requirements. Some of them are discussed next.

6.3.6.1 Spectrum Scenarios

Taking the future technologies and system concepts into account, it is possible to identify scenarios about the possible spectrum configurations for future mobile and wireless communications. They can offer useful overall guidelines towards more detailed planning. The scenarios need to take into account the current bands for mobile communication and their usage, possibilities to make new bands available and the preferred spectrum utilization schemes of the future technologies.

6.3.6.2 Spectrum Arrangements

When the future air interfaces and network topologies are being designed, it is also possible to define preferred spectrum arrangement schemes, where more detailed views about the actual spectrum utilization can be presented. This allows an analysis of the spectrum usage and overall efficiency. The results should be used for choosing the best options.

The traditional approach for mobile networks has been to have a pair of bands to accommodate a number of operator networks, each in a dedicated portion of the bands, next to each other, separated by appropriate guard bands. For WLANs, unpaired bands have been in use and, typically, the operation has been licence exempt and capability for sharing with other licence exempt usage has been important. The preferred arrangements for future systems may be different.

6.3.6.3 Minimum Spectrum

The minimum bandwidth in which the future systems can operate, needs to be taken into account when candidate bands are sought. There will be a minimum bandwidth for the link, and there will be a theoretical minimum for network deployment. Additionally, there may be a practical minimum for a viable network to be deployed. All these need to be studied carefully.

There are already studies about the required bandwidth at the link level [89, 90], but more research is needed.

Future systems are likely to be somewhat flexible in their spectrum use, but this needs to be investigated. It can still be assumed that the number of bands identified should be small, e.g. there should be a few bands wide enough to accommodate the new systems.

6.3.6.4 Frequency Ranges

The increase in data rates will require wideband carriers and a large total spectrum. Today, most of the spectrum allocated for mobile services is below 3 GHz, due to propagation reasons, and partly due to historical reasons. Unfortunately, the spectrum below 3 GHz is heavily in use, therefore finding and making any relatively wideband available from there in any reasonable time does not seem feasible. The new spectrum will most probably be above today's frequency ranges for mobile communications, above 2.5 GHz but still preferably below 6 GHz to enable mobility. It may be that spectrum for low mobility capabilities need to be even higher. 17 GHz and 60 GHz ranges have been indicated as possibilities for low mobility and short-range applications.

6.3.6.5 Total Amount

This is the most visible and tangible of the spectrum requirements. It can be determined from the information about future traffic, its characteristics, required capacity and future system properties by applying a calculation methodology. In practice, it is a very demanding task, due to the expected evolution of the present technologies and services, and due to their yet unpredictable relationship with each other and with the future technologies.

6.3.6.6 Methodologies

Two ITU-R methodologies for calculating spectrum requirements need to be mentioned here: Recommendation ITU-R M.1390 [91], which was developed before the WRC-2000 for calculating how much additional spectrum will be needed by terrestrial IMT-2000 by the year 2010, and Recommendation ITU-R M.1651 [92] for calculating the spectrum needed for RLANs in the 5 GHz band. The M.1651 was derived from the original IMT-2000 methodology M.1390, and it is more focused on packet-based services and applications.

The M.1390 was based on calculating the offered bit quantity (OBQ) per service per environment in the busy hour, and on deriving the total spectrum requirements by calculating the carried bit quantity from the network characteristics, blocking, overhead, spectrum efficiency, etc. The OBQ was calculated from the potential user densities, penetrations, service classes, service characteristics and user behaviour.

Today, it can be seen that M.1390 has some weaknesses and it cannot be used as such for future work. However, it is likely that the new methodology PDNR IMT.METH to be developed by the ITU-R will be influenced by both M.1390 and M.1651.

There are also other activities, outside the ITU, which have either addressed, or are to address, the future spectrum requirements. One important activity was a project called MIND, which was part of the EU's IST Programme. MIND studied, among other things, the possible methodology for estimating the spectrum requirements for future mobile use [93]. The work has now ended and the results have been presented to forums like the UMTS Forum, the WWRF and CEPT. The results have also been contributed through CEPT PT1 to the WP8F as one example of a possible methodology. Other contributions will be needed about the methodology and about the other spectrum requirements before any decisions can be taken on a solid basis by the ITU.

Summary of MIND Methodology
The methodology takes into account the new technologies for wireless communications and the movements towards convergence of different services (see Figure 6.22). Such convergent services are, for example, mobile communications, internet and broadcasting. From a user's viewpoint, the information flows may be requested regardless of the means of delivery.

Spectrum resources are required in the medium and longer term. Bands most suitable for mobile communications are limited to frequencies below 3 GHz due to propagation conditions. In the past, estimations of spectrum requirements have been considered in a static framework. With the advent of convergence, such a simple approach is no longer suitable. Hence, new statistical simulation models, based on the Monte Carlo method, have to be developed in order to allow the consideration of spatial and temporal distributions of the market requirements and network deployment scenarios.

The Monte Carlo simulation method is based upon the principle of taking snapshots of random situations. This flexibility is achieved by defining parameters as statistical distribution functions, which allows one to model complex situations by relatively simple elementary functions. Hence, results provided by such methodology will not give simple numbers of required bandwidth. Ranges of required spectrum resources for different radio access technologies under consideration will be the result, depending on the assumptions defined within the simulation process. On the other hand, consideration of economical

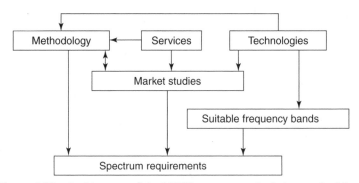

Figure 6.22 Architecture of the MIND spectrum calculation methodology

conditions of network operation as a function of available frequency resources is possible as well. Hence, new spectrum management methods, like enhanced spectrum sharing, may be supported by such results. However, results in terms of frequency requirements can realistically be achieved only if market expectations are available for application.

The result of the work was a proposal for an enhanced spectrum calculation methodology framework. It was not a complete description of how to calculate spectrum requirement figures, but rather a collection of ideas on how to overcome the drawbacks of the methodologies applied in the past.

In conclusion, it seems that the determination of the future spectrum requirements will be extremely difficult to perform by simple spreadsheet types of calculation alone, as was done for the WRC-2000, as some form of simulation is likely to be needed to support the calculations.

However, methodology that may be agreeable to the ITU may need to be largely simplified, perhaps even down to a level where the contents can be verified by traditional calculation methods, without running computer algorithms and simulations. Even in this case, it is likely that there will be a need to perform more detailed calculations using complex methods to get more precise data as background information, and to verify any simpler methodologies.

6.3.7 Conclusions

A wide range of research work is needed to ensure that the requirements for the future mobile communications can be defined and justified, and to ensure that the future spectrum requirements are realistic and can be based on agreeable forecast traffic conditions.

On the other hand, future technologies need to be able to fulfil the future requirements, not only the performance requirements, but also the requirements coming from the realities of spectrum availability and the need for sharing. The level of flexibility and its impact on the spectrum needs to be studied, as well as the impact of convergence.

Applicable results need to be communicated to the international regulatory process in due time to allow regulations to be adapted for the future spectrum requirements and possibly new spectrum utilization schemes.

6.4 References

[1] Walke, B., Pabst, R., *et al.*, 'Relay-based Deployment Concepts for Wireless and Mobile Broadband Cellular Radio', WWRF White paper, 2003. available from http://www.wireless-world-research.org/.

[2] Gosse, K., *et al.*, 'Short Range Communication with New Radio Air Interfaces based on Next Generation of WLANs, WPANs, and WBANs', WWRF White paper, 2003, available from http://www.wireless-world-research.org/.

[3] EU IST Advisory Group, *'Scenarios for Ambient Intelligence'*, February 2001, available from http://www.cordis.lu/ist/istag.htm.

[4] Perkins, C.E., *'Ad-hoc Networking,'* Addison Wesley, December 2000.

[5] Frodigh, M., Johansson, P. and Larsson, P., 'Wireless Ad Hoc Networking–The Art of Networking Without a Network,' *Ericsson Review*, **4**, 248–263, 2000.

[6] Akyildiz, I.F., Su, W., Sankarasubramaniam, Y. and Cayirci, E., 'Wireless sensor networks: a survey,' *Computer Networks*, **38**, 2002.

[7] Maltz, D.A., Broch, J. and Johnson, D.B., *'Experiences Designing and Building a Multi-hop Wireless Ad-hoc Testbed,'* Technical Report CMU, CMU School of Computer Science.

[8] Desilva, S. and Das, S.R., 'Experimental Evaluation of a Wireless Ad-Hoc Network,' *Proceedings of the 9th International Conference on Computer Communications and Networks (IC3N)*, Las Vegas, October 2000.

[9] Dupcinov, M., Jakob, M., Murphy, S. and Krco, S., 'An Experimental Evaluation of AODV Performance,' in *Proceedings of the Medhoc 02*, Sardinia, Italy, September 2002.

[10] Broch, J., Maltz, D.A., Johnson, D.B., Hu, Y.C. and Jetcheva, J., 'A Performance Comparison of Multi-Hop Wireless Ad-Hoc Network Routing Protocols,' in *Proceedings of the Mobicom 98*, Dallas, Texas, USA, October 25–30 1998.

[11] Perkins, C.E., Royer, E.M. and Das, S.R., '*IP Address Autoconfiguration for Ad Hoc Networks*,' Internet Draft: draft-ietf-manet-autoconf-00.txt, expired, 2000.

[12] Perkins, C., '*IP Mobility Support*,' IETF RFC 2002, 1996.

[13] Xi, J. and Bettstetter, C., 'Wireless Multi-hop Internet Access: Gateway Discovery, Routing, and Addressing,' in *Proceedings of the World Wireless Congress*, San Francisco, USA, 2002.

[14] Weniger, K. and Zitterbart, M., 'IPv6 Autoconfiguration in Large Scale Mobile Ad-Hoc Networks,' in *Proceedings of European Wireless 2002*, Florence, Italy, February 2002.

[15] MANET Working group, http://www.ietf.org/html.charters/manet-charter.html.

[16] Perkins, C.E., Royer, E.M. and Das, S.R., '*Ad-hoc On Demand Distance Vector (AODV) Routing*,' IETF RFC 3561, July 2003.

[17] Johnson, D.B., Maltz, D.A., Hu, Y.-C. and Jetcheva, J.G., '*The Dynamic Source Routing Protocol for Mobile Ad-Hoc Networks (DSR)*,' IETF Internet Draft, draft-ietf-manet-dsr-09.txt, April 2003 (work in progress).

[18] Clausen, T. and Jacquet, P., '*Optimized Link State Routing Protocol*,' IETF Internet Draft, draft-ietf-manet-olsr-11.txt, January 2003 (work in progress).

[19] Xu, Y., Heidemann, J. and Estrin, D., '*Adaptive Energy-Conserving Routing for Multi-hop Ad-Hoc Networks*,' Research Report 527, USC/Information Sciences Institute, October 2000.

[20] Singh, S., Woo, M. and Raghavendra, C.S., 'Power aware routing in mobile ad-hoc networks,' in *Proceedings of MOBICOM*, 181–190, 1998.

[21] Ramanathan, R. and Rosales-Hain, R., '*Topology control of Multi-hop Wireless Networks using transmit power adjustment*,' IEEE INFOCOM 2000, 404–413, 2000.

[22] Roduplu, V. and Meng, T., 'Minimum energy mobile wireless networks,' *IEEE Journal on Selected Areas in Communication* 17(8) 1333–1344, 1999.

[23] Hirt, W., *et al.*, 'Ultra Wideband Technology', WWRF White Paper, 2003, available from http://www.wireless-world-research.org/.

[24] Capkun, S., Hamdi, M. and Hubaux, J.P., 'GPS-free Positioning in Mobile ad-hoc Networks,' in *Proceedings of the 34th Annual Hawaii International Conference on System Sciences*, January 2001.

[25] Legrand, F., Bucaille, I., Héthuin, S., De Nardis, L., Giancola, G., Di Benedetto, M.-G., Blazevic, L. and Rouzet, P., '*UCAN's Ultra Wideband System: MAC & Routing protocols*,' International Workshop on UWB systems, Oulu, Finland, June 2003.

[26] Karn, P. 'MACA – A new channel access protocol for packet radio,' in *Proceedings of ARRL/CRRL Amateur Radio 9th Computing Networking Conference*, September 22, 134–140, 1990.

[27] Stojmenović, I. '*Handbook of Wireless Networks and Mobile Computing*,' Wiley-Interscience, 2002.

[28] IEEE 802.11. '*Draft Supplement to Standard for Telecommunication and Information Exchange Between Systems – LAN/MAN Specific Requirements – Part11: Wireless Medium Access Control (MAC) and physical layer (PHY) specification: Medium Access Control (MAC) Enhancements for Quality of Service (QoS)*,' Draft Supplement to Standard IEEE 802.11, IEEE, New York, November 2001.

[29] Mangold, S., Choi, S., May, P., Klein, O., Hiertz, G. and Stibor, L. '*IEEE 802.11e Wireless LAN for Quality of Service*,' European Wireless 2002.

[30] IEEE '*Wireless Medium Access Control (MAC) and Physical Layer (PHY) Specifications for High Rate Wireless Personal Area Networks (WPAN)*,' Draft P802.15.3, February 2003.

[31] Krco, S. and Dupcinov, M. 'AODV Improvement,' accepted for publication in *IEEE Communication Letters*, May 2003.

[32] Buchegger, S. and Le Boudec, J.-Y., '*The selfish node: Increasing routing security in mobile ad-hoc networks*,' Research Report RZ 3354, IBM Research, Zurich Research Laboratory, 8803 Ruschlikon, Switzerland, May 2001.

[33] Buttyan, L. and Hubaux, J.-P., '*Nuglets: a virtual currency to stimulate cooperation in self-organized mobile ad-hoc networks*,' Technical Report DSC/2001/001, Institute for Computer Communications and

Applications, Department of Communication Systems, Swiss Federal Institute of Technology, January 18 2001.

[34] Hubaux, J.P., Buttyan, L. and Capkun, S., 'The quest for security in mobile ad-hoc networks,' *Proceedings of the ACM Symposium on Mobile Ad-hoc Networking and Computing (MobiHOC)*, Long Beach, CA, ACM, October 2001.

[35] Kuhn, T., '*Security architecture for future mobile terminals and applications,*' Interim Report SHA/DOC/ SAG/WP2/D03/2.0, IST-2000-25350 - SHAMAN, available online at: http://www.ist-shaman.org, November 2001.

[36] Perrig, A., Szewczyk, R., Wen, V., Culler, D. and Tygar, J.D. 'SPINS: Security protocols for sensor networks,' *7th ACM International Conference on Mobile Computing and Networking*, Rome, Italy, **1**, 189–199, 2001.

[37] Schmitz, R. '*Results of review, requirements and reference architecture,*' Interim Report SHA/DOC/ SAG/WP1/D02/1.1, IST-2000-25350 - SHAMAN, available online at: http://www.ist-shaman.org, December 2001.

[38] Stajano, F. and Anderson, R. 'The resurrecting duckling: Security issues for ad-hoc wireless networks,' 7th International Workshop on Security Protocols (Eds M. Roe and B. Crispo), *Lecture Notes in Computer Science*', Springer Verlag, 1999.

[39] Yi, S., Naldurg, P. and Kravets, R. '*Security-aware ad-hoc routing for wireless networks,*' Technical Report UIUCDCS-R-2001-2241, UIUC, Urbana, IL USA, August 2001.

[40] Gavrilovska, L. (Ed.), '*State-of-the-art of the WPAN networking paradigm,*' PACWOMAN Project report, July 2002.

[41] Pottie, G.J. and Kaiser, W.J., 'Wireless Integrated Network Sensors,' *Communications of the ACM*, **43**(5), 2000.

[42] Bonnet, P., Gehrke, J. and Seshadri, P., 'Querying the physical world', *IEEE Personal Communications* 10–15, October 2000.

[43] Mehta, V. and El Zarki, M., 'Fixed Sensor Networks for Civil Infrastructure Monitoring–An Initial Study,' in *Proceedings of the Medhoc 02*, Sardegna, Italy, September 2002.

[44] Krco, S., 'Bluetooth Based Wireless Sensor Networks – Implementation Issues and Solutions,' in *Proceedings of Telfor 2002*, Belgrade, Serbia and Montenegro, November 2002.

[45] Intanagonwiwat, C., Govindan, R. and Estrin, D., 'Directed Diffusion: A Scalable and Robust Communication Paradigm for Sensor Networks,' in *Proceedings of the Sixth Annual International Conference on Mobile Computing and Networking (MobiCOM '00)*, Boston, Massachussetts, August 2000.

[46] Braginsky, D. and Estrin, D., 'Rumor Routing Algorithm For Sensor Networks,' in *Proceedings of the International Conference on Distributed Computing Systems (ICDCS-22)*, November 2001.

[47] Yu, Y., Govindan, R. and Estrin, D., '*Geographical and Energy Aware Routing: A Recursive Data Dissemination Protocol for Wireless Sensor Networks*,' UCLA Computer Science Department Technical Report UCLA/CSD-TR-01-0023, May 2001.

[48] Esseling, N., Vandra, H.S. and Walke, B., 'A Forwarding Concept for HiperLAN/2,' in *Proceedings of European Wireless 2000* Dresden, Germany, 13–18, 2000.

[49] Walke, B., Esseling, N., Habetha, J., Hettich, A., Kadelka, A., Mangold, S., Peetz, J. and Vornefeld, U., 'IP over Wireless Mobile ATM – Guaranteed Wireless QoS by HiperLAN2,' in *Proceedings of the IEEE*, **89**, 21–40, 2001.

[50] Walke, B. and Briechle, G., 'A Local Cellular Radio Network for Digital Voice and Data Transmission at 60 GHz,' in *Proceedings of the International Conference on Cellular and Mobile Communications*, London, November 1985, online publication, pp. 215–225, available from www.comnets.rwth-aachen.de/publications/~walke.

[51] Walke, B., 'On the Importance of WLANs for 3G Cellular Radio to Become a Success,' in *Proceedings of the 10th Aachen Symposium on Signal Theory – Mobile Communications*, Aachen, Germany, 13–24, 2001.

[52] Yanikomeroglu, H., '*Fixed and mobile relaying technologies for cellular networks,*' Second Workshop on Applications and Services in Wireless Networks (ASWN'02), Paris, France, 3–5 July 2002, pp. 75–81.

[53] Wu, H., Qiao, C., De, S. and Tonguz, O. 'Integrated Cellular and Ad Hoc Relaying Systems: iCAR,' *IEEE Journal on Selected Areas in Communication*, **19**(10), 2105–2115, 2001.

[54] Sreng, V., Yanikomeroglu, H. and Falconer, D.D., '*Relayer selection strategies in cellular networks with peer-to-peer relaying,*' IEEE Vehicular Technology Conference Fall 2003 (VTC'F03), Orlando, Florida, USA, October 2003.

[55] Zadeh, A.N. and Jabbari, B., 'Performance Analysis of Multihop Packet CDMA Cellular Networks,' in *Proceedings of IEEE Globecom 2001*, **5**, 2875–2879, San Antonio, Texas, November 2001.

[56] Aggélou, G.N. and Tafazolli, R., 'On the Relaying Capability of Next-Generation GSM Cellular Networks,' *IEEE Personal Communications*, **8**(1), 40–47, 2001.

[57] Habetha, J. and Walke, B., 'Fuzzy rulebased mobility and load management for self-organizing wireless networks,' *Journal of Wireless Information Networks, Special Issue on Mobile Ad Hoc Networks (MANETs): Standards, Research, Applications*, **9**, 119–140, 2002.

[58] Walke, B., Pabst, R. and Schultz, D., 'A Mobile Broadband System based on Fixed Wireless Routers,' in *Proceedings of International Conference on Communication Technology 2003*, Beijing, China, pp. 1310–1317.

[59] Esseling, N., Weiss, E., Krämling, A. and Zirwas, W. 'A Multi Hop Concept for HiperLAN/2: Capacity and Interference,' in *Proceedings of European Wireless 2002*, **1**, 1–7, Florence, Italy, February 2002, available from www.comnets.rwth-aachen.de/cnroot_engl.html/~publications.

[60] Mohr, W., Lüder, R. and Möhrmann, K.-H., 'Data rate estimates, range calculations and spectrum demand for new elements of systems beyond IMT-2000,' in *Proceedings of the 5th International Symposium on Wireless Personal Multimedia Communications (WPMC'02)*, Honolulu, Hawaii, October 2002.

[61] ETSI and 3GPP, '*Selection Procedures for the Choice of Radio Transmission Technologies of the Universal Mobile Telecommunication System UMTS* (*UMTS 30.03, 3G TR 101 112*),' European Telecommunications Standards Institute, Sophia Antipolis, France, Technical Report, April 1998.

[62] Irnich, T., Schultz, D., Pabst, R. and Wienert, P., 'Capacity of a Multi-hop Relaying Infrastructure for Broadband Radio Coverage of Urban Scenarios,' in *Proceedings of VTC-Fall*, Orlando, Florida, 2003.

[63] Schultz, D., Pabst, R. and Irnich, T., 'Multi-hop based Radio Network Deployment for Efficient Broadband Radio Coverage,' in *Proceedings of WPMC*, Yokosuka, Japan, October 2003.

[64] Walke, B., '*Mobile Radio Networks – Networking and Protocols,*' 2nd edition, John Wiley & Sons, Ltd, Chichester, 2001.

[65] Xu, B., '*Self-Organizing Wireless Broadband Multihop Networks with QoS Guarantee, Protocol Design and Performance Evaluation*,' PhD Dissertation, Aachen University (RWTH), 2002.

[66] Khun-Jush, J., Schramm, P., Wachsmann, U. and Wenger, F., 'Structure and Performance of the HIPERLAN/2 Physical Layer,' in *Proceedings of VTC-Fall*, 1999.

[67] Lott, M., Weckerle, M., Zirwas, W., Li, H. and Schulz, E., 'Hierarchical cellular multihop networks,' in *Proceedings of the 5th European Personal Mobile Communications Conference (EPMCC 2003)*, Glasgow, Scotland, 22–25 April 2003.

[68] Herhold, P., Zimmermann, E. and Fettweis, G., '*On the Performance of Cooperative Amplify-and-Forward Relay Networks*,' 5th International ITG Conference on Source and Channel Coding (SCC), Erlangen, Germany, January 2004.

[69] Zimmermann, E., Herhold, P. and Fettweis, G., '*On the Performance of Cooperative Diversity Protocols in Practical Wireless Systems*,' 58th Vehicular Technology Conference, Orlando, Florida, Fall 2003.

[70] Zirwas, W., Giebel, T., Rohling, H., Schulz, E. and Eichinger, J., 'Signaling in distributed Smart Antennas,' in *Proceedings of the 7th OFDM-Workshop*, pp. 87–91, Hamburg, September 2002.

[71] Telatar, E., 'Capacity of multi-antenna Gaussian channels,' *European Transactions on Telecommunication*, **10**, 585–595, 1999.

[72] Dohler, M., Said, F., Ghorashi, A. and Aghvami, H., '*Improvements in or Relating to Electronic Data Communication Systems*,' Publication No. WO 03/003672, priority date 28 June 2001.

[73] Dohler, M., Gkelias, A. and Aghvami, H., 'A Resource Allocation Strategy for Distributed MIMO Multi-Hop Communication Systems,' *IEEE Communications Letters*, June 2003, submitted.

[74] Dohler, M., Gkelias, A. and Aghvami, H., 'The Capacity of Distributed PHY-Layer Sensor Networks', *IEEE Journal on Selected Areas in Communication on Sensor Networks*, July 2003, submitted.

[75] Dohler, M., Gkelias, A. and Aghvami, H., '2-Hop Distributed MIMO Communication System,' *IEE Electronics Letters*, **39**(18), 1350–1351, 2003.

[76] Hares, S., Yanikomeroglu, H. and Hashem, B., '*Diversity and AMC (adaptive modulation and coding)-aware routing in TDMA multihop networks*,' IEEE Globecom 2003, San Francisco, USA, 2003.

[77] Viswanathan, H. and Mukherjee, S., 'Performance of cellular networks with relays and centralized scheduling,' in *Proceedings of the IEEE VTC Fall 2003*, Orlando, Florida, October 2003.

[78] Zirwas, W., Lampe, M., Li, H., Lott, M., Weckerle, M. and Schulz, E., '*Radio Resource Management in Cellular Multihop Networks*,' MoMuC 2003, Munich, Germany, 2003.

[79] ITU-R, '*Framework and overall objectives of the future development of IMT-2000 and systems beyond IMT-2000*,' Recommendation M.1645, 2003.

[80] ITU-R, '*Methodology for calculation of spectrum requirements for future development of IMT-2000 and systems beyond IMT-2000 from the year 2010 onwards*,' 8F/TEMP/379-E, revision 1, 2003.

[81] ITU-R, '*Proposed methodology for calculation of spectrum requirements for systems beyond IMT-2000*,' 8F/866-E, 2003.

[82] UMTS, '*Spectrum for IMT-2000*,' Forum Report, 1997.

[83] UMTS, '*Minimum spectrum demand per public terrestrial UMTS operator in the initial phase*,' Forum Report 5, 1998.

[84] UMTS, '*UMTS/IMT-2000 spectrum*,' Forum Report 6, 1998.

[85] Fitzek, F., Zorzi, M., Seeling, P. and Reisslein, M., 'Video and Audio Trace Files of Pre-encoded Video Content for Network Performance Measurements', IEEE Consumer Communications & Networking 'Consumer Networking: Closing the Digital Divide' January, 2004.

[86] Berlemann, L., Walke, B. and Mangold, S., 'Behavior Based Strategies in Radio Resource Sharing Games', Proc. of The 15th IEEE PIMRC, Barcelona, Spain, September 2004.

[87] WWRF, '*New Air Interface Technologies – Requirements and Solutions*,' White Paper, 2003.

[88] Walke, B. and Kumar, V. 'Spectrum Issues and New Air Interfaces,' *Computer Communications*, **26** (1), 2003.

[89] Mohr, W., Lüder, R. and Möhrmann, K.-H., '*Data Rate Estimates, Range Calculations and Spectrum Demand for New Elements of Systems Beyond IMT-2000*,' IEEE, The 5th International Symposium on Wireless Personal Multimedia Communications, Honolulu, Hawaii, USA, October 2002.

[90] Mohr, W., '*Spectrum Demand for Systems Beyond IMT-2000 Based on Data Rate Estimates*,' *Wiley Wireless Communications and Mobile Computing*, John Wiley & Sons, 2003.

[91] ITU-R, '*Methodology for the calculation of IMT-2000 terrestrial spectrum requirements*, Recommendation M.1390, 1999.

[92] ITU-R, '*A method for assessing the required spectrum for broadband nomadic wireless access systems including radio local area networks using the 5 GHz band*,' Recommendation M.1651, 2003.

[93] IST Project IST-2000-28584 MIND Deliverable D3.3, *Methodologies to identify spectrum requirements for systems beyond 3G*, 2002.

6.5 Credits

The following individuals have contributed to the contents of this chapter: Srdjan Krco (Ericsson Systems Expertise, Ireland); Frank Fitzek (acticom, Germany); Bernhard H. Walke, Ralf Pabst and Daniel Schultz (Aachen University, Communication Networks, Germany); Patrick Herhold and Gerhard P. Fettweis (Dresden University, Germany); Halim Yanikomeroglu and David D. Falconer (Carleton University, Ottawa, Canada); Sayandev Mukherjee and Harish Viswanathan (Lucent Technologies, USA); Matthias Lott and Wolfgang Zirwas (SIEMENS ICM, Munich, Germany); Mischa Dohler and Hamid Aghvami (King's College London, UK) and Pekka Ojanen (Nokia, Finland).

7

New Air Interface Technologies

Edited by Bernhard Walke and Ralf Pabst (Aachen University, ComNets)

An improved spectrum efficiency is achievable through improved signal design, i.e. modulation and channel coding. Well-known methods for a vast majority of wireless channels are available. The theoretical limit of channel capacity (Shannon Limit) can be very closely approached through the use of existing mechanisms of channel coding in single input single output (SISO) channels. The system performance in multiple input multiple output (MIMO) channels can be greatly enhanced by use of space–time coding, multiuser detection and smart antenna techniques. Moreover, signal design for very high bit transmission (100 Mb/s) in difficult wireless channels (e.g. for fast moving mobiles) is still an open issue. So is the case of providing throughput like 1 Gbps for the 'hot spots'.

Information compression, i.e. signal processing and source coding techniques useful to reduce the source output bit rate so that a larger number of simultaneous calls can be accommodated per unit bandwidth, is well mastered for voice sources. More efficient algorithms for multimedia services are currently being developed.

The ability to accommodate several users accessing a given service simultaneously in a communication channel can go a long way to improving the situation of spectrum usage. The presently deployed 'contention free schemes' are well understood. Further improvements through the use of 'contention-based (or tolerating) multiple access schemes' for packet mode communication seem possible. Simultaneous implementation of adaptation mechanisms at different layers of the protocol stack should be helpful in improving the system performance, all the while allowing for fine granularity of spectrum usage for different mixes of services (channel or packet switched).

This chapter elaborates on the following topics.

Broadband Multicarrier Techniques and Mixed OFDM plus Single Carrier Techniques

It is a possibility that we will not see a full blown new mobile system generation after 3G. Maybe we will only witness the emergence of a new radio access scheme that can then be connected to a 3G network.

Technologies for the Wireless Future. Edited by R. Tafazolli
© 2005 John Wiley & Sons, Ltd   ISBN: 0-470-01235-8

There are several reasons why we should set challenging beyond-3G research targets of up to 100 Mbps and 1 Gbps peak data rates for wide-area coverage full mobility and local area coverage low mobility cases, respectively. They include:

- 3G will go towards 10/100 Mbps (wide/local area)–beyond-3G should be clearly better.
- No application may need such high continuous bit rates but the system may need them in order to serve many high bit rate users simultaneously, maximize throughput/capacity and minimize latencies.
- There may be an optimum bandwidth which will maximize the spectral efficiency of a wireless system: the research target must be set high to 'capture' that optimum.
- Short distance radio bit rates will go towards 1 Gbps and users expect a wide area coverage service level to be fairly close.

Given the very challenging goals and expected long evolution path of 3G, schedule wise we should target year 2010 or later for a possible commercial launch of any beyond-3G air interface.

While an increase in data rate, and maybe to some extent also QoS, could be achieved by combining techniques in the form of multimode terminals (e.g. UMTS and wireless LAN), this has the disadvantage that for each new scenario another mode has to be added. The combination may also not be suitable for the mobility requirements of wireless radio systems. The OFDM section of this chapter proposes using a broadband multicarrier-based air interface for both uplink and downlink. It also includes a discussion of the enabling technologies for multicarrier mobile systems and the technology trends for future research. The multicarrier technology offers the desired high data rates for the 4G mobile environments and also has advantages for spectral efficiency and low-cost implementation.

Future generation wireless systems, transporting high bit rates, must be robust to severe frequency selective multipath. Frequency domain-based modulation or equalization provide cost-effective multipath compensation. One proposal is that the downlink air interface should use orthogonal frequency division multiplexing (OFDM) and the uplink air interface use single carrier (SC) modulation, with adaptive frequency domain equalization (FDE) carried out in the base station receiver. This arrangement has the advantage that the mobile terminal's power amplifier cost is minimized, and most of the signal processing complexity is concentrated at the base station. A number of research issues relevant to this proposal are described in Section 7.2.

Smart Antennas and MIMO Systems

Smart antennas are essential to increase the spectral efficiency of wireless communication systems. They can be realized by an antenna array at the base station and sophisticated baseband signal processing. Thereby, adaptive directional reception is achieved on the uplink and adaptive directional transmission on the downlink. Hence, an increased antenna gain or an increased diversity gain are realized towards the desired user. At the same time, less interference is received from the other directions on the uplink or transmitted in the other directions on the downlink. Therefore, more users can be accommodated by the system and a corresponding increase of the spectral efficiency is achieved.

Due to the fact that the uplink and the downlink operate on the same frequency in time division duplex (TDD) systems, channel parameters (e.g., spatial covariance matrices) estimated on the uplink can also be used to calculate the weights for downlink beamforming techniques. This is more difficult in frequency division duplex (FDD) systems, since the uplink and the downlink operate on different frequencies and some kind of frequency transformation or feedback from the mobile might become necessary. If no channel knowledge is available at the transmitter, space–time coding techniques can be used to increase the diversity gain.

Even higher spectral efficiencies can be achieved if antenna arrays are not only used at the base station but also at the mobile to create multiple input multiple output (MIMO) systems. The achievable spectral efficiency, however, depends on the propagation and interference environment. If a rich scattering environment is available, an enormous spectral efficiency can be obtained via spatial multiplexing. Current research efforts focus on idealistic channel models and simplified interference scenarios, which do not properly describe real world channels. The spatial dimension of the propagation environment is particularly important to the suitability and success of different space–time multiplexing techniques, as well as to the mapping between the uplink and downlink channels. Moreover, the time-varying properties of the MIMO space–frequency channel are also crucial to the design of efficient MIMO transceivers. Therefore, MIMO channel-sounding campaigns are required at the frequencies and in the environments of interest. Depending on the chosen radio interface, suitable space–time processing techniques should be developed and evaluated by using realistic channel and interference models.

Short-range Wireless Technologies

WLANs are increasingly being perceived as a natural complement to the cellular third generation (3G) mobile communication system. This can be interpreted as the fact that it will be unavoidable to witness in the near future a superposition of radio access networks of various sizes and topologies: macrocellular systems (3G), microcellular systems (based on further developed WLANs), and picocellular systems such as wireless personal area networks (WPANs).

It appears highly probable that a radio interface for a B3G mobile communication system cannot be specified without taking the WLAN/WPAN/WBAN worlds into account.

Ultra-Wideband Radio Technology

Ultra-wideband (UWB) systems as defined by the Federal Communications Commission (FCC) in the United States shall have a fractional bandwidth exceeding 20% or at least an absolute bandwidth larger than 500 MHz referred to the -10 dB band edges. By this definition, pulse-based systems as well as PRBS-, chirp-, noise- and other modulation techniques pertain to the UWB methods. The European Conference of Postal and Telecommunications Administrations (CEPT) and European Telecommunications Standards Institute (ETSI) have created study groups, which are investigating the regulatory framework for UWB radio devices in Europe to ensure that UWB radio systems can coexist with other radio services. The U.S. based FCC and the CEPT/ETSI groups are interested in building a regulatory framework for peaceful radio resource sharing between UWB devices and existing services. An industrial interoperability standard for a functional UWB radio system is instead the objective of Institute of Electrical and Electronic Engineers (IEEE) Group 802.15.3a. UWB radio technology (UWB-RT) holds the promise for a large array of new or improved (short-range) radio devices and services that could have enormous benefits for public safety, consumers as well as businesses alike, through potentially sharing spectrum space with other radio services. UWB technology provides the potential for implementing short-range radio devices that can inherently and effectively support applications based on data communication (potentially at high data rates) as well as precise measurement of distances for location tracking. Thus, UWB-RT provides the opportunity to develop and define (standardize) new short-range wireless systems capable of combining the functions for (low-rate) data communication and location tracking. This feature might also be useful for network management and data routing in ad hoc networks or to support location-aware applications and services or both. These capabilities make UWB-RT also interesting for hand-held devices (e.g., cell phone, PDA, laptop computer) to enable high data rate access and location tracking in areas with high user density ("hot-spot"scenario).

7.1 Broadband Multicarrier-based Air Interface for Future Mobile Radio Systems

7.1.1 Introduction

The success of future mobile radio systems depends on meeting, or exceeding, the needs, requirements and interests of users, and society as a whole. It seems likely that this will require an increase in spectral efficiency to allow high data rates and high user capacities far beyond those of second or third generation systems. Moreover, flexible resource allocation will play a key role in future mobile radio networks. In recent years, much research has been carried out on increasing the performance and efficiency of various air interface components, such as coding or detection. Also, new air interface concepts based on either single carrier or multicarrier transmission have been proposed. These show promising performance results. One future adaptive air interface which could cover the relevant mobile radio scenarios would ease the introduction of a new generation of mobile radio systems. Thus, a comprehensive survey of new air interface technologies which might be able to meet these requirements is necessary.

Motivated by recent results of different system comparisons, this text highlights the flexibility and efficiency of a new broadband multicarrier-based air interface for application in future mobile radio systems. Moreover, different multicarrier-based access technologies are discussed and realization issues are pointed out, OFDM especially offers a low-cost multicarrier implementation. The key advantages of multicarrier modulation realized in the form of orthogonal frequency division multiplexing (OFDM) [1–5] are the high spectral efficiency, simple implementation by an FFT, low receiver complexity due to the elimination of ISI and high flexibility in subcarrier allocation. This is the reason why many standards developed for wireless high rate data transmission in recent years have been based on multicarrier modulation, i.e., OFDM. Among the earliest of these were the digital audio broadcasting standard (ETSI DAB) [6] and the terrestrial digital video broadcasting standard (ETSI DVB-T) [7] which where completed in the 1990s. They where followed by the wireless local area network (WLAN) standards ETSI HIPERLAN/2 [8], IEEE 802.11a [9] and MMAC and the recently developed wireless metropolitan area network (WMAN) standards ETSI HIPERMAN and IEEE 802.16a [10].

Of interest in the WLAN and WMAN standards are the access schemes multicarrier time division multiple access (MC-TDMA) and orthogonal frequency division multiple access (OFDMA).

The multicarrier code division multiple access (MC-CDMA) scheme proposed in 1993 [11, 12] is a candidate for the downlink of future mobile radio systems since it combines the advantages of OFDM and CDMA. The principle is to map the chips of the spreading code on different subcarriers, which makes the system robust against ISI and allows for simple single- and multiuser detection techniques.

More recently, multiplexing of user data followed by a pure OFDM scheme with adaptive modulation has been suggested for downlinks in systems where channel state information is available in the transmitter [13]. Such a system significantly increases the spectral efficiency at the expense of some additional complexity, since it may require both channel prediction and a feedback link.

MC-DS-CDMA was proposed in 1993, this is more suitable for the asynchronous uplink, since it is more related to classical DS-CDMA [14]. Here, the high rate serial data symbols are mapped to multiple parallel low rate data symbols. Each data symbol is spread on a single subcarrier in the time direction, i.e., MC-DS-CDMA can be seen as multiple more or less narrowband parallel DS-CDMA systems, which are less sensitive to ISI and allow less complex receivers.

The research on the combination of OFDM and multiple access was in its infancy during the specification of 3G mobile radio standards and, thus, OFDM did not become part of 3G systems. However, this situation has changed and today a lot of research has been carried out and is ongoing on using OFDM in multiple user systems like 4G. Thus, it is time to apply it for broadband 4G mobile radio systems [15]. In Japan, NTT DoCoMo has identified a hybrid MC-CDMA scheme with variable spreading factor as the most promising to meet their requirements for future cellular systems [16].

To design the next generation of mobile radio systems, a clear understanding of the requirements of these systems is necessary. This section starts with the identification of requirements which are of importance for a future air interface. These requirements have to be specified more precisely on the way towards 4G standardization. In the following, the strengths and weaknesses of individual multicarrier-based air interface concepts are highlighted, and future research trends in multicarrier communications are identified. Moreover, the section emphasizes that future research should complement the various approaches relying on the increase of the throughput capitalizing on the higher capacity provided by using a MIMO channel.

7.1.2 Requirements

The requirements identified in this section give a comprehensive list of parameters which have to be taken into account in the design and evaluation of new air interfaces. Figures are given for key requirements in order to have a starting point for the system comparisons. These parameters will be defined more precisely when the user requirements become better known on the way towards standardization, e.g., from other WWRF working groups dealing with user requirements and applications, or from outside. Since hot spots are becoming more and more significant for very high data rate wireless links, such scenarios are also covered by these requirements. Cellular and Hot Spots radio systems may be considered as two modes in one future broadband wireless system.

7.1.2.1 Data Rate per User/Link

The peak data rate and the scalability of the data rate per user supported by the system have to be specified. Since a single user can establish different links simultaneously, the data rate per link has also to be specified. Under discussion are future data rates per link of the order of 10 Mbit/s to 100 Mbit/s for cellular applications and up to 1 Gbit/s for hot spots.

Also, the variation in data rate (peak vs. typical) must be taken into account. Efficient support for low data rates is also important, as this seems likely to continue to form a significant part of traffic – although it may be possible to support this through the existing 2G and 3G air interfaces.

7.1.2.2 Asymmetric Traffic

Since data traffic is becoming more and more important compared to speech transmission in the future, the demands for data rate between downlink and uplink are becoming asymmetric. These asymmetric traffic requirements have to be taken into account in the system design and can result in different access scheme proposals for up- and downlink. Asymmetry both across individual links and also across cells/hotspots must be taken into consideration. According to current thinking relating to the evolution of cellular communications, and the emergence of mobile clients accessing Internet-like systems, data rates of around 20 Mbit/s to 40 Mbit/s in the uplink seem to be sufficient, while in the downlink, peak data rates of 100 Mbit/s and higher are under discussion. Other types of usage, services and traffic may create different requirements on asymmetry. Dynamic flexibility of asymmetry is likely to be required in order to efficiently meet traffic requirements.

7.1.2.3 Packet-switched and IP-based Traffic

For easy interfacing to core networks based on Internet system architectures, means should be provided to facilitate IP data transmission structures. Since operators have already undertaken significant investments into building an IP core network, these investments have to be taken into account in order to make a 4G system attractive from an economical point of view.

Efficient support of IP data will also ease the integration of fixed and wireless networks. This can encompass the efficient integration of packet-switched connections and end-to-end TCP transmission chains, but also considerations of the special effects and environments of mobile wireless transmission.

7.1.2.4 Quality of Service (QoS)

QoS is the parameter which is of interest for the mobile user and is defined by several parameters:

- Bit error rate (BER) and packet error rate (PER);
- End-to-end delay;
- Link delay;
- Outage probability/packet loss;
- Packet delay variation;
- Real-time/nonreal-time traffic.

At the current stage, as guidelines for the system design, it can be assumed that at least the requirements of 2G and 3G systems have to be fulfilled, this holds especially for the BER and PER requirements. The requirements on the delay can be expected to be stricter.

7.1.2.5 Mobility

The maximum expected velocity of the mobile user is important for the system design, especially for the frame structure, synchronization, channel estimation and predictability, and data detection. The maximum velocities are expected to be in the same order or higher as for 3G systems, except for hot spots, which are by nature scenarios with low mobility. Moreover, it is necessary to think about additional measures to support high velocities such as for example, high-speed trains travelling at more than 300 km/h.

7.1.2.6 Coverage

The coverage within next generation mobile radio systems might depend on the chosen service, i.e., typical cellular services will be guaranteed a wide-area coverage, while very high data rate services might be restricted to hot spots. Thus, the coverage requirements of a 4G system have to be defined per service.

7.1.2.7 System Scalability

The 4G mobile radio systems should be scalable with respect to bandwidth and cell site deployment (macro/micro/pico). These radio systems should also be scalable in terms of the service requirement, including both symmetrical and asymmetrical services. To meet this objective, the physical layer design, including the slot/frame structure hierarchy, might be adaptable to the different bandwidths, deployment scenarios, traffic conditions and a wide range of future application requirements.

7.1.2.8 Frequency Spectrum and Bandwidth Allocation

There is no frequency spectrum allocated yet for the next generation mobile radio. Nevertheless, it is known from 3G that spectrum is very limited and, thus, very expensive, from which it follows that effort is necessary to increase the spectral efficiency of the system.

It may also be useful to consider alternative methods of spectral allocation and use, rather than the current 'fixed block' allocations normally considered. Air interfaces which could identify and use any available spectrum (spectrum sharing), without interfering with other systems, could provide much higher capacities, but there would also be many issues to be solved.

7.1.2.9 Sector/Cell Capacity and Number of Users per Sector/Cell

The capacity and number of users per cell and per sector are of importance for the system design sand will strongly influence the choice of multiple access scheme and frequency reuse factor. Within the International Telecommunications Union (ITU) work on Systems Beyond IMT-2000, they have identified possible data rate requirements for wide-area 4G systems according to total cell capacity, rather than individual links. These requirements suggest up to 100 Mbit/s in the cellular environment, and up to 1 Gbit/s in hot spots, as targets for research.

7.1.2.10 Dynamic Resource Allocation and Scheduling

The system must be able to dynamically change the allocated resources according to changing user requirements and available capacities. A good allocation scheme has to be developed. To this end, the scheduling of services with different QoS requirements and the fairness between different users/services have to be examined carefully. It may be necessary to prioritize certain data, e.g. real-time services or ARQ retransmissions.

7.1.2.11 Medium Access Control (MAC) Design

The new air interface should be designed to carry different types of traffic, including real time traffic and nonreal-time traffic. This will require the design of an efficient MAC that maximizes the system throughput and minimizes the overhead. The MAC design should carry, in addition to data, control information that requires more robustness to errors. With the vision that the 4G system may have different physical layer concepts, the MAC should be generic to support different physical layers, such as those for cellular and hot spot scenarios. This poses the question of whether one MAC is efficient for the support of different 4G air interfaces.

7.1.2.12 Power Consumption and Radiation Limits

Regulatory authorities are specifying mandatory limits for the maximum power consumption and radiation for both the base station and the mobile. These limits are not yet specified for 4G systems, however, they may be similar to or even more severe than those for 2G and 3G systems.

7.1.2.13 Coexistence/Cooperation

According to the 'multiple-systems' vision of systems beyond 3G, it seems likely that any new air interface will have to coexist with many other air interfaces, both cellular and of other types. This should occur with minimal mutual interference, and benefit may be obtained from active cooperation.

7.1.2.14 Low Cost

It is important to achieve the requirements at costs which are moderate, and certainly not significantly higher than those of 2G and 3G systems. This is especially important for user terminals. Advances in digital integrated circuit technologies will probably restrain the digital signal processing costs. However, RF analogue circuit costs for power generation, amplifiers, antennas, filters, mixers, etc. will be harder to control. Note that required transmitter power increases with bit rate. Thus, particular attention must be paid to how the air interface might affect terminal, base station and other infrastructure costs. Other cost-sensitive items include spectrum: licensed or unlicensed, shared with other services or not, etc.

7.1.2.15 Location/Context Awareness

Discussion of services for future mobile radio systems often includes talk of context-aware, or location-based services. In order to provide these services, it is necessary to have a mechanism to derive the user's context. Some of these properties may be extracted from the air interface (e.g. location).

7.1.3 Multicarrier Modulation

The successful deployment of OFDM in various standards and demonstrations, and its high flexibility is a strong motivation for using it in the design of a new broadband air interface for fourth generation (4G) mobile radio systems. The features of OFDM and alternative multicarrier modulation concepts are briefly introduced in the following. More details may be found in [1–5, 15].

7.1.3.1 Orthogonal Frequency Division Multiplexing (OFDM)

The principle of OFDM is to demultiplex a high rate serial data stream with rate $1/T_d$ into multiple parallel lower data rate substreams with rate $1/T_s = 1/(N_c T_d)$. The parallel substreams are modulated on individual subcarriers, which are spaced by

$$F_s = \frac{1}{T_s} \tag{7.1}$$

The complex envelope of the transmitted OFDM signal with rectangular pulse shaping has the form

$$x(t) = \frac{1}{N_c} \sum_{n=0}^{N_c-1} S_n e^{j2\pi f_n t} \tag{7.2}$$

where S_n are the substream source symbols. The N_c symbols S_n are referred to as an OFDM symbol, and its duration is given by T_s. The N_c subcarrier frequencies are located at

$$f_n = \frac{n}{T_s}, \quad n = 0, \ldots, N_c - 1. \tag{7.3}$$

In this way, the signals at different subcarriers become orthogonal to each other after matched filtering in the receiver, when rectangular pulse shaping is used and the channel is frequency flat. Thus, each substream source symbol can be detected without interference from the other subcarrier signals.

With an increasing number of subcarriers, the OFDM spectrum approaches the optimum rectangular spectrum with bandwidth $1/T_s$. The key advantage of OFDM is that multicarrier modulation can be implemented in the discrete domain by using the DFT/IDFT, which for some parameters can efficiently be implemented using FFT/IFFT. When sampling $x(t)$ with rate $1/T_d$ the samples are

$$x_v = \frac{1}{N_c} \sum_{n=0}^{N_c-1} S_n e^{j2m v/N_c} \qquad (7.4)$$

The sequence x_v, $v = 0, \ldots, N_c - 1$, is the IDFT of the source symbol sequence S_n, $n = 0, \ldots, N_c - 1$.

By increasing the OFDM symbol duration by a cyclic extension (guard interval) of duration T_g, the subcarriers' signals can be made orthogonal also on frequency selective channels, which enables simple receivers for high data rate applications. This is, however, at the expense of additional transmit power (reduced power efficiency) and reduced data rate (reduced spectral efficiency). With a sufficiently long cyclic prefix, each subchannel can be considered separately. Furthermore, when assuming that the fading is frequency selective over the total signalling bandwidth but appears flat on each subchannel, a received symbol R_n on subcarrier n is obtained as

$$R_n = H_n S_n + N_n, \quad n = 0, \ldots, N_c - 1 \qquad (7.5)$$

where H_n is the channel coefficient and N_n is the noise of subchannel n. It can be observed from (7.5) that the received signal can simply be pre-equalized in the transmitter or equalized in the receiver by one complex-valued multiplication per subchannel. Moreover, it is straightforward to use different signalling constellations and powers for S_n on the different subcarriers to further improve spectral efficiency of OFDM.

The strengths of OFDM can be summarized as:

- high spectral efficiency due to a nearly rectangular frequency spectrum of width $1/T_s$ with high numbers of subcarriers;
- simple implementation of OFDM by using the fast Fourier transform (FFT);
- reduced receiver complexity due to simple elimination of ISI on frequency selective channels. This is efficiently achieved by cyclically extending each OFDM symbol and thus enabling very high rate data transmission in multipath channels;
- high flexibility in terms of subcarrier allocation strongly supports adaptive data rate and powers, and efficient multiplexing of data from different users on the same OFDM signal;
- low complexity multiple access schemes exist, such as OFDMA, MC-TDMA and MC-CDMA, some of which have already been applied in recent wireless standards.

The weaknesses of OFDM are:

- higher peak-to-average power ratio (PAPR) compared to single-carrier modulation. This makes the OFDM signal more sensitive to nonlinear distortions, which result in performance degradations and enhancement of the out-of-band power;
- sensitivity to synchronization errors.

7.1.3.2 Further Multicarrier Modulation-based Schemes

Multicarrier systems can also be realized by not exploiting the orthogonality properties of OFDM. These systems have typically a much lower number of subcarriers than traditional OFDM schemes. They can achieve a better PAPR compared to OFDM at the expense of higher receiver complexity. Up to now, these schemes have not become as popular as OFDM. The mobile radio standard cdma2000 [17] employs multicarrier transmission with three subcarriers and can be considered as one realization of MC-DS-CDMA [14]. The main difference between OFDM and the other multicarrier modulation schemes is either larger subcarrier spacing or different pulse shaping. The OFDM frequency spectrum and a general multicarrier frequency spectrum are shown in Figure 7.1, which illustrates the spectral efficiency of OFDM.

7.1.3.3 Multicarrier Modulation and Multiple Access

Figure 7.2 shows the fundamental OFDM-based multiple access schemes OFDMA, MC-TDMA and MC-CDMA. Any hybrid combination of these access schemes is also possible.

OFDMA
In an OFDMA scheme, each terminal station is allocated a set of subcarriers which is exclusively used by this station. The different sets of subcarriers assigned to the terminal stations do not overlap, avoiding multiple access interference. The subcarrier allocation can change between OFDM symbols in order to increase the diversity gain.

The OFDMA scheme is, for instance, applied in the WMAN standards IEEE 802.16a and ETSI HIPERMAN, as well as in the return channel of interactive ETSI DVB-T.

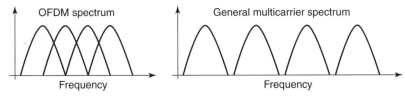

Figure 7.1 Different multicarrier frequency spectra

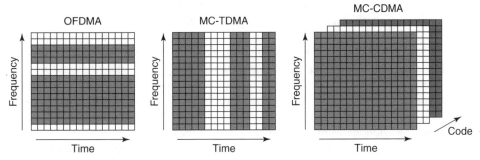

Figure 7.2 OFDM based multiple access schemes. Note that a symbol in MC-CDMA can be spread in frequency only, time only or in both directions

The diversity order of OFDM is one, since each symbol experiences flat fading; thus, its performance is poor in fading channels. Additional measures like FEC coding, antenna diversity, etc., are necessary to increase the diversity order to achieve promising performance results. Alternatively, OFDMA can be combined with code division multiplexing (CDM), resulting in a spread spectrum multicarrier multiple access (SS-MC-MA) scheme [18]. The maximum diversity order now becomes equal to the number of chips per symbol (repetition coding) and is obtained with perfect interleaving of the chips [19]. Due to the avoidance of multiple access interference with SS-MC-MA, it is especially of interest for the uplink, where it also allows for simple channel estimation.

Multiplexing followed by OFDM modulation for a downlink is similar to OFDMA, except that the allocation of user symbols to subcarriers and symbol periods can be made more flexible. The diversity order is, however, improved if subcarriers and symbol periods are allocated based on maximizing the SNR (compare selection diversity in transmitter) [20, 21]. The diversity order then becomes equal to the number of users and can be further improved by traditional diversity schemes.

MC-TDMA

The user separation can be performed by TDMA where each user is assigned its time slot for transmission. A time slot can consist of several OFDM symbols, and a sequence of time slots where each user transmits once is referred to as one MC-TDMA frame.

The principle of MC-TDMA is, for example, applied in the WLAN standards IEEE 802.11a and ETSI HIPERLAN/2.

The performance of MC-TDMA is also poor in fading channels due to the diversity order of one. Traditional diversity schemes (including channel coding) are necessary to achieve promising performance results.

MC-CDMA

The principle of CDMA can be applied in an OFDM system, resulting in MC-CDMA, where all users transmit at the same time in the same frequency band. The users' signals are separated by user-specific spreading codes with good crosscorrelation properties. In the downlink, orthogonal Walsh–Hadamard codes are typically used. In the uplink, depending on the scenario, PN codes like Gold codes are applied. Due to the spreading (repetition coding), MC-CDMA has a maximum diversity order equal to the number of chips per symbol, but requires perfect chip interleaving to obtain it (equal to MC-DS-CDMA).

MC-CDMA can be realized with spreading in the frequency direction, the time direction, or jointly in both directions. Originally, MC-CDMA was proposed with spreading in the frequency direction and the concept with spreading in time direction was introduced as MC-DS-CDMA. Today, various concepts with spreading in both the time and frequency directions exist. As long as perfect interleaving is assumed between the chips in time and frequency, all spreading schemes perform equally since all can exploit a diversity order equal to the spreading code length. These schemes will, however, suffer from multiuser interference since the channel gain is different on different chips due to the channel variation and, more importantly, due to the chip interleaving.

Concepts like variable spreading factor orthogonal frequency and code division multiplexing (VSF-OFCDM) [15] and time–frequency localized CDMA (TFL-CDMA) [22] perform spreading in time and frequency where the chips are located next to each other, so that the variation of the fading gains on the chips of the spreading code can be considered small. Thus, the symbols from different users at the receiver antenna are still almost orthogonal. In the ideal case, multiple access interference can be avoided, when the channel variation is small and all users are synchronized. This also increases the robustness to near–far interference. These schemes have, however, a reduced inherent diversity order. If the whole code experiences the same channel gain, diversity order one is obtained, which is identical to the performance of OFDM without spreading. By relying on redundant block spreading and linear precoding, a so-called multicarrier block-spread CDMA (MCBS-CDMA) transceiver [23] preserves the orthogonality among users and guarantees symbol detection, regardless of the underlying frequency selective fading channels. These properties allow for deterministic (as opposed to statistical) MUI elimination through low-complexity block despreading, and enable full diversity gains, irrespective of the system load.

Various performance comparisons between MC-CDMA and single-carrier DS-CDMA have shown that MC-CDMA can significantly outperform DS-CDMA with respect to BER performance and spectral efficiency [24]. The reason for the better performance with MC-CDMA is that it can avoid ISI and ICI, allowing an efficient, low-complexity user signal separation.

MC-CDMA is considered as a potential candidate for a 4G mobile radio system and has already successfully been tested in the NTT DoCoMo and the DLR 4G test bed.

7.1.3.4 Performance Analysis

This section summarizes some fundamental performance comparison results between the multiple access schemes OFDMA, MC-TDMA and MC-CDMA [15].

- Without FEC coding and other diversity methods, only MC-CDMA has a maximum diversity order larger than one (the maximum equals the spreading code length) in fading channels. Uncoded and unspread OFDMA systems have diversity order one. Thus, the lower bound on the bit error probability for MC-CDMA corresponds to the performance curve for diversity of order L in Rayleigh fading channels, whereas for OFDMA systems without spreading the bit error probability is equal to the performance curve for a system with no diversity [25]. The fading on individual chips in the spreading code must be uncorrelated to obtain the maximum diversity order of MC-CDMA.
- MC-CDMA with chip interleaved spreading as originally proposed in [11, 12] can efficiently exploit diversity at the expense of multiuser interference.
- In isolated cells, coded MC-CDMA requires multiuser detectors to outperform coded OFDMA schemes (when the same FEC is used in both systems) [26]. In cellular environments (with more than one cell), it is an open issue how the performance of the different systems compares to each other, but it is clear that MC-CDMA still needs multiuser detectors to perform well. Thus, there is a clear trade-off between complexity and performance of these systems.
- MC-CDMA systems like TFL-CDMA and VSF-OFCDM with block spreading can reduce the multiple access interference and can use simple single user detectors at the expense of a reduced maximum diversity order [22, 27]. While TFL-CDMA and VSF-OFCDM guarantee low-complexity receivers, the chip interleaved MC-CDMA

may outperform these with multiuser detectors. These trade-offs have to be further investigated in the future.

- MC-CDMA systems with block spreading are more resistant to near–far interference than chip interleaved MC-CDMA systems for single user detection techniques, due to reduced multiuser interference. An open question is what the result will be for multiuser detectors.

- When channel state information is available in the transmitter, the spectral efficiency can be significantly increased with OFDM without spreading [13, 20]. Such a scheme is superior to known results on MC-CDMA. Adaptive modulation and coding has, however, not been investigated much for MC-CDMA and is a topic for future research.

7.1.4 Technology Trends and Research Topics

The focus of future research should include general system aspects and technical aspects of the individual access schemes. An overview of the technology trends summarizing important research results is given here, together with the identification of future research topics.

7.1.4.1 Fundamentals

Duplex Scheme

The air interface design strongly depends on the choice of the duplex scheme. An issue related to this is whether the link should be synchronous or asynchronous. An FDD scheme is applicable for quite static traffic partitioning between up- and downlink and allows asynchronous links, while TDD would allow a flexible support of asymmetric traffic requiring synchronous links (see HIPERLAN/2 and IEEE 802.11a). An additional advantage of TDD is that no feedback channel is needed to make channel state knowledge available in the transmitter, allowing for simpler implementation of adaptive coding, modulation and resource allocation as well as pre-equalization in OFDM-based systems.

The selection of the duplex scheme is crucial for the system design and predefines major characteristics of the system. A 4G MC-CDMA concept based on FDD has already been developed by NTT DoCoMo [16], by the joint DLR-DoCoMo Euro-Labs project [26] and by a national Swedish research project [13]. Within the EU project MATRICE, a TDD-based MC-CDMA 4G concept is proposed [28].

Multiple Access and Multiplexing

This issue is quite different in up- and downlinks. The downlink problem is one of multiplexing data from more than one service and/or user onto one channel, which is OFDM modulated, while the uplink is a combination of multiplexing several services in one terminal with multiple access between more than one terminal.

Even if fundamental performance results are available for multiple access, further research is necessary to show which multiple access method or combination of multiplexing and multiple access methods gives the highest spectral efficiency or system efficiency for the uplink and downlink of an OFDM-based system.

Multiplexing in the downlink (and between services in an uplink) is a complex issue that must take many service dependent parameters like QoS, delay, etc. into account.

Moreover, it is likely that the traffic in the system will be asymmetric and more work is needed to find an appropriate design solution to efficiently handle this.

Relaying Systems
Relaying is an important method to extend the coverage of wireless systems. In such systems, it might not be possible to distinguish between the uplink and the downlink, which means that all links have to be designed in the same way. Some of the special properties of traditional uplinks and downlinks cannot then be utilized. Link designs for relaying systems are very much an open issue and need further study.

Reuse Factor
In cellular systems, it is important that the frequency can be reused as often as possible. The reuse factor in OFDM systems strongly depends on the method used for separating information from different users and services. In average loaded systems, advantages for the MC-CDMA schemes have been shown in cellular environments [29]. In the 4G system design, the reuse factor and maximum cell size have to be carefully selected in order to maximize the area spectral efficiency.

Error Correction Coding and Spreading
It is well known that current OFDM systems require some forward error correction (FEC) coding in order to achieve excellent performance results in fading channels. FEC coding can be classical convolutional coding as used in DAB, DVB-T, IEEE 802.11a or HIPERLAN/2, iterative turbo coding as used in UMTS, or it can be spreading (repetition coding) or a combination of coding and spreading [30]. It is of importance to investigate the trade-off between channel code rate, spreading code length, and symbol allocation on the channel for typical 4G mobile radio scenarios [22].

7.1.4.2 Channel State Knowledge at the Transmitter

Adaptive Coding and Modulation
With channel state information (channel impulse response, received powers, noise levels, interference levels, etc.) available on the transmitting side, as in TDD systems or FDD systems with channel prediction and a feedback link, completely new designs are possible. Now the transmitted power, the coding rate and the modulation constellation on individual subcarriers can be chosen such that the QoS is fulfilled with maximum spectral efficiency [31, 32]. Diversity (through e.g. channel coding or antennas) becomes less important, since this system can utilize multiuser diversity [20, 21], but may be needed when there are few users on very bad channels. A system based on these principles is presented in [13]. Alternatively, it is possible to exploit the redundancy in the prefix or postfix [33].

Pre-Equalization
If channel state information is available in the transmitter, pre-equalization can be applied at the transmitter such that the signal at the receiver appears nondistorted and an estimation of the channel state and equalization are not necessary at the receiver. Various investigations have shown that a high spectral efficiency is achievable with pre-equalization

in MC-CDMA systems, especially in the uplink [34–37]. A new concept is based on combining pre-equalization with conventional equalization, showing additional performance improvements [38]. Further investigations are necessary to show the effects of time variance between TDD slots, which are present in real systems.

Channel Prediction

Large spectrum and system efficiency improvements are possible by making channel state information available to the system and the transmitters [13, 20, 31, 32]. This, however, requires the design of channel prediction algorithms that can predict the channel impulse response well into the future, and assumes that these have reasonably good performance [13]. Furthermore, adaptive coding/modulation schemes, fast scheduling algorithms, and multiplexing and/or multiple access schemes that can use the information (predicted channel impulse response and the accuracy of it) provided by the channel prediction algorithm must be developed. A feedback channel might be needed in the system (depending on the duplex scheme) and this needs to be designed such that it does not consume too much of the bandwidth resources.

7.1.4.3 Channel State Knowledge at the Receiver

Single- and MultiUser Detection

Investigations have shown that the most promising single user detection technique is MMSE equalization. It can be implemented in a suboptimum version with very low performance degradation and with similar complexity to zero forcing or equal gain combining.

Multiuser detection techniques can significantly improve the system performance at the expense of complexity. Various interference cancellation schemes and block linear equalizers, with and without feedback, have been investigated and proposals have been made as to how the optimum maximum likelihood detector can be realized with reasonable complexity in MC-CDMA systems. Results for the up- and downlink have been presented [15].

In future research, it is important to compare all these techniques in a fair way, to consider their imperfections and to investigate their robustness against different kinds of interference, such as inter-cell interference and near–far resistance.

Joint Iterative Detection and Decoding

The iterative combination of detection and decoding has gained increasing interest recently, resulting in novel detection schemes like soft-interference cancellation [26]. These schemes exploit reliability information about the coded bits from the outer decoder and feed this back to the detector, and possibly to the channel estimator. These promising schemes have to be adapted to the requirements of 4G systems. The benefits of these schemes have to be proven also for system imperfections, such as imperfect power control, imperfect channel estimation and interference from neighbouring cells.

Moreover, in order to improve the multicarrier detection scheme, iterative demodulation algorithms (turbo demodulation) of bit interleaved higher order coded modulation enable significant performance enhancement over standard bit metric derivations.

Channel Estimation

Pure OFDM systems can easily apply differential modulation (as used in DAB). In the case of multiple access schemes like MC-CDMA, the coherent detection with pilot symbol assisted channel estimation has gained more interest due to its higher user capacity. Efficient two-dimensional channel estimation concepts have been developed [39–41], which can significantly reduce the overhead due to pilot symbols by exploiting the correlations of the mobile radio channel in both the time and frequency dimensions. These schemes are especially of interest for downlink systems.

For the 4G system, concepts have to be developed for an efficient MC-CDMA uplink channel estimation that can cope with high numbers of active users. Alternatives or extensions of pilot symbol aided channel estimation schemes, including blind or semi-blind estimation algorithms [42, 43] that reduce the overhead due to pilot symbols, have gained much interest in the past years. The suitability of these schemes for 4G broadband systems should also be investigated in detail.

7.1.4.4 General System Aspects

MIMO OFDM

Multiple antennas can be considered as one of the most important contributors to reliable communications, especially in hostile environments, and the need for provision of independently fading replicas of the same data-bearing signal is crucial for systems [44]. The presence of several antennas is usually exploited to provide some sort of diversity. Several diversity techniques exist, including, for example, antenna diversity, polarization diversity, angle diversity, frequency diversity, multipath diversity and time diversity [45].

MIMO OFDM systems provide the capability of exploiting the advantages of both MIMO and OFDM techniques. It has been established [46] that the use of several transmit and receive antennas (MIMO) can provide a significant capacity gain. In MIMO systems, this additional capacity is translated into a throughput and overall performance increase. Commonly, MIMO systems exploit spatial and/or temporal diversity, giving rise to the name 'space–time coding' technique [47, 48]. In a MIMO OFDM system, diversity can be supplied also in the frequency domain, offering an additional coding dimension. Thus, we refer to space–time–frequency coding (STF) schemes [49]. Due to the availability of an additional dimension for coding in MIMO OFDM, better performance than space–time coding is to be expected. However, for this to be achieved, novel STF codes have to be developed that will maintain a transmission rate at least equal to that offered by a SISO OFDM system [50].

Since the conditions of maximum capacity and maximum diversity cannot be achieved simultaneously, a trade-off between them must be reached when designing a MIMO OFDM system. Furthermore, the complexity degree must be considered in the design so that possible implementations can be carried out [51].

The channel estimation for MIMO OFDM systems constitutes a real challenge, since instead of one channel, a multidimensional matrix is envisaged that includes all the possible paths between antennas [52–54]. Moreover, as an OFDM system, MIMO OFDM faces synchronization impairments such as timing, carrier frequency and sampling frequency offsets, which must be taken into account [55].

Achieving synchronization in a MIMO OFDM system would be an extremely complicated task if some of the transmitter (receiver) blocks composing the transmitter (receiver) side differ one with the other. However, from the designer point of view, this is a very improbable situation and the modules at either side are identically implemented [56]. Hence, synchronization complexity reduces significantly to that of a SISO OFDM system. Furthermore, the offset estimates can be enhanced by averaging among the receive antennas.

In conclusion, MIMO OFDM systems provide the following advantages:

- capacity and performance amelioration;
- enhanced robustness versus deep multipath fades;
- granularity between throughput and performance;
- improved performance of synchronization schemes.

Efficient implementations and performance results of MIMO OFDM and MIMO MC-CDMA systems can, for instance, be found in [23, 37, 49, 57].

Synchronization

OFDM inherently offers the possibility to use the guard interval for synchronization [58] and thus can reduce, or even completely avoid, additional overhead due to reference symbols required for synchronization. Investigations are necessary to show whether these algorithms are sufficient to perform acquisition and tracking in a 4G system or whether additional measures as shown in [15, 59] are necessary. As a rule of thumb, the synchronization error in frequency and time should not exceed a few percent of the subcarrier spacing and of the OFDM symbol duration, respectively [15].

Peak-to-Average Power Ratio Reduction

Multicarrier signals have a nonconstant envelope which reduces the efficiency of the power amplifiers in the transmitter. The nonlinear amplifier characteristics distort the transmitted signal and enhance the out-of-band power. A multitude of measures have been proposed to reduce the peak-to-average power ratio (PAPR) or to apply predistortion to increase the power efficiency of the transmitter. It is important to study the PAPR reduction techniques to conclude which methods best fit the requirements in the uplink and in the downlink.

Deriving digital strategies, either in the frequency or time domain, to reduce the peak-to-average power allows a significant decrease of the backoff for the power amplifier [60]. Moreover, spreading codes with low PAPR can be chosen in MC-CDMA systems [61]. PAPR reduction would significantly impact the battery life of wireless terminals.

More generally, the study of digital means for reducing the OFDM constraints on the RF front-end, and transceiver architectures that target lower power operation, will be important for reaching high data rates and preserving true wireless operation.

Spectral Efficiency

OFDM modulation is well known for its robustness to multipath time-varying propagation channels. This is mainly due to the insertion of a cyclic prefix (guard interval), a very efficient and simple way to combat multipath effects. However, adding a guard interval is adding redundancy, which decreases the spectral and power efficiency. It is interesting to study alternative OFDM modulation schemes, which can provide the same robustness without requiring a guard interval, i.e. offering a better spectral and power efficiency. For this purpose, the prototype function modulating each subcarrier must be very well localized in time and frequency to limit ISI and ICI. Functions having these characteristics exist, but the optimally localized ones only guarantee orthogonality on real values [62]. The corresponding multicarrier modulation is OFDM/OffsetQAM [63]. Among these functions, the localization is optimal with the IOTA (Isotropic Orthogonal Transform Algorithm) function [62, 64]. It would be interesting to investigate the potential of OFDM/OQAM modulation for the new OFDM-based air interfaces addressed, as has already been done in the 3G context [65].

7.1.4.5 Joint Optimization of Layers 1 and 2

Advanced ARQ Protocols

A possible area to investigate in order to increase the efficiency and robustness of transmission systems is the request for retransmissions of incorrectly received packets [56]. Clearly, the most efficient ARQ schemes require more sophisticated technology to be implemented. It is therefore important to study which approaches are practical and efficient for a 4G system. This includes different ARQ protocols (e.g. stop-and-wait, selective-repeat, go-back-N) as well as different hybrid ARQ techniques (e.g. code combining, incremental redundancy) and their efficient linkage to channel coding schemes. The reliability and processing delay of an ARQ return channel has also to be considered.

In some applications ARQ is needed to fulfil the QoS requirements. It may even be necessary to have ARQ protocols involved both at the physical layer, to make it good enough for the upper layers, and on the link layer [13]. The protocols need to be designed jointly with coding and modulation on the link layer [32].

The 4G mobile radio system should include optimal error correction with highly flexible dynamic ARQ schemes, especially for the support of future multimedia applications. These algorithms have to evolve from simple repetition to more intelligent redundancy splitting methods which allow for flexible and dynamically increasing error correction capabilities. Hence, retransmissions of data packets can be deployed more efficiently, while occupation of radio resources can significantly be reduced.

The choice between FEC, ARQ, or a balance between the two, may depend on QoS requirements. Whilst ARQ can achieve a higher throughput, by only transmitting overheads when required, FEC can provide a more consistent delay performance.

Scheduling of Resources

In order to increase spectral efficiency in a multiuser system, channel resources have to be efficiently scheduled between users in such a way that spectral efficiency is maximized given the requirements on QoS (BER, delay, etc.) [13, 20, 32]. In this way, resources are given to those that can best use them, e.g. a user that can transmit at high rate with low power is allowed to transmit instead of a user that cannot transmit with reasonable performance, or can only transmit with high power and low rate. New MAC protocols and efficient signalling between the physical and MAC layers will have to be developed and carefully engineered to accomplish this. In such a scheme, the overall generated interference, and thus the radio exposition of humans, is reduced, increasing the conformance to tighter future emission regulations and the public acceptance of a system.

Spatially Differentiated Air Interface Access

With the application of multiple element antennas as linear or two-dimensional arrays with flexible on-demand beamforming in 4G networks, radio access protocols should take into account the spatial separation of users simultaneously requesting service in a certain area. The ability to direct multiple antenna beams to specific sectors and map spatial differentiated data streams to them increases the spectral efficiency of the radio air interface, while also reducing the interference emitted from the network. The MAC should be aware of the individual locations of other communicating stations by making use of positioning techniques such as measured angle of arrival statistics from the uplink direction. With this information, a multidimensional access with respect to temporal, frequency, subcarrier and spatial domains could be established with self-organizing dynamic re-use of radio resources in hierarchical network topologies.

7.1.5 Conclusions

The future demands on data rate, mobility and QoS require the design of completely new air interfaces for the next generation of mobile radio systems. Multicarrier technology has shown its suitability for very high rate systems in many digital broadcasting and wireless LAN standards and seems to be an excellent candidate to fulfil the requirements on 4G mobile radio systems. Future research is expected to show that multicarrier will be a superior technology for 4G.

7.2 A Mixed OFDM/Single Carrier Air Interface

Future generation wireless systems, transporting high bit rates, must be robust to severe frequency selective multipath. Frequency domain-based modulation or equalization provide cost-effective multipath compensation. We propose that the downlink air interface use orthogonal frequency division multiplexing (OFDM) and the uplink air interface use single carrier (SC) modulation, with adaptive frequency domain equalization (FDE) carried out in the base station receiver. This arrangement has the advantage that the mobile terminal's power amplifier cost is minimized, and most of the signal processing complexity is concentrated at the base station. We also describe several research issues relevant to this proposal.

7.2.1 Introduction

Future generation wireless access systems will probably be required to provide peak user data rates of the order of 20 Mb/s, and aggregate data rates of the order of 50 to 100 Mb/s [66, 67]. Furthermore, the bit rates and performance should be adaptable to the subscribers' respective applications and requested quality of service, for whatever environment – indoor, outdoor, static, mobile, urban, etc. – each subscriber terminal is in. A very important additional requirement is low cost: low terminal cost and power consumption for subscribers, low spectrum costs, high spectral efficiency and coverage, and ease of deployment for service providers. The choice of air interface modulation and coding schemes can significantly influence performance and cost for a given range of bit rates. These systems may be required to operate in non-line-of-sight (NLOS) environments, in which the radio channel's impulse response could span upwards of 10 to 20 ms [68, 69] in extreme cases.

7.2.2 OFDM and SC-FDE Systems

For an aggregate bit rate of 100 Mb/s and a 10 ms delay spread, the intersymbol interference (ISI) span could be as much as 1000 data bits. While this may be an extreme example, it indicates that traditional methods of dealing with ISI, such as adaptive time-domain equalization, will probably prove to be impractical in future generation wireless systems, since the number of signal processing operations per bit in traditional time domain equalization methods is proportional to the ISI span. A better equalization performance/complexity trade-off is obtained by doing transmission and reception operations on a block by block basis in the frequency domain, using fast Fourier transform (FFT) processing, for which the signal processing complexity per bit rises only logarithmically with ISI span. Orthogonal frequency division multiplexing (OFDM) and its variants, and single carrier (SC) with frequency domain equalization (SC-FDE) are in this category. In this text, we advocate a hybrid air interface – with OFDM in the downlink from base to subscribers, and SC-FDE in the uplink from subscribers to base. As will be seen in the following, this arrangement benefits from the advantages of both approaches, while minimizing their disadvantages.

7.2.2.1 OFDM

OFDM, or multicarrier modulation, is a well known frequency domain approach [70, 71]. IEEE 802.11a, 802.16, HiperLAN2, and digital video and audio broadcast standards adopt it completely or partially [72, 73], and it can also be considered for a future generation mobile broadband wireless air interface [74].

In OFDM, coded data symbols are transmitted in parallel on narrowband subcarriers, separated in frequency so as to be orthogonal. This multicarrier approach has long been known as a natural and ideal approach to transmitting data over a frequency-selective channel, since the parallel orthogonal subchannels approximate the eigenfunctions of the linear channel [75]. The transmitter's modulation process is carried out on blocks, or frames, of (typically quadrature amplitude modulation (QAM)) data symbols, using a computationally-efficient inverse FFT (IFFT) operation. At the receiver, an FFT operation decomposes each received signal block into the narrowband modulated subcarriers, which are independently equalized and passed on to the decoder.

A cyclic prefix (CP), which is a copy of the last part of the transmitted block, is prepended to each block. The length of the CP is at least the maximum expected length of the channel impulse response. In both SC and OFDM receivers, the received CP is discarded, and FFT processing is done on each M-symbol block. The CP transmitted at the beginning of each block has two main functions:

1. It prevents contamination of a block by ISI from the previous block.
2. It makes the received block appear to be periodic with period M. This produces the appearance of circular convolution, which is the basis of the FFT operation.

Equalization is carried out by a simple complex gain multiplication on each received subcarrier, and can be incorporated within the decoding process. On a frequency-selective channel, the signal to noise ratios (SNR) of the equalized subchannels vary widely in a manner similar to the time-varying SNR observed on a fast fading channel. For this reason, coding and intrablock interleaving are essential, and provide inherent frequency diversity for OFDM systems used on frequency-selective channels. Decoding that uses channel state information about the gain of each OFDM subcarrier is generally employed, since it significantly enhances bit error rate performance [76]. Adaptive loading, mentioned later, can yield a significant spectral efficiency and performance enhancements to OFDM.

7.2.2.2 SC-FDE

Equalization in the frequency domain, using computationally-efficient FFT and IFFT operations, can also be applied to traditional single carrier modulation schemes, in which data symbols are serially modulated (typically with QAM) with a high symbol rate, onto a single carrier. As in OFDM, data is transmitted in blocks, or frames, preceded by a CP. A nonblock version of SC-FDE, without inserted CPs, can also be implemented by using overlap-save or overlap-add filtering at the receiver [77].

The FFT operations in SC are all done at the receiver: an FFT in followed by simple independent equalization of each frequency component, which is followed by an IFFT to restore the now equalized serial data stream [78, 79]. Adaptation of the equalizer parameters is straightforward and efficient, and can use similar techniques as OFDM [80]. For example, the CP can be replaced by one or more short time-domain training sequences or 'unique words', which are processed and extrapolated in the frequency domain to estimate the required equalizer frequency-domain parameters [73, 80–85]. Like OFDM systems, equalized SC systems exploit frequency diversity, but through the equalization process rather than relying entirely on coding. If the channel remains stationary during a block transmission, the linearly equalized channel appears to the decoder as an additive white Gaussian noise channel with constant SNR. The SNR is approximately averaged over the frequency band. Intrablock interleaving is unnecessary.

There is an interesting hybrid of SC-FDE and OFDM which was proposed recently: a carrier interferometry (CI) approach, in which, block by block serial modulation is used, but the individual pulses which transport the data symbols are multicarrier signals [86]. These multicarrier pulses are delayed and cyclically-wrapped versions of one another. This system can be considered and implemented as an SC system in which the transmitted signal is generated in the frequency domain, and then inverse-FFTed to the time domain before transmission.

7.2.2.3 Discussion and Comparison of OFDM and SC-FDE

SC-FDE and OFDM air interface signals can be converted into each other by an FFT or IFFT operation. For example, a block of an SC or of an OFDM signal appears to be one period of a periodic waveform, thanks to the effect of the CP. Thus, the transmitted spectrum during a block is a line spectrum with discrete spectral lines at intervals of the reciprocal of the block length. An SC signal therefore could be interpreted as a particular version of a multicarrier waveform. Likewise, OFDM signals could be generated as SC signals with the transmitted data symbols replaced by their FFT components. However, it is important to bear in mind that SC and OFDM signals transport data symbols on different sets of basis waveforms, and hence have some important differences.

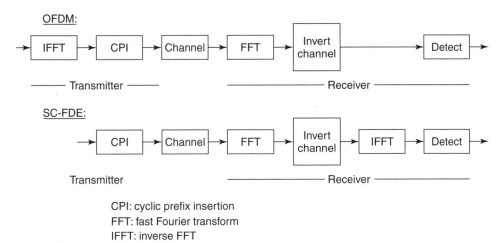

CPI: cyclic prefix insertion
FFT: fast Fourier transform
IFFT: inverse FFT

Figure 7.3   OFDM and SC-FDE – signal processing similarities and differences

For comparison purposes, Figure 7.3 shows block diagrams of an OFDM system and of a single carrier system with FDE and CP insertion (CPI). Note that similar FFT and other signal processing operations are present in these two systems, albeit with different orders of appearance. In each of these frequency domain systems, data is organized in blocks whose length, M, is typically at least eight to ten times the maximum expected channel impulse response length. In the case of OFDM, each transmitted block is processed by an inverse IFFT to implement multicarrier modulation. In the SC case, the IFFT operation is at the output of the receiver's equalizer.

The serially-modulated and coded data symbols in an SC system spectrally occupy the entire transmission bandwidth, but each has a relatively short time duration. Thus, different users' signals are separated by time slots if time division multiple access (TDMA) is the multiple access scheme, by carrier frequency if frequency domain multiple access (FDMA) is used, by spreading codes if code division multiple access (CDMA) is used, or by appropriate packet scheduling in packet radio systems. For TDMA or packet systems, variable user bit rate requirements can be accommodated by varying the number of time slots or the packet length allocated to each user. For SC systems which use overlap-save processing to avoid transmission of a CP, FFT processing block lengths impose no constraint on packet or slot duration. Adaptive modulation and coding for a given user's signal in SC transmission amounts to (generally slow) changes in the signal constellation size and code rate in response to changes in the output signal to interference plus noise ratio (SINR) of a received training signal.

The coded data symbols transported by an OFDM system modulate parallel frequencies. Thus, OFDM systems offer the possibility of another degree of flexibility: separating different users either in time, spreading code, packet time, or by subcarrier. The latter mode, in which different users are assigned different subcarriers, or groups of subcarriers, is called OFDMA (orthogonal frequency division multiple access). In this way, very high bit rate users and very low bit rate users can be frequency multiplexed efficiently and simply. Furthermore, the spectral occupancy of OFDM signals is easily controlled by choosing whether or not to data modulate each subcarrier. Careful allocation of subcarriers, or coded subcarrier frequency hopping, can be used to avoid problems arising from channel frequency selectivity.

Among the other potential strengths of OFDM is its unique flexibility to 'adaptively load' subcarriers; i.e. to optimize the allocation of coded bits to each subcarrier, according to its channel gain or SINR. If properly implemented, this can be a more flexible and powerful form of adaptive modulation and coding than is practically possible for SC systems. Adaptive loading requires feedback from the receiver to the transmitter, and accurate tracking of time varying channel frequency responses. It is applied in wireline discrete multitone systems, but has so far proven to be too much of a challenge to implement in the nonstationary wireless environments encountered by current OFDM-based wireless systems, such as the digital video broadcast (DVB), IEEE 802.11a, HiperLAN2 or OFDM 802.16a. However, for frequency-selective channels, adaptively loaded OFDM has been shown theoretically to offer significant spectral efficiency improvements over linearly equalized SC modulation and nonadaptive OFDM (in which bit allocation is uniform) [87–90]. The flexibility to assign bits to subcarriers also extends to OFDMA.

Both OFDM and SC-FDE systems use efficient FFT and IFFT operations and simple one-tap equalization operations on each frequency component. Their equalization signal processing complexities are similar: of the order of log2M operations per data symbol, for block length M. For channels with multipath delay spreads spanning more than about ten data symbols, this complexity is far less than that typically required by a conventional time domain equalization approach [79, 91]. Figure 7.4, from [83], shows a comparison

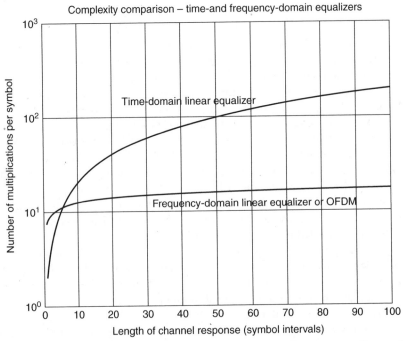

Figure 7.4 Complexity comparison of time and frequency domain linear equalizers with LMS (least mean square) adaptation

of the complexities of time domain and frequency domain (SC or OFDM) linear equalizers as a function of the length of the channel impulse response, measured in symbol intervals. Complexity here is gauged by the number of complex multiplications per transmitted data symbol, including both the filtering operations and LMS adaptation operations (see [92] for a description of the latter). The frequency domain equalizer is assumed to use an FFT block length equal to eight times the channel length.

Comparisons of coded linear SC-FDE systems with non-adaptively loaded coded OFDM systems have shown that the former offer similar, and in high code rate/high SNR cases, better, bit error rate performance to coded OFDM [78, 79, 93–97]. Figure 7.5(a) shows such a comparison for coded linearly equalized SC-FDE and OFDM systems in indoor frequency-selective channels with a maximum delay spread of 0.66 ms and bit rates from 60 to 300 Mb/s. For each system, the FFT block size was 512 symbols. For code rates of 1/2, non-adaptively loaded coded OFDM shows a small (~0.5 to 1 dB) average SNR gain over coded linearly equalized SC-FDE. Higher code rates tend to favour SC-FDE. SC-FDE systems' performance can be further enhanced by the addition of full-length [90] or sparse [80, 83] time-domain decision feedback equalization. By employing iterative block decision frequency-domain DFE techniques [98, 99] or turbo equalization [100, 101], we can also improve the performance of the SC-FDE system.

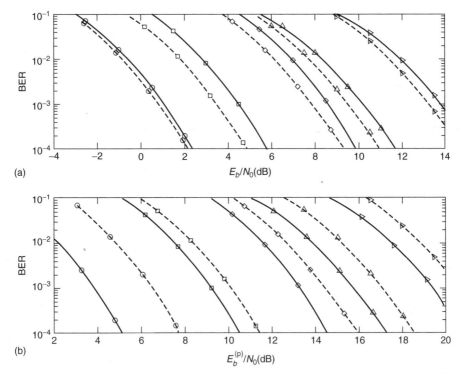

Figure 7.5 Bit error rate performance for pragmatic TCM-coded SC-FDE solid lines and BICM (bit interleaved coded modulation)-coded OFDM (dashed lines); ○: QPSK, rate 1/2; ◇: 16QAM, rate 3/4; △: 64QAM, rate 2/3; ▷: 64QAM, rate 5/6. (a) shows BER versus average E_b/N_0; (b) shows BER versus peak E_b/N_0

However, OFDM has an important hardware implementation drawback: since a transmitted OFDM signal consists of many parallel modulated subcarriers, it has a relatively high peak to average power ratio (PAPR). The amplitude distribution of a transmitted OFDM signal approaches that of random Gaussian noise, while that of an SC signal is closer to a uniform distribution over a finite range. Peak output instantaneous power exceeding a power amplifier's linear range causes significant spectral regrowth [102, 103], and also bit error rate (BER) degradation resulting from nonlinear distortion [104]. These problems necessitate several dB of transmitter power backoff [89, 105–108]. As a result, OFDM systems require more expensive power amplifiers than do comparable SC systems with the same average power output.

Figure 7.6 illustrates the spectral regrowth that occurs with 10 dB power backoff at the output of a typical power amplifier for a QPSK OFDM system with a 256-point FFT block size, with 25 % of the subcarriers not used (Figure 7.6(b)), and for a QPSK SC system with 25 % excess bandwidth. Also shown in this figure is the transmitted power spectrum for an ideal (infinite backoff) power amplifier (Figure 7.6(a)), and also (with straight lines) the FCC spectral mask for multichannel microwave distribution systems (MMDS) with 6 MHz bandwidth (Figure 7.6(c)). Clearly, the OFDM system's output power must be backed off more than 10 dB to comply with the FCC mask in this example.

For a given power amplifier specification, the larger the power backoff the lower the cell coverage. This power backoff penalty is especially important for subscribers near the edge of a cell, with large path loss, where lower-level modulation, such as binary phase shift keying (BPSK) or quaternary phase shift keying (QPSK), must be used.

There are signal processing or clipping techniques for reducing the PAPR and spectral regrowth of OFDM systems [109, 110]. However, they add significant complexity and may degrade performance or bandwidth efficiency. If bit error rate performance is evaluated as a function of the average peak bit energy to noise ratio, as in Figure 7.5(b) [31], coded SC-FDE is consistently superior to coded OFDM over a wide range of code rates and QAM constellation sizes, even if the transmitted OFDM signal undergoes clipping and filtering to reduce its peak power.

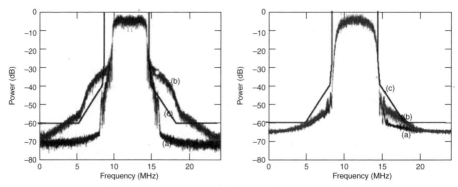

Figure 7.6 Power amplifier output power spectra for a QPSK 256-point OFDM system (left graph) and for a QPSK single carrier system (right). (a) Spectrum with ideal power amplifier (infinite power backoff); (b) spectrum with typical power amplifier with 10 dB power backoff; (c) FCC MMDS (multi-media distribution system) Improvement of resolution is not possible in the short time span available spectral mask

7.2.3 Mixed Mode Air Interface – Downlink OFDM, Uplink SC-FDE

In general, all frequency-domain modulation/equalization approaches – OFDM, OFDMA and SC-FDE – have their own advantages, and can be used singly or together in a complementary fashion. For example, the IEEE 802.16a task group developed a broadband wireless metropolitan area network (WMAN) air interface standard for sub-11 GHz systems, which includes all three as possible modes [73]. In view of the power amplification issues that can particularly impact the cost of the wireless subscriber unit, a very sound approach is to use SC modulation in the uplink and OFDM in the downlink [79, 95]. A block diagram of this approach is shown in Figure 7.7. This arrangement has the following potential advantages:

- The subscriber transmitter is SC, and thus is simple and inherently more efficient in terms of power consumption, due to the reduced power backoff requirements of the single carrier mode. This will reduce the cost of a subscriber's power amplifier.
- More of the signal processing complexity is concentrated at the hub, or base station. Differentially modulated OFDM could be transmitted in the downlink to very low-cost subscriber terminals, obviating their need to do channel estimation. Also, the hub has two IFFT operations and one FFT, while the subscriber has just one FFT operation for receiving the downlink OFDM signal, and transmits with a simple serial modulation scheme. As a result, the subscriber terminals should experience lower battery drain.
- OFDM in the downlink, as well as minimizing the FFT processing in the subscriber unit, has powerful options of adaptive channel loading, coding and multiplexing flexibilities.
- The uplink TDMA single carrier mode is simple and efficient; short media access control (MAC) messages can be transmitted in very short-duration bursts, since serial, SC modulation has a flexible range of burst durations, especially if the CP requirement is removed.
- Interoperability of future generation broadband wireless systems with existing third generation (3G) wireless local area networks (WLAN), WMAN and DVB standards,

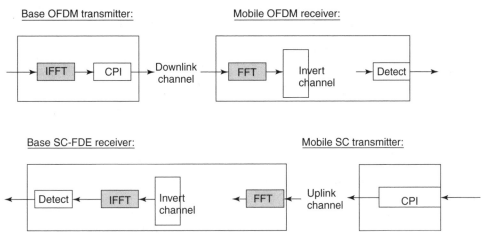

Figure 7.7 Uplink SC-FDE, downlink OFDM architecture. FFT: fast Fourier transform operation; IFFT: inverse FFT; CPI: cyclic prefix insertion

some of which use SC, some of which use OFDM, and some of which include both as options, may be facilitated by the mixed mode.

- Uplink SC and the version of downlink coded OFDM used in DVB systems is a potentially low-cost approach to integrating broadcast digital TV services and broadband wireless two-way services.

7.2.4 System Requirements and Enabling Technologies

Robustness to multipath and spectral efficiency, with minimal hardware cost and signal processing complexity, are the main motivations for proposing a mixed OFDM/SC air interface. However, there are also other requirements that any future generation air interface must satisfy [67].

7.2.4.1 System Scalability and Implications for Multiple Access

New mobile radio systems should be scalable with respect to multimedia user bandwidth requirements, either symmetric or asymmetric, with peak uplink user bit rates ranging from a few Kb/s up to at least several tens of Mb/s. Low-cost user terminals may be limited to low bit rates, while higher grade user terminals will be able to employ the full range of bit rates on demand. Thus, physical layer design, including the slot/frame structure hierarchy should be adaptable to the different deployment scenarios and a wide range of future application requirements. All wireless user terminals should be as efficient as possible – both in spectrum utilization and in transmitted power. Multiple access, as well as modulation methods, must be carefully designed to meet these requirements.

Possible uplink multiple access methods, with roughly equal spectral efficiencies, include TDMA, CDMA and FDMA, or combinations of these. While the average transmitted power requirement is about the same for the above multiple access methods, and is proportional to the user's current average bit rate, the peak power is significantly larger for TDMA than for FDMA or CDMA, especially for terminals transmitting at low bit rates. While this difference has only moderate cost significance for terminals which must handle both high and low bit rates, it could add a significant cost factor to very low bit rate terminals, which have low duty cycles in a TDMA system, but whose power amplifiers must have the high peak rating required for a high aggregate bit rate. These considerations would argue that CDMA- or FDMA- based SC techniques should be employed in the uplink to minimize terminal cost.

7.2.4.2 High-speed Data Rate Coverage

The existing OFDM-based IEEE 802.11a WLAN system has the limitation of small area coverage and uneven high-speed data rate distribution. The new mobile radio system should deliver uniform data rate coverage and ubiquitous service. It should also be scalable with respect to cell site deployment (macro/micro/pico). A study of means to achieve high and uniform coverage includes research on the access architecture, access control, such as rate control and power control, and mesh/multihop network architectures.

7.2.4.3 Soft Hand-off and Macro/Network Diversity

Soft hand-off is a proven technique for 3G, it is a fundamental technology to improve the coverage and reduce the outage. It is important to include the soft hand-off technique in new mobile radio systems to exploit the macro/network diversity.

7.2.4.4 Application to High-speed Mobility

To support the high-speed mobility application for the new mobile radio system, the physical layer needs to include fast parameter estimation and tracking. Efficient frequency-domain parameter estimation methods using training words and/or pilot tones are known for OFDM and SC-FDE systems [83]. However, high bit rates and multipath delay spreads call for large FFT block lengths; there is then a question of the ability of receivers to estimate and track the resulting large number of frequency-domain parameters over rapidly time-varying channels.

For the OFDM downlink, if adaptive loading is used, the rate of updating and feeding back channel estimates, and their required accuracy, needs to be determined as a function of the rate of change of the channel.

7.2.4.5 Support Universal Frequency Reuse

To achieve high capacity of future mobile radio systems based on the mixed mode air interface, it is essential that the new air interface be deployable in a frequency reuse one network. This requires the new air interface to have a strong interface resistant capability, and the cochannel interference cancellation and avoidance techniques should be embedded into the new air interface design.

7.2.5 Enhancements and Relevant Research Issues

7.2.5.1 Enhancements to Basic OFDM in the Downlink

Coding, Modulation and Interleaving

OFDM requires powerful coding and channel-state-aware decoding techniques, especially if adaptive loading is not used. Convolutional codes, in some cases concatenated with Reed–Solomon (RS) codes, are used in several existing and draft OFDM-based air interface standards (802.11a, HiperLAN, 802.16a, DVB). Convolutional or block turbo codes are also included as powerful options. Further study of the optimization and implementation of coded modulation and interleaving techniques should be made for nonadaptive and adaptive OFDM. Bit-interleaved coded modulation (BICM) [111, 112] has been found to be effective for OFDM, since a coded OFDM block presents to it the appearance of a fast fading channel with channel state information. Iterative detection can also be combined with BICM decoding. It remains to be seen if other coded modulation methods may be better.

Adaptive Loading and OFDMA

As mentioned earlier, adaptive loading of OFDM subcarriers has been shown to offer superior performance and capacity. Currently no wireless OFDM systems implement adaptive loading, although it is implemented in wireline discrete multitone systems. The main implementation issue for wireless is the amount of feedback overhead required for effective adaptive loading in typical time-varying wireless channels. A related topic is the efficiency and implementation in time-varying channels of OFDMA in the downlink; i.e. multiplexing different users in both time and frequency.

Multicarrier CDMA
There are various versions of multicarrier CDMA (MC-CDMA), formed by combining CDMA and OFDM [71, 113, 114]. MC-CDMA offers simple and flexible MAC for multimedia uplink traffic. The pros and cons of employing it in the downlink should be studied.

Power Efficiency and PAPR
While power amplifier efficiency and cost may be somewhat less of a concern at base stations than in subscriber terminals, overall system cost would still benefit from signal processing which lowers PAPR in downlink OFDM signals. Combinations of DSP and RF signal processing to reduce transmitted peak to average power ratio could benefit both OFDM and SC systems [115]. Also, the power amplifier provisioning and thermal mounting cost dependence on required peak power should be studied, as in [102].

MIMO Techniques and Smart Antennas
The emerging space–time coding and smart antennas have also to be taken into account in the design of a new air interface. The effects of MIMO schemes on the system capacity of an OFDM-based air interface are of high importance for the design of the detection and decoding scheme.

MIMO-OFDM is an enabling technology to fundamentally improve the spectral efficiency of a new mobile radio system. The advantage of OFDM over conventional CDMA to implement MIMO technology is significant. There should be a study of OFDM-specific MIMO and smart antenna schemes to allow the antenna technology to be optimized and embedded into the new air interface. In particular, space–time coding and multiplexing using multiple transmitting antennas for OFDM should receive attention, since multiple antennas are easier to deploy at the base station than at the subscriber terminal.

Cochannel Interference Cancellation
In a frequency reuse one cellular network, the cochannel interference impact on the overall system is significant. Interference mitigation and cancellation techniques should be developed to increase the spectral efficiency. This is important technology for new mobile radio systems. In general, existing OFDM-based systems, such as IEEE 802.11a, are not operated in a frequency reuse one environment.

7.2.5.2 Enhancements to Basic SC-FDE in the Uplink
DFE
DFE gives better performance for frequency-selective radio channels than does linear equalization [116]. In conventional time-domain DFE equalizers, symbol-by-symbol data decisions are made, filtered, and immediately fed back to remove their interference effect

from subsequently detected symbols. Because of the delay inherent in the block FFT signal processing, this immediate filtered decision feedback cannot be made with low complexity in a frequency domain DFE, which uses frequency-domain filtering of the fed-back signal.

A hybrid time–frequency domain DFE approach, which avoids the abovementioned feedback delay problem, uses frequency-domain filtering only for the forward filter part of the DFE, and conventional transversal filtering for the feedback part [79, 80, 90]. It is preferable to replace the CP of data with a fixed unique word (UW) or training sequence to facilitate receiver adaptation and synchronization, and to allow the DFE to start up properly in each block [85, 90, 117]. The transversal feedback filter is relatively simple in any case, and it could be made as short or long as is required for adequate performance. Complexity is minimized by making the feedback taps few in number and sparse, corresponding to the largest channel impulse response echoes [79, 80]. This also tends to minimize possible DFE error propagation problems. There is a significant performance enhancement over linearly equalized SC-FDE systems [79, 80, 90] – enough to yield similar or better performance than coded non-adaptively loaded OFDM.

It has also been demonstrated [90] that nonadaptive SC-FDE with a sufficiently long decision feedback filter approaches the performance of adaptively loaded OFDM, confirming the earlier theoretical results of [118]. Alternative enhancements to DFE, which avoid error propagation problems include Tomlinson–Harashima transmitter filtering and receiver maximum likelihood sequence estimation.

7.2.5.3 Iterative Block Decision Feedback Equalization

A promising IB-DFE scheme (iterative block decision feedback equalization) was proposed in [90], where both the feedforward and feedback filtering operations are done in the frequency domain. For this purpose, the feedback loop uses the blockwise hard decisions from the previous iteration. The particular case where only one iteration is employed corresponds to a conventional, linear FDE scheme, and the subsequent iterations with the IB-DFE scheme are shown to provide increasingly improved performances. Dinis and Gusmão compared the performances of linearly equalized SC-FDE, SC with IB-DFE (three iterations) and OFDM systems [99], when QPSK constellations and convolutional codes were employed (for the OFDM system the coded bits are interleaved before being mapped, as with BICM schemes). For code rates of 1/2, the performance of SC with IB-DFE was similar to the OFDM performance, and about 0.5 dB (for BER = 10^{-4}) better than with the linearly equalized SC-FDE. For code rates of 3/4, the linearly equalized SC-FDE is about 1.5 dB better than the OFDM; an additional gain of 1.5 dB is observed for the SC with IB-DFE. It was shown [99] that these techniques can be easily modified to include receive space diversity, as well as two-branch, Alamouti-like transmit diversity [119, 120]. The feasibility of using soft decisions on the feedback loop, before and/or after the channel decoder, should be studied.

Turbo Equalization
Conventional linear or decision feedback equalization techniques for SC, and conventional coded OFDM techniques for multicarrier broadband signaling, combine channel multipath components in a suboptimal way to achieve path diversity improvement. Recent advances

in turbo signal processing techniques have achieved closer to optimal SC broadband signalling for mobile communications [121]. A turbo equalization technique, a soft canceller followed by an MMSE filter (SC/MMSE), can achieve close to the matched filter bound performance without requiring prohibitively large computational effort. For example, the performance of a MIMO SC system with turbo equalization has been verified through simulations using field measurement data [100]. The SC MIMO system concept can easily be introduced onto the mixed mode system, when a MIMO turbo equalizer is in place, and hence it can be seen as an extension of the mixed mode new air interface. Furthermore, complexity of a time-domain MIMO turbo equalizer algorithm can be reduced to a practical level [122] in the same way as in [121], and the algorithm can easily be converted into the frequency domain. Frequency-domain turbo equalization significantly reduces the complexity burden [101], and should be examined further.

Coding, Modulation and Interleaving

Uncoded SC-FDE systems significantly outperform uncoded nonadaptive OFDM systems on frequency-selective channels [78, 83]. However, coding still enhances SC-FDE performance further. After equalization, an SC-FDE decoder sees an additive white Gaussian noise (AWGN) channel with an SNR which is approximately constant over an FFT block. Symbol-based interleaving and coded modulation is therefore appropriate; e.g. pragmatic trellis coded modulation (TCM) for high-SNR channels [123]. The BER performance comparisons shown in Figure 7.5 used BICM coding for the OFDM system and pragmatic TCM for the SC system.

The gain in performance available from convolutional or block turbo codes, and even more advanced coding methods, such as low density parity check codes, in comparison to TCM should also be weighed. Adaptive modulation and code rate also maximizes the efficiency of SC-FDE, although it does not offer the subcarrier flexibility that is possible with adaptively loaded OFDM.

Eliminating the CP

The overhead created by the CP in SC-FDE systems could be eliminated by using well known frequency domain overlap-save or overlap-add processing methods [77]. It can be shown that the performance (MMSE or BER) of the overlap-save approach, with $2N$ coefficients equals that of the corresponding time-domain equalizer with N optimal coefficients, but the complexity varies logarithmically with N, rather than linearly. On the other hand, block frequency domain processing, where each block is preceded by a CP, avoids any inter-block interference. Furthermore, block frequency domain processing with CP achieves the equalization performance expected from using an infinite number of time-domain equalizer taps; e.g. a linear FDE with CP could perfectly invert the cyclic frequency response of a channel, whereas a time-domain equalizer, in general, could achieve this feat only with an infinite number of taps [124].

An overlap-save equalizer cannot exceed the BER performance of a corresponding CP FDE (apart from the small E_b/N_0 gain resulting from the CP overhead elimination. This has been verified in simulation studies comparing BER performance of SC-FDE systems with CP and with M frequency-domain taps against time domain equalizers with M time-domain taps (the equivalent of overlap-save equalizers with $2M$ frequency-domain taps) [125].

Computationally efficient estimation of the overlap-save equalizer's parameters is a research issue. A pragmatic approach was proposed in [125].

CDMA and FDMA

Minimal PAPR for all uplink transmitters could be achieved with SC direct sequence CDMA, with each uplink low bit rate user transmitting with a common chip rate on a single modulated carrier using a different orthogonal multispreading factor code [126]. Such a system resembles the wideband CDMA schemes used in 3G systems for variable bit rate users. A different masking code would be used in each cell to separate interfering users in different cells. Uplink SC CDMA terminals would transmit continuously, or nearly continuously, in a common bandwidth, each with its own current bit rate. The PAPR would be low, and the average power would be proportional to the terminal's bit rate. Frequency-domain processing at the base station could perform the Rake receiver function or act as a minimum mean squared error (MMSE) filter which suppresses interference and exploits multipath [127–129]. High bit rate users would have low, or even unity, spreading gains [129].

Unfortunately, this approach appears unattractive for systems with large spreading gains, since the complexity of the adaptation/training process of either the MMSE time domain or frequency domain multiuser equalizer turns out to increase as the square or cube of the spreading gain [129], depending on the parameter estimation algorithm employed.

A promising approach to this problem is to use chip interleaving and frequency-domain orthogonal signature sequences (FDOSS) [130, 131]. An FDOSS-coded transmitted block of M data symbols, preceded by a CP, turns out to have a line spectrum, with M spectral lines at frequency intervals corresponding to the spreading gain K. As illustrated in Figure 7.8, up to K such uplink transmitted line spectra can be interleaved and remain disjoint. This means that the system can support up to K uplink FDOSS user signals,

Spacing between successive lines of a given user = K

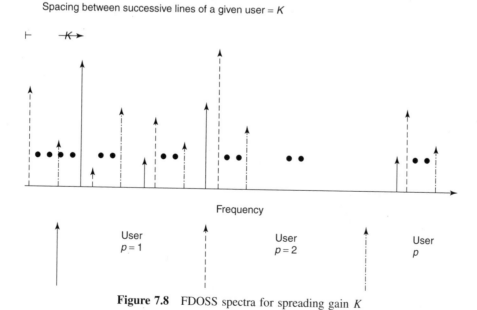

Frequency

User
$p = 1$

User
$p = 2$

User
p

Figure 7.8 FDOSS spectra for spreading gain K

which remain mutually orthogonal, independent of channel frequency selectivity and delay conditions. They are separated at the base station receiver by a simple decimation operation on the FFT of the received composite signal [132]. It is clear from this spectral interpretation of FDOSS signals that they in fact have attributes of both CDMA and FDMA signal formats. FDOSS can also accommodate multirate users and MIMO techniques. Issues being investigated include sensitivity to frequency errors and channel time variability.

FDOSS may also be regarded as an SC version (with low peak to average power ratio) of OFDMA. It may also be regarded as an SC version of the multicarrier SS-MC-MA scheme introduced by Kaiser [114]. A particular version of this idea, called 'OFDM-FDMA' was presented by Galda and Rohling [133]. Efficient implementation of such an FDMA system would probably involve transmitter signal processing and filtering in the frequency domain, using FFT and IFFT techniques.

7.2.5.4 MIMO Enhancements to Both OFDM and SC-FDE

Transmitter and Receiver Diversity
Receiver diversity combats fading, cochannel interference and multipath dispersion. It is equally applicable and effective for OFDM systems [134] and SC-FDE systems [92, 135, 136]. In combating cochannel interference, it facilitates SDMA and multiplies the spectral efficiency of wireless systems. Space–time coding with multiple transmitting antennas and single or multiple receiving antennas also applies equally well to time-domain or frequency-domain SC [117, 120, 135, 137, 138], or OFDM [134, 139, 140], and it offers substantial advantages in reliability and capacity. For example, a MIMO SC system's performance has been verified through simulations using field measurement data [141]. The SC MIMO system concept can easily be introduced into the mixed mode system, when the SC/MMSE turbo equalizer is in place, and hence it can be seen as an extension of the mixed mode air interface. Because of the signal processing similarities between SC-FDE and OFDM, implementations of single input multi output (SIMO), multi input single output (MISO) and MIMO structures are also very similar. Figures 7.9(a) and 7.9(b) show examples of SIMO and MIMO SC-FDE and OFDM block diagrams.

Further Research Topics
Channel Estimation and Adaptation to Time-Varying Channels
Efficient frequency-domain parameter estimation methods using training words and/or pilot tones are known for OFDM and SC-FDE systems [76, 80–82, 125, 142]. However, high bit rates and multipath delay spreads call for large FFT block lengths; there is then a question of the ability of receivers to estimate and track the resulting large number of frequency-domain parameters over rapidly time-varying channels. Investigations should include adaptation using training and blind adaptation, as well as the use of UW for channel tracking. For the OFDM downlink, if adaptive loading is used, the rate of updating and feeding back channel estimates, and their required accuracy, needs to be determined as a function of the rate of change of the channel. The use of a known UW sequence offers several enhancements to the concept of SC-FDE. Since the UW is known at the receiver it can effectively be used for tasks like synchronization (carrier frequency, clock frequency, etc.), channel estimation, and preventing the error propagation in the DFE [117].

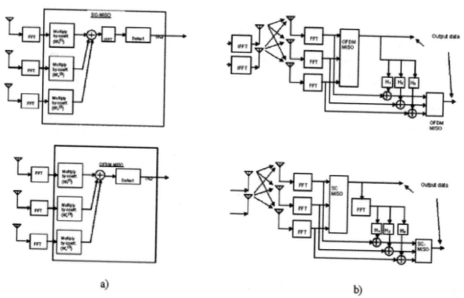

Figure 7.9 (a) Examples of SIMO systems: SC-FDE (top), OFDM (bottom); (b) Examples of MIMO V-BLAST systems: OFDM (top), SC-FDE (bottom)

MAC Layer Issues and Radio Resource Management

The impact of the proposed mixed mode on the MAC layer needs to be investigated. For example, feedback information must be exchanged for adaptive modulation, power control, synchronization, parameter control for coding, modulation and antennas. The MAC contained in the draft IEEE 802.16a broadband WMAN standard, which has OFDM, OFDMA and SC modes, may be a possible starting point model [73], especially in consideration of possible interoperability synergies between future mobile broadband systems and fixed broadband WMAN systems standardized by 802.16. The mobile systems' MAC should be modified to take into account the higher expected mobility and rate of channel change in future mobile broadband systems. The MAC should be as efficient and as simple as possible. For example, omitting the CP through the use of overlap-save processing means that the SC uplink is not necessarily constrained by a fixed block length structure; very short and efficient uplink MAC messages can then be accommodated.

While FDMA or CDMA uplink approaches appear attractive from the viewpoint of power and spectrum efficiency, the resulting constraints imposed on the MAC layer must be carefully assessed. Most broadband wireless system architectures, such as IEEE 802.16, HiperAccess, HiperMAN, DVB, IEEE 802.11 and HiperLAN, are based on sharing in time, rather than simultaneously in frequency or code. It may be possible to find ways of combining SC TDMA for high bit rate users with some version of CDMA or FDMA for low bit rate users. Interference and interference cancelling issues between the two types of multiple access scheme would then have to be addressed.

Resource allocation functions, distributed between uplink and downlink, is another issue: e.g. distribution of coded bits among downlink OFDM or OFDMA subcarriers, and adaptive modulation and code rate assignment in the SC uplink. The new mobile radio system should support unpaired spectrum by means of time division duplexing (TDD). The mobile radio system should also support multicast and broadcast services.

Operation in Unlicensed Environments with Intermittent, Uncontrolled Interference
With the integration and interoperability possibilities of cellular systems with wireless LANs and other unlicensed systems, interference is likely to be much less predictable and controlled than in pure cellular systems. Smart antennas [143] or the use of DS-CDMA or frequency hopping (FH)-CDMA are technologies which can mitigate self or external interference in uncontrolled interference environments. The performance and implementation trade-offs of these techniques should be studied, as well as other approaches such as transmission etiquette.

Mesh Systems
Adequate coverage and performance of very high bit rate future generation wireless systems will probably require mesh or multihop system architectures [66]. The impact of the mixed mode air interface on multihop systems should be evaluated. Multihop systems can use digital or analogue relaying, or combinations of these.

Detailed Performance Evaluation Over Realistic Fourth Generation (4G) Channel Models
Any proposed air interface must be evaluated against suitable channel models that are considered realistic for future generation wireless systems. These models are likely to be MIMO versions of 3G [69], WMAN [68] or WLAN models.

Interoperability
The interoperability between 4G systems and existing systems has been identified as a major issue. Interoperability of mixed mode systems with existing 3G, WLAN, WMAN [144, 145] and DVB standards should be explored. Some of these use single carrier, some use OFDM, and some include both as options.

7.2.6 Summary

The proposed mode of OFDM in the downlink and SC in the uplink, with SC-FDE, provides a beneficial mobile terminal cost/performance trade-off, compared to pure OFDM in both directions. The OFDM and SC-FDE technologies are well understood, as are their reduced signal processing complexities for channels with large delay spread, relative to classical time-domain equalization methods. The proposed mode is quite compatible with MIMO and space diversity, and also with uplink SC CDMA. This mixed OFDM/SC mode may offer the best combination of high performance and capacity with low terminal costs. Further research topics related to the mixed OFDM/SC mode have been suggested.

7.3 Smart Antennas and Related Technologies

The adoption of smart antenna techniques in future wireless systems is expected to have a significant impact on the efficient use of the spectrum, the minimization of the cost of establishing new wireless networks, on the optimization of the service quality provided by wireless networks and the realization of transparent operation across multitechnology wireless networks. Nevertheless, the design of future generation smart antenna systems involves a number of challenges, such as reconfigurability to varying scenarios in terms of propagation conditions, traffic models, mobility, transceiver architectures, mobile terminal resources (i.e. battery lifetime), QoS requirements for different services and interference conditions. The design of robust solutions matched to the reliability of the available channel information is required in order to account for the impact of channel estimation errors and feedback quantization and delay. The techniques addressing these challenges need to be assessed by adequate performance evaluation based on realistic modelling assumptions, both at the link and the system level. Moreover, in the design of smart antenna techniques, system architecture, implementation and complexity limitations need to be taken into account. Finally, the requirement for future generation systems operating in multitechnology networks introduces a further set of challenges associated with the design of smart antenna systems, which enhance the performance and facilitate the interoperability across different wireless technologies.

This section provides an overview of the most recent advances in smart antenna algorithms. System level performance optimization strategies, channel and interference modelling, realistic performance evaluation and implementation issues for the deployment of smart antennas in future systems will be presented.

7.3.1 Introduction

The development of a truly personal communications space will be based on next generation wireless communication systems, the dominant means of providing a whole new concept of mobility and new service paradigms. The adoption of smart antenna techniques, i.e., of multiple antenna processing schemes at the transmitting and/or the receiving side of the communication link, is indeed expected to have a significant impact on the efficient use of the spectrum, the minimization of the cost of establishing new wireless networks, the enhancement of the quality of service, and the realization of reconfigurable, robust and transparent operation across multitechnology wireless networks. To this end, current research efforts in the area are focusing on:

- The design and development of advanced and innovative multiple input multiple output (MIMO) processing algorithms that allow for increased network capacity, adaptivity to varying propagation conditions and robustness against network impairments.
- The design and development of innovative smart antenna strategies for optimization of performance at system level and transparent operation across different wireless systems and platforms.
- The realistic performance evaluation of the proposed MIMO algorithms and smart antenna strategies, based on the formulation of accurate channel and interference models, introduction of suitable performance metrics and simulation methodologies.
- The analysis of the implementation, complexity and cost efficiency issues involved in the realization of the proposed smart antenna techniques for future generation wireless systems.

The latest trends and future directions in the area of smart antennas are analyzed in Section 7.3.2, an overview is presented of how smart antennas can improve the performance of mobile communication systems. Section 7.3.3 addresses MIMO transceivers that support reconfigurability and robustness to varying propagation and traffic conditions. Smart antenna system level strategies are discussed in Section 7.3.4 by investigating novel radio resource management techniques based on reconfigurable scheduling schemes and cross layer optimization designs, as well as network optimization exploiting contextual information. MIMO channel modelling issues are summarized in Section 7.3.5, before strategies for a realistic performance evaluation of smart antenna systems are provided in Section 7.3.6. Implementation issues for the deployment of smart antennas are treated in Section 7.3.7. Finally, Section 7.3.8 summarizes the main conclusions.

7.3.2 Benefits of Smart Antennas

Smart antennas are generally used at the base station for uplink reception and downlink transmission with multiple antennas. They allow spatial access to the radio channel by means of different approaches, e.g., based on directional parameters or by exploiting the second order spatial statistics of the radio channel. Thus, space–time processing reduces interference and enhances the desired signal. Moreover, adaptive antennas can exploit long-term and/or short-term properties of the mobile radio channel to achieve improved channel estimation accuracy and reduced computational complexity [146]. Some antenna array processing techniques are classified in Figure 7.10. This classification is based on the propagation channel properties, i.e. on the structure of the spatial correlation matrix at the antenna array.

Spatial antenna processing is seen as a key technology for future high-performance 4G wireless systems. Future system designs are likely to embody a combination of beamforming, MIMO, interference cancellation and other related technologies in order to maximize system capacity and high data rate coverage across the network. To this end, we therefore propose the study of the intelligent use of multiple antennas at the base station and user terminal. Antenna array processing will provide the basis for improved system operation and performance, and result in a substantially lower cost per user within the network. The multiple antenna concepts should be targeted at scalable, tiered cell deployments, with the emphasis on providing high capacity, high data rate wide-area coverage to mobile and nomadic users. Optimized spatial processing concepts will be identified for both uplink and downlink aspects of the system design.

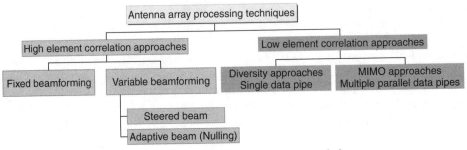

Figure 7.10 Taxonomy of smart antenna techniques

The gains achievable with multiple antenna systems on the transmit, as well as the receive, side can be classified as follows [147].

- The array (or beamforming) gain is the average increase in signal power at the receiver due to a coherent combination of the signals received at all antenna elements. It is proportional to the number of receive antennas. If channel knowledge is available at the transmitter, the array gain can also be exploited in systems with multiple antennas at the transmitter.
- The achievable diversity gain depends on the number of transmit and receive antennas, as well as the propagation channel characteristics, i.e. the number of independently fading branches (diversity order). The maximum diversity order of a flat fading MIMO channel is equal to the product of the number of receive and transmit antennas. Transmit diversity with multiple transmit antennas can, for instance, be exploited via space–time coding and does not require any channel knowledge at the transmitter.
- The interference reduction (or avoidance) gain can be achieved at the receiver and the transmitter by (spatially) suppressing other cochannel interferers. It requires an estimate of the channel of the desired user.
- The spatial multiplexing gain can be obtained by sending multiple data streams to a single user in a MIMO system or to multiple cochannel users in an SDMA (space division multiple access) system. These techniques take advantage of several independent spatial channels through which different data streams can be transmitted.

Multiple antennas at the base station and also at high performance terminals achieve significantly higher data rates, better link quality, and increased spectral efficiency [148]. Hence, more users can be accommodated by the system and a corresponding capacity increase achieved. To obtain very high data rates, multiple input multiple output (MIMO) processing techniques, such as spatial multiplexing and space–time coding, can be used [149]. Alternatively, MIMO techniques can be used in order to reduce the total transmitted power, while preserving the data rate. This in turn reduces the overall system interference. Overall, the adoption of MIMO processing techniques in future wireless systems is expected to have a significant impact on the efficient use of the spectrum, the minimization of the cost of establishing new wireless networks, the optimization of the service quality provided by those networks and the realization of transparent operation across multitechnology wireless networks.

Performance enhancements achieved by the use of multiple transmit and/or receive antennas in wireless telecommunications can be summarized as follows:

- Increase of channel (and hence system) capacity. Investigations show that in the case of uncorrelated Rayleigh fading, the channel capacity limit (Figure 7.11(a)) grows logarithmically (Figure 7.11(b)) when adding an antenna array at the receiver side (single input multiple output system – SIMO). In the MIMO case, it grows as much as linearly, with min (M_R, M_T), where M_R and M_T denote the number of antennas at the receiver and the transmitter, respectively (Figures 7.11(c) and 7.12).
- Decrease of the bit error rate (BER) without any bandwidth expansion or transmit power increase when receive (Rx) diversity and space–time linear processing or coding, i.e. transmit (Tx) diversity, are used jointly, or alternatively cell range expansion if performance is traded off for coverage.

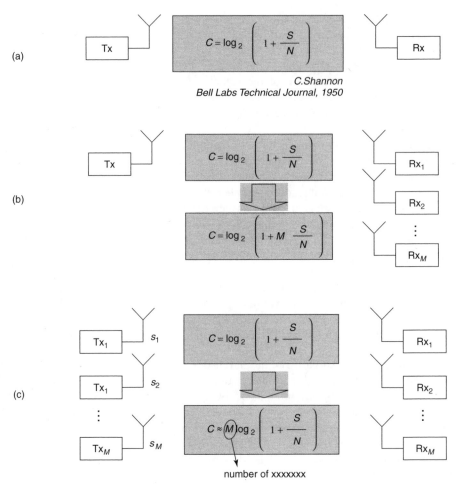

Figure 7.11   (a) Single input single output (SISO) capacity limit label; (b) adding an antenna array at the receiver (single input multiple output – SIMO) provides logarithmic growth of the bandwidth efficiency limit; (c) multiple input multiple output (MIMO) provides linear growth of the bandwidth efficiency limit

- Decrease of the impact of fading effects, this potentially leading to enhanced robustness and reliability in detected data.
- Decrease of total transmitted power/system interference for the same data rates.
- Increase of the packet call and cell throughput at the system level.
- Improvement of coverage for high data rates.

7.3.3 MIMO Transceivers

7.3.3.1 MIMO Transceiver Design

In a multiple transmit multiple receive antenna system such as the one illustrated in Figure 7.13, the data block to be transmitted is encoded and modulated to symbols of a

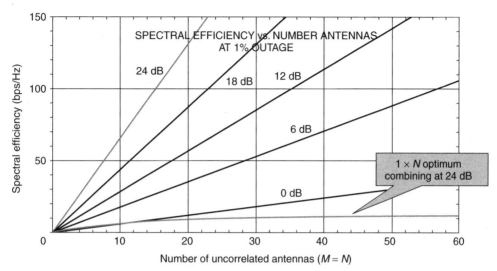

Figure 7.12   Spectral efficiency vs. number of antennas

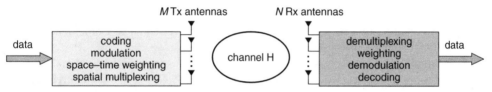

Figure 7.13   MIMO transceiver

complex constellation. Each symbol is then mapped to one of the transmit antennas (spatial multiplexing) after spatial weighting of the antenna elements or a linear space–time precoding technique is applied. After transmission through the wireless channel, at the receiver, demultiplexing, weighting, demodulation and decoding is performed in order to recover the transmitted data.

The transmission schemes over MIMO channels, designed to maximize spectral efficiency, typically fall into three categories, corresponding to the maximization of the following criteria:

- Diversity;
- Data rate; and
- SINR.

In the case of *maximization of diversity*, joint encoding – space–time coding–is applied and thereby the level of redundancy between transmit antennas is increased, as each antenna transmits a differently encoded fully redundant version of the same signal. Space–time codes were originally developed in the form of space–time trellis codes (STTC) [150], which required a multidimensional Viterbi algorithm for decoding at the receiver. These codes can provide diversity equal to the number of transmit antennas as well as coding gain, depending on the complexity of the code without loss in bandwidth efficiency. Space–time block codes (STBC) [151, 152] offer the same diversity as the STTCs but do not provide coding gain. However, STBCs are often the preferred solution against STTCs, as their decoding only requires linear processing.

The *maximization of data rate* is achieved by performing spatial multiplexing, i.e. by sending independent data streams over different transmit antennas. BLAST (Bell labs Layered Space–Time) technology takes advantage of several independent spatial channels through which different data streams can be transmitted [149, 153–156]. The receiver must demultiplex the spatial channels in order to detect the transmitted symbols. Various techniques have been used for this purpose, such as zero-forcing, which uses simple matrix inversion but gives poor results when the channel matrix is ill conditioned; MMSE, which is more robust in that sense but provides limited enhancement if knowledge of the noise/interference is not used; and maximum likelihood, which is optimal in the sense that it compares all possible combinations of symbols but can be too complex, especially for high-order modulation.

The *maximization of SINR* is achieved through focusing energy into the desired directions and minimizing energy towards all other directions. Beamforming [157] allows spatial access to the radio channel by means of different approaches, e.g., based on directional parameters or by exploiting the second order spatial statistics of the radio channel (Figure 7.14). Thus, space–time processing reduces interference and enhances the intended signal. Moreover, adaptive antennas can exploit long-term and/or short-term properties of the mobile radio channel to achieve improved channel estimation accuracy and reduced computational complexity.

These transmission strategies require efficient techniques to separate the signals of multiple users sharing the spectrum resources at the receiver, and to cancel interference under various interference scenarios [158, 159]. Some of them use pilot signals known by both the transmitter and receiver (non blind techniques), whereas other ones only employ *a priori* knowledge of the received signals (blind/semi-blind techniques). Another class of techniques assumes that the transmission channel is known at the transmitter, this knowledge being obtained either by feedback or by the TDD channel reciprocity assumption.

Depending on the multiple access technique, different receiver strategies have been proposed, from theoretically optimal strategies to practical ones like parallel or successive

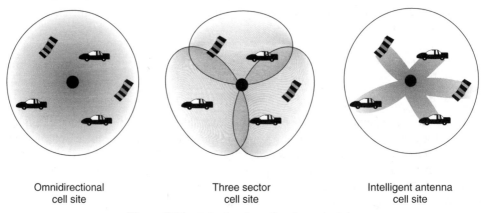

| Omnidirectional | Three sector | Intelligent antenna |
| cell site | cell site | cell site |

Figure 7.14 Adaptive beamforming principle

interference cancellation (PIC or SIC), decorrelation, and joint detection. Reduced complexity schemes may use turbo techniques for performance enhancement [160]. The channel estimation process can also be incorporated within the turbo mechanism, which can enhance the channel throughput [161].

Technologies using *a priori* channel knowledge at the Tx have been recently proposed, which allow for independent data streams to be sent and received without interference, relying on the computation of the channel matrix SVD [162, 163].

The key parameters to be taken into account when designing MIMO transceivers are:

- Channel knowledge: what are the assumptions (narrowband, frequency-selective, long-term properties, short-term properties, open loop or closed loop systems, feedback data rate in closed loop systems), how reliable is channel state information (CSI) and how robust are algorithms to potential CSI mismatches?
- Deployment environment: picocell, macrocell, etc.
- Fading correlation seen by the antenna elements on both transmit and receive antennas.
- Dispersion in the channel.
- Characteristics of the transmitted signal: modulation format, multiplexing of traffic and training information, etc.
- Sample size in number of observations with respect to the dimension of the array.
- Level of desired computational complexity and its partitioning between Tx and Rx.

7.3.3.2 Reconfigurability and Robustness

Wireless systems beyond 3G require signal processing techniques that would be capable of operating in a wide variety of scenarios [164–167] with respect to:

- propagation environment (indoor/outdoor, rich-scattering/specular etc.);
- traffic environment (hotspot/spotty, uniform/directional, dense/sparse etc.);
- interference environment (intra-cell/inter-cell, same system/other system etc.);
- user mobility (static/mobile users/users in bullet train, speed of interference etc.);
- antenna configuration (number of antennas at the terminal/base, antennas correlation/bandwidth, antenna characteristics etc.);
- radio access technology (single or multiple parallel technologies, frequency, modulation/waveform etc.);
- reliability, availability or unavailability of information on the channel prior to transmission (feedback channel, TDD vs. FDD duplex mode).

There are two main approaches to allow wireless communication transceivers to operate in a multiparametric, continuously changing environment:

- reconfigurable, adaptive techniques for adjusting the structure and parameters of the transceivers to allow them to demonstrate the best performance in a variety of situations;
- robust techniques, which can demonstrate reasonable (required) performance in a variety of unspecified situations.

The first approach assumes that the particular scenario can be identified, the optimal solution is known and the required transceiver configuration can be provided. MIMO receivers capable of reconfiguring themselves by switching automatically between 'beam-forming' and 'spatial multiplexing' can be considered as an example of the first approach [168]. It is currently known that there is a fundamental trade-off between spatial diversity and multiplexing in MIMO channels [169], but still there is no general approach to achieve this limit in a wide range of scenarios. Novel approaches have been proposed [164], which achieve reconfigurability to antenna correlation and channel state information reliability by employing a transmission scheme adaptive to channel conditions, as illustrated in Figure 7.15.

The second approach can be illustrated by 'short-burst' systems, which allow avoiding nonstationarity tracking. Other examples are MIMO and interference cancellation techniques based on semi-blind estimation algorithms without explicit estimation of the propagation channel for all signals received by an antenna array [158, 159]. A general problem of these techniques is the fact that they must operate with a low number of observations. These situations are especially unsuitable for traditional array processing tools, which are designed to give a good performance when the number of observations is high. General statistical analysis tools based on random matrix theory can be very useful in order to operate in low sample size situations.

The current vision of systems 'beyond 3G' (everything is connected everywhere at anytime with high frequency reuse and high mobility) suggests that complicated system/interference scenarios will be important, this cannot be reduced to a fixed set of

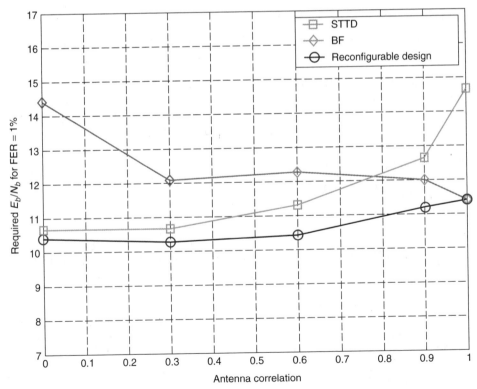

Figure 7.15 Reconfigurability to antenna correlation

situations with known optimal solutions. Thus, a combination of reconfigurable and robust techniques is recommended. In this context, algorithms able to gracefully adapt to varying degrees of knowledge of, for example, channel state information, are expected to achieve optimal performance.

Furthermore, user and environmental (contextual, location-based, etc.) information can be used to enhance MIMO processing performance. More specifically, location-related information, such as user position, velocity and orientation, surrounding propagation environment, interference parameters, traffic distribution, and region-specific statistics, can be used to centrally optimize multiple antenna wireless networks by making the best use of wireless resources, aiding user portability/transparency and network reconfigurability across heterogeneous networks and environments, and for a variety of services and applications, while preserving target QoS requirements.

The overall system performance can be enhanced further by finding ways for the upper layers in the OSI stack, e.g., resource management functions like scheduling, admission control and H-ARQ, to benefit from the array processing algorithms at the physical layer. Such a tighter interaction with upper layer functionalities would not only bring remarkable technical benefits but it could also entail further adoption of array processing schemes in future wireless communication systems. That is, a layer-isolated approach where no cross-layer interaction is envisaged could possibly prevent most of the physical layer innovations from being extensively recognized at the standardization level.

Overall, an integrated system design approach must be employed if one is to achieve the best overall level of system performance. This means that the antenna array processing techniques are developed in close collaboration with the PHY and MAC layer designs, rather than attempting to optimize the designs in isolation of one another.

7.3.4 Smart Antenna System-level Strategies

The overall system performance can be enhanced further by finding ways to account for the upper layers in the OSI stack. For instance, resource management functions like scheduling, admission control and H-ARQ can benefit from the array processing algorithms at the physical layer. In other words, an integrated system design approach must be employed if one is to achieve the best overall level of system performance. This means that the antenna array processing techniques are developed in close collaboration with designs at the physical, link (MAC, DLC, scheduling, etc.) and network layers (radio resource management, routing, transport, etc.), that is, in a cross-layer fashion rather than attempting to optimize the designs in isolation of one another. In subsequent paragraphs, several important issues to be addressed when dealing with cross-layer designs will be further detailed.

As for the information to be exchanged among those functionalities residing in different OSI layers, a three-fold classification can be established:

1. Channel state information (CSI), that is, estimates for channel impulse response, location information, vehicle speed, signal strength, interference level, interference modelling, etc;
2. QoS-related parameters, including delay, throughput, bit error rate (BER), packet error rate (PER) measurements, etc; and

3. PHY-layer resources made available in the corresponding node, such as multiuser reception capabilities, spatial processing schemes, number of antenna elements, battery depletion level, etc.

It is also important to carefully consider the optimization criteria. When applying MIMO techniques to wireless packet switched networks, it is common practice to select transmission parameters with the goal of maximizing channel capacity. However, the specific coding schemes being used, MAC strategies, scheduling policies, or even the performance of protocols in the upper layers of the protocol stack (such as TCP) also have a major impact on link quality. Thus, all these constituent blocks should alternatively be designed so as to attain the highest possible throughput (i.e. maximum number of data packets successfully delivered), instead of the highest data rate (capacity-maximizing approaches).

For delay-tolerant services, the combination of MIMO architectures, such as V-BLAST, with hybrid ARQ schemes (go-back N, selective repeat, stop & wait) turn out to be a fertile field for research. Cross-layer designs encompassing hybrid-ARQ methods (i.e. combinations of FEC with ARQ strategies) have also attracted considerable attention, in particular, type-II HARQ methods which address soft joint decoding of successive packet retransmissions. By undertaking cross-layer designs, novel strategies for mapping transmit data on specific streams, alternative methods to adjust frame size (as is the case for the HSDPA standard), or new antenna selection methods, either on the transmit or receive side, can be derived. Apart from that, the identification of methods aimed at dynamically adjusting rates and powers on transmit antennas is also needed.

In a multiuser context, the so-called opportunistic approaches have recently attracted considerable attention. In short, this is based on the idea of multiplexing users by granting the channel to those with higher chances of completing a successful transmission. In the end, an overall (rather than individual link) throughput maximization is pursued. For highly correlated space channels, opportunistic beamforming approaches will decide pointing at the user with the highest SNIR out of those present in the system. On the contrary, in rich-scattering scenarios where space channels exhibit low correlation (e.g. indoors), opportunistic approaches will implicitly exploit fading by granting access to those with highest instantaneous capacity (while possibly maintaining fairness among users, i.e. not causing excessive delay). This clearly departs from traditional strategies aimed at stabilizing individual links against channel (or interference) fluctuations by using multiple antennas. Moreover, artificial fading has to be induced for slowly time-varying channels on the transmit side by randomly changing transmit weights. More sophisticated multiple access can be derived by combining space and code diversities (e.g. V-BLAST with DS-SS). In this case, additional issues like efficient user grouping arise.

In summary, the approaches mentioned in the last paragraph go beyond the link level and jointly exploit multiuser diversity as a complement to code, time, frequency or space diversity. This clearly has an impact on the design of MAC protocols, which are forced to abandon the collision-avoiding paradigm (CSMA, Aloha, etc.) and evolve towards multiuser MAC schemes. Multiuser capabilities of the system will clearly depend on specific physical layer technique (optimum combining, transmit beamforming, SVD-based, etc.) this, again, reinforces the need for cross-layer designs.

In previous paragraphs, it has been shown how a tighter interaction of smart antenna techniques with upper layer functionalities can bring remarkable technical benefits. Furthermore, this could also entail further adoption of array processing schemes in future wireless communication systems. That is, a layer-isolated approach where no cross-layer interaction is envisaged could possibly prevent most of the physical layer innovations in the array-processing field from being extensively recognized at the standardization level.

7.3.5 MIMO Channel Modelling

A good understanding and a realistic modelling of the multipath propagation structure of the mobile communication channel are prerequisites for the design and realistic performance evaluation of smart antenna systems, in particular MIMO architectures. Based on MIMO channel sounder measurements, statistical channel models, as well as parametric channel models, can be developed.

The most common statistical model is the spatially uncorrelated Rayleigh fading model for the flat fading MIMO channel. It is commonly used as a benchmark in the performance evaluation of MIMO algorithms and models the non line-of-sight case. A line-of-sight component can be added to get Rician fading. Moreover, spatial correlations at the transmit and the receive array may be introduced by defining appropriate spatial correlation matrices [147]. In the frequency selective case, each tap in the matrix-valued FIR channel can be modelled as described for the flat fading case.

Parametric channel models, on the other hand, can be further divided into three groups [170]:

- *Full-wave models.* These models accurately reproduce the physics involved in the propagation by solving the Maxwell equations. Most methods discretize the space on a lattice and calculate the field values at each lattice point. In order to obtain useful models the lattice must be dense and the inhomogeneities present in the space must be fully described in their geometry and physical properties. This approach is associated with a significant computational complexity.
- *GTD-based models.* GTD (Geometrical Theory of Diffraction) postulates the existence of direct, diffracted and reflected rays only. As the wavelength approaches zero, this approximation becomes increasingly accurate. The possibility to interpret the propagation as a sum of diffracted and reflected rays simplifies greatly the computational effort required. Ray-tracing tools can identify the rays that actually reach the receiver. These tools still require an exhaustive description of the objects surrounding the antennas, and in the case where multiple reflections are considered, a fairly high computational complexity also has to be taken into account.
- *Measurements-based parametric models.* These models are directly derived from measurements and rely on parameter estimation techniques. From these parameters it is possible to extend the measurements to similar scenarios (for example, changing the antenna arrays) without having to repeat the measurement campaign.

7.3.6 Realistic Performance Evaluation

Realistic performance evaluation, both at the link and the system level, of the techniques described in Section 7.3.1 will be based on measurement data-based simulations as well as model-based simulations in different deployment scenarios (macrocellular, microcellular, in-building, into building, ad hoc, etc.).

The measurement data-based performance evaluations should rely on link-level simulations and system-level simulations. Both are offline simulations using multidimensional channel sounding field measurement data, but their objectives are different [171, 172]:

1. Link-level simulations provide an assessment of the implementation losses, and evaluate the ST-equalizer link performance in terms of bit error rates (BER) in various propagation environments. They focus on analysing/evaluating the impact of ST-processing and/or coding configurations, parameters and algorithms on performance.
2. System-level simulations provide an assessment of the overall multicell system performance, and evaluate performance metrics such as overall capacity, throughput, signal-to-interference plus noise power ratio (SINR) and BER in terms of either geographical distributions in the area of interest or outage probability [173]. The main focus is on interference scenarios, which depend on cell design, including frequency reuse and user distribution.

The efficiency of smart antenna techniques depends on the characteristics of a highly variable environment in terms of propagation, antenna array configuration, user behaviour, terminal resources, minimum QoS demands, reliability of channel state information (CSI), traffic patterns, service profiles, and interference scenarios. It is therefore critical to develop realistic spatio-temporal models for the characterization of the MIMO channel in a highly variable environment [170, 174] and of representative interference models for adequate evaluation of baseband signal processing techniques in multiple antenna, multiuser, multiservice, multitechnology radio networks. Furthermore, it is necessary to extend the operation of these models to enable their use in simulating multitechnology radio environments, including ad hoc networks, such that it is possible to utilize differing models from different systems in a consistent manner. Of particular importance are any time-correlated effects (e.g. shadowing), which need to be applied across the models in a coherent manner. Since frequency diversity is the traditional tool to combat these phenomena, it is necessary to develop hybrid technology processing schemes where spatial diversity processing is combined with frequency, time and code.

MIMO channel modelling is required in order to sufficiently assess the performance of smart antenna techniques at a link level, but Quality of Service evaluation in a multiuser, multiservice and multitechnology environment requires adequate modelling of the interference as well. An appropriate characterization of the spatio-temporal properties of these issues is important, not only to obtain realistic simulation results, but also to establish a theoretical framework describing the link level behaviour from an analytical perspective. In this sense, new mathematical tools such as random matrix theory could be the key to a wider understanding of the MIMO channel characteristics [175, 176]. These novel techniques provide a very simple way of analysing and studying the performance of MIMO channels with fading correlation at the transmitter or the receiver. Despite the fact that these models are asymptotic by nature, a good agreement with the actual (nonasymptotic) channel response is observed in practice.

As for the characterization of the global performance of the communications link, system level simulators have usually been employed. These simulators generate certain traffic patterns within a network of cells and statistically measure interference in terms of signal-to-interference ratio (SIR).

The realistic performance evaluation of smart antenna techniques will therefore rely on the following modelling issues [177–179]:

- MIMO channel modelling based on statistical formulation of the scatterers' distribution, and real-time MIMO measurement campaigns followed by link-level simulations for a variety of antenna array configurations, air interfaces, user mobility patterns and service profiles.
- Realistic interference modelling based on system level simulations taking into consideration the intra- and inter-cell impact of smart antenna techniques, nonuniformity of traffic (e.g. hot spots), mixed services scenarios and interoperability between different air interfaces [165].
- Accurate mapping between link- and system-level results based on more realistic interference models [162, 177, 180, 181].
- Realistic modelling of implementation losses such as channel estimation errors, feedback quantization and delay etc.

7.3.7 Deployment of Smart Antennas in Future Systems – Implementation Issues

According to recent studies, smart antenna technology is now deployed in one of every ten base stations in the world. The study predicts that the deployment of smart antenna systems will grow by a factor of two in the next five years. It was shown in the same study that smart antenna technology has been successfully implemented for as little as 30 % more cost than a similar base station without the technology. However, implementation costs can vary considerably and cost-effective implementation is still the major challenge in the field.

The application of sophisticated smart antenna processing techniques in future systems will impose stringent demands upon both base station and terminal implementations, and this increased functionality will also be expected to be achieved at low cost. It is therefore important that the viability of the proposed smart antenna processing techniques is addressed, and novel technologies applied wherever appropriate.

At the base station, of particular importance is the development of improved antenna structures (possibly employing MEMS technology, e.g., microswitches, or left-handed materials), improved cabling structures, and efficient low cost RF/DSP architectures.

At the terminal, the application of smart antenna techniques can have a significant impact, not only in terms of system performance, but also in terms of cost and the terminal's physical size. It is therefore important to examine the viability of such terminals, which are likely to be both multimode and multiband in nature, and available in a wide variety of forms. Particular areas that need to be examined are: efficient MIMO/diversity antenna system designs, small, low-power RF structures, use of RF combining techniques [182], and viable low power digital signal processing implementations. Terminal cost requirements may lead to use of nonperfect RF/analogue components and DSP algorithms for compensating should be investigated.

In addition, for both the base station and the terminal, an important consideration is how it may be possible effectively to develop the antenna structures, RF architectures, and DSP implementations to enable them to operate efficiently within a wide variety of air interface scenarios, both separately and in parallel. To this end, innovative development flow methodologies jointly covering the RF and the baseband parts of a complex wireless system on chip should be studied. This methodology effort will allow a short time development of integrated smart antenna systems starting from high-level algorithm descriptions and going up to efficient implementation optimized for the scenarios through reusability of IP blocks. This should allow advantage to be taken of a high integration level of the latest digital technology to improve 4G terminals in terms of complexity, cost and power consumption, thus allowing a mass introduction of new smart devices. In general, the implementation feasibility of different concepts and algorithms should be evaluated, considering analogue and digital component technology that evolves with time and finally leads to a 'feasibility roadmap', which indicates the viability of different concepts over time.

A key output of this area of study is an understanding of the base technologies that are required to make the future use of smart antennas viable. The financial impact of the deployment of smart antenna technologies in future wireless systems has been studied [183] for the cases of CDMA2000 in the United States and UMTS in Europe. The results of this study showed that 'smart antenna techniques are key in securing the financial viability of the operator's business, while at the same time allowing for unit price elasticity and positive net present value. They are hence crucial for operators that want to create demand for high data usage and/or gain high market share'. Based on this type of analysis, technology roadmaps, along with their associated risks, can be concluded, which will enable appropriate technology intercept points to be determined, resulting in the development of technologies appropriate for each application area.

7.3.8 Conclusions

The adoption of smart antenna technologies in future wireless systems will have a critical impact on the achievement of higher data rates in a number of challenging scenarios (varying propagation conditions, multiservices, multitechnology networks) and it is expected to play a significant role in the realization of end-to-end reconfigurability and transparent operation across networks of different technologies. The array processing techniques have to be taken into account in the original system design to provide the basis for scalable, high data rates, as well as high capacity system solutions, by achieving an optimum combination of array gain, diversity gain, interference reduction gain and spatial multiplexing gain on the receive, as well as the transmit, side.

7.4 Short-range Wireless Networks

7.4.1 Motivation

Wireless Local Area Networks (WLANs) are nowadays increasingly perceived as a natural complement to the cellular third generation (3G) mobile communication system. This can be interpreted as the fact that it will be unavoidable in the near future to witness a superposition of radio access networks of various sizes and topologies: macrocellular systems (3G), microcellular systems (WLANs) and picocellular systems, such as Wireless Personal Area Networks (WPANs) or even Wireless Body Area Networks (WBANs). This somehow reflects and mimics the overlap of the body area/personal area/local area/wide area radio spheres of human interactions. Using this approach and similarities between how users communicate, it is now clear that no radio interface for the next generation mobile communication system can be considered without capturing inherently the WBAN/WPAN/WLAN worlds.

Moreover, for ease of deployment and economy of scale, the next technical challenge is to design a comprehensive fourth generation (4G) system that would be in a position to fulfil the user requirements for all the possible environments the user will face. Each of these being characterized by performance/mobility/coverage/QoS/topology constraints such as, for instance, a wide area with vehicular speed, or hot spots, home and enterprise environments with at most pedestrian speed.

It is clear that this cannot be realized without a high degree of flexibility and adaptivity, so that a variety of wireless access modes can be integrated into a common structure, or even a common system concept, that encompasses next generation cellular mobile systems, broadband wireless LAN and inexpensive wireless PAN/BAN. These wireless access modes will complement each other with respect to services and radio environments, allow seamless interworking, and provide access to an IP-based backbone network enhanced with QoS and mobility management support. In this respect, it is foreseen that WLAN/WPAN/WBAN will be an essential part of the overall 4G landscape by providing radio access capabilities that complement and strengthen those of cellular networks, so that throughput and coverage can be enhanced in areas where short-range scenarios are the most relevant, especially for the home, enterprise and very dense urban environments.

In a future wireless communication environment there is another type of short-range communication system that will play a significant role: Wireless Sensor Networks (WSNs) [184]. WSNs are composed of hundreds to thousands of tiny low-cost, low-power smart sensor nodes, which can cooperate among themselves by dynamically establishing short-range communication links between neighbouring nodes in order to forward raw or aggregated sensor data to a task manager node. Efficient integration of this type of network into a global mobile communication system is a long-term research goal (see Figure 7.16).

Given the importance of short-range communication systems in the overall landscape of wireless communication networks, as highlighted above, the intent of this text is to present key technologies to be explored in order to fulfil 'needs' in terms of wireless connection

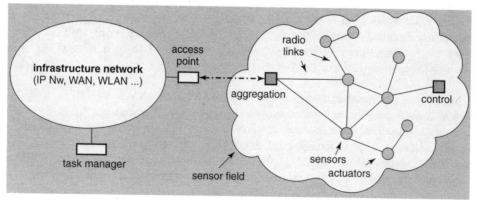

Figure 7.16   Wireless sensor network connected to an IP-based core network

capabilities, as well as the most up-to-date research results in the area. Thus, in a first step, we give an overview of the short-range communication world covered here, i.e. including WLANs, WPANs, WBANs and WSNs. Typical user deployment and scenarios relevant for short-range communication networks are presented, relevant spectrum bands and propagation aspects are described and finally, as a result, technical requirements derived from user cases and propagation constraints are listed.

In a second step, the technologies identified as key in the development of short-range communication systems are detailed, both at PHY (physical) and MAC (medium access control) layers, and the dedicated research tasks that need to be performed in order to exploit to the maximum extent what this environment can offer are listed. In addition, research results highlight the potential of the approaches selected.

Finally, to conclude, standardization aspects are tackled in Section 7.4.4.

7.4.2 Requirements and Constraints for Short-range Wireless Network Deployment

7.4.2.1 User Deployment Scenarios and Potential Application Space of WLAN/WPAN/WBAN/WSN Technologies

The main advantage of short-range communication in environments where mobility is restricted to pedestrian speed is the enabling of higher cell throughput in a restricted area. However, other advantages can be found in different trade-offs, such as achieving very low power consumption for lower data rates for instance. Thus, it is easily understood that this technology is very well suited for three types of deployment scenario: microradio, picoradio limited to body neighbourhood, and hot spots.

Microradio: Private Indoor Operations (WLANs WPANs)

- *Enterprise environment.* WLANs are already well integrated in enterprise, where wired Ethernet has met its replacement. The workers are no longer attached to their desks and a new world of interactions and way of working has now been opened, thanks to WLANs. However, we are far from reaching the 100baseT, or even 1000baseT, rates required to completely get rid of wires, and current systems did not integrate enough quality of service (QoS) for providing the wireless phone capabilities using Voice over IP (VoIP) that are desirable in enterprise.
- *Home environment.* The home is still a white space to be conquered. The requirements in terms of quality of service for multimedia consumer services (VoIP, MPEG4 streaming, etc.) demand more technology developments in order to revolutionize the home consumer landscape.

Picoradio and Microradio: Connecting Wireless Sensors (WSNs)

Typical usage scenarios of wireless sensor networks are characterized by the presence of numerous small low-cost, battery-powered sensing and computing devices, which are densely scattered over a geographical area. These sensors are equipped with a radio transceiver, which enables them to communicate and collaborate with each other. Wireless sensors are operated in a pico- and microradio environment within a growing number of diverse civil and military applications. The commercial application space covers areas such as industrial process control, home automation, automotive networks, and personal health care. Remote metering, managing of inventory, monitoring of product quality, and surveillance of premises for providing security are well known classical applications which can use either wired or wireless sensors. However, there are also applications on the horizon which are enabled by integrating a radio transceiver into sensors and supporting self-configuration in wireless sensor networks, for instance, for collecting, processing and disseminating wide ranges of complex environmental data for detecting environmental hazards and monitoring remote terrain and disaster areas. The potential future markets for WSNs will thus include transportation and shipping, fire fighting and rescue operations, home automation, and even interactive toys.

Picoradio: Connecting Electronic Devices Worn on the Human Body (WBANs)

- *Electronic health assistance.* In their main user application, WBANs provide a basic infrastructure element for control and monitoring of medical sensors and actors (e.g. implants). A wide range of diseases could be prevented, treated and managed more effectively today – for the benefit of both patients and the health system – using wireless multiparameter monitoring technology that requires a reliable, low-power wireless communication network for linking several sensors together to a central monitoring station like, for example, a watch or PDA.
- *Context-aware assistance.* Using motion sensors or localization systems, movements of a person can be observed, leading to automatic reactions of devices in the WBAN. For example, it is possible to exchange business cards automatically if sensors observe a handshake, or an alert can be given if a craftsman forgets a work step. Higher data rates will be necessary if multimedia devices like cellular phones, mp3 players or video glasses join a WBAN. With such devices, a speech controlled navigation system can be integrated into the WBAN, the map being displayed on the video glasses.

Hot Spots: Offering Fast Download Public Access in an Operator Friendly Way
WLANs are already deployed in very dense urban environments for providing fast public download capabilities. Due to the absence of regulation and unlicensed nature of the bands where WLANs operate, the unexpected emergence of these new services is shaking the conventional centralized way of distributing data in the outdoor environment. This new capacity will seed new needs and business opportunities. Even though this scenario is at an early experimental stage, projections of throughput needs, if it becomes a success, are tremendous: no existing technology would cover this demand, and the design of more advanced air interfaces able to cope with these new requirements is required.

7.4.2.2 Spectrum Landing Zones and Propagation Constraints
Spectrum Usage Approaches
Extensions of short-range communication standards need to be envisaged first in light of available spectrum resources. Without enumerating all the possible strategies in that field it is still worth mentioning the following approaches.

Conventional Spectrum Allocation:

- 2.4/5 GHz bands are the frequency bands already allocated for WLAN operation. They are used by IEEE 802.11x, ETSI HIPERLAN/2 and ARIB MMAC standards, and are considered as landing zones for providing extended throughput/coverage, for instance by the newly created IEEE 802.11n Task Group for High Throughput WLAN.

- The 17 GHz band was assigned by CEPT (CEPT/T/R 22-06 [185]) for very high-speed wireless LAN use on a nonprotected and noninterference basis. Besides that, CEPT [185] also recommended, among others options, the use of 17.1–17.3 GHz for short-range wireless connectivity. A contiguous extension of this available spectrum was proposed by the ITU study group JRG 8A-9B [186], for further extension from 17.3 to 17.7 GHz. For USA and Japan, similar bands are generically allocated for radio communications. Given the clear need for wireless systems to move to upper carrier frequency ranges in order to offer additional capacity and higher spatial efficiency, and the fact that 17 GHz allocation does not provide any harmful interference to other wireless systems, it is expected that FCC and global regulatory approval of this band will not be an issue, assuming industry support of such bands.

- Or the 60 GHz band, a tremendous amount of spectrum is already allocated worldwide (3 GHz!) for wireless broadband communications. The inherent Doppler spread present due to the high carrier frequency restricts the usage of this band to fixed or low mobility applications. The Japanese standardization body (ARIB) is ahead in this domain, even though 60 GHz has already been partially addressed at the IST level through the BroadWay project [187].

Break Away Approach

Ultra Wide Band (UWB) communication systems can be superimposed on the frequency axis with conventional narrower band systems through frequency spreading of transmitted signals and appropriate receiver processing, thus introducing a new way to use and share the spectrum. The potential of UWB technologies for short-range communication is addressed elsewhere in this book, and in [188]. It is worth noting that UWB seems to have the potential to support the requirements related to the WPAN (low and high data rates) and WBAN 'picoradio' scenarios listed subsequently.

Propagation Characteristics and Channel Models

In order to assess properly the various technologies that will be proposed for addressing the considered scenarios, an important task that is required is the propagation characterization at the carrier frequency selected. Many propagation studies have already been conducted in the bands allocated to WLAN operation (see for instance [189–191]); however, the introduction of new system parameters (bandwidth, antenna configuration, multi-antenna set-up, mobility, etc.) requires extending them accordingly. Also, depending on the type of analysis conducted (link, system or MAC level), adapted channel models need to be derived. In addition, new scenarios, such as the picoradio one in the framework of WBANs, require specific attention, and the 'human body' channel characterization is a new field for propagation studies that has just started to be addressed. First BAN channel measurements in the frequency range from 3–6 GHz can be found in [192].

Tap-delay line models, widely used in the 5 GHz WLAN area for single antenna link-layer simulations, are based on [193]. The indoor environment, rich in scattering and multipath, is very well suited for multi-antenna approaches that can exploit the spatial diversity. However, there is no widely accepted 5 GHz multi-antenna channel model yet. Several models have been developed in European projects, and the IEEE 802.11n standardization body is currently working on the definition of a multi-antenna geometric channel model (see [194] for latest progress) that would serve as a basis for contribution comparisons.

When using a low power transmission, which is an absolute requirement for devices operating in a wireless LAN system, the wall-confined propagation in the 17 GHz band allows very dense frequency reuse and increased spatial efficiency over covered areas, and provides intrinsic security in both office/house neighbourhoods. The impact of 17 GHz indoor channel models on OFDM transceiver performance is analysed in [195] At 60 GHz, the propagation losses are even more severe; channel models are available from several European projects (see for instance the IST BroadWay project [187] deliverable WP3-D7).

7.4.2.3 Technical Requirements for WLAN/WPAN/WBAN/WSN Communication Systems

WLANs

Next generation WLANs are expected to meet very challenging requirements. Their exact nature, as well as the extent to which they will have to be met, is determined by the scenario (e.g. home, enterprise or hot spot environment) in which the wireless access system will be used. In general, they are characterized by a low mobility vs. high throughput trade-off, supporting a high percentage of static communications.

The home environment offers a large number of applications for which WLANs are useful, and the WLAN requirements are thus very demanding. They have to:

- enable communication among several devices, playing the role of an Ethernet connection;
- provide data rates that may vary from 100 kbps (e.g. sending a file to the printer) to 100 Mbps (e.g. video streaming) depending on the application;
- use low transmitting power levels, both to limit human exposure to electromagnetic radiation and to reduce power consumption;
- coexist with other devices, in particular WPANs, since the home environment is prone to the proliferation of wireless devices, not always based on the same technology.

The interest of enterprises in acquiring WLANs is, to a large extent, due to the simple way changes and moves can be dealt with. In the corporate environment, next generation WLANs will have to:

- provide peak data rates of at least 100 Mbps in order to be attractive, as opposed to 1 to 10 Gigabit Ethernet;
- ensure full QoS support to guarantee other service attributes, such as throughput, PER, latency and jitter, for time-sensitive applications such as real-time conferencing;
- consider backward compatibility since it justifies a gradual migration to next generation without requiring the entire investment to be concentrated at one point in time.

A key driver in the development of public-area access has been, and remains, the corporate requirement to increase workforce mobility and productivity. As a consequence, as far as hot spots are concerned, WLANs should be able to provide the applications that are demanded by business travellers, which translates into the following requirements:

- to enable high-speed access to the Internet, thus allowing access to the company's intranet, as well as providing lower rate modes with wider cell range;

- to fulfil strict QoS requirements, and also QoS differentiation;
- to enable adjacent, uncoordinated, network deployment with interference control;
- to address the mobility or nomadicity objective more consistently, in particular by working on handoff capabilities;
- to introduce new functions for secure authentication and accounting through the network.

WPANs

A wireless personal area network is a wireless short-range communication system specifically targeted to provide very low power consumption, low complexity and wireless connectivity among devices within a personal operating space (POS). In this context, a WPAN-enabled device should:

- interoperate with other devices without user intervention;
- coexist with other technologies in the same space (e.g. WLANs);
- provide low or high data rates, depending on the application, as well as QoS support as required by multimedia applications.

WSNs and WBANs

User requirements for WSNs and WBANs translate into unique technical specifications and constraints, which need to be taken into account when selecting adequate technologies for radio interface design. WBANs can be regarded as a special case of a heterogeneous WSN, in which the closely positioned nodes range from very simple, extremely energy-efficient and low-cost simplex nodes that can only sense and transmit, to nodes that have high computational capabilities. Thus, technical requirements for WSNs and WBANs are compared in Table 7.1.

A comparison of the requirements for noninvasive WBANs and other short-range systems is shown in Figure 7.17.

7.4.3 Technology Trends to be Explored

7.4.3.1 Baseband Technologies for Short-range Communication

The fulfilment of the technical requirements listed below relies on new air interface designs. New scientific developments enable one to challenge existing limits provided by the specified standards, and the most promising ones are listed in this section.

Technologies Exploiting Signal Processing Principles
Waveform Formatting at the Transmitter
New Modulation Paradigms

Most recent developments in short-range communication systems design rely on a multicarrier modulation called OFDM (orthogonal frequency division multiplexing). Several recent studies aim to renew this now classical OFDM approach, and the first direction for further study is the combination of the strength of multicarrier OFDM with CDMA: OFDM-CDMA, single carrier OFDM, etc.

For instance, [196] presents a mixed OFDM plus single carrier mode air interface. The idea is to preserve the smaller peak to average ratio of single carrier systems, while

Table 7.1 Comparison between WBAN and WSN distinctive features

	WSN features	WBAN features
Data rates	Low data rates (tens of kbps)	Modest data rate requirements regarding medical applications, higher rates for context-aware applications
Network topology	Unlimited number of network nodes (from hundreds to millions), scattered over any distance within the short-range sphere (less than 100 metres): • Scalable protocols are required, to support the network size and density, in sharp contrast to what can be encountered in traditional networks	Dense distribution of a large number of network nodes but limited by the size of the body (Maximal distance between nodes is about 2 m)
	Random placement of nodes, no predetermined position of the master node(s); positions may change frequently: • Protocols and algorithms providing self-organizing and self-initializing capabilities are necessary	Deterministic placement of nodes possible, the shape of the human determining the network layout, thus enabling partial *a priori* channel knowledge. Network topology also characterized by simultaneous birth and death of larger batch of nodes (e.g. when putting off/on a jacket). In general, only one master node is considered
Power consumption	Autonomous operation (access to sensors may be considerably costly or dangerous), and thus multimonth to multiyear battery life. • low-power transceivers and power-aware MAC protocol have to be developed so as to reduce the power consumption incurred by radio transmission. It is also crucial to reduce the messaging overhead and the link set-up delay, while allowing a node to spend a large fraction of time in sleep mode	Batteries and nodes can usually be replaced, but power constraints are still crucial, especially for medical devices like implants (small batteries, multimonth to multiyear battery life)
Electromagnetic exposure	No *a priori* constraint on electromagnetic exposure	Electromagnetic exposure to the human body has to be very low, thus constraining the maximum transmit power to be very low, and favouring multihop connections from one side of the body to the other (noninvasive WBAN). Such noninvasive WBANs cause only low interference to other networks. • Efficient support of extremely low-power simplex nodes is required
Propagation	Position of obstacles is unknown	The body is always an obstacle in at least one direction

(continued overleaf)

Table 7.1 (*continued*)

	WSN features	WBAN features
Cost	Small nodes cost for profitable network deployment, thus implying severe constraints on the hardware (and especially radio transceiver) complexity	
Location	Position location capability to be used for sensing and routing tasks	
Coexistence	Very diverse situations	Coexistence with other BAN and longer range wireless systems (such as Bluetooth, 802.11a/b, etc.)

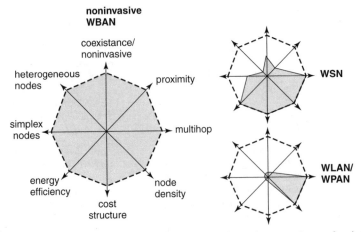

Figure 7.17 Requirements for noninvasive WBANs (left) and comparison of existing wireless technologies (right)

still performing a low-cost (for channel delay spread exceeding about ten data symbols) frequency-domain equalization on the receiver side.

A research theme thus consists of extending and enhancing the concept of mixed OFDM plus a single carrier air interface [196] to deal with typical user deployment scenarios of WLAN/WPAN/WBAN/WSN technologies.

The cost/performance trade-off of downlink OFDM, uplink single carrier frequency-domain equalization (SC-FDE) offers the possibility of having low-cost, minimal power consumption (and reduced power amplifier backoff) subscriber terminals. The potential of this complexity trade-off needs to be revisited when considering peer-to-peer, terminal-to-terminal connections, typical in short range networks. In addition, the mixed mode approach allows a seamless migration path from existing 3G, wireless metropolitan area networks (WMANs) and WLANs to the next generation. The IEEE 802.16a WMAN standard has both modes of SC-FDE and OFDM, while 802.11a/g use OFDM. FDE capabilities are mainly a receiver-based technology, that allows upgrading of existing wireless protocols through implementing extra signal processing at network nodes.

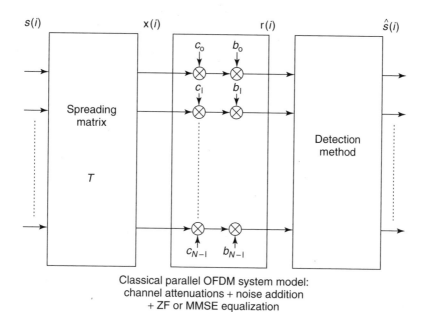

Classical parallel OFDM system model:
channel attenuations + noise addition
+ ZF or MMSE equalization

Figure 7.18 Block diagram of a linearly precoded OFDM scheme, with diagonal OFDM model

This mixed mode approach has been generalized with linearly precoded OFDM schemes, depicted in Figure 7.18. Performance enhancements are discussed in [197] depending on various nonlinear decoders [198–200]. Possible complementary paths to be explored are the design of a smarter guard time that would enhance the robustness of the modulation scheme against multipath propagation compared to the usual cyclic prefix (granting channel invertibility in the absence of noise, irrespective of channel nulls location). Based on the recent zero-padded OFDM transmission scheme introduced in [201, 202], a pseudorandom postfix modulator, as depicted in Figure 7.19, was shown [203] to be very attractive in terms of robustness against channel variations. New extensions have been derived in the case of multi-antenna transmission schemes.

Other potential improvements could be brought about by the use of differential modulation schemes, shortening the receiver start-up time and simplifying equalization (see, for instance, [204]). In addition, current short-range radio interfaces rely on several modulation and coding schemes and adaptively select one for transmission. Adaptive modulation and coding is thus an important feature that is further discussed in the section below.

Use of Channel State Information
The link performance can be improved by applying transmission techniques that require feedback from the receiver on the monitored state of the radio channel. These transmission techniques are adaptive rate and power control and adaptive bit loading (see, for instance, [205, 206]).

Adaptive bit loading is the process of modulating a different number of bits on each OFDM subcarrier based on the monitored SNR of each subcarrier. Here, the channel state information enables one to exploit opportunistic approaches: instead of struggling

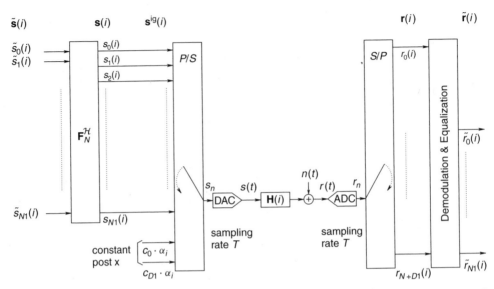

Figure 7.19 Blockw diagram of a constant postfix OFDM modulator (for zero-padded OFDM, $\alpha = 0$)

for a constant data rate for any instantaneous channel and interference condition, one should benefit from the time-varying nature of radio propagation by transmitting at high rates when conditions allow, and by reducing transmission speed when conditions suggest doing so. It has been shown that opportunistic transmission may provide gains of up to 150% with respect to other schemes that do not take into account the channel [207].

An interesting related field of research is the exploitation of the reciprocity characteristic of the time division duplex (TDD) propagation channel that enables one to derive some knowledge of the transmission channel from receiver measurements. Indeed, WLAN standards, such as IEEE 802.11a, rely on a TDD-TDMA access scheme and could benefit from such techniques. In addition, multi-antenna techniques, relying on *a priori* channel information, have demonstrated their strengths and are being implemented in 3G. These should be studied for short-range communication purposes as well.

New Multiple Access Schemes

Besides the well known multiple access schemes such as TDMA, FDMA and CDMA, advanced techniques can be employed to provide radio channel access for several users. These schemes are orthogonal frequency division multiple access (OFDMA), multicarrier CDMA (MC-CDMA), and spatial division multiple access (SDMA). They have been known and studied for some time now [208–210], but conclusions and effective combinations are still to be found in the specific context of short-range communications. The selection of multiaccess techniques also has an impact on interference scenarios that need to be carefully evaluated in order to design appropriate interference reduction/cancellation approaches whenever necessary.

Spatial Processing

Approaches Exploiting Microdiversity

New spatial processing techniques exploiting spatial diversity or spatial multiplexing have been shown to improve link reliability or error rate, to increase network capacity, cell throughput or peak data rate, and to extend the cell coverage range, so that they are now considered as key in new air interface designs.

Diversity techniques are known as a powerful approach to combat fading effects, or even to turn them into benefits, by providing to the receiver a number of uncorrelated (or weakly correlated) replicas of the transmitted signal. In the indoor 5 GHz WLAN context, the propagation environment appears to be sufficiently rich in scattering effects to guarantee that significant improvements can be obtained thanks to spatial diversity. Among the well known diversity techniques are space–time coding methods that are, for instance, compared in [211] in the WLAN context, highlighting the robustness of space–time block coding solutions for two transmit antennas over space–time trellis codes. Cyclic delay diversity is an alternative to space–time coding that provides a flexible and simple transmit diversity solution for WLANs [212]. Spatial multiplexing leads to an increase in capacity by splitting the stream to be transmitted into substreams that are transmitted simultaneously in the same frequency band over the scattering radio channel using multiple transmit and receive antennas. The related processing relies on multiple input multiple output (MIMO) signal processing methods. Zhuang *et al.* [213] compare MIMO solutions with space–time block coding ones, and propose a hybrid STBC/MIMO design providing a good trade-off between robustness to fading and capacity.

Such approaches have been extensively studied in recent years, and the purpose here is not to detail the wide range of techniques devised for short-range communication interfaces, and more specifically for OFDM-based systems. In the multicarrier context, the three dimensions of time/frequency/space can be exploited simultaneously, and space–frequency (SF) or space–time–frequency (STF) codes, such as the ones reported in [214], have to be designed.

In general, the reader should also refer to [215] for an extensive overview on smart antenna techniques. Note, simply, that substantial gains in overall network capacity using such techniques have also been reported [216]. This underlines the necessity to take into account the multi-antenna feature at all levels of the radio interface design, including the MAC layer protocols.

Beamforming

In addition to MIMO/diversity techniques, more conventional spatial processing techniques, such as fixed and adaptive beamforming, can be applied at both access point and terminal to provide significant performance improvements in wireless systems. The adaptivity inherent within these techniques is particularly suited to WLAN systems, where users are mobile and infrastructure deployment is unplanned. The ability of spatial processing techniques to provide coverage improvements will also help to reduce the costs of infrastructure in large buildings. Thus, several start-ups are already proposing WLAN products implementing such technologies.

A practical study considering the application of a switched sector circular array to WLANs is given in [217], which concludes that two users can be employed in a single time slot for 54 % of the time when a switched sector circular array is employed.

In some proposed techniques, beamforming is associated to transmit diversity for improved detection reliability [218]. Feasibility of the implementation and antenna configurations are major issues in this context and will be treated according to the scenario considered.

On the other hand, beamforming has also been deployed at the receiver in WLANs, where its use is standard compliant [219]. The authors proposed several alternatives for the spatial processing of the signal at the receiver in both domains of OFDM, namely time and frequency. Substantial gain was achieved in the temporal domain.

Virtual Antenna Arrays
In scenarios where spatial limitations preclude the use of physical antenna arrays, virtual antenna arrays can be spanned by cooperating terminals. This method has been shown to provide diversity gains when other forms of diversity are not available.

In a virtual antenna array (VAA), a transmitter sends a data stream to a receiver via a given number of relaying nodes, which perform distributed space–time encoding at a single relaying stage. VAAs are hence a subset of distributed communication systems which are introduced in [220]. The deployment of VAAs in xAN (i.e. WLAN, WPAN, WBAN) would greatly enhance the achievable data rates, reduce delays and improve power efficiency. Thus, into, inter and intra xAN communication scenarios are of interest. Related design challenges include synchronization, scheduling and resource allocation, as well as low duty cycle feedback mechanisms for communicating suitable decision metrics.

The deployment of VAAs creates various challenges that need to be addressed. A generic system-independent problem arising is the ability of the terminals to relay simultaneously and thus to operate in full duplex mode. The duplex communication problem is simply solved by assuming that the frequency bands for the main link and the relaying links differ. There is a related strong constraint on the radio frequency (RF) chain of the terminals, which needs to ensure an efficient separation between transmitting and receiving branches.

Of further importance is the actual relaying process. Similar to satellite transponders, the signal can be retransmitted using a transparent or regenerative relay. A transparent relay is generally easier to deploy since only a frequency translation is required. However, major changes in current standards are required. For a simpler adaptation of VAAs to current standards, regenerative relays have to be deployed, which generally require more computational power.

Advanced Receiver Algorithms
The accurate reception of the signals, and thus the increased robustness of the devices, relies on the design of efficient mechanisms at the receiver, such as:

- Semi-blind yet simple channel estimation techniques enhancing mobility support (i.e. robustness against Doppler) [221], possibly coupled with decoding [222] through a well-suited expectation maximization (EM) procedure tailored to OFDM. This can be achieved through a special embedding of the *a priori* knowledge of the OFDM modulation format (i.e. channel confined in the cyclic prefix duration) or Doppler spectrum in a modified EM cost function leading to a new channel estimation update formula.
- Investigation of signal processing algorithms for decoding the received SF or STF encoded signals in an OFDM-based WLAN system by:

- — estimation of achievable diversity/capacity gains through SF/STF coding and appropriate modulation [223];
- — studying the implications of multi-antenna systems on channel estimation, such as robustness and appropriate training sequence issues [224];
- — investigating schemes with reasonable signal processing/decoding complexity [213];
- Elaboration of more robust synchronization algorithms tailored to the new modulation formats with reduced false alarm/detection failure probabilities. This is especially necessary for the OFDM modulator, known to be very sensitive to carrier and clock frequency offsets (FO), which generate inter-carrier interference and parasitic rotations of the subcarriers [225]. Several techniques already exist for estimating and cancelling a carrier FO based on time correlations (e.g. [226]). A frequency-domain estimator of the carrier and clock FO is also proposed in [227] based on linear least squares fitting. Unfortunately, this approach assumes that the channel attenuation is the same for all subcarriers, which leads to a performance loss. New and more robust joint maximum likelihood frequency-domain estimators of the carrier and clock FO, relying on a model that explicitly takes into account the channel frequency selectivity, have been proposed [228] for coherent/differential modulation schemes relying, or not relying, on the pilots (data-aided or blind) in order to overcome this issue.
- Active and passive peak-to-average power ratio (PAPR) reduction techniques, either in digital or analogue. A very large literature exists on this subject, since in OFDM, the time domain samples generated after modulation are approximately Rayleigh distributed, and thus large sample amplitudes occur from time to time. In the main, two problems are encountered: (i) the corresponding power amplifiers need a large backoff which makes them inefficient in terms of power consumption and cost; (ii) the analogue-to-digital (ADC) and digital-to-analogue (DAC) converters have to provide a high conversion precision which leads again to power consuming and costly devices. These issues motivate the application of techniques limiting the PAPR value of OFDM symbols.
- Powerful nonlinear equalization schemes capitalizing on latest multiuser detection and successive/serial [229] or parallel [230] interference cancellation techniques (SIC/PIC) for linearly precoded OFDM systems (SC-OFDM, CDMA-ODFM, spread OFDM). The potential of these new nonlinear detection schemes is high, but the area still deserves research in order for the schemes to be applied and implemented in short-range radio systems.
- Digital compensation of RF front-end impairments (e.g. phase noise, I/Q imbalance etc.) [231]. Indeed, OFDM-based systems are highly sensitive to the phase noise which is present in the up-conversion and down-conversion oscillators, and that can reduce significantly the capacity of the system, as depicted in Figure 7.20. The amount of the degradation depends on the system parameter, such as carrier frequency, constellation size and subcarrier spacing. High quality RF components can be used, but they increase the system cost drastically. Another approach is thus to design signal processing algorithms which make the system more robust to RF imperfections. Currently, a variety of algorithms for combating phase noise is being investigated [232–234]. The results suggest that the phase noise can be tackled at baseband by reducing the amount of inter-carrier interference between the carriers, and that the capacity can be significantly increased compared with the case when no phase error correction algorithm is used.

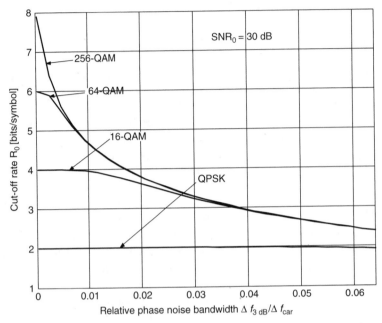

Figure 7.20   Capacity degradation as a function of the oscillator quality

Technologies Exploring Information Theory Advances

In the field of error correction techniques, two main promising directions are identified:

1. The application of the latest error correction strategies (among which are the well known turbo codes, low-density parity check codes) and the study of their interaction/integration with the new above-mentioned techniques. The application of LDPC codes to OFDM is studied, for instance, in [235]. An emphasis will be placed on suboptimal, low-complexity receiver architectures which enable the selection of algorithms that are exploitable in future product generations.
2. The improved reliability of the decoding process of traditional bit-interleaved coded modulations, such as the ones used in the WLAN standards, by applying turbo demodulation [236].

Implementation Feasibility and Realization

Evaluating the implementation feasibility of various technology proposals before product realization is key for new radio interface introduction. This applies both to digital hardware design and to radio frequency front-end architecture. Such studies need first to look at the complexity evaluation level, together with the modelling and simulation of analogue and digital hardware impairments, in order to enable efficient feedback on the second path in the algorithm design. They comprise a vital overall sanity check for the systems proposed, before going into effective implementations.

In addition, WBANs and WSNs set very specific constraints on the hardware design, which need to be carefully taken into account. WBANs use very low radiated power to just match the range of human body dimensions, which translates into low transmit power, poor receiver noise figure and low receiver dynamic. This sets very specific constraints on the design of the RF front-end, which need to be studied carefully. For WSNs, simple and low-power transceivers for sensor nodes need to be designed that can still cope with the adverse radio signal propagation effects. The hardware design should lead to tiny, low-power, low-cost transceiver, sensing and processing units that also implement strategies for efficient power management and processor workload prediction [237, 238].

7.4.3.2 Data Link Control (DLC), Medium Access Control (MAC) and Network Topology

The requirements imposed on future short-range communication systems in terms of capacity, user data rate, and quality of service also demand enhancements in DLC, MAC and network topology. In particular, the MAC has to control the physical layer transmission scheme so that end users obtain the best QoS in terms of throughput, delay and jitter. It also needs to operate in different environments (LAN, PAN, BAN) for a large scale of applications, which puts additional requirements on the system.

New MAC for High-speed WLANs

The CSMA/CA-based random access protocol used in 802.11-based WLANs has a throughput limit given by a set of protocol parameters and the payload size. Therefore, increasing the physical layer data rate does not necessarily lead to a higher throughput. It has been shown that capacity is wasted by the 802.11 MAC protocol in cases needing data rates above 54 Mbps. Other approaches based on TDMA have been investigated for WLANs (Hiper-LAN/2) and WPANs (802.15.3). Although the problem of capacity wastage is not as severe as for the CSMA/CA-based approach, the protocol introduces some overhead that would need to be reduced for a high-speed WLAN. A comparison of several MAC alternatives for future generation WLANs can be found in [239].

New MAC for Heterogeneous Applications

Having an efficient MAC layer for bandwidth demanding applications is not sufficient when considering environments like hot spots, where the typical scenario includes a large number of users using mainly low-bandwidth applications like VoIP. Hence, the MAC protocol should also be able to efficiently share the common resource among users. This research topic does not involve the MAC layer only. A consistent combination of physical multiaccess schemes, such as OFDMA, MC-CDMA and SDMA in particular, and MAC design has to be studied.

Low power Protocols

True wireless operation is required to allow the design of terminals that are not only plugged laptops. This is a key enabler for the hot spot scenario, where multimode terminals will have to support both cellular and short-range air interfaces with maximum battery life. It is the *sine qua non* condition for convincing the cellular industry to invest significantly in the short-range technologies. A few studies have already addressed this area, such as [246], but this is still an open space for further research.

For WBANs, where radiation and transmission power consumption are minimized to just match the range of human body dimensions, low-power protocols are also key. Indeed, the low-power protocol has to be capable of connecting several devices and sensors and–depending on application–to reserve individual link capacity and long sleep periods for sensor communication. The latter is a vital feature for low-power operation of the system.

QoS-Aware Protocols

With the increasing demand for support of multimedia services: VoIP, wireless high quality video streaming, etc., there is now a market need for built-in QoS support. Thus, one of the main goals is to challenge the chaotic IEEE 802.11 CSMA/CA MAC to allow clean integration of various levels of QoS, satisfying the various multimedia applications under consideration and supporting the resulting heterogeneous traffic flows simultaneously. ETSI BRAN HIPERLAN/2 has partially already addressed this need. The goal is to leverage and extend this technology in order to transform it into a viable standard that will be built on a clean basis. However, succeeding in defining such a QoS-aware MAC is not sufficient: the application needs to be translated into MAC characteristics. This piece of the protocol stack, often referred to as the convergence sublayer, should be able to list and maintain available resources and to collect applications requirements. Then it should be able to constantly perform the adequate mapping and to apply any predefined QoS policy. QoS support in a wireless and mobile environment is, in itself, a huge and hot research topic.

Better Management of the PHY Layer, Auto Planning, Zero Configuration

Link Adaptation and Transmit Power Control

One of the major roles of the DLC is to exploit at its full potential the data rate that is provided by the PHY layer. This cannot completely be addressed when adopting a modular layered approach for protocol design; therefore, a cross-layer design approach has to be applied, in which the transmission scheme and the protocols are no longer designed independently. Due to the variability of the wireless environment, it is necessary to constantly dialogue and dynamically reconfigure the PHY layer in terms of modulation order (constellation size), coding rate and transmit power. These procedures, known as dynamic link adaptation (LA) and transmit power control (TPC), impact greatly the overall throughput achievable (see, for instance, [241]). Moreover, they interact with other mechanisms such as auto frequency allocation and dynamic frequency selection that aim to reduce automatically the interference levels and to perform an easy network deployment.

Scheduling

Scheduling and QoS

In order to exploit multiuser diversity, transmissions of different users should only be scheduled when their instantaneous channel quality is good. When dealing with such mechanisms, the challenge is to conceive clever resource allocation and scheduling mechanisms that exploit all available dimensions in order to reach a higher system spectral efficiency. An example of a scheduling technique guaranteeing QoS in 802.11 WLANs can be found in [242].

Scheduling and Multi-antenna Processing

Multiple antennas can be deployed to increase the data rate by the simultaneous communication of several users. A good alternative is the use of multiple antennas at the centralized controller (base station or access point) while the terminals have a single one. This configuration is called multiuser MISO [243]; the problem at hand is the scheduling of the multiple users in this SDMA configuration, scheduling that is affected by the channel variations, by the users' mobility and by the usually scarce battery life [244].

The fairness of the resource allocation is then a key issue [245]: it is well known that techniques optimizing the global performance of the cell might not take into account the individual needs of the users having a lower channel quality, whereas schemes that assign an equal share of the resources might not achieve a good performance in terms of global optimization [246]. In [247], the authors propose and evaluate several scheduling alternatives, partitioning the total available power among the users. It appears that techniques maximizing the sum rate tend to distribute the resources asymmetrically, whereas if the users are granted the same rate, there is a lack of efficiency since the central controller uses a lot of power for the poorest users. Besides, the selection of the users served simultaneously deeply impacts on the capacity of the system. It is shown in [248] that the optimum number of users to be served is always lower than the number of antennas. Bartolomé *et al.* [243] show that, in terms of power, serving at the same time users that have correlated channels requires a higher amount of power than if the users have orthogonal channels.

Concerning the choice of the beamforming design criterion, optimality [249] might be sacrificed for the sake of complexity since real-time operation is needed for schedulers. In [250], it is shown that the zero forcing criterion is a good option.

Finally, the distribution of the several users in the temporal axis is not an easy task. Ideas borrowed from opportunistic communication [251] could be useful, but further research regarding how it can be implemented for SDMA is needed.

Hybrid ARQ

Another topic of interest with high potential sits again at the border of L1 and L2: hybrid ARQ. A first idea is to send, in the case of an unsuccessful codeword transmission, a number of retransmissions of the coded data packet and let the decoder combine all received packets prior to decoding. This is the Chase combining concept [252], which was recently applied to CSMA/CA-based WLANs in [253].

Transmitting incremental redundancy is another promising hybrid ARQ method. The concept of incremental redundancy was proposed in the framework of rate compatible punctured convolutional codes by Hagenauer *et al.* [254], but has also been applied to turbo codes. Several studies have been performed on the combination of hybrid ARQ with adaptive modulation schemes (see [255]). The performance evaluation of a hybrid ARQ scheme with adaptive modulation is influenced by system parameters such as the MAC protocol overhead or the channel and interference fluctuations (see [256] for an application to IEEE 802.11a), and this topic needs further investigation.

Coexistence with Other Wireless Systems and, In Particular, IEEE 802.11e

Most WLANs, WPANs and WBANs operate in frequency bands that do not require an application for a license (unlicensed bands). Hence, the frequencies used by 802.11 and its extensions are not exclusively used by WLANs. Thus, it is important that short-range systems provide means to coexist, not only with other WLAN systems, but also with systems using a totally different protocol. Dynamic frequency selection (DFS), present in

802.11h, 802.15.3 and HiperLAN/2, is a first attempt to address this issue, but needs to be taken further.

The upcoming standard for QoS support in the IEEE 802.11 WLAN family, 802.11e, discusses centralized and decentralized mechanisms for QoS. The centralized form of 802.11e, based on the hybrid coordination function, uses a single coordination instance that has full control over its BSS and the frequency used therein. This concept is useless if two or more overlapping coordination instances have to share the same frequency: for instance, if none of the hybrid coordinators performs a backoff, the immediate access on the channel leads to repeated collision and increasing delay, and QoS cannot be guaranteed any more. Therefore, a new method to support competing centrally coordinated WLANs is needed.

In general, establishing a general policy based on a set of rules for sharing a wireless channel is a superior solution; in [257], such a solution, based on the application of game theory, promises benefits to all users, guaranteeing QoS and an efficient share of scarce resources among independently competing stations. In particular, the application of cooperating strategies is of mutual benefit and helps to increase the payoff of all stations.

Figure 7.21 presents a normalized space where the gained payoff of a hybrid coordinator is a function of the demanded throughput and tolerated delay. Below a certain throughput,

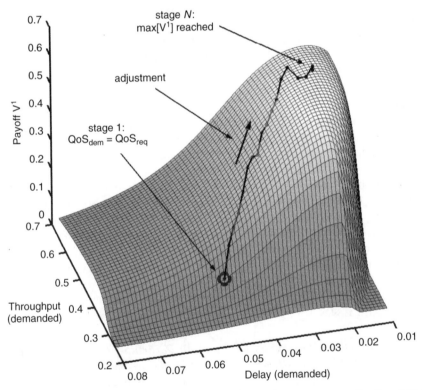

Figure 7.21 The values for throughput and delay presented here are normalized to 100% of a single superframe. Based on demand, a player can react to the opponent's behaviour and adapt its scheduling of transmission opportunities to maximize its payoff

there is no or little usage of data transmission (e.g. a video stream which has a lower bound on the transmission rate). Also, the delay must not be too high to be acceptable. As both players independently try to maximize their payoff, an optimum sequence of transmission opportunities exists which maximizes the sum of payoff for both players. It is the task of game theory to reach that maximum.

Multihop Communications

In order to increase the coverage/cell range in short-range communication systems, limited by propagation losses and low maximum transmission power, the installation of numerous access points, that are low cost with respect to cellular base stations, is possible. Since a wired line to the fixed backbone is not available everywhere, full coverage can only be achieved by forwarding access points that offer network connectivity thanks to multihop connections. On the other hand, mobile devices with sufficient computing and electrical power to carry additional data flows of other wireless devices will be able to provide forwarding capabilities. This can be helpful for ad hoc scenarios where not all the stations are in reception range of a central access point. Figure 7.22 presents a scenario where connectivity to the network is realized by forwarding access points without connection to the fixed network.

Unfortunately, existing WLAN systems do not provide any means for multihop communication. WLANs based on 802.11 are even very inefficient when used as forwarding terminals. They suffer from the 'neighbourhood capture effect' [258] that has a huge

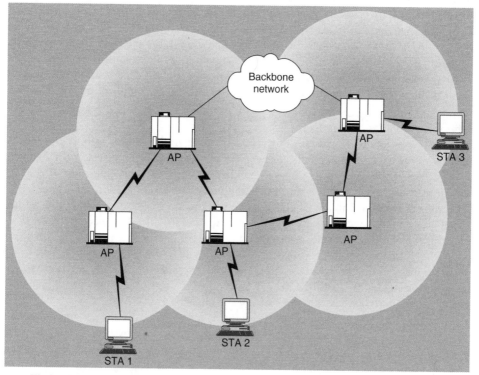

Figure 7.22 Forwarding access points increase the range using multihop procedures

impact on delay and throughput when using 802.11 for multihop communication (this was validated by simulations at ComNets, Aachen). Only neighbouring stations are in reception range. The stations having only one neighbour are more likely to grab the channel than the bottleneck stations 2 and 3, although most communications in the network need to be bidirectional.

In consequence, the 802.11 protocol needs to be modified or replaced by a new self-organizing, ad hoc, MAC protocol with embedded support for QoS, enabling one to avoid collisions, and optimizing the usage of channel capacity. As every collision in 802.11 increases the backoff duration, and therefore decreases the achievable throughput, another collision avoidance mechanism is also needed. In [259] a distributed reservation mechanism is proposed. It allows scheduling of transmission requests in a distributed network and offers support for QoS by prioritizing data transmissions. Since it is an addition to the 802.11e standard its implementation may easily be carried out. Simulations already show the advantages of such a distributed scheduling procedure. This indicates that, although the existing protocols are not capable of multihop communication, they can be enhanced by new MAC procedures to increase the achievable throughput and to decrease the delay on multihop routes, while supporting QoS. This is an important area for future research as it can significantly increase short-range system coverage areas.

Ad hoc Networking

Wireless networks can be divided into infrastructure-based and self-organizing networks. Traditionally, radio networks have always been infrastructure-based. However, interest in self-organizing networks has recently grown, owing to the possible ad hoc deployment of the systems without any assistance from the existing infrastructure. The ad hoc network architecture has two main goals: to respond rapidly and correctly when adapting to the network's behaviour, and to minimize the consumption of the network's transmission. Ad hoc networks must also be capable of rapid deployment. Two classes of ad hoc network can be distinguished: centralized (also called clustered) and decentralized ad hoc networks.

Centralized Networks

In cluster-based networks, nodes autonomously organize themselves into interconnected clusters, all of which contain a cluster-head (or master), one or more gateways, and zero or more ordinary nodes (slaves) that are neither cluster-heads nor gateways. The slaves become part of the cluster after an association procedure.

There are several definitions of cluster in network topology literature. In a fully meshed cluster architecture concept, every device is able to establish direct connection to any other device of the cluster. Typically, a terminal belongs to a cluster if it is in the transmission range of the cluster-head. Transmission from one cluster to another can be performed from one gateway (or bridge) to another, without any aid from the cluster-head; this is the operation mode for HIPERLAN/2 [185]. Most of the time, the cluster-head schedules transmissions and allocates resources within the clusters.

To establish a link-clustered control structure over a physical network, the nodes discover neighbours to whom they have bidirectional connectivity by broadcasting a list of those neighbours they can hear and by receiving broadcasts from neighbours, elect cluster-heads and form clusters, and agree on gateways between clusters. The cluster-head election procedure ensures that the master is the device having the best position within the cluster (for example, the node with the highest number of neighbours, or the node positioned in the centre of the cluster). Obviously in a wireless LAN system the position of the nodes, and thus the status of the link, can change frequently, so this procedure must be repeated periodically.

To ensure a node's maximum availability for communications, cluster membership must be updated accordingly when nodes move in the network. This requires that all nodes are capable of executing the cluster-head and gateway functions if elected to do so. In a high-traffic environment, the performance of a centralized network is much better than that of a distributed network, since resources must be controlled efficiently and with great rapidity, so as to avoid reaching unacceptable levels of throughput and delay.

Such a cluster-based architecture also offers QoS guarantees very easily, such as fairness in medium access, maximum loss of data and maximum delay. Furthermore, it is possible to reuse infrastructure-oriented centralized protocols and equipment. A major drawback for cluster-based networks is the significant complexity of the architecture, which is undesirable in ad hoc networks. The other major issue with this approach is its limited flexibility. Even if nodes agree that a new master should be elected every time the master exits the network (an operation used in many centralized wireless networks), it is easy to understand that this kind of strategy is not ideal in an environment in which nodes can be mobile and the number of nodes often changes. In fact, it could be said that centralization is antithetical to ad hoc networking.

Decentralized Networks
Distributed architecture is based on a decentralized control of the network, now equally distributed among the network nodes that all have the same amount of information on the network. Hence, necessary information must be as little as possible so as not to overload the nodes with immense quantities of topological data. In some cases, the overhead can be reduced by reducing the maximum number of hops needed for a node to reach all other nodes. This reduces the routing information stored in the nodes. When nodes need to transmit, they simply sense the medium and determine whether it is free or not.

All of these features give rise to an architecture that is very simple to implement, and to a fast access to the medium. The related drawbacks are increased delays and packet collisions, which could be a disadvantage in real-time applications, or heavily loaded networks. Moreover, one of the appealing advantages of a noncentralized topology when direct links are possible, is a tremendous saving of bandwidth when intra-cell communications are predominant.

The distributed architecture also presents increased robustness against failures, since the damage of a given node does not affect the functioning of the overall network. This is critical, especially for military or disaster recovery applications. When considering a military scenario, mobile ad hoc networks are both security sensitive and easily exposed to security attacks [260]. Not only the information passing from node to node is confidential, but also the wireless traffic itself can reveal the location of a target to the enemy. Thus, the exchange of control information must be reduced as much as possible and the possibility that some of the nodes remain silent for tactical reasons while they can continue receiving data must be envisaged. Also, if public key cryptography is involved, a central, and thus vulnerable, certification authority is problematic.

In conclusion, the two sections above illustrate two scenarios that require two radically different network topologies. The applications targeted do not all fit in one of these two categories. Thus, intermediate levels of centralization/ad hoc networking need to be considered and investigated further.

MAC, Network Layer and Management Protocols for Wireless Sensor Networks

The MAC protocol in a multihop self-organizing WSN must create communication links between sensors for data transfer. This forms the basic infrastructure needed for wireless hop-by-hop communication and gives the sensor network the self-organizing infrastructure. In addition, the MAC protocol needs to ensure that communication resources are fairly and efficiently shared between sensor nodes. New MAC protocols have to be designed which perform well, not only in static environments but also in the case of moving sensors.

The networking layer of WSNs must be designed so that power efficiency is guaranteed, data packet routing can be data-centric, and data aggregation and location awareness are supported. Protocols that support these features have to be developed. A system administrator can interact with WSNs using a sensor management protocol (SMP). This protocol allows the introduction of rules for data aggregation, the exchange of data related to location finding algorithms, the querying of sensor status, and the distribution of keys for authentication and data security. An efficient SMP has to be designed.

Interpenetration of Various Network Topologies

The inherent coexistence of centralized and unsupervised networks deserves more study on how they can interoperate in order to allow seamless vertical handover, to ensure security and to guarantee QoS. Thus, in order to go beyond coexistence, these two networks need to be efficiently bridged for full complementarity. Seamless interworking of the various types of access network can be achieved by using an IP-based backbone network with QoS and mobility management support.

Part of this work can also be dedicated to transforming the autonomous cell operations of WLANs into a picocellular infrastructure.

7.4.4 Standardization Impact for WLANs/WPANs/WBANs/WSNs

Future WLANs will have to provide a larger range of capabilities in terms of range/bit-rate/etc., as discussed in a previous section. The WLAN industry has recognized this need, and various organizations in the US and Japan–namely IEEE 802.11 and ARIB MMAC–are working on the standardization of these future short-range communication systems. However, at the European level, we can notice a slow down in terms of WLAN leadership, with uncertainties related to the success of ETSI BRAN HIPERLAN/2 that is superseded by the IEEE 802.11a WLAN and its various extensions.

The main driver for the use of WLAN remains, for the moment, the enterprise requirement to enable flexible workplace mobility, which penalizes the European standard that has the vocation to embrace both the home and office environments. This fact leads to a general perception that IEEE is the main standard body leading the specification of WLANs worldwide, and the creation of a new task group IEEE 802.11n, devoted to the standardization of a next generation high throughput WLAN air interface, reinforces this leadership. In order to properly take into account European specificities in the WLAN landscape, it is important to form a pole of European interest that would have enough strength for leveraging IEEE standards to answer local European requirements.

Moreover, Europe is today completely absent in the WPAN landscape that is fully driven by the US through the IEEE 802.15.x workgroup (.1 Bluetooth,. 2 Bluetooth 2,. 3 high data rate,. 4 low data rate). Some European initiatives, such as the EU-funded projects on Ultra Wide Band (UWB), are trying to partially fill the gap.

The IEEE 1451 working groups in the US are standardizing smart sensor networks. Wireless functionality can be added to this family of standards by adopting parts of the IEEE 802.15.4 physical layer/MAC specification and the networking/application part of the open standard 'ZigBee'.

7.5 Ultra-Wideband Radio Technology

The commercialization of wireless devices based on the principles of ultra-wideband (UWB) radio technology (UWB-RT) is widely anticipated given the recent endorsement by US regulators and corresponding efforts in Europe and Asia. The possible use and benefits of UWB-RT can be significant owing to its applicability to communications, imaging, ranging, sensing and public security, as well as its potential to alleviate problems related to finite spectrum resources. It is anticipated that the future *Wireless World* will include large numbers of "intelligent devices" in the form of wireless transceivers built into appliances, sensors, beacons as well as identification tags and the like. The envisaged scenarios of such networked "ambient intelligence" imply that many of these devices will interact among themselves and with their environment by exchanging various types of information, including their geographic positions in support of location-aware applications and services as well as intelligent routing mechanisms. The prospect of a shortage of spectrum resources combined with the growing number of active wireless devices favors the introduction of license-free systems based on spectrum overlay concepts.

As indicated in this section, UWB-RT has been identified as a potential technology to enable the development of dynamic, short-range networks that combine robustness and resistance to intentional attacks, with communication and location tracking features. UWB devices will help fill the current technical gaps towards realizing the vision of a future *Wireless World*, where a variety of high data rate- and/or location-centric applications interact seamlessly. The commercial prospects as well as the regulatory and technical challenges to be overcome to develop and deploy UWB radio systems are described, with a look to future activities within the wireless industry and research institutions. The potential impact of UWB-RT on the future *Wireless World* that builds on interoperation and coexistence of the various systems is substantial, but equally numerous are the challenges of and the need for dedicated research and development to move this technology towards reality.

After an introduction in the following section, the three subsequent sections deal with regulatory issues, UWB signal characteristics as well as standardization of UWB physical layers (PHY) and medium access control (MAC) functions. Three further sections discuss coexistence, systems issues and application scenarios and a final section concludes the UWB discussion giving an outlook into the future.

7.5.1 Introduction

The increasingly "pervasive" nature of communications is expressed by the vision that the future *Wireless World*, i.e., wireless systems beyond the third generation (B3G), will enable connectivity for "everybody and everything at any place and anytime." This ambitious view assumes that the future *Wireless World* will be the result of a comprehensive integration of existing and new wireless systems, including second-generation (2G) cellular systems and their enhancements (2.5G), 3G/UMTS wide area networks (WANs), wireless or radio local area networks (WLAN or RLANs), wireless personal area and body area networks (WPANs and WBANs) as well as novel ad hoc networks that link devices as diverse as portable and fixed appliances, personal computers and entertainment equipment. Realization of this vision requires the creation of new wireless technologies and system concepts based on easy-to-use interfaces with a focus on users' needs. This user-centric scenario forms the basis of the concepts of *Ambient Intelligence* [261] and – in a very similar sense – *Pervasive Computing* [262]. In the scenarios envisaged for future smart environments, the user needs to manage information easily, while increasingly relying on the electronic creation, storage and transmission of personal and other confidential information by having complete access to time-sensitive data, regardless of physical location. Users will expect various devices – such as personal digital assistants, mobile phones, office and mobile computers as well as home entertainment equipment – to access information from anywhere by working together in one seamless, integrated system. This automatic device connection enables users to manage information exchange and interaction with the environment efficiently and effortlessly, for example, through interfaces that understand people's natural speech and movements.

Pervasive communication enables people to accomplish an increasing number of personal and professional transactions using a new class of intelligent and portable devices to access relevant information stored within powerful networks, allowing them to easily take action anywhere and anytime. To be widely accepted, however, these devices must be very simple to use and their communication links need to be stable, fast and very reliable. Figure 7.23 illustrates how short-range wireless technology will play a key role in a scenario where "everybody and everything" is connected by different types of communication links, that is human-to-human, human-to-machine, machine-to-human and machine-to-machine. While the majority of the human-to-human information exchanges still take place via voice, a rapid increase in data transfers is observed in other types of links as manifested by the rising need for high-volume video transfer capability within the home and office environments[263]. But in the future *Wireless World* we will see a continuing need for ever higher data rates developing jointly with a flourishing increase of large numbers of low-rate devices in the form of transceivers embedded in common devices and appliances, sensors, beacons as well as radio frequency identification (RFID)

Figure 7.23 A variety of pervasive computing scenarios create dense and rich environments of ambient intelligence.

tags and the like. The envisaged scenario of networked "ambient intelligence" implies that many of these devices will interact among themselves and with their environment by exchanging various types of information, including their geographic positions in support of location-aware applications or efficient data routing protocols.

7.5.2 Spectrum is a Finite Commodity

The prospect of a future shortage in spectrum resources in view of a rapidly growing variety of active wireless devices within given coverage areas (macro- or picocells) favors the introduction of license-free radio systems using spectrum overlay concepts. The apparent spectrum squeeze is already evident, for example cell phones cannot access the network in environments where there are too many cellular users in close proximity and short-range wireless systems may have difficulties to coexist. For example, Bluetooth, Wi-Fi (IEEE802.11b/g) and Home RF devices in the 2.4 GHz ISM band interfere with one another and may cause a link loss if used in close proximity to each other. Moreover, the explosive deployment of pagers, cell phones and other wireless services has generated intense competition over any remaining spectrum.

UWB-RT has been widely identified as a potential novel key technology in the realization of the future *Wireless World*, for one thing because of the possibility of efficient (re-use) of existing spectrum that has been allocated but often only intermittently used [264]. UWB-RT offers a compelling – if not somewhat disruptive – solution to many of the challenges facing today's wireless (short-range) industry, where the likely squeeze in radio frequency (RF) spectrum may become a bottleneck for the further evolution of wireless technologies. In addition, the numerous differences in RF spectrum assignments from one country to another often hinder or may even prohibit interoperability of RF-based devices on a global scale. Given that a global regulatory framework is not yet in place for UWB-RT, this novel radio technology remains unburdened by such RF constraints and thus offers the rather unique opportunity to achieve a regulatory situation that supports worldwide compatibility and interoperability. On the other hand, RF spectrum is already so extensively allocated that there is no free RF band available to accommodate the bandwidth requirements of UWB devices. However, the key differentiator of UWB radio lies in the fact that its emissions occur at power spectral density levels that do not exceed the "noise floor" specified for unintentionally radiating electronic devices (FCC Part 15 rules). Thus, UWB radio emissions can operate in a spectral overlay mode without inflicting undue interference – if any – onto incumbent narrowband systems that share the same band.

The commercial prospects and the technical challenges to be overcome when developing and deploying UWB radio systems have spurred rapidly growing activities within the wireless industry and research institutions, initially mainly in North America and more recently also in Europe and beyond. Given the potential impact of UWB-RT on the finite amount of RF spectrum and its use and management, there exists a definite need for a globally harmonized regulatory framework for the operation of UWB radio devices and systems.

7.5.3 Historical Background of UWB Radio

Many have claimed to have spurred the idea of (pulse-based) UWB radio. However, it is widely recognized that Dr. Gerald F. Ross first demonstrated the feasibility of utilizing

UWB waveforms for radar and communications applications in the late 1960's and early 1970's; in 1973, Ross was awarded the first UWB communications patent [265]. Ross' insight into the value and potential applications of this technology some 30 years ago has been fundamental in shaping the development of UWB-RT and its applications, initially mainly for military purposes, for which UWB radio was particularly well suited, i.e., for RADAR and highly secure wireless links, also known as "low probability of intercept and detection (LPI/D)" communications. In the late 1980's, this wideband technology was alternately called *baseband, carrier-free* or *impulse communication;* the term "ultra-wideband" was only introduced approximately in 1989 by the U.S. Department of Defense. Mainly in the 1990's, UWB-RT became the focus of commercial interests, leading the U.S. Federal Communications Commission (FCC) in 1998 to recognize the significance of this technology. A lengthy process of regulatory review began back then culminating in May 2000 in the FCC *Notice of Proposed Rule Making* (NPRM) to collect comments from all interested parties on this new radio technology [266]. Comments and review reports were received by early 2002, when on February 14 the FCC formally adopted the rule changes to permit the deployment and operation of certain types of commercial UWB radio devices [264]. Since then, the business interests for UWB-RT have grown exponentially, bringing this technology closer and closer to the markets[1].

7.5.4 UWB Regulatory Issues

One of greatest advantages of the UWB technology, namely the unlicensed reuse of existing radio spectrum, can also be considered its greatest liability. The concept of overlaying different radio emissions represents a radical departure from the historical practice of splitting the spectrum into separate (narrow) bands and allocating them to a unique licensee. In contrast, the novel approach used for UWB radio is based on sharing the existing radio spectrum resources rather than allocating new bands. This means emitting radio signals in spectrum areas that are already allocated to other services. In such a resource-sharing situation, coexistence is guaranteed by the low-power spectrum density allowed for UWB devices [264], so that their effect can be compared with that of added white noise at a level typically below the ambient noise and interference level of a close-by narrowband system.

For this critical relation between UWB and narrowband systems, the regulation authorities have to take appropriate steps to analyze in detail, simulate and measure the possible impact of emissions from UWB signal sources to gain the expertise and technical background to prepare new rulings. Based on the common goal of seeking an efficient reuse of spectrum by allowing the coexistence of existing narrowband services with UWB radios, the authorities of the main regions of the world are assessing the compatibility of UWB radio and are taking steps towards defining a set of applicable rules, as outlined below.

7.5.5 United States of America

The recent process for changing radio regulations in the United States of America was very complex and involved various parties in the decisions. In particular, two organizations played a significant role, namely the

---

[1] Note – An excellent review with key references documenting the history of UWB-RT has been compiled by Dr. Robert Fontana, available at URL: http://www.multispectral.com.

- Federal Communications Commission (FCC) and the
- National Telecommunications and Information Administration (NTIA).

These organizations are the principal administrations dealing with spectrum management-related issues in the United States. The FCC is an independent U.S. government agency, directly responsible to the Congress. It is charged with regulating interstate and international communications by radio waves as well as by wire, satellite and cable.

The NTIA is an arm of the Department of Commerce and the President's principal advisor on spectrum management issues. It is co-responsible with the FCC to enable new radio-based technologies to enter the market place. In simple terms, the FCC has jurisdiction over non-federal users (commercial, private, state and local public safety), whereas the NTIA is responsible for federal government users. There is a Memorandum of Understanding in place between the two organizations to allow joint use of the spectrum, and to give reciprocal notice of any proposed action and early notification with priority of commenting. In their final actions, the two agencies are independent (not requiring each other's approval); typically however, they strive to reach a consensus.

Other parties and federal agencies are also involved in the regulation process, as spectrum regulation is an interactive and collaborative process. An important part of any ruling is played by a very broad range of interested parties, who submit public comments to aid the agencies in their evaluation process of a new technology.

The FCC process for revising regulations includes the following major steps:

- *Notice of Inquiry (NOI):* The FCC releases an NOI to gather information about a broad subject or to generate ideas on specific issues. NOIs can be initiated either by the FCC itself or by an outside request. In the case of UWB, the FCC issued an NOI on ultra-wideband regulations in September 1998, after years of industrial lobbying.
- *Notice of Proposed Rulemaking (NPRM):* After collecting the necessary material and reviewing pertinent suggestions from the public, the FCC may issue an NPRM, which proposes changes to the current FCC rules and seeks public comment on the new proposals. The FCC issued an NPRM on ultra-wideband regulations in May 2000.
- *Further Notice of Proposed Rulemaking (FNPRM):* After reviewing public comments on the NPRM, the FCC may also choose to issue an FNPRM regarding specific issues raised in comments. The FNPRM provides an opportunity for further public commenting on the specific proposal.
- *Report and Order (R&O):* After considering comments to an NPRM or FNPRM, the FCC may issue an R&O to develop new rules or amend existing rules, or it may decide not to change existing rules. The FCC emitted the R&O on UWB technology in April 2002, based on a deliberation made public on 14th February 2002. An entirely new legislation was introduced to legalize the use of unlicensed UWB radio devices.
- *Changes after the R&O:* Further changes are possible after the R&O. In particular there can be:
  - *Petitions for Reconsideration:* Those dissatisfied with the way an issue is resolved in the R&O can file a *Petition for Reconsideration* within 30 days from the date the R&O appears in the Federal Register. Several petitions for reconsideration were filed in the case of UWB technology, which led only to minor adjustments formalized in the Memorandum Opinion and Order (MO&O), which the FCC made public on March 13, 2003.

— *Memorandum Opinion and Order (MO&O):* In response to a Petition for Reconsideration, the FCC may issue a *Memorandum Opinion and Order* (MO&O) or an Order on Reconsideration amending the rules or stating that they will not be changed. This can be accompanied also by official statements from the FCC chairman or Commissioners, if deemed appropriate. The FCC issued in parallel an MO&O and an FNPRM on UWB technology on March 13, 2003. This MO&O amends Part 15 in response to 14 petitions filed after the First R&O on UWB. The new MO&O introduced no major changes in the technical operating parameters for UWB devices, but it did relax some of the restrictions on the operation of through-wall imaging systems and ground-penetrating radar.

Current US regulations, as established in the R&O [264], list six classes of UWB devices (ground-penetrating radars, wall imaging systems, medical systems, surveillance systems, communication devices, vehicular radar systems). For the first wave of products, the mainstream consumer electronics and semiconductor industries will be concentrating on systems that support communication applications, whose defined FCC and – at this time still very speculative – "sloped" draft ETSI power limits are shown for indoor use in Fig. 7.24. The main band (i.e., the -10 dB bandwidth) of operation will therefore be within the 3.1–10.6 GHz region with an average EIPR (effective isotropic radiated power) not exceeding -41.25 dBm/MHz (only approved by the FCC).

In the meantime, the FCC is progressing with its continuous assessment of UWB technology and plans to conduct further tests and measurements with UWB radio prototypes. The vision pursued by the FCC, as stated by Ed Thomas (FCC Chief, Office of Engineering and Technology), is in fact the one first expressed by Lao-Tzu (604–531 BC): "A journey of a thousand miles must begin with a single step." The FCC regulations

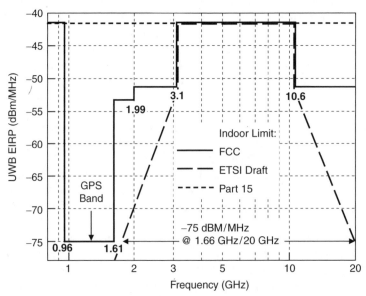

Figure 7.24 Power spectral density masks for indoor UWB communication as defined by the FCC's First Report and Order and as discussed within ETSI ("sloped mask" draft, July 2003)

on UWB are therefore expected to evolve with time in the course of future technology developments.

7.5.6 Europe

The European regulatory process started in early 2000 inspired by the FCC's NPRM for UWB devices adopted on May 10, 2000. Several interest groups and industrial parties have been active in the past few years to assess and verify potential interference issues between UWB devices and existing radio services. Two organizations are particularly relevant to the discussions and rulings on radio communications in Europe:

- CEPT (European Conference of Postal and Telecommunications Administrations) is focused on harmonization of telecommunications regulations across 44 European countries. The groups within CEPT interested in UWB radio matters are assigned to the Electronic Communications Committee (ECC). Among those, the most active working groups (WG) are:
  — WGFM: Frequency Management
    • Short-range Devices Maintenance Group (SRDMG)
  — WGSE: Spectrum Engineering
    • SE 21 (unwanted emissions)
    • SE 24 (short-range devices) until early 2004
  — Task Group 3 (TG3) since early 2004
  — WGRR: Radio Regulations
- ETSI (European Telecommunications Standards Institute) was created by the CEPT in 1988 as a new organization to which all its telecommunication standardization activities were transferred. ETSI has been promoting and regulating telecommunications in Europe since then (examples of standards set by ETSI include: GSM, POCSAG, HYPERLAN).

Within ETSI, UWB radio issues are dealt with under TC-ERM (Technical Committee EMC and Radio Spectrum Matters), in TG31A (Communication and Radar in the bands <10 GHz) and TG31B (automotive applications in the bands >10 GHz, e.g., 24 GHz). The activities of the groups that have been particularly involved in technical studies and discussions towards the introduction of UWB radio in Europe are described in some more detail:

- *ETSI TG31:* Selected by the European Commission (Enterprise Directorate) as the group that should respond to the mandate[2] of February 25, 2003, calling for *Harmonized Standards Covering Ultra-wideband (UWB) Applications*, which derives from the *R&TTE Directive* (1999/5/EC) defining the essential requirements that equipment must meet to be placed on the market and to be put into service for its intended purpose. The mandate directed at ETSI points out that "*... as a technology, UWB radio does not fit in the classical radio regulatory paradigm, which bases itself on a subdivision of*

---

[2] Mandate forwarded to CEN/CENELEC/ETSI in the field of information technology and telecommunications, DG ENTR/G/3, February 25, 2003; URL: http://portal.etsi.org/public-interest/Documents/mandates/M329EN.pdf.

the spectrum in bands and allocated for specific usage(s)." A stipulation of the mandate is that the work be aligned with equivalent activities within the ITU's (International Telecommunication Union) Radiocommunication Task Group dealing with UWB radio matters (ITU-R TG 1/8) and ISO/IEC, while due account should be taken of (draft) regulations adopted in other economies to ensure a global market for UWB devices.

- *CEPT SE24:* Until early 2004, this group was responsible for the technical analysis of the possible impact of UWB radio signals on the spectrum currently allocated to other services. Attended by national regulators and radio administrations, the SE24 group has been conducting studies of UWB radio deployment and signal propagation to derive required separation distances between UWB devices and victim receivers and to define signal power margins, based mainly on theoretical models and simulations.

- *CEPT/ECC/Task Group 3 (TG3):* As of March 2004, the terms of reference for this newly formed group are:

  1. To develop the draft ECC responses to the European Commission mandate[3] to CEPT to harmonize radio spectrum use for ultra-wideband systems in the European Union.
  2. To liaise and consolidate the outputs from all relevant ECC groups, particularly, WGSE (SE21, SE24) and WGFM.
  3. To provide a *Draft Final Report* by April 2005 (after issuing a Draft First report followed by a Draft Interim Report).
  4. To coordinate European positions in preparation for ITU-R TG1/8 on ultra-wideband issues.
  5. To consult with relevant European organizations, in particular ETSI.

  TG3 has approved the creation of three *Working Groups* (WGs) to accomplish the required work:
  - WG1: UWB Characteristics and Measurement Techniques;
  - WG2: UWB Compatibility;
  - WG3: Frequency Management and Regulatory Issues.

The mandate[2] to CEPT is an order by the EC to identify the conditions relating to the harmonized introduction in the European Union of radio applications based on UWB radio, i.e., "...pursuant to Art. 4 of the *Radio Spectrum Decision*, CEPT is mandated to undertake all necessary work to identify the most appropriate technical and operational criteria for the harmonized introduction of UWB-based applications in the European Union." Among other reasons, the European Commission's *Directorate General for the Information Society* (DG INFSO) provides the following justification for this mandate: "*UWB technology may provide a host of applications of benefit for various EU policies. However, its characteristic broad underlay over spectrum already used by other radio services may also have an impact on the proper operation of radio services of significance for the successful implementation of EU policies. It is therefore important to establish conditions of the use of radio spectrum for UWB which will allow UWB to be introduced on the market as commercial opportunities arise, while providing adequate protection to other radio services.*"

The EC mandate orders CEPT further to "...*report on actual or planned real-life testing within the European Union [and] consider the possible benefits of experimental rights*

---

[3] Mandate forwarded to CEPT to harmonize radio spectrum use for ultra-wideband systems in the European Union, DG INFSO/B4, February 18, 2004; URL: http://europa.eu.int/information_society/topics/radio_spectrum/docs/pdf/mandates/rscom0408_mandate_uwb.pdf.

to use radio spectrum (or licenses) for UWB applications." The mandate also encourages CEPT "*...to make use of ongoing activities and know-how of EU RTD (Research and Technology Development) projects on UWB,*" such as those participating within the European UWB Cluster.

There is a *Memorandum of Understanding* (MoU) in place between CEPT and ETSI, which commissions ETSI with addressing technical definitions and measurement techniques to be used by CEPT members. The ETSI work includes defining the market space for UWB radio devices, and promoting a technical standard that is not application-specific.

The progress made within ETSI in the past three years has resulted in a very early draft-discussion proposal for a "sloped" power spectral density (PSD) mask for indoor communication devices using UWB principles (see Fig. 7.24, dotted lines). This PSD mask is still debated but already includes a number of additional protection measures for the existing radio communication services in Europe, compared to the corresponding FCC mask. For example, the "sloped" PSD mask includes more than five additional dB of protection in the 2.4 GHz ISM band, more than 10 dB of additional protection at the 3 G/UMTS frequencies, and more than 15 additional dB at the GSM 1800 frequencies. The original schedule of ETSI as mandated by the European Commission called for delivery of the Harmonized Standards by December 2004; however, there have been delays in the process and it is expected that regulations for UWB radio will not be in place in Europe before 2005.

7.5.7 Asia

Many administrations on the Asian continent have been careful and rather pragmatic observers of the developments and decisions related to UWB radio regulation in the western world and within ITU-R TG 1/8. However, two initiatives have been particular noteworthy up to now:

- In Japan, a technical consortium formerly led by the *Communications Research Laboratory* (CRL) and now called the *National Institute of Information and Communications Technology* (NICT), has been established to promote R&D initiatives and to study appropriate radio regulations for the commercial use of UWB radio in Japan. Currently, more than 40 companies have joined the consortium. Four working groups have been formed to investigate compatibility between UWB radio devices and other radio [communication] systems, i.e., the *Compatibility Model Working Group*, the *Fixed-Broadcasting System Working Group*, the *Radar-Aviation and Maritime Systems Working Group*, and the *Satellite Low- Power Systems Working Group*. The work thus far has led to a Draft Interim report, for which 22 comments had been received by February 22, 2004. This work is expected to lead to a draft proposal for UWB radio in Japan sometime in 2005.

- The Infocomm Development Authority of Singapore (IDA), a regulator but also a supporter of economic and technical development, is "... fostering a competitive world-class 'Infocomm' industry, preparing residents for living and working in the 'New Economy', spearheading the delivery of citizen-centric e-government services, and building and operating the Government's IT infrastructure." In February 2003, IDA launched an innovative UWB radio experimentation and evaluation phase with

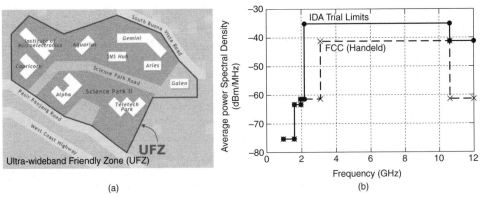

(a)   (b)

Figure 7.25   The innovative approach to UWB radio regulation of Singapore's IDA: The UWB Friendly Zone (left) and the experimental UWB mask (right).

the opening of a so-called UWB Friendly Zone (UFZ); the program will last until February 2005. The UFZ (see Fig. 7.25) is an area located within the Science Park II in Singapore, where a series of tests will be conducted to establish the real effects of UWB radio signal interference on existing radio systems. To encourage experimentation and attract innovative platforms, IDA permits within the UFZ transmissions at 6 dB above the FCC Part 15 level (i.e., up to a total of −35.3 dBm/MHz) between 2.2 GHz to 10.6 GHz (i.e., with 0.9 GHz additional bandwidth with respect to the FCC's lower limit of 3.1 GHz). However, a (draft) ruling on UWB radio cannot be expected before 2005.

7.5.8 International Telecommunication Union (ITU)

In July 2002, the *Study Group 1* (SG 1) of the *Radiocommunication* standardization sector of the ITU (ITU-R) decided (document 1/95) to form a *Task Group 8*, named ITU-R TG 1/8, to carry out relevant studies to answer two questions (ITU-R 226/1 and ITU-R 227/1) adopted by SG1 on ultra-wideband technology[4]. These questions (see Table 7.2) concern the proposed introduction of UWB devices and the implications of compatibility with radiocommunication services. The work load was divided among four working groups as shown in Table 7.3.

It is foreseeable that the four originally planned meetings of TG 1/8 until the end of 2004 will not suffice to complete the work as anticipated. At its third meeting from 9–18 June, 2004, in Boston, MA, USA, the leadership of TG 1/8 decided that up to a total of seven TG 1/8 meetings will be necessary to produce the final deliverable in the form of a *Proposed Draft New Recommendation* (PDNR). Therefore, this PDNR cannot be expected before the end of 2005.

7.5.9 Characteristics of UWB Signals

Historically, UWB devices have been designed to transmit sequences of information carrying pulses of very short duration (e.g., 0.1–2 ns) that are widely spaced such that the

---

[4] International Telecommunication Union (ITU), *Report of the Second Meeting of Task Group 1/8*, Geneva, 27–31 October, 2003.

Table 7.2 Questions assigned to ITU-R TG 1/8

Question ITU-R	Title	Category (*)
226/1	Spectrum management framework related to the introduction of ultra-wideband (UWB) devices	S1
227/1	Compatibility between ultra-wideband (UWB) devices and radiocommunication services	S1

(*) S1: Urgent studies intended to be completed within two years.

Table 7.3 Working Groups of ITU-R TG 1/8

Working Group	Title	ITU-R Question/Decision
WG 1	UWB Characteristics	Question 226/1 and 227/1
WG 2	UWB Compatibility	Question 227/1
WG 3	Spectrum Management Framework	Question 226/1
WG 4	Measurement Techniques	Decision 1/95

waveform's duty cycle is up to several orders less than unity (e.g., 1/10–1/1000) [267, 268]. The FCC's recent First Report and Order, however, does not require the exclusive use of these extremely low-duty-cycle signals, thereby allowing for a flexible approach to generating UWB signals for communication applications in the spectrum range 3.1–10.6 GHz, where the average power spectral density (EIRP) is limited to −41.25 dBm/MHz [264]. In addition, the peak power is limited to 0 dBm measured within a 50 MHz bandwidth[5] centered on the frequency of maximum emission. While it is anticipated that Europe will also have adopted a similar ruling by 2005 at the earliest [269–271], the present FCC rules represent the regulatory reference framework. According to the FCC's definition, an UWB radio device is defined as any device with a fractional bandwidth (ratio between the signal's absolute bandwidth and its centre frequency) greater than 0.20 or with an absolute bandwidth occupation of at least 500 MHz.

The formula and conditions for calculating fractional bandwidth are defined as

$$B_F = B/f_C = 2(f_H - f_L)/(f_H + f_L) > 0.2,$$

[5] The FCC allows peak power measurements with a resolution bandwidth (RBW) between 50 MHz and 1 MHz according to the scaling law $(RBW/50 \text{ MHz})^2$ mW [264].

where f_H is the upper frequency of the -10 dB emission point and f_L is the lower frequency of the -10 dB emission point. The centre frequency of the transmitted signal is therefore defined as

$$f_C = (f_H + f_L)/2,$$

and the signal bandwidth is lower bounded as

$$B = f_H - f_L \geq 500 \text{ MHz}.$$

It is further required that the bandwidth is determined using the antenna that is designed to be used with the UWB device. Furthermore, systems should not be precluded from the UWB definition simply because the bandwidth of the emission is due to a high data rate instead of the width of the pulse or impulse; various modulation types are permitted as long as the products comply with all of the technical standards adopted in the FCC proceeding. Thus, as long as the transmission system complies with the fractional bandwidth or minimum bandwidth requirements at all times during its transmission, the FCC permits systems to operate under the UWB regulations. However, this first ruling precludes certain types of modulations, such as swept frequency (e.g., FMCW), stepped frequency, and certain types of frequency hopping systems, since the FCC's current measurement procedures require that measurements of swept frequency devices are made with the frequency sweep stopped, rendering it unlikely that swept frequency (linear FM or FMCW) or stepped frequency modulated emissions would comply with the fractional bandwidth or minimum bandwidth requirements. Furthermore, it is unlikely that any known, conventional frequency hopping systems would comply, unless extremely wide bandwidth hopping channels are employed [264].

7.5.10 Pulsed Monoband Signals (Wavelets)

The original approach to generating UWB radio signals is based on the use of very short pulses, ideally often described in the form of Gaussian monocycles as

$$p(t) = \frac{1}{\tau}e^{-(1/\tau)^2} \Leftrightarrow P(f) = -jf\tau^2 e^{-(f\tau)^2},$$

where $p(t)$ and $P(f)$ represent the pulse in the time and frequency domain, respectively. The parameter τ determines the pulse duration while the frequency where $P(f)$ is maximum, f_{max}, is inversely proportional to τ, i.e., $f_{max} \propto 1/\tau$. Higher-order derivatives of the Gaussian monocycle are often also applied for shaping the spectrum of the pulse. The actual emission spectrum of a transmitter is also determined by the response characteristics of the transmitter's antenna, which tends to differentiate the pulse. A number of more or less effective methods can be applied to generate such pulses and to modulate them, e.g., pulses amplitude modulation (PAM), on-off keying (OOK), various forms of pulse position modulation (PPM), and polarity (bi-phase) modulation; the latter is the most efficient method of those mentioned.

An early-proposed modulation method that can today be considered a near-classic approach is time-hopping pulse position modulation (TH-PPM) or time-modulated UWB

(TM-UWB). With this method, the interval between individual pulses is jointly varied based on the information data and a defined, overlaid code sequence. The latter serves two purposes, for one thing it allows a kind of code division multiple access (CDMA) operational mode and it smoothes (whitens) the emission spectrum, thereby avoiding excessive discrete spectrum lines that are typically spaced at intervals dictated by the average pulse repetition frequency (PRF). With TH-PPM, a data symbol is often represented by several pulses to achieve a higher processing gain by applying "pulse integration" methods at the receiver, in return for a reduced data rate [272].

Recent developments in UWB signal design and modulation deviate considerably from classic pulse shaping and TH-PPM. There are various reasons for this. For example, although the precise generation of Gaussian monocycles is perhaps still of academic interest, the practical challenge when designing UWB pulse generators stems from the requirement for a highly flexible spectrum-shaping ability to meet regulatory constraints (spectral mask) and/or to mitigate potential or actual interference constellations in a heterogeneous network setting. One of the implementation challenges of TH-PPM is the need for extremely precise timing signals at the receiver to within picoseconds. In addition, although lower-rate (binary) TH-PPM can be implemented by using a relatively low PRF, this classic monoband signaling method is not well suited to support higher-order modulations or high PRFs, which would be required for TH-PPM to achieve data rates in excess of some 100 Mb/s. Increased data rates can be more easily achieved, for example, by applying simple polarity (bi-phase) modulation or – as shown in the following section – by dividing the available bandwidth into multiple bands, where each band supports a lower data rate, to obtain a high compound data rate.

A variation of modulated, pulsed monoband UWB signals (wavelets) is called direct-sequence, code-division multiple access (DS-CDMA) UWB, or DS-UWB, and is currently under discussion within the IEEE 802.15.3a working group. This method partitions the spectrum between 3.1 and 10.6 GHz into two parts by excluding the band around 5.5 GHz allocated to WLAN systems (IEEE 802.11a). In this way, three modes of operation are possible: (a) low band (3.1 to 5.15 GHz) operation, (b) high band (5.825 to 10.6 GHz) operation, or (c) both. In the latter case, by combining the two sub-bands, an aggregated data rate can be achieved or, with an appropriate diplexer, independent reception and transmission in the two sub-bands can be supported, thereby enabling full-duplex operation.

7.5.11 Multi-Band Signals

Even if the origins of UWB technology lie in the use of impulse radios, many further techniques have recently been proposed for UWB radio signaling which fully satisfy the FCC definitions as described at the beginning of this chapter. One of the most interesting alternative approaches is based on a combination of the best elements of frequency hopping, multiple frequency bands and *Orthogonal Frequency Division Multiplexing* (OFDM).

This approach – under discussion also within the IEEE 802.15.3a working group and within the *Multi-Band OFDM Alliance Special Interest Group* (SIG) – subdivides the frequency spectrum in a number (up to 14 with a minimum of 3 mandatory) of sub-bands of about 528 MHz each and uses each sub-band independently from the others, one at a time. Within each sub-band an OFDM micro system is built, generating via Fast Fourier Transforms (FFT) a large number (typically 122) of tone frequencies-also called *sub-carriers*- which carry *Quadrature Phase Shift Keying* (QPSK) or *Dual Carrier Modulation* (DCM) information at a frequency spacing of around 4.125 MHz. For data rates up to 200 Mb/s, time spreading is performed with a spreading factor of two to improve frequency diversity and the performance of *Simultaneous Operating Piconets* (SOP). Convolution coding is also applied to the generated sequences with a mother code of rate R = 1/3 from which further codes (R = 1/2, 5/8, 3/4) are derived by puncturing according to the desired output data rate. The resulting generated payload varies from 53 Mb/s to 480 Mb/s using the three lower frequency bands (Band Group 1).

In a multi-band OFDM system, the transmitted signals can be described (as in [273]) using a complex baseband signal notation, with the actual RF transmitted signal related to the complex baseband signal, as follows:

$$r_{RF}(t) = \text{Re}\left\{ \sum_{k=0}^{N-1} r_k(t - kT_{SYM}) \exp(j2\pi f_k t) \right\},$$

where Re($\cdot$) represents the real part of a complex variable, $r_k(t)$ is the complex baseband signal of the k-th OFDM symbol and is nonzero over the interval from 0 to T_{SYM}, N is the number of OFDM symbols, T_{SYM} is the symbol interval, and f_k is the centre frequency for the k-th band. The exact structure of the k-th OFDM symbol will depend on its location within the packet (preamble, header or data), but all of the OFDM symbols $r_k(t)$ can be constructed using an inverse Fourier transform with a certain set of coefficient C_n, where the coefficients are defined as either data, pilots, or training symbols:

$$r_k(t) = \begin{cases} 0, & t \in [0, T_{CP}], \\ \displaystyle\sum_{n=-N_{ST}/2}^{N_{ST}/2} C_n \exp(j2\pi n \Delta_f)(t - T_{CP}), & t \in [T_{CP}, T_{FFT} + T_{CP}], \\ 0, & t \in [T_{FFT} + T_{CP}, T_{FFT} + T_{CP} + T_{GI}]. \end{cases}$$

The parameters Δ_f and N_{ST} are defined as the subcarrier frequency spacing and the total number of subcarriers used, respectively. The resulting waveform has a duration of $T_{FFT} = 1/\Delta_f$. Shifting the time by T_{CP} creates the "circular prefix", which is used in OFDM to mitigate the effects of multipath. The parameter T_{GI} is the guard interval duration. At the receiver end, after filtering and digital conversion, the big task of recovering data from the correct tones will be given to a fast FFT engine connected to a Viterbi decoder. A typical multi-band OFDM radio will have a minimum sensitivity of at least -83 dBm, a predicted (CMOS 090) die size on the order of 5 mm^2, and power consumption of about 100 mW (Tx)/170 mW (Rx) when in active mode.

There are numerous advantages implicit in the multi-band approach. The lower bandwidth of each operating sub-band means a relaxation in the specifications for ADCs, LNAs and precision on the oscillators (lower timing requirements). The OFDM signaling has also good inherent robustness to multi-path channel environments, the technology is mature and has already been tested for other radio systems (for example 802.11a), the system is easily scalable with additional bands and can facilitate coexistence with other systems by tone dynamic selection. Working fully in the frequency domain also enables the use of high percentages of digital processing, hence it is highly synthesizable with

CMOS chip processes and subject to large successive shrinks, with the consequent cost, area and power savings implicit in the continuous evolution of the CMOS technology forecast by Moore's law.

7.5.12 UWB Standards Issues

As with any other wireless technology that has high potential for widespread use, the eventual application of systems based on UWB-RT will greatly depend on the availability of suitable physical layer (PHY) and medium access control (MAC) standards. In addition, such standards need to be in strict compliance with the rules imposed by the regulatory authorities and must be backed – as well as technically realizable – by a majority of the interested parties. Thus, broadly supported PHY/MAC standards will be a prerequisite for successful deployment. Figure 7.26 illustrates that the regulatory framework and standardized PHY/MAC functions together with well-supported applications are also key building blocks for the successful deployment of systems based on UWB-RT. Moreover, standardization efforts will likely also be necessary at the (ad-hoc) network management level to enable a flexible integration of a variety of participating mobile nodes and to ensure interoperability between different devices. Although one can argue over the degree to which the need for standardization depends on the intended application, the emergence of UWB-RT should be considered a unique opportunity to develop applications and standards that build on combined data transmission as well as location tracking (LT) capabilities. It is anticipated that low data rate (LDR) devices, because of their larger link range, will more likely support the LT feature, making them so-called LDR-LT devices.

Within the IEEE, the recently established Task Group 3a (TG3a) focuses on defining an alternative physical-layer (Alt-PHY) that is to be based on UWB-RT, complementary to the existing 802.15.3 PHY standard. The emerging Alt-PHY standard will support high data rate (HDR) consumer applications for multimedia distribution, typically in a home, and is intended to work essentially with the existing 802.15.3 MAC to provide a unique combination of standard features and new technology. The adoption of a predefined MAC

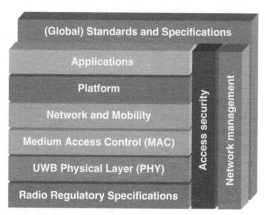

Figure 7.26 The regulatory framework and broadly supported standards for PHY/MAC, network management functions and applications will be key building blocks of systems based on UWB-RT.

layer on top of an alternative PHY (i.e., the IEEE 802.15.3 MAC over the 802.15.3a Alt-PHY) might initiate the first phase of commercial development of consumer UWB-RT applications. Although such a rather pragmatic approach will likely reduce the overall system efficiency and limit the level of achievable QoS (Quality of Service), it appears to be an acceptable compromise to expedite the commercial deployment of UWB-RT.

IEEE P802.15.3a (TG3a) considered the following initial set of system requirements [274, 275]:

- Coexistence with all IEEE 802 wireless PHYs,
- Target aggregate data rate of 110 Mb/s and support of 3 to 4 parallel links at a range of up to 10 m for embeddable consumer applications; possible extensions to 200 Mb/s (range up to 4 m),
- Robust multipath performance,
- Location awareness enabling applications such as range-dependent authentication,
- Anticipation of using additional unlicensed spectrum for high-rate WPANs to relieve possible spectrum congestion.

It is evident that these requirements can be met collectively (!) almost only if the Alt-PHY system incorporates the principles of UWB-RT. Further applications envisaged under this future standard are very high data rate (VHDR) links replacing short-range cable connections (e.g., up to 480 Mb/s data rate as in IEEE1394 or USB2.0). In particular, the TG3a group aims at keeping as much as possible from the MAC as defined in the IEEE 802.15.3 standard; the corresponding access scheme is based on TDMA. However, there exists no real evidence that this MAC scheme is performing well for an UWB PHY. Without appropriate modifications, the resulting standard could thus exhibit a weaker performance than one would expect from an UWB system.

In parallel with the standardization effort and following the successful path of branding a wireless standard as demonstrated by the WiFi® association, a sister association – called WiMedia® – has been formed (2002) to build a brand image and to establish test and interoperability compliance procedures for the emerging IEEE 802.15.3 standard suite. In addition, the recently established IEEE 802.15.4a Task Group (TG4a) for the development of an Alt-PHY for the already adopted (2003) low-rate, low-power wireless personal area (WPAN) 802.15.4 PHY/MAC standard has also adopted this approach. The proponents of this Alt-PHY justify its development with the need for low power LDR-LT devices to support location-aware network features and applications (e.g., asset and/or people tracking, protection and security systems, smart sensor networks). In fact, these combined LDR-LT capabilities are poorly supported by today's conventional narrow-band systems and much more effort needs to be spent to develop and standardize these essential capabilities for future short-range wireless applications.

7.5.13 Coexistence

The issue that was debated the most during the work that led to the approval of the UWB radio technology in the USA was the capability of UWB devices to coexist with other systems already operating in the same frequencies without causing harmful interference. From the point of view of a classical narrowband receiver, the UWB signals can

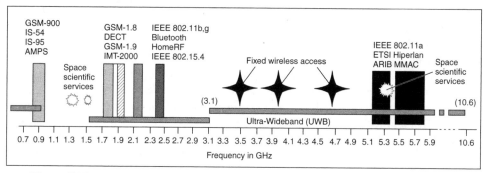

Figure 7.27 Incumbent European services with overlaid UWB (FCC) spectrum mask.

in principle be seen as an additional source of Gaussian noise and therefore can have an effect on a system's link budget. Given this potential issue, careful attention has been given to limit the power allowed for UWB signals to such a level (same level as allowed by FCC/Part 15 rules for unintentional radiation in the USA), that they will not cause any harm to existing (narrowband) signals. The variety of services already in use across frequencies allocated to UWB radio is extensive (see Fig. 7.27 for a simplistic representation of the situation in Europe) and each incumbent has concerns about the potential (aggregated) effects that a large quantity of UWB radio devices could cause. Discussions and complete analysis on each single service and link are fundamental to ensure that the coexistence is peaceful. But it is of paramount importance that models accurately represent real radio links as closely as possible. Even with extremely high densities of UWB devices (i.e., many thousands per km^2) it has been demonstrated that is possible to have peaceful coexistence (see [276] for a description of coexistence UWB with Fixed Wireless Access systems).

Regulatory guidelines and spectrum masks are not defined to prevent interference under any arbitrary operating condition; rather, they are designed to prevent interference under "reasonable" deployment scenarios. Thus, an important point to note is that, while designing simulation scenarios, it is fundamental to consider all the characteristics of UWB systems and to introduce the so-called mitigation factors, i.e., characteristics of UWB signals that will allow a reduction in the effective power captured by close-by narrowband receivers. Independent of the details of specific modulation schemes, i.e., whether UWB systems operate with wideband DSSS or frequency hopping, pulse position modulation or multi-band OFDM, there are a number of generic mitigation techniques that can be considered:

- **Power Control:** The principle of using power control techniques is based on the assumption that not all the radio devices can be at the edge of the serving UWB piconet coverage area. Instead the devices will be distributed (almost) randomly and will accomplish a variety of services at different required bit rates and with different Quality of Service (QoS) requirements. Under these conditions, an effective way to reduce the total power emissions (and therefore the aggregate interference at a given distance) is to distribute a number of master units around a serving area. These units are responsible for controlling the QoS requirements and commanding the slave units to use the minimum possible energy to guarantee the link requirements. The master unit might also be the piconet controller in a master-slave configuration. In a peer-peer situation, one of the devices (if enabled with master functionalities) has to take control.

- **Activity of the Radio Device:** A second important mitigation factor is the activity of the UWB radio device. UWB systems are designed with optimization of power consumption in mind and will need to run also on battery-powered devices. Keeping them actively transmitting all the time would be inappropriate. In fact typical devices utilizing UWB radio technology will need to be active only for a fraction of the total time available during the day. The activity factor can be defined as the fraction of transmission time divided by (transmission + non-transmission time) for the active UWB device in its coverage area. UWB devices should be considered like any other radio local area network (RLAN) device. For these devices the industry target is an aggressive minimization of the power consumption, obtained with a combination of sleep modes and activity strictly limited to service provision. Therefore, any UWB device introduced in coexistence models should be simulated taking this factor into consideration. Recent studies from the Wireless Ethernet Compatibility Alliance (WECA) and others show that the average activity factor for a typical RLAN device is estimated to between 5 % and 10 % (references in [276]).
- **Effective Path Link Dynamics:** A third element needed for an accurate coexistence investigation is a careful model of the propagation (the path loss as a first-order approximation) between a potential UWB device and the "victim" device. Using simple free space path losses for the indoor propagation path is in most cases far too conservative, even in open-space office environments. In fact, within a typical open office we find low-height partitions (or even cubicles) between working areas, walls, closed (tele)conference rooms, lifts, stairs, books, furniture, people. A typical PC or laptop UWB card add-on will suffer severe path attenuation while operating in such *Non-Line of Sight* (NLOS) conditions. Estimated decay factors in these conditions are between 3 and 4 (see [277]). Considering free space attenuation models will result in an excessive overestimation of UWB interference. Furthermore, in the case of services operating under outdoor conditions, care needs to be taken also concerning the propagation conditions indoor-to-outdoor where severe attenuation is incurred by walls and windows, which are often of composite materials and form thick (and highly attenuating) barriers to the outdoor world. Finally the outdoor path, again often simplified as free space, should be analyzed in the particular scenario under investigation. Non free-space models (ITU-R P.1411-1, Cost231-Hata) might be more appropriate especially in the presence of scatterers (trees, other buildings, people, cars) and a nonlinear path between the UWB interference radiation and the victim receiver.

Many additional mitigation techniques can be identified, either specific to the particular UWB technology implementation, such as multi-band static/dynamic allocation (band or sub-band suppression) or selective introduction of narrow-band notches in the spectrum, admission control techniques, antenna polarization mismatches and multiple trough-wall indoor losses. A careful coexistence analysis will help identify any potential problem related to the overlay approach offered by UWB radios.

7.5.14 System Design Issues

UWB radios are about to become commercially viable, stimulated by the appearance of a regulatory framework and because of accelerated decrease of costs through recent advancements in the development of fast semiconductor and chip technologies. UWB radio technology is driven towards the consumer market by the prospect of building UWB circuitry with standard CMOS technology. Therefore, as new CMOS technologies scale down from, say, 0.18 to 0.13 and 0.09 micrometres and eventually even lower, so does UWB circuitry. As a result, UWB radios will particularly benefit from a trend that is sometimes called the "Moore's Law Radio" or the "Silicon Radio". Until recently, the design of radio circuitry – for UWB radios particularly depending on the design approach and architecture used – was constrained by power and form factor limits; with the prospect that entire UWB radios can eventually be implemented in standard CMOS technology, this restriction appears to vanish.

Short-range wireless systems based on narrowband-modulated carriers are often inadequate or incapable of providing both sufficiently high data rates and precise information – derived from their narrowband signals – about a terminal node's location to support location-aware applications or network routing mechanisms. On the other hand, we see a growing need for these capabilities in today's marketplace; given the large available signaling bandwidth and associated potential to resolve multipath signal components at the receiver, UWB radios are inherently capable of supporting precise indoor location determination in 2D and 3D space. It remains a further objective of ongoing as well as future research (e.g., see [278]) to determine the practical limits of achievable spatial capacity – measured in terms of a network's aggregate data rate (b/s) per cell area (m^2) – and to introduce other relevant parameters to characterize system performance and spectral efficiency of UWB radio devices. A sensible "performance measure" (M) has recently been proposed by comparing the ratio of a system's spatial capacity (C_S) and the product of DC power drawn from the battery (P_{DC}), cost ($P_\$$) and volume size (form factor) of a device (V), as indicated by the formula in [279]:

$$M = \frac{C_S}{P_{DC} P_\$ V} \cdot \left[\frac{(b/s)/m^2}{W \$ m^3} \right].$$

Technical challenges also exist in the areas of modulation and coding techniques suited for UWB radio systems. Originally, UWB-RT has been applied for military purposes, where the support of a large number of users was not an immediate objective. However, a large multi-user capacity now becomes important for both military and commercial applications. Coding and modulation are known to be some of the most effective means to improve a system's multi-user capacity. Adaptive modulation methods – for example, dynamic change of modulation used in each sub-band when a multi-band approach is employed – and channel coding schemes will have to be devised that take into account the specific time-domain properties of the UWB radio channel. Also, although the average EIRP (equivalent isotropic radiated power) appears very low in UWB radios (strictly less than 0.56 mW [264]), the required peak power in a given short time interval might become relatively large for certain pulse-based modulations. Thus, adequate characterization and optimization of transmission techniques (e.g., adaptive power control or duty cycle optimization) will be required. To cope with difficult signal-propagation environments, such as in industrial and manufacturing or commercial areas, advanced technologies such as UWB MIMO (Multiple-Input/Multiple Output) systems may be able to provide the necessary high degree of link reliability and rate adaptation capability [280]. Moreover, unlike narrowband radio systems, UWB systems suffer much less from signal-fading effects because the extremely narrow pulses propagating over different paths cause a large number of independently fading signal components that can be distinguished owing to the high temporal resolution, resulting in a significant multipath diversity. UWB MIMO systems also cope better with adverse intersymbol interference (ISI) and inter-channel interference (ICI) in the time domain, because of the very favorable auto-/cross-correlation properties of the received signals and the capability to simply adapt the pulse repetition frequency to the prevailing channel delay spread.

Today, an interesting research area at the PHY level appears to be the antenna design and implementation for UWB radio devices; the FCC, for example, considers the antenna an integral part of an UWB radio [264]. Generally, portable communication devices require small, unobtrusive antennas that can be integrated into small-form-factor devices and operate effectively under varying environmental conditions, often in near-field propagation conditions (e.g., near objects or carried on or close to the body). For UWB radio systems, design and implementation of effective antennas are more challenging than for conventional narrowband systems given the large bandwidths, linearity requirements and variety of anticipated operating conditions. A further aspect that has not yet been fully investigated relates to the deteriorating effects of in-band interference in UWB receivers originating from other radio signals, be they in near- or far-field proximity. The problem of nearby interference is not only one of academic interest, considering that UWB devices might be integrated into mobile platforms that make simultaneous use of a variety of other radios. Thus, the very advantage provided by the fact that UWB devices emit an extremely low power spectral density (PSD) – as a result of the excessive signal bandwidth – potentially gives rise to increased susceptibility to noise and interference in the UWB receiver. Similar effects may occur in areas with a large concentration of active UWB devices. This raises questions concerning harmful compound effects of multipath propagation and cross-device interference phenomena as well as on how to initiate and maintain synchronization at the receiver and network levels.

To improve adaptive modulation methods, it would be necessary to identify methods for measuring prevailing noise levels and interference characteristics "on the fly" to be able to apply suitable interference-rejection schemes. For example, narrowband signals tend to interfere at levels tens of decibels above the level of the received UWB signal, leading to challenging requirements for the dynamic range of the UWB receiver. The impact of such a high dynamic range on the analogue front-end and/or the digital baseband of an UWB receiver needs analyzed to realize adequate tradeoffs between hardware cost, dynamic range requirements, power consumption, and receiver performance. The excessive bandwidth of UWB is likely to require dedicated custom designs for specific circuit components (e.g., wideband LNA and power amplifier, very fast analogue-to-digital converters and sampling devices).

Finally, the use of alternative semiconductor technologies in UWB system realizations needs to be explored, such as MEMS (micro-electro-mechanical systems) and SOI (silicon on insulator) techniques as well as non-linear, analogue circuit and component design techniques (e.g., efficient integration of tunnel diodes on silicon is an open problem). These techniques could potentially provide interesting solutions for problems such as excessive clock speed, synchronization latency and power consumption; their successful exploitation may become a crucial factor in future developments and applications of UWB-RT.

7.5.15 Application Scenarios

The choices for possible deployments and user scenarios when considering UWB systems for the enterprise and consumer markets are numerous. Figure 7.28 indicates the two potential and complementary application areas of UWB radio devices. It is thus imperative that the important scenarios from which the key PHY/MAC system requirements and specifications should be derived are considered carefully. Common elements among different scenarios will have to be identified and then investigated to minimize the potentially large number of components and functional modules necessary to realize these scenarios. Based on the outcome of such investigations, the pertinent requirements on the PHY and MAC layer functions must be clearly identified, e.g., link budgets, range, data rate, location precision, battery burden, level of adaptability to channel conditions, multi-user scalability, network topologies, to name only a few.

While the commercialization of UWB-RT is just beginning, the technology offers significant potential for the deployment of short-range communication systems supporting high-rate applications and lower-rate, intelligent devices embedded within a pervasive and personal *Wireless World*. For example, FCC-compliant UWB radio systems – using simple modulation and appropriate coding schemes – can transmit information rates in excess of 100 Mb/s over short distances, an operational mode previously defined as HDR [281].

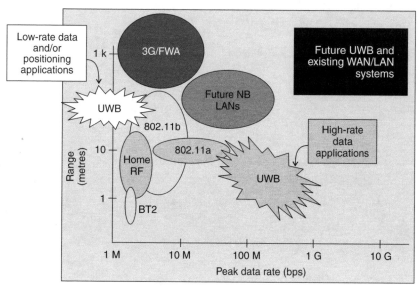

Figure 7.28   Complementary application areas of UWB radio devices.

Alternatively, UWB radios can trade a reduced information rate for increased link range, as is also shown in [281], potentially combined with accurate LT capabilities, thus offering an operational-mode-defined LDR-LT. The two complementary application spaces are unique to UWB radio systems as they can be implemented based on very similar architectures with an unprecedented degree of scalability.

A number of practical usage scenarios well suited to exploit the potentials of UWB-RT can be identified, see Fig. 7.29. In these scenarios, system implementations based on UWB-RT could be beneficial and potentially welcome by industry and service providers alike. While numerous other usage models can be thought of, those shown in Fig. 7.29 illustrate some possible choices of human-to-machine or machine-to-machine interactions:

- "High Data Rate" Wireless Personal Area Network (HDR-WPAN),
- Wireless Ethernet Interface Link (WEIL).
- Intelligent Wireless Area Network (IWAN),
- Outdoor Peer-to-Peer Network (OPPN),
- Sensor, Positioning and Identification Network (SPIN).

The first three scenarios assume a network of UWB devices deployed in a residential or office environment, mainly to enable wireless video/audio distribution for entertainment, control signals or high-rate data transfers. The fourth scenario presents a deployment in outdoor peer-to-peer situations, whereas the fifth takes industry and commercial environments into account. The identification of common elements among the scenarios listed and the optimization of system cost, coverage range, data rate, localization precision, battery

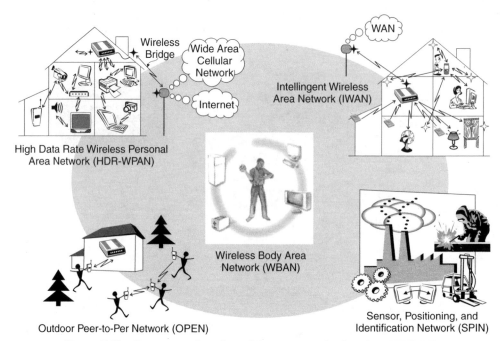

Figure 7.29   Some examples of possible user scenarios based on UWB-RT.

burden, and level of adaptability to channel conditions still are tasks ahead. But assuming adherence to FCC regulations in principle, some preliminary considerations about the individual scenarios can be made:

"High Data Rate" Wireless Personal Area Network (HDR-WPAN)

We define HDR-WPANs as networks with a medium density of active devices per room (5–10 active devices) transmitting at up to 100–500 Mb/s data rate over a distance between 1 and 10 m, mainly based on a peer-to-peer topology and using a relay/bridge to the outside world based on existing (either wireless or cable) standards. The interface and adaptation between "local" and "remote" modes need to be defined carefully as the outside world may be limited to accept only lower data rates (e.g., wireless WAN).

Wireless Ethernet Interface Link (WEIL)

This is an extension of the concept of HDR transmission to extremely high data rates (e.g., 1 Gb/s, 2.5 Gb/s), probably only over rather short distances (i.e., a few metres at most). The WEIL concept could satisfy a specific demand (i) from PC manufacturers that calls for a direct wireless replacement for Ethernet cables, or (ii) from consumer-electronics firms asking for a high-quality wireless video-transfer capability between a PC and an LCD screen, such as for wireless DVI (Digital Video Interface). This latter application is the most demanding one, and much research is still needed to determine whether it is feasible at the current transmission power limits.

Intelligent Wireless Area Network (IWAN)

These systems are characterized by a high density of devices in a domestic or office environment, covering distances over 30 m (see Fig. 7.29). The main requirements for the devices are very "low cost" (no more than 1 dollar/euro per unit) and very low power consumption (e.g., 1 – 10 mW) to provide users with access to intelligence distributed around the home/office (e.g., automated smart appliances). Device capabilities will include accurate location tracking to support context-aware services (e.g., child and/or asset tracking, alarm zones, phone auto-modes, electronic virtual guides) that is not readily realizable with the current generations of narrowband short-range networks. In this scenario an eventual wireless last mile and/or other interconnections to the outer world could be used to send alarms and control signals, and/or to check the status of sensors around the home remotely.

Outdoor Peer-to-Peer Network (OPPN)

This is a network of UWB devices deployed in outdoor areas, mainly to respond to new market demands for PDA (Personal Digital Assistant) link-up and information exchange, digital kiosks for fast download of newspaper text, photographs, automatic video rental or sale distribution systems. To what degree the specifics of an application scenario chosen will determine whether the OPPN architecture is centralized or distributed is an open research problem. It should also be noted that today's UWB regulations in the USA and the envisaged "generic UWB standard" to be adopted in Europe severely constrain any deployment of UWB devices supporting outdoor scenarios. However, this situation might change, because it is anticipated that future UWB usage regulations will likely follow an evolutionary path to an even greater extent than that experienced for other wireless services in the past – provided of course that relevant and currently still open coexistence questions and issues can be resolved.

Sensor, Positioning and Identification Network (SPIN)

SPIN is a system characterized by a high density (e.g., 100 s per floor) of devices (intelligent sensors or tags) in industrial factories or warehouses transmitting low-rate data combined with position information (e.g., data rate greater than several tens of kb/s and position accuracy well within 1 m). SPIN devices operate over medium-to-long distances (typically ~100 m) between many individual devices and a master station with a typical master-slave topology, e.g., as in widely anticipated smart sensor networks. In such industrial applications, SPINs require a high level of link reliability and adaptive system features to react to the dynamically changing and very challenging interference and propagation environment.

Note that both WPAN and IWAN scenarios may include elements of the wireless body area networks (WBANs) introduced in Fig. 7.29. However, current UWB regulations applicable in the USA and the envisaged (initial) rules to be adopted in Europe severely constrain the deployment of UWB devices supporting the two outdoor scenarios listed above. On the other hand, networks of UWB devices deployed in outdoor areas appear desirable for surveillance and control of a variety of sensitive key areas and functions (e.g., school yards, hospitals, car parks, pool area, etc.). For certain applications under these outdoor scenarios, long-distance coverage will be essential and might be realized, for example, by using multi-hop and dynamic routing. While these usage models still are restricted today, it is anticipated that UWB usage regulations in this area will also follow an evolutionary path.

Finally, an important role that UWB technology will play is the provisioning of effective services in response to user demands. Keeping this in mind, the schematic breakdown of scenarios and development of separate networks covering each of the cases under analysis are not sufficient to satisfy user expectations and provide concrete advances over narrowband systems. A strategic target of future investigations and development is therefore the capability for seamless coexistence, interoperability and integration among different scenarios as well as with incumbent wireless protocols (e.g., IEEE 802.11, cellular WANs) to achieve truly heterogeneous networking. Thus, the design of efficient bridges, automatic roaming mechanisms, and adaptation of the data link will be essential aspects in any future R&D effort in this area.

7.5.16 Conclusions and Outlook

The commercialization of wireless devices based on the principles of ultra-wideband (UWB) radio technology (UWB-RT) is widely anticipated given the recent endorsement by US regulators as well as the current European and Asian efforts in the same direction. The prospect of a shortage in spectrum resources – in part solely due to established spectrum allocation and management procedures – combined with the growing number of active wireless devices favors the introduction of license-free systems based on spectrum overlay concepts. UWB-RT has been identified as potential key technology in the realization of the future *Wireless World* by enabling novel data- and/or location-centric applications.

It was assumed that the regulations that govern the spectral and power constraints (to be) imposed on the use of UWB radio devices will provide sufficient flexibility to allow the deployment of technically as well as commercially viable systems; this assumption is backed by the recent regulatory action in the USA, which the FCC considers to be a cautious and conservative first approach. The desirability of – if not the very need for – global regulation and standardization has been emphasized, recognizing that the necessary level of standardization may well depend on the intended application. Similar past standardization efforts for WLANs have clearly shown, however, that too expeditious and uncoordinated an introduction of standards bears the risk of creating (too) many standards, with the consequence that few enjoy a sufficiently broad (global) market support. The commercial deployment of UWB-RT and its applications is only on the verge of happening, and the development of devices towards form and power factors suitable for widespread integration into mobile platforms and appliances will remain a challenge for some time to come. On the basis of the promises as well as the challenges of the future Wireless World, it was argued that there exists ample and significant opportunity as well as the need for coordinated research and development actions. Integrating the current European and worldwide research efforts related to UWB-RT towards a synergetic cooperation is an obvious choice all interested parties should contemplate, given the need for a global regulatory framework and standards.

UWB-RT has the potential to become a viable and competitive wireless technology for short-range HDR-WPANs as well as low-power-consuming low-cost LDR-LT devices and networks, with the capability to support a truly pervasive user-centric and thus personal *Wireless World*. UWB-RT's innovative but somewhat disruptive spectral overlay technique can become the basis for new short-range wireless systems, services and applications within the future heterogeneous network world, where seamless transition from one network to another will be transparent to the user. This paper emphasizes some of the merits and challenges related to the use and commercial deployment of UWB-RT and it gives an overview of the current status of worldwide UWB regulations and standardization together with an introduction to potential future usage scenarios and applications.

The technical, economical, and regulatory challenges ahead are probably still as numerous as the promises that this intriguing wireless technology appears to hold. Thus, dedicated research and development efforts combined with a viable regulatory framework will further the chances of UWB-RT to become the enabler of choice for intelligent short-range networking applications and services within the next ten years.

7.6 References

[1] Saltzberg, B.R., 'Performance of an efficient parallel data transmission system,' *IEEE Transactions on Communications*, **15**, 805–811, 1967.

[2] Weinstein, S.B. and Ebert, P.M., 'Data transmission by frequency-division multiplexing using the discrete Fourier transform,' *IEEE Transactions on Communications*, **19**, 628–634, 1971.

[3] Bingham, J.A.C., 'Multicarrier modulation for data transmission: An idea whose time has come,' *IEEE Communications Magazine*, **28**, 5–14, 1990.

[4] Bahai, A.R.S. and Saltzberg, B.R., *'Multi-Carrier Digital Communications: Theory and Applications of OFDM,'* Kluwer Academic/Plenum Publishers, Boston, 1999.

[5] van Nee, R. and Prasad, R., *'OFDM for Wireless Multimedia Communications,'* Artech House, Boston, 2000.

[6] ETSI ETS 300 401, '*Radio Broadcasting Systems: Digital Audio Broadcasting (DAB) to Mobile, Portable and Fixed Receivers*,' European Standard (Telecommunications series), Valbonne, France, February 1995.

[7] ETSI EN 300 744, '*Digital Video Broadcasting (DVB): Framing Structure, Channel Coding and Modulation for Digital Terrestrial Television*,' European Standard (Telecommunications series), Valbonne, France, July 1999.

[8] ETSI TS 101 475, '*Broadband Radio Access Networks (BRAN): HIPERLAN Type 2: Physical (PHY) Layer*,' European Standard (Telecommunications series), Valbonne, France, November 2000.

[9] IEEE 802.11a, '*LAN/MAN Specific Requirements – Part 2: Wireless MAC and PHY Specifications – High Speed Physical Layer in the 5 GHz Band*,' IEEE 802.11, May 1999.

[10] IEEE 802.16a, '*Air Interface for Fixed Broadband Wireless Access Systems – Part A: Systems Between 2 and 11 GHz*,' IEEE 802.16, June 2000.

[11] Fazel, K. and Papke, L., 'On the performance of convolutionally-coded CDMA/OFDM for mobile communication systems,' in *Proceedings of IEEE PIMRC'93*, Yokohama, Japan, pp. 468–472, 1993.

[12] Yee, N., Linnartz, J.-P. and Fettweis, G., 'Multi-carrier CDMA for indoor wireless radio networks,' in *Proceedings of IEEE PIMRC'93*, Yokohama, Japan, pp. 109–113, 1993.

[13] Sternad, M., Ottosson, T., Ahlén, A. and Svensson, A., 'Attaining both coverage and high spectral efficiency with adaptive OFDMA downlinks,' in *Proceedings of IEEE VTC 2003 Fall*, Orlando, USA, October 2003.

[14] Kondo, S. and Milstein, L.B., 'Performance of multicarrier DS CDMA systems,' *IEEE Transactions on Communications*, **44**, 238–246, 1996.

[15] Fazel, K. and Kaiser, S., '*Multi-Carrier and Spread Spectrum Systems*,' John Wiley & Sons, 2003.

[16] Atarashi, H., Maeda, N., Abeta, S. and Sawahashi, M., 'Broadband packet wireless access based on VSF-OFCDM and MC/DS-CDMA,' in *Proceedings of IEEE PIMRC 2002*, Lisbon, Portugal, pp. 992–997, 2002.

[17] TIA/EIA/IS-2000-2, '*Physical Layer Standard for cdma2000 Spread Spectrum Systems*,' Arlington, USA, August 1999.

[18] Kaiser, S. and Fazel, K., 'A flexible spread spectrum multi-carrier multiple-access system for multi-media applications,' in *Proceedings of IEEE PIMRC'97*, Helsinki, Finland, pp. 100–104, 1997.

[19] Anderson, J.B. and Svensson, A., '*Coded Modulation Systems*,' Kluwer Academic/Plenum Publishers, Boston, USA, 2003.

[20] Wang, W., Ottosson, T., Sternad, M., Ahlén, A. and Svensson, A., 'Impact of multiuser diversity and channel variability on adaptive OFDM,' in *Proceedings of IEEE VTC 2003 Fall*, Orlando, USA, October 2003.

[21] Knopp, R. and Humblet, P.A., 'Multiple accessing over frequency-selective fading channels,' in *Proceedings of IEEE PIMRC*, Toronto, Canada, pp. 1326–1330, September 1995.

[22] Persson, A., Ottosson, T. and Ström, E., 'Analysis of direct-sequence CDMA for downlink OFDM systems, Part I,' submitted to *IEEE Transactions on Communications*, 2003.

[23] Petré, F., Leus, G. and Moonen, M., 'Multi-Carrier Block-Spread CDMA for Broadband Cellular Downlink,' *EURASIP Journal on Applied Signal Processing* (to appear).

[24] Hara, S. and Prasad, R., 'Overview of multi-carrier CDMA,' *IEEE Communications Magazine*, **35**, 126–133, 1997.

[25] Proakis, J.G., '*Digital Communications*', 3rd edition, McGraw-Hill, New York, NY, 1995.

[26] Auer, G., Dammann, A., Sand, S. and Kaiser, S., 'On modelling cellular interference for multi-carrier based communication systems including a synchronization offset,' in *Proceedings of WPMC 2003*, Yokosuka, Japan, October 2003.

[27] Persson, A., Ottosson, T. and Ström, E., 'Analysis of direct-sequence CDMA for downlink OFDM systems, Part II,' submitted to *IEEE Transactions on Communications*, 2003

[28] IST-MATRICE (IST-2001-32620), '*Physical Layer Simulation Chain Description*,' D3.1, November 2002.

[29] Abeta, S., Atarashi, H. and Sawahashi, M., 'Forward link capacity of coherent DS-CDMA and MC-CDMA broadband packet wireless access in a multi-cell environment,' in *Proceedings of VTC 2000 Spring*, Boston, USA, 2000.

[30] Kaiser, S., 'OFDM code division multiplexing in fading channels,' *IEEE Transactions on Communications*, **50**, 1266–1273, 2002.

[31] Chung, S.T. and Goldsmith, A.J., 'Degrees of freedom in adaptive modulation: A unified view,' *IEEE Transactions on Communications*, **19**, 1561–1571, 2001.

[32] Falahati, S., '*Adaptive Modulation and Coding in Wireless Communications with Feedback*,' PhD Thesis, Chalmers University of Technology, Sweden, October 2002.

[33] Muquet, B., de Courville, M., Duhamel, P. and Giannakis, G.B., 'OFDM with trailing zeros versus OFDM with cyclic prefix: Links, comparisons and application to the HiperLAN/2 system,' in *Proceedings of IEEE ICC 2000*, New Orleans, USA, pp. 1049–1053, June 2000.

[34] Pu, Z., You, X., Cheng, S. and Wang, H., 'Transmission and reception of TDD multicarrier CDMA signals in mobile communications systems,' in *Proceedings of IEEE VTC'99*, May 1999.

[35] Jeong, D.G. and Kim, M.J., 'Effects of channel estimation error in MC-CDMA/TDD systems,' in *Proceedings of IEEE VTC 2000 Spring*, pp. 1773–1777, May 2000.

[36] Mottier, D. and Castelain, D., 'SINR-based channel pre-equalization for uplink multi-carrier CDMA systems,' in *Proceedings of IEEE PIMRC 2002*, Lisbon, Portugal, September 2002.

[37] Kaiser, S., 'Space frequency block coding in the uplink of broadband MC-CDMA mobile radio systems with pre-equalization,' in *Proceedings of IEEE VTC 2003 Fall*, Orlando, USA, October 2003.

[38] Cosovic, I., Schnell, M. and Springer, A., 'Balanced channel equalization techniques for uplink time division duplex MC-CDMA,' in *Proceedings of MC-SS 2003*, Oberpfaffenhofen, Germany, September 2003.

[39] Höher, P., Kaiser, S. and Robertson, P., 'Pilot-symbol-aided channel estimation in time and frequency,' in *Proceedings of IEEE GLOBECOM'97*, Communication Theory Mini-Conference, Phoenix, USA, pp. 90–96, November 1997.

[40] Li, Y., Cimini, L. and Sollenberger, N.R., 'Robust channel estimation for OFDM systems with rapid dispersive fading channels,' *IEEE Transactions on Communications*, **46**, 902–915, 1998.

[41] Frenger, P. and Svensson, A., 'Decision directed coherent detection in multicarrier systems on Rayleigh fading channels,' *IEEE Transactions on Vehicular Technology*, **48**, 490–498, 1999.

[42] Heath, J. and Giannakis, G., 'Exploiting input cyclostationarity for blind channel identification in OFDM systems,' *IEEE Transactions on Signal Processing*, **47**, 848–856, 1999.

[43] Muquet, B., de Courville, M., Duhamel, P. and Buzenac, V., 'A subspace based blind and semi-blind channel identification method for OFDM systems,' in *Proceedings of SPAWC'99*, pp. 170–173, May 1999.

[44] Haardt, M. and Alexiou, A. (Eds), '*Smart Antennas and Related Technologies (SMART)*,' WWRF/WG4 White Paper, June 2002.

[45] Stein, S., 'Fading channel issues in system engineering', *IEEE Journal on Selected Areas in Communications*, **5**, 68–89, 1987.

[46] Foschini, G.J., Gans, M.J. and Cioffi, J.M., 'On limits of wireless communications in a fading environment when using multiple antennas', *Wireless Personal Communications*, **6**(3), 311–355, 1998.

[47] Tarokh, V., Seshadri, N. and Calderbank, A.R., 'Space–time codes for high data rate wireless communication: Performance criterion and code construction,' *IEEE Transactions on Information Theory*, **44**, 744–765, 1998.

[48] Alamouti, S.M., 'A simple transmit diversity technique for wireless communications,' *IEEE Journal on Selected Areas in Communications*, **16**, 1451–1457, 1998.

[49] Kaiser, S., 'Space time frequency coding in broadband OFDM systems,' in S. Chandran (Ed.) '*Adaptive Antenna Arrays,*' Springer Verlag, 2003.

[50] Pan Yuh Joo et al. '*Novel Design for STBC for OFDM/OFDMA using Frequency Diversity*,' IEEE 802.16, 802.16abc-01/59.

[51] Tarokh, V., Jafarkhani, H. and Calderbank, A.R., 'Space–time block codes from orthogonal designs,' *IEEE Transactions on Information Theory*, **45**, 1456–1467, 1999.

[52] Jeon, W.G., Paik, K.H. and Cho, Y.S., 'An efficient channel estimation technique for OFDM systems with transmitter diversity,' *IEICE Transactions on Communications*, **E84-B**(4), 967–974, 2001.

[53] Li, Y., Seshadri, N. and Ariyavisitakul, S., 'Channel estimation for OFDM systems with transmitter diversity in mobile wireless channel,' *IEEE Journal on Selected Areas in Communications*, **17**, 461–471, 1999.

[54] Li, Y., 'Simplified channel estimation for OFDM systems with multiple transmit antennas,' *IEEE Transactions on Wireless Communications*, **1**, 67–75, 2002.

[55] Mody, A.N. and Stuber, G.L., 'Synchronization for MIMO systems,' in *Proceedings of GLOBECOM 2001*, San Antonio, USA, pp. 509–513, November 2001.

[56] Draft ETSI TS 101 XXX, '*Broadband Radio Access Networks (BRAN) HIPERMAN. OFDM Physical (PHY) Layer*,' September 2002.

[57] Petré, F., Leus, G., Deneire, L., Engels, M., Moonen, M. and De Man, H., 'Space–time block-coding for single-carrier block transmission DS-CDMA downlink,' *IEEE Journal on Selected Areas in Communications*, **21**(3), 350–361, 2003.

[58] Sandell, M., van de Beek, J.-J. and Börjessen, P.O., 'Timing and frequency synchronization in OFDM systems using the cyclic prefix,' in *Proceedings 1995 International Symposium on Synchronization*, Essen, Germany, pp. 16–19, December 1995.

[59] Schmidl, T.M. and Cox, D.C., 'Robust frequency and timing synchronization for OFDM,' *IEEE Transactions on Communications*, **45**(12), 1613–1621, 1995.

[60] Tellado-Mourelo, J., '*Peak to Average Power Reduction for Multicarrier Modulation*,' PhD thesis, Stanford University, September 1999.

[61] Nobilet, S., Helard, J.-F. and Mottier, D., 'Spreading sequences for uplink and downlink MC-CDMA systems: PAPR and MAI minimization,' *European Transactions on Telecommunications (ETT)*, **13**, 465–474, 2002.

[62] Le Floch, B., Alard, M. and Berrou, C., 'Coded orthogonal frequency division multiplex', *Proceedings of the IEEE*, **83**, June 1995.

[63] Hirosaki, B., 'An orthogonally multiplexed QAM system using the discrete Fourier transform,' *IEEE Transactions on Communications*, **29**, 1981.

[64] Siohan, P. and Roche, C., 'Cosine modulated filterbanks based on extended Gaussian functions,' *IEEE Transactions on Signal Processing*, **48**, 3052–3061, 2000.

[65] Lacroix, D., Goudard, N. and Alard, M., 'OFDM with guard interval versus OFDM/OffsetQAM for high data rate UMTS downlink transmission,' in *Proceedings of VTC 2001 Fall*, Atlantic City, USA, October 2001.

[66] Mohr, W., Lüder, R. and Möhrmann, K.-H., '*Data Rates and Range Calculations for New Elements of Systems Beyond IMT-2000*', Contribution to WG4 in WWRF6, London, June, 2002.

[67] Kaiser, S. *et al.*, '*WWRF/WG4/Subgroup on New Air Interfaces White Paper: New Air Interface Technologies – Requirements and Solutions*', draft version, August, 2002.

[68] Erceg, V. *et al.*, '*Channel Models for Fixed Wireless Applications*', Contribution to IEEE 802.16, IEEE 802.16.3c-01/29r4, July 17, 2001. Available under the heading 'contributions' in task group 802.16a section of www.ieee802.org/16.

[69] ETSI, '*Universal Mobile Telecommunications System (UMTS): Selection Procedures for the Choice of Radio Transmission Technologies of the UMTS*', Technical Report TR101 112v3.1.0, November, 1997 (Sections B.1.4, B.1.8, and B.1.9).

[70] Cimini, L.J. Jr., 'Analysis and Simulation of a Digital Mobile Channel Using Orthogonal Frequency Division Multiplexing', *IEEE Transactions on Communication*, **33**(7), 665–675, 1985.

[71] van Nee, R. and Prasad, R., '*OFDM for Wireless Multimedia Communications*', Artech House, 2000.

[72] http://www.ieee802.org/11/

[73] http://www.ieee802.org/16/

[74] Kaiser, S. *et al.*, '*White Paper: Broadband Multi-Carrier Based Air Interface*', Draft Version 1.0, May 6, 2002.

[75] Holsinger, J., '*Digital Communication over Fixed Time-Continuous Channels with Memory – With Special Application to Telephone Channels*', PhD thesis, Department of Electrical Engineering, M.I.T., October 1964.

[76] Sandell, M., Wilson, S.K. and Börjesson, P.O., 'Performance Analysis of Coded OFDM on Fading Channels with Non-Ideal Interleaving and Channel Knowledge', in *Proceedings of VTC '97*, May 1997.

[77] Shynk, J., 'Frequency Domain and Multirate Adaptive Filtering', *IEEE SP Magazine*, pp. 14–37, January 1992.

[78] Sari, H., Karam, G. and Jeanclaude, I., 'Frequency-Domain Equalization of Mobile Radio and Terrestrial Broadcast Channels,' *Proceedings of Globecom '94*, San Francisco, November–December 1994, pp. 1–5.

[79] Falconer, D., Ariyavisitakul, S.L., Benyamin-Seeyar, A. and Eidson, B., 'Frequency Domain Equalization for Single-Carrier Broadband Wireless Systems', *IEEE Communications Magazine*, **40**(4), 58–66, 2002.

[80] Falconer, D., Ariyavisitakul, S.L., Benyamin-Seeyar, A. and Eidson, B., '*White Paper: Frequency Domain Equalization for Single-Carrier Broadband Wireless Systems*', www.sce.carleton.ca./bbw/papers/Ariyavisitakul.pdf.

[81] Deneire, L., Vandenameele, P., van der Perre, L., Gyselinckx, B. and Engels, M., 'A Low Complexity ML Channel Estimator for OFDM', in *Proceedings of ICC 2001*, May 2001, pp. 1461–1465.

[82] Deneire, L., Gyselinckx, B. and Engels, M., 'Training Sequence Versus Cyclic Prefix – a New Look on Single Carrier Communication', *IEEE Communications Letters*, **5**(7), 292–294, 2001.

[83] Koffman, I. and Roman, V., 'Broadband Wireless Access Solutions Based on OFDM Access in IEEE 802.16', *IEEE Communications Magazine*, **40**(4), 96–103, 2002.

[84] Witschnig, H., Mayer, T., Springer, A., Koppler, A., Maurer, L., Huemer, M. and Weigel, R., 'A Different Look on Cyclic Prefix for SC/FDE', in *Proceedings of 13th International Symposium on Personal, Indoor and Mobile Radio Communications, PIMRC, 2002*, October 2002, Lisbon, Portugal.

[85] Witschnig, H., Mayer, T., Springer, A., Maurer, L., Huemer, M. and Weigel, R., 'The Advantages of a Known Sequence versus Cyclic Prefix in a SC/FDE System', in *Proceedings of 5th Symposium on Wireless Personal Multimedia Communications WPMC02*, October 2002, Honolulu, Hawaii, USA.

[86] Nassar, C.R., Natarajan, B. and Wu, Z., 'Multi-Carrier Platform for Wireless Communications. Part 1: High Performance, High Throughput TDMA and DS-CDMA Via Multi-Carrier Implementations', *Wireless Communications and Mobile Computing*, **2**(4), 357–379, 2002.

[87] Czylwik, A., 'Comparison Between Adaptive OFDM and Single Carrier Modulation with Frequency Domain Equalization', in *Proceedings of VTC '97*, Phoenix, May 1997, pp. 865–869.

[88] Willink, T.J. and Wittke, P.H., 'Optimization and Performance Evaluation of Multicarrier Transmission', *IEEE Transactions on Information Theory*, **43**(2), 426–440, 1997.

[89] Carron, G., Ness, R., Deneire, L., Van der Perre, L. and Engels, M., 'Comparison of Two Modulation Techniques Using Frequency Domain Processing for In-House Networks', *IEEE Transactions on Consumer Electronics*, **47**(1), 63–72, 2001.

[90] Benvenuto, N. and Tomasin, S., 'On the Comparison Between OFDM and Single Carrier with a DFE Using a Frequency Domain Feedforward Filter', *IEEE Transactions on Communications*, **50**(6), 947–955, 2002.

[91] McDonnell, J.T.E. and Wilkinson, T.A., 'Comparison of Computational Complexity of Adaptive Equalization and OFDM for Indoor Wireless Networks', in *Proceedings of PIMRC '96*, Taipei, pp. 1088–1090.

[92] Clark, M.V., 'Adaptive Frequency-Domain Equalization and Diversity Combining for Broadband Wireless Communications', *IEEE JSAC*, **16**(8), 1385–1395, 1998.

[93] Sari, H., Karam, G. and Jeanclaude, I., 'Transmission Techniques for Digital Terrestrial TV Broadcasting', *IEEE Communications Magazine*, **33**(2), 100–109, 1995.

[94] Aue, V., Fettweis, G.P. and Valenzuela, R., 'Comparison of the Performance of Linearly Equalized Single Carrier and Coded OFDM over Frequency Selective Fading Channels Using the Random Coding Technique', in *Proceedings of ICC '98*, pp. 753–757.

[95] Gusmão, A., Dinis, R., Conceição, J. and Esteves, N., 'Comparison of Two Modulation Choices for Broadband Wireless Communications', in *Proceedings of VTC 2000 Spring*, Tokyo, May 2000.

[96] Montezuma, P. and Gusmão, A., 'A Pragmatic Coded Modulation Choice for Future Broadband Wireless Communications', in *Proceedings of VTC 2001 Spring*, Rhodes, May 2001.

[97] Springer, A., Koppler, A., Witschnig, H. and Weigel, R., '*Robust and Efficient W-LAN Systems for Very High Data Rates Employing Single Carrier Transmission with Frequency Domain Equalization*', contribution to WG4 in 6th WWRF Meeting, London, June 25–26, 2002.

[98] Benvenuto, N. and Tomasin, S., 'Block Iterative DFE for Single Carrier Modulation', *IEEE Electronic Letters*, **39**(19), 2002.

[99] Dinis, R. and Gusmão, A., 'On Broadband Block Transmission Over Strongly Frequency-Selective Fading Channels', in *Proceedings of Wireless '03*, Calgary, Canada, July, 2003.

[100] Müller, S. and Huber, J., 'A Comparison of Peak Reduction Schemes for OFDM', in *Proceedings of IEEE Globecom '97*, Phoenix, December 1997.

[101] Tüchler, M. and Hagenauer, J., 'Linear Time and Frequency Domain Turbo Equalization', in *Proceedings of VTC Fall* 2001.

[102] Struhsaker, P. and Griffin, K., '*Analysis of PHY Waveform Peak to Mean Ratio and Impact on RF Amplification*', IEEE 802.16 contribution IEEE 802.16.3c-01/46, March 6, 2001.

[103] Struhsaker, P., McKown, R. and Griffin, K., 'IEEE 802.16 TG3 Single Carrier Standard: Application of Block Frequency Domain Equalization', in *Proceedings of WCA Technology Conference*, January 14–17, 2002.

[104] Van den Bos, C., Kouwenhoven, M.H.L. and Serdijn, W.A., 'Effect of Smooth Nonlinear Distortion on OFDM Symbol Error Rate', *IEEE Transactions on Communications*, **49**(9), 1510–1514, 2001.

[105] Martone, M., 'On the Necessity of High Performance RF Front-Ends in Broadband Wireless Access Employing Multicarrier Modulations (OFDM)', in *Proceedings of Globecom 2000*, December 2000, pp. 1407–1411.

[106] Come, B., Ness, R., Donnay, S., van der Perre, L., Eberle, W., Wambacq, P., Engels, M. and Bolsens, I., 'Impact of Front-End Non-Idealities on Bit Error Rate Performances of WLAN-OFDM Transceivers', in *Proceedings of RAWCON 2000*.

[107] Tubbax, J., Come, B., Van der Perre, L., Deneire, L., Donnay, S. and Engels, M., 'OFDM versus Single Carrier with Cyclic Prefix: a System-Based Comparison', in *Proceedings of VTC 2001*, Atlantic City, October 2001.

[108] Ochiai, H., 'Power Efficiency Comparison of OFDM and Single Carrier Signals', in *Proceedings of VTC 2002 Fall*, Vancouver, October 2002.

[109] Li, X. and Ritcey, J., 'Bit-interleaved coded modulation with iterative decoding', in *Proceedings of IEEE ICC'99*, pp. 858–863, June 1999.

[110] Armstrong, J., 'Peak-to-average power reduction for OFDM by repeated clipping and frequency domain filtering', *Electronics Letters*, 38(5), 246–247, 2002.

[111] Caire, G., Taricco, G. and Biglieri, E., 'Bit-Interleaved Coded Modulation', *IEEE Transactions on Information Theory*, May 1998.

[112] Li, X., Chindapol, A. and Ritcey, J., 'Bit-interleaved coded modulation with iterative decoding and 8 PSK signaling', *IEEE Transactions on Communications*, 50(8), 1250–1257, 2002.

[113] Fazel, K. and Kaiser, S. (Eds), '*Multicarrier Spread Spectrum and Related Topics*', Kluwer Academic Publishers, 2000 and 2002.

[114] Kaiser, S., 'MC-FDMA and MC-TDMA versus MC-CDMA and SS-MC-MA: Performance Evaluation for Fading Channels', in *Proceedings of 5th International Symposium On Spread Spectrum Techniques and Applications*, 1, 1998.

[115] Raab, F.H. *et al.*, 'Power Amplifiers and Transmitters for RF and Microwave', *IEEE Transactions on Microwave Theory and Techniques*, 50(3), 814–826, 2002.

[116] Qureshi, S.U.H., 'Adaptive Equalization' in *Proceedings of IEEE*, 73(9), 1349–1387, 1985.

[117] Tubbax, J., Van der Perre, L., Donnay, S. and Engels, M., 'MIMO Communications for Single-Carrier Using Decision-Feedback Equalization', in *Proceedings of ICC 2003*, Anchorage, May 2003.

[118] Zervos, N. and Kalet, I., 'Optimized Decision Feedback Equalizer Versus Optimized Orthogonal Frequency Division Multiplexing for High Speed Data Transmission over the Local Cable Network', in *Proceedings of International Conference On Communications*, June 1989.

[119] Alamouti, S., 'A Simple Transmit Diversity Technique for Wireless Communications', *IEEE Journal on Selected Areas in Communications*, 16, 1998.

[120] Al-Dhahir, N., 'Single-Carrier Frequency-Domain Equalization for Space–Time Block-Coded Transmissions Over Frequency-Selective Fading Channels', *IEEE Communication Letters*, 5(7), 305–306, 2001.

[121] Omori, H., Asai, T. and Matsumoto, T., 'A Matched Filter Approximation for SC/MMSE Iterative Equalizers', *IEEE Communication Letters*, 5(7), 310–312, 2001.

[122] Kansanen, K. and Matsumoto, T., 'A Computationally Efficient MIMO Turbo Equalizer', in *Proceedings of IEEE VTC Spring*, Jeju, Korea, April 2003.

[123] Viterbi, A.J., Wolf, J.K., Zehavi, E. and Padovani, R., 'A Pragmatic Approach to Trellis-Coded Modulation', *IEEE Communications Magazine*, 11–19, July 1989.

[124] Manton, J.H., 'Dissecting OFDM: The Independent Roles of the Cyclic Prefix and the IDFT Operation', *IEEE Communication Letters*, 5(12), 474–476, 2001.

[125] Falconer, D. and Ariyavisitakul, S.L., 'Broadband Wireless Using Single Carrier and Frequency Domain Equalization', in *Proceedings of 5th International Symposium on Wireless Personal Multimedia Communications*, Honolulu, October 27–30, 2002.

[126] Adachi, F. and Sawahashi, M., 'Wideband Wireless Access Based on DS-CDMA', *IEICE Transactions on Communications*, E81-B(7), 1305–1316, 1998.

[127] Madhow, U. and Honig, M.L., 'MMSE Interference Suppression for Direct Sequence CDMA', *IEEE Transactions on Communications*, 42(12), 3178–3188, 1994.

[128] Buehrer, R.M., Correal, N.S. and Woerner, B.D., 'Simulation Comparison of Multiuser Receivers for Cellular CDMA', *IEEE Transactions on Vehicular Technology*, 49(4), 1065–1085, 2000.

[129] Sabharwal, A., Mitra, U. and Moses, R., 'MMSE Receivers for Multirate DS-CDMA Systems', *IEEE Transactions on Communications*, 49(12), 2184–2197, 2001.

[130] Chang, C.-M and Chen, K.-C, 'Frequency-Domain Approach to Multiuser Detection in DS-CDMA Communications', *IEEE Communication Letters*, **4**(11), 331–333, 2000.

[131] Chang, C.-M and Chen, K.-C, 'Frequency-Domain Approach to Multiuser Detection over Frequency-Selective Slowly Fading Channels', in *Proceedings of PIMRC* 2002.

[132] Falconer, D., Dinis, R., Lam, C. and Sabbaghian, M., 'Uplink DS-CDMA Using Frequency Domain Orthogonal Signal Sequences', submitted to *IEEE Communication Letters*.

[133] Galda, D. and Rohling, H., 'A Low Complexity Transmitter Structure for OFDM-FDMA Uplink Systems', in *Proceedings of VTC 2002 Spring*, May 2002.

[134] Li, Y. and Sollenberger, N.R., 'Adaptive Antenna Arrays for OFDM Systems with Cochannel Interference', *IEEE Transactions on Communications*, **47**(2), 217–229, 1999.

[135] Benyamin-Seeyar, A. *et al.*, '*SC-FDE PHY Layer System Proposal for Sub-11 GHz BWA (An OFDM Compatible Solution)*', presentation 802.16.3p-01/3lr2, March 12, 2001.

[136] Kadel, G., 'Diversity and Equalization in Frequency Domain – a Robust and Flexible Receiver Technology for Broadband Mobile Communication Systems', in *Proceedings of VTC '97*, May 1997.

[137] Abe, T. and Matsumoto, T., 'Space–Time Turbo Detection in Frequency Selective MIMO Channels with Unknown Interference', in *Proceedings of WPMC01*, Aalborg, Denmark.

[138] Abe, T. and Matsumoto, T., 'Iterative Channel Estimation and Signal Detection in Frequency Selective MIMO Channels', in *Proceedings of VTC2001 Fall*, Atlantic City, USA.

[139] Raleigh, G.G. and Cioffi, J.M., 'Spatio-Temporal Coding for Wireless Communication', *IEEE Transactions on Communications*, **46**(3), 357–366, 1998.

[140] Sampath, H., Talwar, S., Tellado, J., Erceg, V. and Paulraj, A., 'A Fourth-Generation MIMO-OFDM Broadband Wireless System: Design, Performance and Field Trial Results', *IEEE Communications Magazine*, 143–149, September 2002.

[141] Abe, T., Tomisato, S., Matsumoto, T., 'Performance Evaluations of a Space–Time Turbo Equalizer in Frequency Selective MIMO Channels Using Field Measurement Data', *IEE Workshop on MIMO Communication Systems*, London, December, 2001.

[142] Muquet, B., de Courville, M., Duhamel, P. and Giannakis, G., 'OFDM with Trailing Zeros Versus OFDM with Cyclic Prefix: Links, Comparisons and Application to the Hiperlan/2 System', in *Proceedings of ICC 2000*, New Orleans, June 2000, pp. 1049–1053.

[143] Siddiqui, F., Sreng, V., Danilo-Lemoine, F. and Falconer, D., 'Antennas Array Training and Adaptation Techniques in an Unpredictable and Uncontrolled Interference Environment', in *Proceedings of VTC Fall 2003*, Orlando, October, 2003.

[144] Ran, M., Falconer, D. and Seeyar, A.B., '*Mixed Mode OFDM Downstream and FDE-SC Upstream for HIPERMAN*,' BRAN30d025, European Telecommunications Standards Institute BRAN#30, Sophia Antipolis, France, 1–4 October 2002.

[145] Ran, M. and Bellot, L., '*Hiperman SC-FDE UL Interoperable Options*', BRAN30d025, European Telecommunications Standards Institute BRAN#30, Sophia Antipolis, France, 1–4 October 2002.

[146] Haardt, M., Mecklenbräuker, C., Vollmer, M. and Slanina, P., 'Smart Antennas for UTRA TDD,' *European Transactions on Telecommunications (ETT), special issue on Smart Antennas*, J. A. Nossek and W. Utschick, (Eds) **12**(5), 393–406, 2001.

[147] Paulraj, A., Nabar, R. and Gore, D., '*Introduction to Space–Time Wireless Communications,*' Cambridge University Press, 2003.

[148] Hottinen, A., Tirkkonen, O. and Wichman, R., '*Multi-antennas Transceiver Techniques for 3G and Beyond,*' John Wiley & Sons, 2003

[149] Foschini, G.J. and Gans, M.J., 'On Limits of Wireless Communications in a Fading Environment when using Multiple Antennas,' in *Wireless Personal Communications 6*, 311–335, Lumer Academic Publishers, 1998.

[150] Tarokh, V., Seshadri, N. and Calderbank, A.R., 'Space–Time Block Coding for Wireless Communication: Performance Criterion and Code Construction,' *IEEE Transactions on IT*, **44**(2), 744–765, 1998.

[151] Tarokh, V., Jafarkhani, H. and Calderbank, A.R., 'Space–Time Block Coding for Wireless Communication: Performance Results,' *IEEE SAC*, **17**(3), 451–460, 1999

[152] Tarokh, V., Naguib, A., Seshadri, N. and Calderbank, A.R., 'Combined Array Processing and Space–Time Coding,' *IEEE Transaction on Information Theory*, **45**, 1121–1128, 1999.

[153] Telatar, E., '*Capacity of Multi-Antenna Gaussian Channels*,' AT&T Bell Technical Memorandum, 1995.

[154] Chizhik, D., Rashid-Farrokhi, F., Ling, J. and Lozano, A., 'Effect of antenna separation on the capacity of BLAST in correlated channels', *IEEE Communications Letters*, **4**(11), 337–339, 2000.

[155] Bach Andersen, J., 'Array Gain and Capacity for Known Random Channels with Multiple Element Arrays at Both Ends,' *IEEE Journal on Selected Areas in Communication*, **18**, 2172–2178, 2000.

[156] Farrokhi, F.R., Foschini, G.J., Lozano, A. and Valenzuela, R.A., 'Link-Optimal Space–Time Processing with Multiple Transmit and Receive Antennas,' *IEEE Communications Letters*, **5**, 85–87, 2001.

[157] Liberti, J.C. and Rappaport, T.S., *'Smart Antennas for Wireless Communications,'* Prentice Hall, 1999.

[158] Hatzinakos, D., (Ed.), 'Signal Processing Technologies for Short Burst Wireless Communications', Special issue of *Signal Processing*, **80**(10), 2000.

[159] Kuzminskiy, M. and Hatzinakos, D., 'Multistage semi-blind spatio-temporal processing for short burst multiuser SDMA systems', *32nd Asilomar Conference on Signals, Systems and Computers*, pp. 1887–1891, Pacific Grove, 1998.

[160] Abe, T. and Matsumoto, T., 'Space–Time Turbo Detection in Frequency Selective MIMO Channels with Unknown Interference,' in *Proceedings of WPMC01*, Aalborg, Denmark.

[161] Abe, T. and Matsumoto, T., 'Iterative Channel Estimation and Signal Detection in Frequency Selective MIMO Channels,' in *Proceedings of VTC2001 Fall*, Atlantic City, USA, 2001.

[162] Raleigh, G.G. and Cioffi, J.M., 'Spatio-Temporal Coding for Wireless Communication,' *IEEE Transactions on Communications*, **46**, 357–366, 1998.

[163] Sampath, H., Stoica, P. and Paulraj, A., 'Generalized Linear Precoder and Decoder Design for MIMO Channels Using the Weighted MMSE Criterion,' *IEEE Transactions on Communications*, **49**, 2198–2206, 2001.

[164] IST-FITNESS project (http://www.ist-fitness.org).

[165] IST-FLOWS project (http://www.flows-ist.org).

[166] IST-STRIKE project (http://www.ist-strike.org).

[167] IST-MATRICE project (http://www.ist-matrice.org).

[168] Lozano, A., Farrokhi, F.R. and Valenzuela, R., 'Asymptotically optimal open-loop space–time architecture adaptive to scattering conditions', *IEEE 53rd Vehicular Technology Conference, VTC 2001 Spring*, **1**, 73–77, 2001.

[169] Zheng, L. and Tse, D.N.C., 'Diversity and Multiplexing: a Fundamental Tradeoff in Multiple Antenna Channels,' *IEEE Transactions on Information Theory*, **49**(5), 1073–1096, 2003.

[170] Correia L.M., (Ed.), *'Wireless Flexible Personalised Communications'*, John Wiley & Sons, 2001.

[171] Yamada, T., Matsumoto, T., Tomisato, S. and Trautwein, U., 'Results of Link-Level Simulations Using Field Measurement Data for an FTDL-Spatial/MLSE-Temporal Equalizer,' *IEICE Transactions on Communications*, **E84-B**(7), 1956–1960, 2001.

[172] Yamada, T., Matsumoto, T., Tomisato, S. and Trautwein, U., 'Performance Evaluation of FTDL-Spatial/ MLSE-Temporal Equalizers in the Presence of Co-channel Interference – Link-Level Simulation Results Using Field Measurement Data', *IEICE Transactions on Communications*, **E84-B**(7), 1961–1964, 2001.

[173] Czylwik A. and Dekorsy, A., 'System level simulations for downlink beamforming with different array topologies,' in *Proceedings of the IEEE Global Telecommunications Conference (GLOBECOM 2001)*, San Antonio, pp. 3222–3226, 2001.

[174] COST 259, (http://www.lx.it.pt/cost259).

[175] Müller, R., 'A random matrix model of communication via antenna arrays,' *IEEE Transactions on Information Theory*, **48**(9), 2495–2506, 2002.

[176] Lozano, A., and Tulino, A.M., 'Capacity of multiple-transmit multiple-receive antenna architectures,' *IEEE Transactions on Information Theory*, **48**(12), 3117–3128, 2002.

[177] IST-ASILUM project (http://www.ist-asilum.org).

[178] IST-METRA project (http://www.ist-metra.org).

[179] IST-SATURN project (http://www.ist-saturn.org).

[180] Rehfuess, U., Ivanov, K. and Lueders, C., 'A novel approach of interfacing link and system level simulations with radio network planning,' *GLOBECOM'98*, **3**, 1503–1508, 1998.

[181] Pons, J. and Dunlop, J., 'Enhanced system level/link level simulation interface for GSM,' *IEEE VTS 50th Vehicular Technology Conference*, **2**, 1189–1193, 1999.

[182] ESPRIT-ADAMO project (http://www/cordis.lu/esprit).

[183] Avidor, D., Furman, D., Ling, J. and Papadias, C., 'On the Financial Impact of Capacity-Enhancing Technologies to Wireless Operators,' *IEEE Wireless Communications*, **10**(4), 62–65, 2003.

[184] Akyildiz, I.F., Su, W., Sankarasubramaniam, Y. and Cayirci, E., 'A survey on sensor networks', *IEEE Communications Magazine*, 102–114, August 2002.

[185] CEPT T/R 22-06 *'Harmonized Radio Frequency Bands for High Performance Radio Local Area Networks (HIPERLAN) in the 5 GHz and 17 GHz frequency range.'*

[186] ITU Study Groups, Documents 8A-9B/58-E, *'Spectrum Aspects of Fixed Wireless Access,'* 28 September 1998.

[187] *'BroadWay, the Way to Broadband Access at 60 GHz'*, IST-2001-32686, http://www.ist-broadway.org/.

[188] Hirt, W. and Porcino, D., *'Pervasive Ultra-wideband Low Spectral Energy Radio Systems (PULSERS)'* white paper, WWRF/WG4/UWB-subgroup, contribution to the WWRF7 meeting, Eindhoven, December 3–4, 2002.

[189] Medbo, J., Hallenberg, H. and Berg, J.-E., 'Propagation characteristics at 5 GHz in typical radio-LAN scenarios', *IEEE VTC*, **1**, 185–189, 1999.

[190] Stridh, R. and Ottersten, B., *'Spatial characterization of indoor radio channel measurements at 5 GHz'*, Sensor Array and Multichannel Signal Processing Workshop, pp 58–62, 2000.

[191] Smulders, P., 'Exploiting the 60 GHz band for local wireless multimedia access: prospects and future directions', *IEEE Communications Magazine*, **40**(1), 140–147, 2002.

[192] Zasowski, T., Althaus, F., Stäger, M., Wittneben, A. and Tröster, G., 'UWB for noninvasive wireless body area networks: Channel measurements and results', *UWBST '03*, Reston, Virginia, USA, to appear.

[193] ETSI BRAN Normalization Committee, *'Channel Models for HIPERLAN/2 in Different Indoor Scenarios'*, document 3ERI085B, ETSI, Sophia Antipolis, France, 1998.

[194] IEEE 802.11 Standardization Committee, *'Minutes of the High Throughput Study Group Channel Model Special Committee Teleconference on 19 June 2003'*, IEEE 802.11-03/460r0, June 2003.

[195] Dlugaszewski, Z., Wesolowski, K. and Lobeira. M., 'Performance of several OFDM transceivers in the indoor radio channels in 17 GHz band', *IEEE Vehicular Technology Conference, Spring 2001*, **2**, 825–829, May 2001.

[196] Falconer, D., Matsumoto, T., Ran, M., Springer, A. and Zhu, P., *'A Mixed OFDM plus Single Carrier Mode Air Interface'*, white paper, WWRF/WG4/Subgroup on New Air Interfaces, 2002.

[197] Debbah, M., Hachem, W., Loubaton, P. and de Courville, M., 'MMSE Analysis of Certain Large Isometric Random Precoded Systems,' to appear in *IEEE Transactions on Information Theory*.

[198] Debbah, M., Muquet, B., de Courville, M., Muck, M., Simoens, S. and Loubaton, P., 'A New MMSE Successive Interference Cancellation Scheme for Spread OFDM Systems,' *VTC2000*.

[199] Debbah, M., de Courville, M. and Patrick Maille, M., *'Multiresolution Decoding Algorithm for Walsh–Hadamard Linear Precoded OFDM,'* 7th International OFDM Workshop, September 2002.

[200] Debbah, M., Loubaton, P. and de Courville, M., 'Linear Precoded OFDM Transmissions with MMSE Equalization: Facts and Results,' *ICASSP 2002*.

[201] Muquet, B., Wang, Z., Giannakis, G.B., de Courville, M. and Duhamel, P., 'Cyclic-prefixed or Zero-padded Multicarrier Transmissions?,' *IEEE Transactions on Communications* **50**(12), 2136–2148, 2002.

[202] Muquet, B., de Courville, M., Duhamel, P. and Giannakis, G., 'Turbo Demodulation of Zero-Padded OFDM Transmissions,' *IEEE Transactions on Communications*, **50**(11), 2002.

[203] Muck, M., de Courville, M. and Debbah, M., *'Pseudo-random Postfix OFDM-based Modulator for Next Generation 60 GHz WLAN'*, IST Summit 2003, June 15–18, Aveiro, Portugal, 2003.

[204] Rohling, H. and May, T., *'OFDM systems with differential modulation schemes and turbo decoding techniques,'* International Zurich Seminar on Broadband Communications, pp. 251–255, 2000.

[205] van der Perre, L., Thoen, S., Vandenameele, P., Gyselinckx, B. and Engels, M., 'Adaptive loading strategy for a high speed OFDM-based WLAN', *GLOBECOM '98*, **4**, 1936–1940, 1998.

[206] Barreto, A.N. and Furrer, S. 'Adaptive bit loading for wireless OFDM systems', *IEEE Personal, Indoor and Mobile Radio Communications*, **2**, G–88–G-92, 2001.

[207] Liu, X., Chong, E.K.P. and Stroff, N.B. 'Opportunistic Transmission Scheduling with Resource-Sharing Constraints in Wireless Networks', *IEEE Journal on Selected Areas in Commun.*, **19**(10), 2053–2064, 2001.

[208] Sari, H., Levy, Y. and Karam, G., 'An analysis of orthogonal frequency-division multiple access', *GLOBECOM '97*, **3**, 1635–1639, 1997.

[209] Lindner, J., Nold, M., Teich, W.G. and Schreiner, M., 'MC-CDMA and OFDMA for indoor communications: the influence of multiple receiving antennas', *IEEE International Symposium on Spread Spectrum Techniques and Applications*, **1**, 189–194, 1998.

[210] Vandenameele, P., Van Der Perre, L., Engels, M.G.E., Gyselinckx, B. and De Man, H.J., 'A combined OFDM/SDMA approach', *IEEE Journal on Selected Areas in Communications*, **18**(11), 2312–2321, 2000.

[211] Rouquette-Leveil, S. and Gosse, K., 'Space–time coding options for OFDM-based WLAN,' *VTC Spring 2002*, **2**, 904–908, 2002.

[212] Bossert, M., Huebner, A., Schuehlein, F., Haas, H. and Costa, E., *'On Cyclic Delay Diversity in OFDM Based Transmission Schemes'*, 7th International OFDM Workshop, Hamburg, Germany, September 2002.

[213] Zhuang, X., Vook, F.W., Rouquette-Leveil, S. and Gosse, K., 'Transmit diversity and spatial multiplexing in four-transmit-antenna OFDM', *ICC '03*, **4**, 2316–2320, 2003.

[214] Bolcskei, H. and Paulraj, A.J., 'Space–frequency codes for broadband fading channels', *ISIT 2001*, p. 219.

[215] Haardt, M. and Alexiou, A., *'Smart Antennas and Related Technologies (SMART)'*, white paper, WWRF/ WG4/subgroup on smart antennas, 2002.

[216] Javed, A., *'A View on the Development of Enabling Technologies for Future Wireless Systems,'* Nortel Networks, WWRF7, Eindhoven, 3 December, 2002.

[217] Pesik, L.J., Beach, M.A. and Allen, B.H., 'Performance Analysis of Switched-Sector Antennas for Indoor Wireless LANs,' *IEEE VTC Fall 2001*.

[218] Dammann, A., Raulefs, R. and Kaiser, S., 'Beamforming in combination with space–time diversity for broadband OFDM systems,' *ICC 2002*, **1**, 165–171, 2002.

[219] Bartolomé, D. and Pérez-Neira, A.I., 'MMSE Techniques for Space Diversity Receivers in OFDM-based Wireless LANs,' *IEEE Journal on Selected Areas in Communications*, **21**(2), 151–160, 2003.

[220] Dohler, M., Lefranc, E. and Aghvami, H., 'Virtual Antenna Arrays for Future Wireless Mobile Communication Systems,' *IEEE ICT 2002*, Beijing, China, Conference CD-ROM, 2002.

[221] Muquet, B., de Courville, M. and Duhamel, P., 'Subspace-based Blind and Semi-Blind Channel Estimation for OFDM Systems,' *IEEE Transactions on Signal Processing*, **50**(7), 1699–1712, 2002.

[222] Mazet, L., Buzenac, V., de Courville, M. and Duhamel, P., 'An EM based semi-blind channel estimation algorithm designed for OFDM systems,' in *Proceedings of the 36th Asilomar Conference*, 2002.

[223] Bolcskei, H., Gesbert, D. and Paulraj, A.J., 'On the capacity of OFDM-based spatial multiplexing systems,' *IEEE Transactions on Communications*, **50**(2), 225–234, 2002.

[224] Ribeiro Dias, A., Rouquette, S. and Gosse, K., *'MTMR Channel Estimation and Pilot Design in the Context of Space–Time Block-Coded OFDM-based WLAN,'* IST Summit, Thessaloniki, Greece, June 2002.

[225] Pollet, T., van Bladel, M. and Moeneclaey, M., 'BER sensitivity of OFDM systems to carrier frequency offset and Wiener phase noise,' *IEEE Transactions on Communications*, **43**, 191–193, 1995.

[226] Schmidl, T.M. and Cox, D.C., 'Robust frequency and timing synchronization for OFDM,' *IEEE Transactions on Communications*, **45**(12), 1613–1621, 1997.

[227] Hwang, I., Lee, H. and Kang, K., 'Frequency and Timing Period Offset Estimation Technique for OFDM Systems,' *Electronic Letters*, **34**(6), 520–521, 1998.

[228] Simoens, S., Buzenac, V. and de Courville, M., 'A New Method for Joint Clock and Carrier Frequency Offsets Estimator for OFDM Receivers over Frequency Selective Channels,' *IEEE VTC Spring*, Tokyo, **1**, 390–394, 2000.

[229] Wolniansky, P.W., Foschini, G.J., Golden, G.D. and Valenzuela, R.A., 'V-BLAST: An Architecture for Realizing Very High Data Rates Over the Rich-Scattering Wireless Channel,' *International Symposium on Signals Systems and Electronics*, **4**, 295–300, 1998.

[230] Chan, A.M. and Wornell, G.W., 'A class of asymptotically optimum iterated-decision multiuser detectors,' *IEEE ICASSP Conference*, Salt Lake City, Utah May 2001.

[231] Simoens, S., de Courville, M., Bourzeix, F. and de Champs, P., 'New I/Q Imbalance Modeling and Compensation in OFDM Systems with Frequency Offset,' *PIMRC 2002*, Lisboa, Portugal, 2002.

[232] Petrovic, D., Rave, W. and Fettweis, G., 'Common Phase Error Correction Algorithm in the Presence of Phase Noise,' *International OFDM Workshop*, Hamburg, September 2003.

[233] Wu, S. and Bar-Ness, Y., 'A Phase Noise Suppression Algorithm for OFDM-Based WLAN,' *IEEE Communication Letters*, **6**(12), 2002.

[234] Casas, R.A., Biracree, S.L. and Youtz, A.E., 'Time Domain Phase Noise Correction for OFDM Signals,' *IEEE Transactions On Broadcasting*, **48**(3), 2002.

[235] Lu, B., Wang, X. and Narayanan, K.R., 'LDPC-based space–time coded OFDM systems over correlated fading channels: Performance analysis and receiver design,' *IEEE Transactions on Communications*, **50**(1), 74–88, 2002.

[236] Magniez, P., Muquet, B., Duhamel, P. and de Courville, M., 'Improved Turbo-Equalization, with Application to Bit Interleaved Modulations', *Thirty-Fourth Asilomar Conference on Signals, Systems and Computers*, **2**, 1786–1790, 2000.

[237] *'Wireless Integrated Network Sensors,'* University of California, Los Angeles, http://www.janet.ucla.edu/WINS (1993–1998).

[238] Kahn, J.M., Katz, R.H. and Pister, K.S.J., 'Next century challenges: Mobile networking for smart dust,' in *Proceedings of Mobicom 1999*, pp. 483–492, 1999.

[239] Simoens, S., Pellati, P., Gosteau, J., Gosse, K. and Ware, C., 'The Evolution of 5GHz WLAN Toward Higher Throughputs,' *IEEE Wireless Communications*, **10**(6), 6–13, 2003.

[240] Ben Dhaou, I., 'A novel load-sharing algorithm for energy efficient MAC protocol compliant with 802.11 WLAN,' *VTC 1999 Fall*, **2**, 1238–1242, 1999.

[241] Chevillat, P., Jelitto, J., Noll-Baretto, A. and Truong, H.L., 'A dynamic Link Adaptation Algorithm for IEEE 802.11a Wireless LANs,' *IEEE International Conference on Communications (ICC)*, Anchorage, Alaska, May 2003.

[242] Jacob, L., Radhakrishna, P. and Prabhakaran, B., 'MAC protocol enhancements and a distributed scheduler for QoS guarantees over the IEEE 802.11 wireless LANs,' *VTC 2002 Fall*, **4**, 2410–2413, 2002.

[243] Bartolomé, D., Pascual-Iserte, A. and Pérez-Neira, A.I., 'Spatial Scheduling Algorithms for Wireless Systems,' in *Proceedings of ICASSP'03*, April 2003.

[244] Fattah, H. and Leung, C., 'An overview of scheduling algorithms in wireless multimedia networks,' *IEEE Wireless Communications Magazine*, **9**(5), 76–83, 2002.

[245] Le Boudec, J.Y. and Radunovic, B., 'A Unified Framework for max–min and min–max fairness with applications,' in *Proceedings of 40th Annual Allerton Conference on Communication, Control, and Computing*, October 2002.

[246] Borst, S. and Whiting, P., *'The use of diversity antennas in high-speed wireless systems: capacity gains, fairness issues, multi-user scheduling,'* Bell Labs Technical Memorandum, 2001.

[247] Bartolomé, D., Palomar, D.P. and Pérez-Neira, A.I., 'Real-Time Scheduling for Wireless Multiuser MISO Systems under Different Fairness Criteria,' in *Proceedings of ISSPA'03*, July 2003.

[248] Hochwald, B. and Viswanath, S., 'Space–Time Multiple Access: Linear Growth in Sum Rate,' in *Proceedings of Allerton Conference on Communications, Control, and Computing*, October 2002.

[249] Bengtsson, M. and Ottersten, B., 'Optimal and suboptimal transmit beamforming,' in L.C. Godara (Ed.) *'Handbook of Antennas in Wireless Communications,'* CRC Press, 2001.

[250] Heath Jr., R.W., Airy, M. and Paulraj, A.J., 'Multiuser diversity for MIMO wireless systems with linear receivers,' in *Proceedings of Asilomar Conference on Signals, Systems and Computers*, November 2001.

[251] Viswanath, P., Tse, D.N.C. and Laroia, R., 'Opportunistic beamforming using dumb antennas,' *IEEE Transactions On Information Theory*, **48**(6), 1277–1294, 2002.

[252] Chase, D., 'Code combining. A maximum-likelihood decoding approach for combining an arbitrary number of noisy packets,' *IEEE Transactions on Communications*, **33**, 385–393, 1985.

[253] Gidlund, M., 'An approach for using adaptive error control schemes in wireless LAN with CSMA/CA MAC protocol,' *VTC Spring 2002*, **1**, 224–228, 2002.

[254] Hagenauer, J., 'Rate-compatible Punctured Convolutional Codes (RCPC Codes) and their applications,' *IEEE Transactions on Communications*, **36**, 389–400, 1988.

[255] Falahati, S. and Svensson, A., 'Hybrid type-II ARQ schemes with adaptive modulation systems for wireless channels,' *IEEE Vehicular Technology Conference 1999, Fall*, **5**, 2691–2695, 1999.

[256] Calvanese Strinati, E., Simoens, S. and Boutros, J., 'Performance evaluation of some hybrid ARQ schemes in IEEE 802.11a networks,' *IEEE Vehicular Technology Conference 2003, Spring*, **4**, 2735–2739, 2003.

[257] Mangold, S., Berlemann, L. and Walke, B., 'Radio Resource Sharing Model for Coexistence of IEEE 802.11e WLAN,' *ICCT Conference*, Beijing, China, April 2003.

[258] Benveniste, M., 'Wireless LANs and Neighbourhood capture', *PIMRC 2002 – The 13th IEEE International Symposium on Personal Indoor and Mobile Radio Communications*, Portugal, 2002.

[259] Hiertz, G. and Habetha, J., *'A new MAC protocol for a wireless multi-hop broadband system beyond IEEE 802.11,'* WWRF9 meeting, Zurich, July 1–2, 2003.

[260] Mäki, S., 'Security Fundamentals in Ad hoc Networking,' in *Proceedings of the Helsinki University of Technology Seminar on Internetworking – Ad hoc Networks*, May 2000.

[261] Aarts, E.H.L., 'Ambient Intelligence: calming, enriching and empowering our lives,' *Password*, **8**, Royal Philips Electronics, July 2001.

[262] Zimmerman, T.G., 'Wireless networked digital devices: A new paradigm for computing and communication,' *IBM Systems Journal (Pervasive Computing)*, **38**(4), 566–574, 1999.

[263] IEEE P802.15.3 SG3a, '*Application Summary (summary to the response to the Call for Applications)*,' IEEE Document 02149r0P802-15_SG3a.

[264] Federal Communications Commission (FCC), '*Revision of Part 15 of the Commission's Rules Regarding Ultra-Wideband Transmission Systems*,' First Report & Order, ET Docket 98-153, FCC 02-48; Adopted: February 14, 2002; Released: April 22, 2002.

[265] Ross, G.F., '*Transmission and Reception System for Generating and Receiving Base-band Duration Pulse Signals for Short Base-band Pulse Communication System*,' U.S. Patent 3728,632, 1973.

[266] Federal Communications Commission (FCC), '*Revision of Part 15 of the Commission's Rules Regarding Ultra-Wideband Transmission Systems*,' Notice of Proposed Rule Making (NPRM), ET Docket 98-153, FCC 00-163, May 11, 2000.

[267] Reference list at http://www.aetherwire.com/CDROM/Welcome.html.

[268] Publications in: *United States Patent – Reexamination Certificate, B1 5361,070* (concerning T. McEwan's Ultra-Wideband Radar Motion Sensor, U.S. Patent No. 5361,070), May 16, 2000.

[269] *1st European Workshop on Ultra-Wideband Technology*, Brussels, Belgium, December 12, 2000.

[270] European Radiocommunications Office (ERO), *1st Workshop on Introduction of Ultra Wideband Services in Europe*, RegTP, Mainz, Germany, March 20, 2001.

[271] European Radiocommunications Office (ERO), *2nd Workshop on Introduction of Ultra Wideband Services in Europe*, RegTP, Mainz, Germany, April 11, 2002.

[272] Win, M.Z. and Scholtz, R.A., 'Ultra-wide bandwidth time-hopping spread-spectrum impulse radio for wireless multiple-access communications,' *IEEE Transactions on Communications*, **48**(4), 679–691, April 2000.

[273] Batra, A. *et al.* (Texas Instruments *et al.*), '*Multi-band OFDM Physical Layer Proposal for IEEE 802.15 Task Group 3a*,' IEEE 802.15.3a 03268, September 2003.

[274] IEEE P802.15 Working Group for Wireless Personal Area Networks (WPANs), '*IEEE P802.15SG3a Call for Applications*,' IEEE P802.15-02/027r0; Submitted: November 26, 2001; Released: December 11, 2001.

[275] Porcino, D. and Shor, G., '*Response to CFA – ULTRAWAVES*,' IEEE 802.15 WPAN™ High Rate Alternative PHY Study Group 3a (SG3a) (http://www.ieee802.org/15/pub).

[276] Porcino, D., 'Ultra-Wideband Overlay on Existing Radio Services: Mitigation Techniques for Peaceful Coexistence,' *International Workshop on UWB Systems (IWUWBS) 2003*, Oulu, Finland, 2–5 June 2003.

[277] Hovinen, V., Hämäläinen, M., Tesi, R., Hentilä, L., Laine, N., Porcino, D., and Shor, G., '*A proposal for an indoor UWB path loss model*,' IEEE Document 802.15-02/280, IEEE 802.15.3a, Vancouver, July 2002.

[278] Hirt, W., 'Ultra-Wideband Radio Technology: Overview and Future Research,' *Computer Communications*, **26**(1), 46–52, January 2003.

[279] Gandolfo, P., '*XtremeSpectrum – SG3a CFA response*,' IEEE P802.15 ALT PHY Study Group, Document: 02031r0P802-15_SGAP3-CFAResponseAltPHY.ppt, January 14, 2002.

[280] Weisenhorn, M. and Hirt, W., 'Performance of binary antipodal signaling over the indoor UWB MIMO channel,' in *Proceedings of IEEE 2003 International Conference on Communications (ICC 2003)*, Anchorage, AK, USA, paper CT 19-2, May 11–15 2003.

[281] Porcino, D. and Hirt, W., 'Ultra-Wideband Radio Technology: Potential and Challenges Ahead,' *IEEE Communications Magazine*, 66–74, July 2003.

7.7 Credits

The following individuals have contributed to the contents of this chapter: Stefan Kaiser (German Aerospace Centre (DLR), Germany); Arne Svensson (Chalmers University of Technology, Sweden); Bernard Hunt (Philips Research, UK); Rui Dinis (Instituto Superior Técnico, Lisbon, Portugal); David Falconer (Carleton University, Ottawa, Canada);

Tadashi Matsumoto (University of Oulu, Finland); Moshe Ran (Holon Academic Institute of Technology, Israel); Andreas Springer (University of Linz, Austria); Peiying Zhu (Nortel Networks, Canada); Martin Haardt (Communications Research Laboratory, Ilmenau University of Technology, Germany); Angeliki Alexiou (Bell Labs, Wireless Research, Lucent Technologies, UK); Karine Gosse (Motorola Labs, France); Walter Hirt (IBM Zurich Research Laboratory, Rüschlikon, Switzerland) and Domenico Porcino (Philips Research, UK).

8

Scenarios for the Future Wireless Market

Edited by Ross Pow (Analysys) and Klaus Moessner (University of Surrey)

8.1 Current Status and Prediction for Evolution of the Wireless Market

How might the wireless market develop over the years up to 2010? The forecast developed is based on understanding of the current position of the mobile telecoms industry and the future impact of various drivers of change. The output from this study will help to identify the technological research and development priorities that need to be initiated by the wireless industry.

This research has used the construction of three scenarios – 'Blue', 'Red' and 'Green' – to build alternative possible pictures of the future wireless market. The scenarios focus on the societal, business and regulatory trends that will drive demand for, and use of, wireless communications over the period to 2010, and are based on the different potential outcomes of a range of market drivers.

A range of uncertainties, including customer demand, service innovation and industry structure, lead to three scenarios for the future wireless market. These scenarios for the future of the industry have been created based on the different potential outcomes of a range of market drivers (shown in Figure 8.1), including customer willingness to adopt new services, innovation in service propositions and marketing, technical success of 3G services, convergence, industry structure and regulation.

8.2 The Wireless World of the 'Blue' Scenario

The 'Blue' scenario is a world in 2010 where wireless is the dominant technology in connecting people and machines. Customers have demonstrated an increasing appetite for new wireless technologies and services and use these extensively in all aspects of their lives. The demand for quality digital content has grown steadily and people are

Technologies for the Wireless Future. Edited by R. Tafazolli
© 2005 John Wiley & Sons, Ltd ISBN: 0-470-01235-8

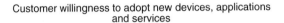

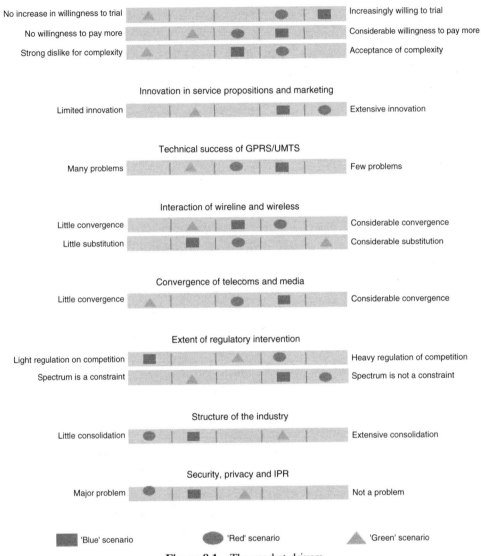

Figure 8.1   The market drivers

keen to access this wherever they can, using simple and reliable devices. Large, strongly branded vendors and service providers have invested heavily in offering well-packaged products and services which customers can use easily, initially subsidizing devices and services to maximize adoption in the early years. Regulators have ensured that there is plenty of spectrum available to meet the requirements of mobile and WLAN networks. Customers have shown that they are willing to pay more for devices and content that meet their day-to-day needs so, following on from the success of 2.5 and 3G technologies, and against a backdrop of stable capital markets, companies are now introducing even higher speed mobile networks that can meet the demand for accessing personal, work and entertainment content while on the move. Bigger operators, vendors and content owners dominate, though niche players do well in selling to these.

8.2.1 The Market Drivers in the 'Blue' Scenario

The state of the market drivers for the blue scenario is defined by the following assumptions: customers demonstrate increasing willingness to adopt, and pay for, new wireless devices, applications and services. They do not like complexity though, and show a strong preference for ease-of-use. Mobile devices have to be multifunctional, as individuals use them to use different services and applications. Customers tend to buy prepackaged equipment and services, with specified applications and content, from larger vendor and operator brands. Big content brands do well and customers want this digital content integrated across different delivery platforms, linking the home, office and public spaces. In the early years these devices and services are subsidized by service providers and vendors, but over time this reduces as customers are willing to pay more for the value they get. On the deployment side, implementation of 2.5 and 3G goes well, and mobile operators invest in innovative service propositions, with marketing that is not focused on technology, but which reflects the needs of different end user segments. WLANs do well where service quality is high, and create increasing demand for true broadband mobility. Growth in revenues from developed markets mean that these markets remain the primary focus of operators and vendors. Customers tend to be 'owned' by operators, working with other big brands, large vendors dominate the technology development process, while innovative smaller players can thrive as suppliers in the equipment and service value chains. There is some equipment specialization, with niche devices emerging, but people would generally rather not have to use multiple devices, so the focus is on providing integrated devices with easy to use software.

Large mobile operators, consumer equipment manufacturers, content owners and software firms collaborate to ensure that all the major device platforms and public mobile networks can interoperate, enabling the seamless delivery of applications, content and services in different locations. Mobile specific content becomes more and more important and the demand for mobile specific applications also grows. Vendors and operators also take advantage of the rapidly increasing machine-to-machine (m2m) market, driven by other sectors such as transport, health and consumer goods. Against an improving economic and financial market backdrop, the mobile industry invests in higher speed technologies. New devices, applications and services emerge to capitalize on the faster data rates. These trends are depicted in Figure 8.2.

8.2.1.1 Customer Willingness to Adopt New Devices, Applications and Services

Mass market business and consumer customers show a rapidly increasing willingness to trial new wireless devices, applications and services that aim to improve their daily life. In the home and office, wireless is used to connect different devices for entertainment, communication, computing and information access, thereby linking digital content either

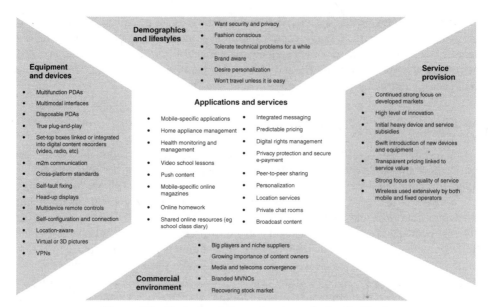

Figure 8.2   Example of demand and supply trends (blue)

created or purchased by the user. People generally avoid long-distance travel wherever possible, following from an increased emphasis on balanced lifestyles, economic pressures and fears over terrorism. Fashion is a strong driver of demand for all wireless devices and services. Customers prefer branded goods, especially those offering the latest designs and innovations. Colour, size and materials are often as important as functionality and cost. Customers also like prepackaged services, where they can get predefined bundles of applications and content. A lot of content is 'pushed' from service providers to customers, some of this becoming broadcast by the end of the period.

Mobile network data services are viewed as a valuable and natural part of lifestyle and are adopted by people of all ages. These range from highly branded, entertainment-focused services targeted at the youth market, to more practical health and life-facilitating services targeted at more mature users.

There is high use of peer-to-peer applications for downloading copyright material, which is tracked using advanced watermark technology and paid for using electronic wallets. Customers are willing to pay for content that is strongly branded and not available through other channels.

8.2.1.2 Innovation in Service Propositions and Marketing

New mobile devices are easy to use, with multimodal interfaces and multimedia presentation, encouraging customers to try them. Operators initially provide heavy subsidies of mobile devices to stimulate take-up and vendors also price aggressively in the home and office networking market. As attractive mobile services are introduced swiftly, and without too many problems, this further inspires demand, and customers soon appreciate the value of the new equipment so that subsidies are reduced and the price paid by the end user increases.

Mobile voice prices continue to fall, but slower than in the period 1998–2002, as operators focus on developing innovative, cash generative mobile data services, reducing their need to compete on the price of voice. Data charges are transparent and easy to understand. Pricing evolves to represent service value, and charging for airtime becomes outdated. Because of a reduction in travel, customers demand being able to access centrally stored information from wherever they are in the world, often using disposable high-speed devices with limited processing power that they buy at their destination. Vendors develop and produce advanced devices to serve the demand for integrated voice and data services within specific customer segments. There is a strong focus on quality of service (QoS) and investment in fault tolerance and self-fixing devices. As demand grows for home and office networking equipment, the focus is on equipment that is easy to install and maintain. Devices are designed to recognize each other and are enabled to 'learn' how to connect to new equipment, either through m2m communication or via instructions over a network. The need to integrate PCs, TVs and similar equipment is accompanied by enormous customer interest in wireless, in preference to wired, solutions. By 2010, most equipment ships without connector wires and 'self-connects' to other devices in its local area. Fixed operators begin to use the available spectrum to deliver broadcast services to households using fibre to the kerb and fixed wireless access (FWA). This proves successful and cost-effective and becomes one of the fastest growth areas in developing markets.

8.2.1.3 The Technical Success of GPRS/UMTS

Network sharing and efficient deployment lead to lower than anticipated investment costs. This enables operators and service providers to be more flexible on pricing, which is used as a key tool in stimulating the market. The available technology readily supports the advertised services, fulfilling the requirements of users. Where technical problems do occur, user tolerance is reasonably high and there is relatively low churn from the new services. The early introduction of personalized content and services goes well.

8.2.1.4 Interaction of Wireline and Wireless

Initially, both the fixed and mobile broadband markets will enjoy steady growth. WLANs will be largely run by large mobile operators who use these to complement their 2.5, and then 3G, networks. Over time, customers are enabled to move easily between the different networks. Ambient technologies are developed to allow customers to make use of their personal service and preference information, depending on which access network they connect to. A range of applications are designed that are able to take advantage of either medium. In the mobile environment, applications and services have been designed with mobility in mind, making them more valuable to the end customer.

Ultimately, there is a steady trend towards substitution of fixed by mobile in the public environment. In the home and office, short-range wireless also becomes the dominant technology for connecting a wide range of devices to the fixed network. Migration of voice traffic from fixed to mobile continues but does not accelerate, the business use of virtual private networks (VPNs) grows strongly, and voice over internet protocol (VoIP) starts to make an impact in the corporate segment.

Convergence of telecoms and mediaWLAN networking speeds increase and business and consumer customers make use of this to connect a wide range of devices. Generally, telecoms and media entertainment devices become more interoperable. There is an increasing demand for entertainment services while on the move. Mobile operators offer attractive packages based around information and entertainment services, and these are adopted by users through the introduction of broadcast-like capabilities to mobile networks and WLANs. Initially, customers are happy to accept lower quality sound and images, but towards the end of the period, the higher speed access networks are able to deliver them content much closer to that obtained from fixed broadcast platforms. Content owners gain an increasingly strong position and take a growing proportion of customer spend on mobile technologies.

8.2.1.5 Extent of Regulatory Intervention

Regulators remain largely hands-off in terms of mobile service competition because high customer adoption rates encourage mobile operators to lower prices and meet their roll-out obligations. Regulators do not attempt to enforce unbundling of service provision from network businesses, and development is driven by market needs. The regulators, industries and consumer groups work together to introduce acceptable and practical digital rights management solutions. Regulators realize the demand for wireless applications and services so act to ensure, through consultation with the appropriate industries, that sufficient spectrum is made available. An increasing proportion of licence-exempt spectrum comes under regulatory control. Regulators harmonize the rules and conditions for these licences and this becomes similar across all frequencies. Regulators decide to introduce charges for the commercial use of WLAN spectrum, reducing the cost difference with the use of 3G licences.

8.2.1.6 Structure of the Industry

There is little consolidation in the mobile market. Each country continues to support four to five network operators, resulting in fierce competition. MVNOs with a strong brand do well, usually focusing on delivering services at niche customer segments. Different players in the mobile value chain work to develop effective revenue-sharing mechanisms, increasing the viability of small innovative players and resulting in a more rapid introduction of new content and services, with richer and more varied content. Many different large brands fight for consumers' attention, resulting in many formal and ad hoc partnerships between device manufacturers, content owners and network operators.

8.2.1.7 Security, Privacy and IPR

Various interest groups successfully lobby governments and regulators to introduce strict security and privacy legislation. This legislation reinforces the position of major players and trusted brands and does not reduce the willingness of customers to adopt new services. Users trust wireless networks sufficiently to experiment with new data services such as business VPNs, location-based applications, m-health and m-commerce, and entrust service providers with personal data to facilitate push services.

As a result of international conflict and terrorism, governments place increasing pressure on service providers and those with access to personal information to share that for security purposes. Increased restrictions are placed on carrying electronic goods in planes as airlines respond to concerns about interference with flight systems, as well as possible security threats.

8.2.2 A Day in the Life (2010) – Blue Scenarios

Two scenarios, considering the outlined drivers, describe the days of individuals in the 'blue' wireless world of 2010.

Helene, Student, Aged 15, Developed Country

07.00   Helene wakes up. After showering, she checks the latest downloads of chart songs on her X-Life (which she got for Christmas from one of the big stores in town) and set her alerts and ringtones for the day from these.

07.30   She checks multimedia messages from friends on her X-Life, then goes down to breakfast. Helene's brother is already downstairs playing Maniac Builders on his Gamebox3 using their 120 cm TV as a screen, while downloading new characters from the internet as he plays. He's looking forward to playing it at school with his friends, although he can't get it to work as fast there.

08.30   She sits at the kitchen table and checks her class diary to see which of her friends are working at school and which are working from home, and to remind her of any scheduled lessons during the day.

09.00   She stays online and meets three of her friends in their private Smart-Zone where they select some background music, chat about boys and arrange lunch.

09.45   Helene browses the latest fashion pages from a selection of online magazines and invites her friend to join her in a virtual shopping trip that evening.

10.15   She sends a remastered video clip of Kylie to her Dad, who has also left for work, then asks him for EUR30 to go into her electronic wallet.

10.30   Helene makes herself comfortable in the garden and looks up her homework from the SchoolWeb.

11.00   She attends her maths class via the new videolesson service introduced nationally by the education department the previous year.

12.30   Her X-Life reminds her about lunch and she sets off to meet her friends. On the bus she puts on her Adidas sunglasses and switches over the X-Life to the wearable display so that she can watch a kids TV programme.

13.30   Helene and her friends receive vouchers offering 2 for 1 at a nearby VitaBar so they vote to eat there instead.

14.15   Her father's face appears on her X-Life to say it's time she was getting back to school – how embarrassing!

16.00   She notices from the locator that her father is in Paris but her mother is nearby so she makes sure she is home on time.

19.15   After finishing schoolwork, Helene's friend comes round and they try on clothes at the virtual mall.

21.25   Helene's friend goes home but they chat to each other while watching the 15th series of Big Brother.

Robert, Businessman, Aged 50, Developed Country

06.00 Robert wakes up. He downloads his daily programme to his exercise equipment and does his workout.

06.30 After showering, Robert uses the remote PC control pad to request a summary of his national property and share portfolios while he is in the shower. While shaving, he listens to his messages – mostly email but a few calls from his colleagues in the US – before switching to the AxleGrease website to listen to some classic Kylie tracks. Much more reliable than Morpheus ever was and it is legal!

06.45 According to the readings taken through sensors in his shaver, Robert's pulse appears to be slightly higher than normal so a heart rate and body fat reading is sent to Dr Online. Dr Online returns an all-clear but suggests that Robert should watch his diet today and sends a menu for lunch to Robert's CPS (complete personal system) and a shopping list of ingredients for dinner to the my@home management system.

07.45 Over breakfast, Robert browses his personalized news round-up on the TV, then asks his CPS to download various files from his office.

08.30 A videomail comes in from one of his big customers in Paris and he decides he has to go there in person to close the sale they are negotiating. This is a nuisance as he hates travel and the security at the airport has made flying less attractive than it used to be. However, a sale is a sale and so he purchases airline tickets and a portable CPS using the ufly.sec site and pays using his corporate e-wallet. So reassuring to have these highly secure. sec websites following the rise in fraud.

09.15 He works on his CPS as he travels to the airport and then leaves that at the deposit centre for picking up when he returns.

13.10 On arrival at the airport he picks up the disposable CPS he paid for and selects a taxi based on its recommendation of service. He receives a message from his daughter, including rare old Kylie footage – superb! Robert transfers EUR30 into his daughter's e-wallet. Then he accesses the details and photos of his meeting, and of the attendees who have all input their status; and obtains an up-to-date company profile of his client.

14.00 Just before the meeting starts, Robert checks his daughter's whereabouts using location tracker and agrees with his wife to chase her home to do her schoolwork.

16.30 Meeting goes well, so his sales manager videoconferences in to talk money and close the deal. They need to decide when to meet again, so they ask their CPSs to select a time and location convenient to them all, which takes less than a minute.

17.30 Robert leaves the meeting and heads to a seafood restaurant in Montmartre which he picked on the basis of it offering most of the food suggested by Dr Online.

19.30 Robert finishes his meal then moves to a pavement café to attend to the rest of the day's work. He conferences in to a number of meetings in the US, selecting office style wallpaper for his video images.

23.00 After catching the plane home Robert and his wife relax with a glass of wine and try out some virtual holiday venues.

8.2.3 Event Timeline and Market Outcomes

The event timeline for the blue scenario is shown in Figure 8.3.

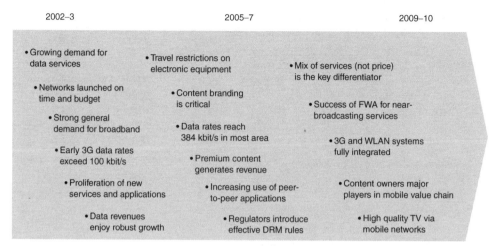

2002–3   2005–7   2009–10

• Growing demand for
  data services
                        • Travel restrictions on
                          electronic equipment
                                                        • Mix of services (not price)
                                                          is the key differentiator
  • Networks launched on
    time and budget
                          • Content branding
                            is critical
                                                          • Success of FWA for near-
                                                            broadcasting services
    • Strong general
      demand for broadband
                            • Data rates reach
                              384 kbit/s in most area
                                                            • 3G and WLAN systems
                                                              fully integrated
      • Early 3G data rates
        exceed 100 kbit/s
                              • Premium content
                                generates revenue
        • Proliferation of new
          services and applications
                                • Increasing use of peer-
                                  to-peer applications
                                                              • Content owners major
                                                                players in mobile value chain
          • Data revenues
            enjoy robust growth
                                  • Regulators introduce
                                    effective DRM rules
                                                                • High quality TV via
                                                                  mobile networks

Figure 8.3   Event timeline blue

The markets of a 'blue' scenario will be based on:

- vendors and global operators focusing on large and developed markets as main drivers of profit growth;
- voice revenues driven largely by new customers in non-European markets;
- the 2.5/3G devices that dominate in Europe and North America, and some devices operating on faster 3G+ networks;
- WLAN devices becoming more prevalent, but customers generally preferring integrated devices – a proportion of devices can use both 2.5/3G and WLANs;
- data revenues growing quickly worldwide as a wide variety of entertainment and other content-related services are used, in addition to basic and advanced messaging;
- devices and services that are either priced aggressively or subsidized in the early years but, by 2010, customers, recognizing the value they gain from such devices and services, will be prepared to pay higher prices.

A summary of some key figures relating to the blue scenario is shown in Table 8.1.

8.3 The Wireless World of the 'Red' Scenario

The 'Red' scenario is a world in 2010 where customers are highly experimental, intent on finding and trying out the applications and services that meet their needs best. The great success of fixed broadband has resulted in people becoming very familiar with the open nature of the internet and the ready access to a hugely diverse range of content and applications. Users know that not all the information is of good quality but they also know that they can lay their hands on lots of content and applications without having to pay much, if any, money. They apply this thinking to mobile networks and generally shy away from overly prepackaged services, preferring to use mobile devices simply to extend their access to content from the internet. New wireless networking products are enthusiastically embraced in the home and office, where wireless connectivity becomes the

Table 8.1   The blue scenario

	2002	2010	CAGR[a]
Mobile customers by region (millions)			
Western Europe	287	323	1.5 %
North America	122	232	8.4 %
Asia Pacific	397	650	6.4 %
RoW	221	369	6.6 %
Total mobile customers	1026	1575	5.5 %
Mobile devices split by technology			
2G	97 %	26 %	
2.5/3G	2 %	53 %	
WLAN	1 %	21 %	
End customer expenditure on mobile devices and services (EUR millions)			
Devices	38 983	70 215	7.6 %
Voice services	329 931	410 955	2.8 %
Data services	37 802	243 302	26.2 %
Total end customer expenditure on mobile	406 716	724 472	7.5 %

[a] CAGR = compound annual growth rate

default edge of network solution. Customers like owning a range of specialist devices that meet their specific needs and more of these are introduced to the market by new equipment vendors that take advantage of the general trend towards open technical platforms. Mobile service providers and equipment vendors, faced with strong competition, have to price their products and services aggressively. The rapid early adoption of WLANs has resulted in a profusion of ad hoc networks, including in developing markets, where short-range wireless networks overlap to provide cheap voice and data services to a rapidly growing number of customers. The fragmentation in network ownership, device manufacture and application development results in an emphasis on ambient technologies to enable device collaboration and m2m communication. This is also required by users who constantly want to swap information. New equipment vendors and service providers enter the market, while established MVNOs and system integrators increasingly challenge vertical operators.

8.3.1 The Market Drivers in the 'Red' Scenario

Users gradually increase in their willingness to try out new wireless devices, applications and services. However, they are heavily influenced by their experience of wireline services and do not want to pay a lot extra for content or applications. Prepackaged services do not tend to flourish.

   Implementation of 2.5 and 3G networks face some delays and quality problems, but the initial adoption rates look promising. Operators still tend to focus their marketing too much on technology, and innovation in service propositions is limited. Competitors, in the form of MVNOs and specialist service providers, emerge to challenge the vertically integrated operators, taking ownership of customers and competing through differentiation on niche service bundling.

Users are increasingly comfortable with complex devices, applications and services, buying the best solution that meets their needs, even if this needs effort on their part to set up. Exposure to wireline broadband services and the growing penetration of home film and music entertainment products results in users demanding technology that can help them integrate devices, especially in the home, through wireless networking. Wireless devices become quite specialized, with people content to carry a number of devices, each of which is very good at meeting a particular need. Service providers and equipment vendors, faced with strong competition, have to price their products and services aggressively. The demand for m2m devices grows, as manufacturers in a wide range of sectors integrate wireless devices to improve functionality, performance and durability.

WLANs grow fast in the business market for both data and voice, and low-cost devices begin to penetrate the home market. Ad hoc networks proliferate, and increasingly these are able to interlink to provide alternatives to public fixed and mobile networks. Ambient technology is used to assist interoperability based on open platforms.

Customers are prepared to spend time getting the digital content they want, and prefer that to consuming the material that is pushed at them by mobile network operators. Ambient technology enables users to track their own preferences. A lot of this relies on device storage and processing power, as the mobile network bandwidth is limited, and users prefer the applications they have sourced to the ones that mobile operators offer them. Users also like to share a lot of information with each other and individual device manufacturers work together to make this possible.

System integrators begin to increase their share of the corporate market, delivering services that encompass wide area networking as well as LANs. Operators suffer as customers expect prices to decline and their services become more commoditized. New entrants in the vendor and service provider markets emerge to offer niche devices and services. Limited revenue growth in developed markets means operators look to developing countries to increase the penetration of their basic voice and messaging services.

These trends are depicted in Figure 8.4.

8.3.1.1 Customer Willingness to Adopt New Devices, Applications and Services

Business and consumer customers gradually show a growing willingness to trial new wireless devices, applications and services. Users grow more used to complexity and can cope with an increasing amount of self-configuration. There is very strong demand for fixed internet services. This is accompanied by very strong growth in the use of peer-to-peer applications for downloading copyright material, much of which is done illegally.

Customers are initially discouraged from using mobile data services by per minute and per kilobyte charging, which they associate with high costs. Customers also refuse to pay for mobile content that they think should be free, and there is little initial demand for paying a premium for content over mobile networks when this can be acquired via wireline networks and uploaded to a growing range of wireless devices. Wireless extensions to fixed networks become more common in the workplace from 2003, and in the home from 2006. This facilitates access to internet services – including access to 3D games and online education at home, and high-speed data transfer and videoconferencing at

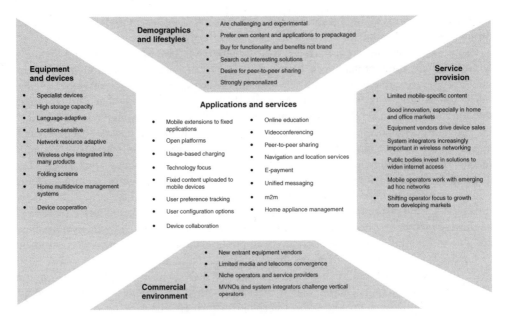

Figure 8.4 Example of demand and supply trends (red)

work – from a broader range of locations, such as people's gardens and hot desk meeting rooms. Wider WLAN coverage enables greater consumption of mobile data and content services in hotels, airports and coffee shops.

Application development initially focuses on adding mobility to fixed network applications, rather than the development of new types of mobile-specific content, but the access to a wider range of content and applications on the move via ad hoc networks leads, over time, to more demand for applications and services that require higher speed mobile networks.

8.3.1.2 Innovation in Service Propositions and Marketing

Niche devices are developed to meet the specific needs of different customer segments, including portable home-and-away entertainment consoles for the youth market and unified message managers for business users. These devices require ever-increasing amounts of memory, processing power and battery life to store and manipulate the huge amount of information carried around by users. Wireless devices become more adaptive to their situation, responding to differences in language, network resources and location.

Equipment vendors enjoy great success in meeting the demand for home and office wireless networking, either through selling their own branded equipment or by white labelling to other big organizations, but both vendors and service providers and equipment vendors have to price their products and services aggressively faced with strong competition. System integrators are increasingly important in putting in place large wireless networks for corporate customers that fully integrate voice and data services. Public authorities and other government bodies also invest in high speed WLAN solutions to make the internet more accessible to local people.

Mobile network operators acknowledge the emergence of alternative ad hoc networks that are providing connectivity to people on the move. They work with equipment vendors to offer devices that can operate across a wide range of access technologies. Billing and operational support system (OSS) vendors develop systems to track IP traffic across diverse networks. This enables network operators to deliver core voice and messaging services to end users, while also deriving revenues from their intermittent use of high-speed access while on the move. However, many users find out how to make use of alternative network capacity without paying for it, and a range of service discovery software becomes available for locating and gaining access to these networks. This impacts on developing markets where interlinking short-range networks emerge, on a structured and ad hoc basis, to meet the voice and data needs of a fast growing customer base. There is rapid and wide-ranging innovation in m2m applications and services (e.g. enabling easier home management through performing remote diagnostic checks on the central heating/utility systems, or introducing intelligent inventory management systems for the fridge, linked directly to online purchasing systems). This scenario relies on the following assumptions:

- the technical success of GPRS/UMTS;
- services being introduced incrementally;
- there being a number of technical glitches, but as initial user numbers are low, long-term damage to the market's development is prevented;
- by the time demand for broadband mobility starts to emerge around 2009, network problems have been ironed out;
- network costs mostly meet expectations;
- low initial demand for services inspires network operators to limit deployment wherever regulators allow;
- networks do not need to be upgraded to cope with high user numbers as early as initially anticipated;
- as network costs follow expectations, operators are not induced to charge higher prices, nor are they offered great freedom to stimulate the market with low-priced service launches.

8.3.1.3 Interaction of Wireline and Wireless

Fixed broadband is the primary tool for delivery of content and applications over the period 2002 to 2007. However, while initial problems with deployment of new mobile data services induce operators to focus on voice services to increase revenues, public WLANs become an increasingly important part of the wide-area broadband infrastructure, and enjoy strong growth during the period 2004 to 2008.

As a result of the proliferation of public WLAN services, consumers and businesses get used to accessing a wider range of content and applications while on the move. This gradually begins to inspire users to see the benefits of applications and content that require higher speed mobile networks. Competition to generate voice traffic leads to continued migration of voice from fixed to mobile networks. As broadband mobile data starts to interest customers from 2009 onwards, operators develop new ways of differentiating their offerings and stimulating usage. Price competition on voice becomes less important.

8.3.1.4 Convergence of Telecoms and Media Over Time

Telecoms and media converge in terms of services, content and devices. Initially, media is consumed at home, or downloaded while at home, over a fixed broadband connection. Convergence is instigated by the desire amongst customers to access their fixed internet content, and towards the end of the scenario period, while on the move. Demand grows for greater integration and interoperability between different technical platforms (mobile data, fixed broadband, broadcast, etc). Mobile services continue to be dominated by person-to-person communications (e.g. voice, email or text messaging). Within the mobile value chain, most of the value stays in transport.

8.3.1.5 Extent of Regulatory Intervention

Regulators force unbundling of mobile network operation and service provision to encourage a wider range of services and lower retail prices. Spectrum policy ensures plenty of frequencies are available to meet the needs of those developing public and private applications and services.

8.3.1.6 Structure of the Industry

Different players in the value chain work to develop effective revenue share mechanisms, increasing the viability of small innovative players. This is forced by the fact that no single platform (fixed or mobile) dominates. Opportunities emerge for new players to offer fixed wireless solutions for the home and office. Companies able to deliver an integrated suite of services, based on a range of technologies (wireline, WLAN, mobile, broadband), flourish and these tend to be service providers that buy their network capacity from wholesale operators. Many new companies appear to develop, manufacture and market equipment and applications.

8.3.1.7 Security, Privacy and IPR

Governments and regulators find it impossible to reach consensus on how to address concerns related to privacy and the protection of intellectual property. Those laws and regulations that are introduced are largely ineffectual, and customers are forced to accept responsibility for making their own decisions on the use of new services and available content.

8.3.2 A Day in the Life (2010) – Red Scenarios

Two scenarios, considering the outlined drivers, describe the days of individuals in the 'red' wireless world of 2010.

David, Blue-collar Worker, Aged 50, Developed Country

07.45 Before leaving the house, David downloads some football videos over his home WLAN to his MVP (mobile video player) so that he can show them to his workmate, Colin. His new home was preinstalled with a WLAN, but in previous houses David had added first wired, and then wireless, extensions to his internet connection.

08.15 He gets into his car ready to drive to work. The in-car navigation and traffic system notifies him that there has been an accident on his usual route and provides alternative directions. It's a good service but he initially didn't use it as the usage charges were so high. Now, a new company seems to have taken over the service and the flat-rate package they offer is much more cost-effective.

08.45 David arrives at work on time and takes his place on the computer-controlled production line. He is responsible for correcting faults on cars that come off the production line, where they self-diagnose and report faults of assembly.

14.00 David attends a company meeting. The sales director updates the production staff on the company's profitability. He is currently working away from the main premises so he uses the videoconferencing facility.

15.30 Later in the afternoon, David and Colin decide that they will avoid the rush-hour traffic by going home early. David takes some of his paperwork home with him.

16.15 The two friends drive home and, while David watches the road, Colin watches the videos David downloaded earlier, doing this via the car's entertainment system so that he can watch it on the bigger screen. While driving past a cinema, a film preview downloaded over the cinema's public promotional network pops up on the screen menu.

16.35 Feeling peckish, they use Colin's latest food book to locate the best local noodle restaurant by connecting that to the car's net connection. Two such places are online at that moment and they order a takeaway for home delivery in 40 minutes.

17.00 At David's home, the two friends log into the company's shared resources via David's WLAN.

18.00 When the food arrives, they eat and watch some old 1980s bratpack films on one of the cable channels.

Eric, White-collar Worker, Aged 22, Developed Country

09.00 Eric logs into his university class. He switched to an American university to improve the quality of his education, but he did not have to move to the USA. The US universities discovered that there was much more demand than they could cope with, so they deliver high-quality courses and lectures through a 'virtual university'.

09.15 Eric downloads the multimedia lecture notes to the standard student T-pac issued to him by the university. He has been delighted with the ethernet to his home since 2004 and now he can get 100 Mbit to the apartment using the new wireless HERO (home electronic resource organizer), which gives super-fast internet access plus a whole range of new broadcast-type services. He has some questions to ask his lecturer but these can wait until later.

11.30 He sets off for his part-time job at a big international company and on the way Eric emails his lecturer. The signal was all right outside the FastDonut place but the network near the posh Mellanger Hotel was a bit patchy – too many businessmen eating up the capacity!

12.00 At the railway station ticket counter, Eric uses his e-wallet to pay for the weekly train ticket and it reminds him of the time he was mugged late at night at the same place and lost all his credit cards.

13.00   On the train, Eric receives a notification telling him that his boss is having to travel away and would like to meet up with him at the train station when he arrives, so that he can brief him on the project they are working on. Sometimes it is a bit tiring when people always know where you are.

13.20   The message specified which café to meet at and he spends 20 minutes with his boss, who transfers several large files to his T-pac. Eric does all his work on this, since it has quite a big folding screen, and it was wonderfully cheap from the big supermarket.

14.45   At work, Eric connects his T-pac to the office network. It's great these days how machines do so much of the connection and security business without you having to do anything.

18.00   When Eric gets home he downloads and updates some of his 3D games and virtual reality applications. He has just bought a new steering wheel and is looking forward to joining the live Formula One race in Japan, along with other virtual participants – with much less hassle and cost than having to go there.

8.3.3 Event Timeline and Market Outcomes

The event timeline for the red scenario is shown in Figure 8.5.

The scenario for the market looks different to the current situation, while vendors benefit from strong demand for new devices in developed markets, price pressures from customers mean that volume growth is much greater than increases to revenue. Operators seek to offset the effects of competition in the larger markets by investing in smaller developing countries. After some initial problems, 2.5/3G devices take virtually all of the Western European and North American markets, but networks of higher speed have not yet been launched. There is continued substitution of mobile for fixed voice and this helps maintain some voice revenue growth in mature markets, in addition to the expanding revenues from new customers in developing markets.

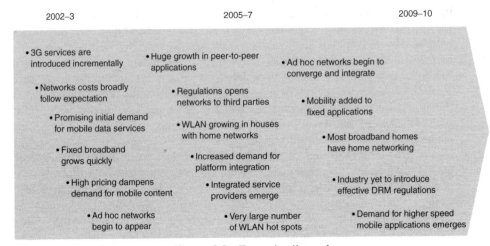

Figure 8.5   Event timeline red

Table 8.2 The red scenario

	2002	2010	CAGR[a]
Mobile customers by region (millions)			
Western Europe	287	302	0.7%
North America	122	217	7.5%
Asia Pacific	397	866	10.3%
RoW	221	492	10.6%
Total mobile customers	1026	1877	7.8%
Mobile devices split by technology			
2G	97%	30%	
2.5/3G	2%	42%	
WLAN	1%	28%	
End customer expenditure on mobile devices and services (EUR millions)			
Devices	38 983	77 286	8.9%
Voice services	329 931	523 729	5.9%
Data services	37 802	219 111	24.6%
Total end customer expenditure on mobile	406 716	820 126	9.2%

[a] CAGR = compound annual growth rate

WLAN devices are extremely popular, sometimes integrated into 2.5/3G handsets, but often used in specialized short-range devices. Data revenues grow slowly in the early years, but then accelerate as customers want to get access to their content on the move. Device prices fall in real terms over the period as customers are not willing to spend more, and service providers and vendors compete intensely for market share.

A summary of some key figures relating to the red scenario is shown in Table 8.2.

8.4 The Wireless World of the 'Green' Scenario

The 'Green' scenario is a world in 2010 where customers primarily want to meet their basic personal, mainly speech-based, communication needs. Users do not tend to be experimental and adopt only simple low-cost devices that enable them to keep in touch with their family, friends and colleagues. Problems in the deployment of 2.5/3G networks, and the high cost of the initial services on these networks, seem to have deterred customers from trying out more sophisticated applications and content services. Financial constraints mean that vendors and operators have not been able to subsidize new devices or services to stimulate take-up, while for established products and services, customers, supported by regulators, demand ever lower retail prices.

Short-range WLANs have not been taken up in the home market and are restricted to the business environment. There is, though, strong demand for video telephony and videoconferencing, and service providers have worked hard to deliver such services in a basic, low-cost way. New applications have been adopted where they offer enhancements to managing communication, like integrated messaging, or make life easier, like car or appliance management. The difficulties for 2.5/3G in the early years, and the resulting negative investor sentiment, has led to substantial industry consolidation amongst both mobile operators and vendors. Those that survive have turned their attention to the developing markets, where new subscribers interested in getting basic mobile services are being served by new cost networks and devices.

8.4.1 The Market Drivers in the 'Green' Scenario

Customer willingness to adopt new wireless devices, applications and services does not increase. Users want simple, low-cost devices offering easy-to-use applications in the home, office and on the move. 2.5 and 3G suffer delays in their implementation, and this causes a lot of frustration amongst early adopting customers. Mobile data services gain a bad name and most mobile customers generally only use enhanced voice and messaging services, which generate most of the content that passes across these networks. The economic backdrop is poor and financial support for industry remains low, meaning that vendors and operators are not able to subsidize new devices or services to stimulate take-up. There is consolidation in the operator and vendor markets and mobile operators focus on 'sweating' their assets and avoid investment in future higher speed technologies. Smaller application and equipment suppliers struggle and many fail to survive. This tends to stifle technical and service innovation and the rate of investment slows. Few people make use of WLANs in the public environment, with this technology largely restricted to businesses.

The pressure on mobile operators to grow revenues results in a focus on growing voice minutes. Significant reductions in mobile voice prices result in dramatic substitution of wireline telephony, first amongst businesses and then residential customers. Capacity enhancements are needed as substitution of fixed voice takes off. Fixed operators fight back and, while few households have home networks, either wireline or wireless, there is fast-emerging demand for wireless telephony, first through DECT, with devices integrated with mobile handsets, and then its successors.

The continuing demand for enhanced voice and messaging services provides the main area of growth. Voice recognition, unified messaging, video telephony and videoconferencing are emerging mass market applications that customers are willing to pay for, although operators need to be able to deliver these at low cost. Service providers invest heavily in making it much easier for people to manage their communication and, by doing so, greatly increase the time people spend on these services. Easy-to-use devices assist in this, while ambient technology enables value-added services to be offered by other companies through either the mobile network or device-to-device.

The lack of interest from customers in data and multimedia applications means even more emphasis is placed on using wireless in m2m communication, especially in relation to transport, health and security. High levels of development investment can be justified on 'background' applications that automate many of life's regular activities.

With a focus on growing voice minutes and messaging, operators invest to increase the capacity on their existing networks. There is also a need to cut costs, and this leads to cost-saving innovations in network equipment and end user devices that enable vendors to sell more effectively into developing countries, driving the adoption of basic mobile services faster in those countries.

These trends are depicted in Figure 8.6.

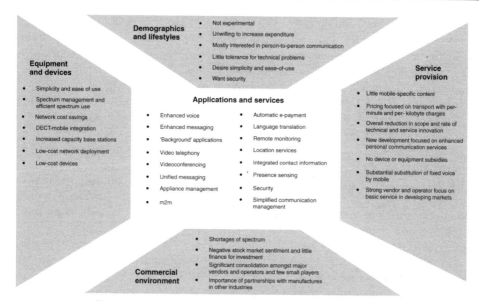

Figure 8.6 Example of demand and supply trends (green)

8.4.1.1 Customer Willingness to Adopt New Devices, Applications and Services

Customers' willingness to adopt new wireless devices, applications and services does not increase. Users show a strong preference for simple, low-cost devices offering easy-to-use applications in the home, office and on the move. There is relatively little growth in demand for home networking products as the poor economic climate has dampened the growth in sales of consumer electronic products, especially those related to advanced home entertainment. There is also very little willingness to pay more for mobile content or services and the services that are adopted meet very basic needs to do with personal security (e.g. parents knowing where their children are) and managing voice and message communications. Consumers tend to download entertainment-related broadband content via fixed connections, uploading them to mobile devices for use when they are on the move. Few people make use of WLANs in the public environment, with this technology largely restricted to businesses.

8.4.1.2 Innovation in Service Propositions and Marketing

Customers demand simplicity, reliability and cost-effectiveness and are discouraged from using data mobile services by the per-minute and per-kilobyte charging, which they associate with complexity and high costs. The continuing demand for enhanced voice and messaging services provides the main area of growth. The pressure on mobile operators to grow revenues results in a focus on growing voice minutes and there are sharp price reductions. Voice recognition, unified messaging, video telephony and videoconferencing are emerging mass market applications that customers are willing to pay for. Mobile operators invest heavily in making it much easier for people to manage their communication and, by doing so, greatly increase the time people spend on these services. Customers are inherently conservative and do not trust newer companies. They tend to gravitate towards bigger, well-known brands, and strong brand marketing from the bigger players makes it harder for new entrants to compete.

The difficult economic climate and problems with 3G deployment puts pressure on operators, large vendors and smaller application and equipment suppliers, resulting in slowing rates of investment and a reduction in the overall scope and rate of technical and service innovation. Financial constraints mean that vendors and operators have not been able to subsidize new devices or services to stimulate take-up, while for established products and services there is constant pressure from customers, supported by regulators, to reduce retail prices.

The lack of interest from customers in data and multimedia applications means even more emphasis is placed on using wireless in m2m communication, especially in relation to transport, health and security. High levels of development investment can be justified on 'background' applications that automate many of life's regular activities. Easy-to-use devices assist in this, while ambient technology enables value-added services to be offered by other companies (e.g. ticket payment, presence identification and language translation), enabled either through the mobile network or directly device-to-device.

The pressure to cut costs leads to innovations in network equipment and end user devices that simplify deployment and cut the cost of ownership. These developments enable vendors to sell higher volumes into developing countries, driving the adoption of basic mobile services faster in those markets.

8.4.1.3 The Technical Success of GPRS/UMTS

Service launches are delayed by several years and beset by difficulties, such as finding sufficient base station sites and handset availability. This causes a lot of frustration amongst early adopting customers and inhibits longer-term demand growth. User tolerance for initial poor QoS is low, causing high churn rates.

As network costs are higher than expected, this leads mobile operators to seek to restrict geographic roll-out, often coming into conflict with regulators. High bandwidth only emerges in pockets within mobile networks and mobile operators struggle to deliver the hoped for data rates. These problems lead network operators to charge higher prices for 3G data services. Enhanced voice and messaging services generate most of the content that passes across these networks and capacity enhancements are needed as large-scale substitution of fixed telephony by mobile takes off.

8.4.1.4 Interaction of Wireline and Wireless

The fixed broadband market enjoys steady growth. Wireline technologies remain the primary means for delivery of content and applications to the home and business. Customers buy fixed and mobile services separately and there is little demand for bundled offerings. Significant reductions in mobile voice prices result in dramatic substitution of wireline telephony, first amongst businesses and then residential customers. Fixed operators fight back with new service propositions and, while few households have home networks, either wireline or wireless, there is fast-emerging demand for wireless telephony. This initially comes in the form of rejuvenated marketing of DECT-based products which have been designed to integrate easily with mobile handsets (e.g. through the transfer of phonebook information), and then successor telephony and messaging technologies, delivering both voice and video.

8.4.1.5 Convergence of Telecoms and Media

There is very little telecoms/media convergence. Broadcast content does not migrate in any form to telecoms networks and the PC remains a separate tool to the TV and radio, used primarily for browsing and ecommerce transactions. Demand for wireless devices to integrate different equipment in the home and office is low. Most value in mobile communications remains related to transport and little is derived from content. Consumers are cost-conscious and capture content at home, uploading to mobile devices for consumption when they are on the move.

8.4.1.6 Extent of Regulatory Intervention

Regulators attempt to enforce licence terms and roll-out schedules by issuing penalties and revoking licences. There is considerable friction with the operators, who attempt to slow down widespread deployment of 3G networks in the face of apparently limited customer demand. Regulators, faced with a consolidating mobile industry, push for low-cost wholesale services that can be taken up by third parties.

While licence exempt bands do continue to exist, the disputes over the use of 3G spectrum mean that operators have to invest in technology to overcome capacity constraints on their 2/2.5G networks, something that becomes even more of a priority as they focus on driving up voice minutes to generate revenue growth.

8.4.1.7 Structure of the Industry

The economic backdrop is poor and financial support for industry remains low. The failure to generate significant revenue from 3G services leads to dramatic industry consolidation. The market is populated by a few, very large mobile specialists. Even with the benefits of scale, these companies struggle to survive. Consolidation amongst operators is accompanied by a focus on 'sweating' their assets and avoiding investment in future higher speed technologies. Smaller application and equipment suppliers struggle, and many fail to survive.

The market moves towards separation of the network and service provision functions and there are a few niche operators providing commoditized voice and data services, but most focus on the wholesale market. In an attempt to generate additional revenues from their telecoms investments, the mobile operators start to form partnerships with manufacturers in other sectors to develop m2m applications and services.

8.4.1.8 Security, Privacy and IPR

Demand for mobile commerce services is limited for most of the decade because users are slow to trust wireless services, with concerns about security, reliability, abuse of personal data and spamming. Eventually, industry-wide standards, supported by governments and regulators, are implemented to address security and privacy concerns.

8.4.2 A Day in the Life (2010) – Green Scenarios

Two scenarios, considering the outlined drivers describe the days of individuals in the 'green' wireless world of 2010.

Jane, Parent, Aged 42, Developed Country

07.00 Jane gets up and gets her children ready for school. She makes sure that the children's mobile devices are charged and that their GPS tags are attached.

08.15 Jane hurries the children into the car to go to school. The red light on the car's management screen flashes again to warn her of the broken exhaust. The car has already notified the garage and they send her a text message with options for times to book it in. Jane talks into the car's in-built microphone and it confirms in real time with the garage the slot she selects, before updating the diary in her phone.

08.20 As she passes the road toll, the car's id-unit debits her account with the local authority. The children play games and watch video clips in the back of the car that they downloaded earlier onto their phones. The screens are small but it keeps them happy.

08.45 After dropping off the children, Jane phones her fridge to check whether there is any milk and to place an order for more. She gets a message reminding her that her favourite TV programme is on today and she hits the button that ensures the video records it in the afternoon.

09.15 She gets a voicemail from her Utilicom account and authorizes the necessary payment by saying yes to the appropriate question. She has only recently begun to trust these ecommerce services, but a lot of recent legislation and technical developments seem to have made them more secure – and it so much easier than it used to be.

13.00 Jane has lunch out. ChildSafe send her the regular 30 minute update on her children's location. Last week she had such a shock when they sent a warning message when one of them strayed out of the school grounds during lunchtime – the alarm sound used by the phone was so loud!

13.45 She checks her phone diary and is reminded that her daughter has a drama lesson, and that her son has football practice.

16.00 Jane picks up her daughter from school and takes her to drama; her son will stay on at school to play football, and Jane can track his walk home using GPS.

19.00 Jane watches the programme she recorded earlier, then goes out to pick up her daughter.

Paula, Teenager, Aged 15, Developed Country

07.15 Paula wakes up with her digital radio alarm.

07.30 She hears an advertisement from E3Fone offering a great price on the latest U?R handset with videoconferencing. She'd love to have that and decides to ask her dad about it tonight.

07.45 She listens to the chart run-down while getting ready, and downloads a workout video to her laptop. She has promised her physical education teacher that she will bring it in for use at lunchtime on the main hall projector – so many of the kids are overweight these days!

08.15 She gets on the bus and Ticketchip automatically pays for another week of travel passes without her taking it out of her bag. She spends the time on the bus calling and messaging friends who are also on their way to school on her FunFone. Messaging is still best, what with all the graphics and sounds you can attach. The videophone service is good but expensive and her dad doesn't like her using it too much. Beside her, her brother is playing the new 3Dlizards game that he has downloaded to his mobile phone.

10.30 During break, Paula receives notification of a special programme that afternoon on psychological well-being. She scrolls down the details and then forwards the time to her video recorder at home.

16.00 Paula is picked up from school by her mother and taken to her drama class.

20.00 When she gets home, she downloads a pop programme over the internet.

21.10 Paula goes to ask her dad about getting a new phone. He doesn't seem all that keen and says she will have to wait until a cheaper version comes out.

8.4.3 Event Timeline and Market Outcomes

The event timeline for the green scenario is shown in Figure 8.7.

Problems with the introduction of 2.5/3G put pressure on both vendors and operators to focus on developing markets for the bulk of their growth. In developed markets, a reliance on voice revenues pushes prices down and this leads to substantial volume growth, much of it from the substitution of fixed voice. Mobile data services grow slowly, with a heavy emphasis on low-cost video services and m2m applications.

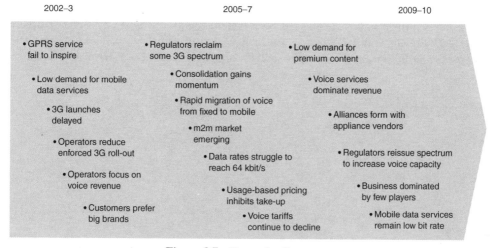

Figure 8.7 Event timeline green

Table 8.3 The green scenario

	2002	2010	CAGR[a]
Mobile customers by region (millions)			
Western Europe	287	296	0.4 %
North America	122	212	7.2 %
Asia Pacific	397	1039	12.8 %
RoW	221	590	13.1 %
Total mobile customers	1026	2138	9.6 %
Mobile devices split by technology			
2G	97 %	53 %	
2.5/3G	2 %	40 %	
WLAN	1 %	7 %	
End customer expenditure on mobile devices and services (EUR millions)			
Devices	38 983	61 046	5.8 %
Voice services	329 931	635 456	8.5 %
Data services	37 802	133 494	17.1 %
Total end customer expenditure on mobile	406 716	829 996	9.3 %

[a] CAGR = compound annual growth rate

WLAN devices do not experience rapid take-up outside the business environment and are not generally integrated into 2.5/3G handsets. End customer prices for devices fall rapidly over the period as a result of customer and operator pressure, although this is offset to some extent by growth in volumes in developing markets.

A summary of some key figures relating to the green scenario is shown in Table 8.3.

8.5 Credits

The study described in this chapter has been undertaken jointly by Analysys and partners of the IST Project WWRI (Wireless World Research Initiative, www.ist-wwri.org). The support of the European Commission and all partners is gratefully acknowledged.

9

Reference Model and Technology Roadmap

Edited by Andreas Schieder (Ericsson)

A reference model for the wireless world has been proposed by the IST Wireless Strategic Initiative (WSI). The model is composed of communication elements consisting of four basic building blocks and reference points between building blocks and communication elements. The work of WSI on the model should be regarded as a starting point. Issues where further thought and work are still required are also described.

This reference model proposal has also been contributed to the Wireless World Research Forum (WWRF, http://www.wireless-world-research.org/), the global forum to develop and maintain a harmonized vision of the wireless world. The WWRF is an outcome of earlier WSI activities.

9.1 Introduction

The definition and development of complex systems requires structuring of different kinds, static system block structuring, dynamic transaction behaviour, project organization, and so on. Dividing such systems into different modules, which might be constructed of modules themselves, is an approach often used. Such an approach allows the definition of reference points, or interfaces, between such blocks.

A mistake sometimes made in the past was to take reference models as 'build plans', or system designs. This resulted in unwanted inefficiencies and distortions in products. As an example, the OSI layered model is probably the most important structuring scheme used in communication engineering, but only rare products are direct implementations of this model.

The Wireless Strategic Initiative proposes a structuring scheme for wireless communication that is expected to be able to support the definition of complex mobile communication concepts and the structuring of the research work on the wireless world.

Technologies for the Wireless Future.   Edited by R. Tafazolli
© 2005 John Wiley & Sons, Ltd   ISBN: 0-470-01235-8

9.2 The Starting Point

The Wireless Strategic Initiative developed a vision of the wireless world by thorough assessments of usage scenarios. Their potential benefits for the end user were evaluated and requirements on the network infrastructure were identified. Based on these analyses, a vision was developed that reflects the key characteristics of the envisaged communication scenarios.

The main aspects of the vision are the recognition of an increased diversification of communication partners, and the possibility to logically group them into spheres representing the relation of the communication partner to the end user [1]. A graphical presentation of the sphere model is presented in Figure 9.1.

9.2.1 From the Vision of the Wireless World to a Reference Model

The vision was used to derive a reference model supporting the definition and design of the wireless world infrastructure. The WW infrastructure, in turn, is supposed to realize the envisaged service scenarios. The services and communication scenarios, as well as their characteristics, are thus the input parameters that determine the structure of the reference model.

The most influential aspects derived from the WSI sphere model are:

- an increased number of devices interconnected and participating in the wireless world;
- multiple, diverse access technologies;
- recursive structures of communication paths (connections routed via relay devices mainly situated in the spheres 'immediate environment' and 'instant partners');
- an omnipresent interconnectivity plane;
- ubiquitous access to services;
- gathering, transportation and utilization of ambient information for service personalization and adaptation.

Figure 9.1   The WSI sphere model

In the following sections we present the reference model, as an outcome of the considerations on the above-mentioned key aspects of the WW communication scenario.

9.3 The Reference Model

9.3.1 Design Goals and Purpose

The specification of a reference model has been used as the means to structure the design and the work towards the specification of the targeted system, i.e. a future generation mobile system. This model does not make any technology choices nor restrict any implementation decision, but aims to guide research and development such that the envisaged system characteristics can become realized.

9.3.2 The Basic Element of the Reference Model

The vision of the wireless world is partly based on the assumption that the number of interconnected devices will increase, and their characteristics and appearance will become more diverse than today. Still, it is assumed that all these devices can be described by a common structure. This structure, representing the physical devices, is the common denominator and is referred to as a *communication element* (CE), see Figure 9.2.

The functionalities integrated in the communication element are provided by different building blocks. The basic assumption is that the reference model should separate content processing, control and management functions into their own end-to-end planes and subsystems. The architecture specification does not allow a mixed use of these three

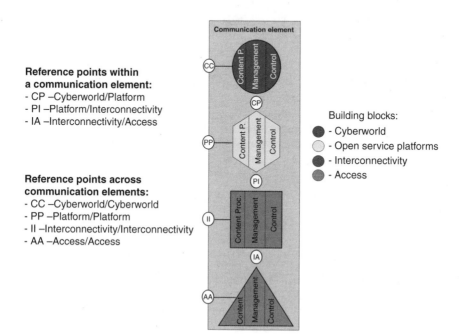

Figure 9.2   The wireless world communication element and its building blocks

functions. To facilitate this, the subdivision into these three functions is mirrored in all four building blocks of the communication element. The various functions include:

- *Content processing functions.* These deal with the processing, transformation, adaptation and end-to-end delivery of application data.
- *Management functions.* These include those functions responsible for both horizontal and vertical communication management. The horizontal management takes care of functions inside one certain block of the communication element, while the vertical management coordinates the cooperation between the different building blocks.
- *Control functions.* These handle all signalling associated with the content processing functions. They take care of negotiation and agreement of QoS parameters and need to span all building blocks in order to provide true end-to-end QoS for end users.

A wireless world communication element consists of four basic building blocks (as depicted in Figure 9.2):

- *Cyberworld.* This hosts all application-specific functionality. It relies on a generic service infrastructure provided by the 'open service platforms' and exploits these platforms to implement applications and services. Cyberworld implementations have to possess means to generically describe and explain their characteristics and requirements to ensure that the underlying infrastructure is used efficiently and the services are provided to the user's satisfaction.
- *Open service platforms.* These are responsible for providing a flexible and generic service infrastructure to the Cyberworld. Implementations of this platform facilitate the creation of new services, according to both user's and operator's needs. The restrictions imposed on the service creator have to be reduced to a minimum. This implies the use and provision of reusable generic service functions.
- *Interconnectivity.* This may also be referred to as the networking part of the wireless world reference model. The functions located in this building block take care of the logical links between CEs located within different spheres. Interconnectivity functions also maintain and manage these links, even in cases when they are subject to change of network topologies or to changing access networks.
- *Access.* This implements all aspects of the physical connection(s) between different CEs. This includes radio or other types of physical connection. Due to the hierarchical structure of the reference model, a connection between two higher spheres could use multiple connections in underlying spheres, relying on the connection services provided by the 'interconnectivity' functional block.

9.3.3 Spheres

The aforementioned CEs (see Figure 9.2) define a common structure of all elements forming the WW. However, the physical appearance of a CE can be very different, depending on the type of sphere it (logically) belongs to (Figure 9.3). Elements of the personal or local sphere, corresponding to the spheres 'immediate environment' and 'instant partners' of the original sphere model (see Figure 9.1), may be single devices, while CEs of the global sphere may represent complete networks or access systems. For the latter case, the

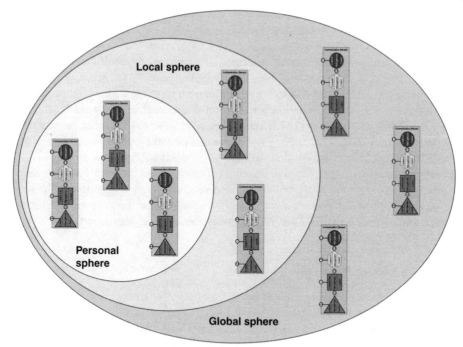

Figure 9.3 Logical grouping of communication elements according to communication spheres

recursive nature of a CE becomes important. A communication element may comprise a whole structure of interconnected CEs, i.e. complete networks. This makes it possible to specify networks or subnetworks in the same way as the complete wireless world infrastructure. The CE thereby represents the complete network, and the actual details of the network structure and capabilities may be encapsulated and hidden in the CE description.

9.3.4 The Building Blocks

A CE comprises a maximum of four building blocks, which structure it according to functional layers ranging from the physical connection means, providing the basis for connectivity, up to the level of services and applications. A building block relies on the presence of a building block located logically below it. A CE, though, does not need to include all four building blocks.

The four building blocks are now described in more detail.

9.3.4.1 Cyberworld

Today, the World Wide Web is chaotic. So much unstructured data exists that most people have a hard time finding information relevant to the situation at hand. This problem is particularly acute for mobile users, who increasingly want to access web services while on the move. At the same time, media are increasingly going digital, and more and more material is being made available every day. Digital imaging, desktop video, loop-based music software and web diaries are examples where both professionals and amateurs are producing content with computer-based tools.

The current nonmobile web offers little help for mobile contexts. The mobile user faces another kind of chaos in the form of information overload. As positioning systems and short-range communications are becoming commonplace, mobile terminals will be flooded with increasing amounts of on-the-spot information. Despite the added local aspect, there will be abundant data available, but relevant information will still be hard to find.

The aim of the wireless world will be to make the relevant information available to people in their daily lives. Considering the numerous possible contexts (not just location), it becomes evident that common descriptions for context information are needed. Similarly, mechanisms for managing and handling dynamic context information within the semantic web and devices on the move are also required. New kinds of mobile service, based on virtualization of presence and a digitally augmented ambient environment, are supported via advanced mechanisms of presence, identity and interaction.

9.3.4.2 Open Service Platforms

The introduction of new services, and convergence between networks, will create a highly diversified value network with new roles and key players (e.g. content providers, application providers, portal providers, mobile virtual network operators, etc.) in the area of network and service provisioning. Flexibility, in terms of supported business models, will reduce the time to market for new services. Standard open interfaces on top of generic service functions will foster the deployment of such service delivery chains.

Despite the underlying network complexity and heterogeneity, the service delivery will be, from the user perspective, simple, uniform and seamless. Advanced techniques for delivering mobility, service continuity, end-to-end security and QoS have to be provided.

The introduction, provision and management of numerous, probably highly complex, services will be associated with factorization of key information, such as user knowledge and context awareness. A set of advanced key features, such as profiling, contextual information delivery, filtering, billing, user privacy guarantee, etc. will be provided by the open service platforms.

9.3.4.3 Interconnectivity

The WW will offer seamless integration of all means of access, as well as far reaching scalable support for the user's communication and information needs. A unified control space will make these resources usable and conveniently accessible. The interconnectivity block is defined to support all capabilities of flexibly attaching and detaching terminal networks (e.g. PANs or BANs). Similarly, interconnectivity provides support for the open service platforms and assists those functions by providing generic functions for mobility, connectivity and delivery, etc. The functions of interconnectivity are of particular importance in a mobile wireless world, where the users may constantly change their point of access, and even the actual network topology might change. New challenges are raised by networks moving within other networks (e.g. BANs/PANs, etc.) and also by reconfigurable cooperative network infrastructures.

In future, communication systems will extend into the user domain, featuring both personal area networks (PANs) and privately owned and operated access networks. This will lead to private networks that move within wider area public networks; such moving networks should be connected in a flexible, simple, secure and rich manner to public short-range and wide-area networks. Body area networks (BANs), PANs, car networks and public hot spots are precursors of this development and will need the capability to seamlessly access visited network infrastructure in a convenient and transparent way for the user.

The WW assumes an increasingly diversified value chain with new key players and new roles in the area of network and service provision. This new paradigm requires a more flexible, reconfigurable and cooperative network environment. Networks have to provide the required flexibility to support different types of business model, and nodes within networks will need to be reconfigurable to fully enable cooperation between fixed and flexibly attached communication equipment.

To summarize, the 'interconnectivity' block acts as the networking part of the wireless world reference model and its functions take care of the logical linking of communication elements within different spheres. Interconnectivity establishes, maintains and manages these links, even when the underlying network topologies or attachment to access networks are being changed.

9.3.4.4 Access

The goal aimed at by all WW efforts is to facilitate easy, natural and intuitive communication for everybody, everywhere. 'Access' will be a key issue in this context, because the future will witness a wide variety of different access technologies, employing a plethora of technical solutions, ranging from intelligent medium access algorithms to smart antennas (a comprehensive set of envisaged technologies is described in [2]). The 'access' building block groups all functionality and technologies which are directly related to the transmission of content (data) via a physical medium, thereby paying respect to the wide range of requirements imposed by users, operators, services and the environment (including governmental regulation). To support the 'multiaccess vision', special emphasis will have to be put on structures supporting the interworking of multiple access technologies. The 'access' block will have to provide a flexible interface towards 'interconnectivity' to facilitate easy negotiation for physical bearers and their characteristics.

The generic components, which cooperatively form the 'access' block, and provide its functionalities, based on the named goals, are identified as (also see [3]):

- *Transceivers and antennas.* Aiming to provide small, and unobtrusive, cost-efficient solutions with a preferably low degree of complexity and low energy consumption, while being reconfigurable and adaptive to a maximum extent.
- *Transmission schemes.* Tailored to the scenarios in different spheres, but compatible to reduce development efforts. They further need to be inherently spectrum-efficient, support coexistence with other transmission schemes and provide high transmission quality.
- *Medium access.* Strategies and related protocols that aim to optimize the exploitation of the physical medium, and energy efficiency. They need to be QoS-aware and support the vision of ubiquitous medium availability.

9.3.4.5 Reference Points

The building blocks that form CEs are connected by reference points. The early identification and specification of these reference points will enable more flexible communication
systems than we will have with 3G systems. There are 'vertical' and 'horizontal' reference points. The vertical reference points are the interfaces between the building blocks of
the communication elements. A connection can also take place between communication
elements that reside in different spheres using 'horizontal' reference points.

The reference points between the building blocks are crucial elements for the precise
technological description of the model. The functionalities which the different blocks have
to provide at these reference points will have to be well-defined, complete and generic,
in order to assure the proper functioning of the model and to allow treating the building
blocks as 'black boxes' from the viewpoint of the adjacent blocks.

The reference points represent well-specified points of contact between the building
blocks. This specification will cover so-called 'generic vertical functions' that have to be
provided by all reference points to ensure a proper functioning of basic services of the
wireless world. Vertical functions provide certain functionality through all the building
blocks by addressing the dedicated problems and technologies of each building block. At
the moment, the WSI project has identified nine different functions (see Figure 9.4) that
have to be provided.

Note that the list of vertical issues is not claimed to be complete, since totally new
services, probably unknown as of today, may raise totally new 'vertical issues'.

The service architectures for the wireless world will have to cope with things like
numerous service providers, always connected users, automatic service adaptation, context awareness and new IP devices. Aspects like dynamic service discovery and service
provisioning in (for users) unknown environments and the personalized service usage
require new mechanisms.

The generic functions will enable providers to make their products and services available
in a flexible way. This will equally make it possible for users to discover and use the

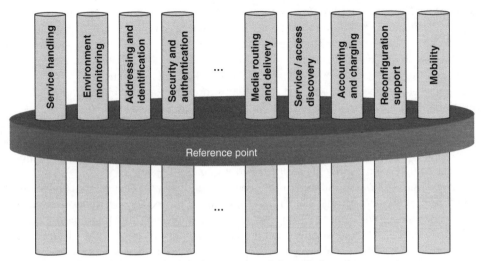

Figure 9.4   Generic functions supported through the reference points

desired service. Assembly and configuration of contexts composed of various service offers is needed to achieve the requirements stated explicitly or implicitly by the user.

The detailed definition of the reference points will be the work of future standardization activities. Before such work can start, however, the principles of how communication works at reference points, and how such communication is to be specified, must be defined.

Definitions and assumptions made on reference points include the following:

- Communication among building blocks is established dynamically, i.e. in the normal case there are no pre-established configurations.
- Communication across reference points is asynchronous and uses messages exclusively.
- Function distribution among building blocks follows the principles of minimizing message traffic and delay.
- The first message to request a service and start a new transaction normally carries no destination, but the destination is determined by a selector function included in each building block's control function. Further messages needed to conclude the transaction are directed to particular instances in the system.
- Communication across reference points is capable of supporting several protocols without requiring knowledge about the protocol's specifics.

9.4 Background

The WSI project started in May 2000 as a joint activity of four large wireless equipment manufacturers (Alcatel, Ericsson, Nokia and Siemens), with the main objective to develop visions of the wireless world (WW), a term given to the wireless communication systems that may follow third generation systems, and that would become operational after 2012. The project decided to make the process to develop visions totally open, and to invite everybody who wanted to contribute. The open approach resulted in a major success and the creation of the WWRF in 2001.

It became quickly apparent that structuring principles were needed to maintain momentum in the work of the forum and its working groups. It was therefore decided to devote the last year of the WSI project to the definition of a reference model, i.e. a structural framework for the definition of, and research on, the wireless world [4–6].

9.5 Conclusions

This chapter describes an attempt at constructing a useful reference model for the wireless world. The work is well advanced and based on a good understanding of the requirements, and of the working of future communication systems. There is, however, considerable research work remaining to be undertaken that exceeds the boundaries of this limited study.

In addition to the static view presented here, the dynamic behaviour of the model needs to be addressed in more detail. Some further issues for future work include:

- a formal description of the semantics of the reference points;
- a methodology to define communications via the reference points;

- definition of master–slave relationships between the building blocks of the model;
- dynamic representations of major transactions.

9.6 References

[1] Wireless World Research Forum, http://www.wireless-world-research.org.
[2] Wireless Strategic Initiative, '*Book of Visions 2000*,' http://www.wireless-world-research.org/Bookofvisions/ BookofVisions2000.pdf.
[3] Wireless Strategic Initiative, '*Book of Visions 2001*,' http://www.wireless-world-research.org/BoV1.0/BoV/ BoV2001v1.1B.pdf.
[4] IST-WSI Project, Deliverable D9, '*The WSI Reference Model*,' http://www.ist-wsi.org/D9_final.pdf, December 2002.
[5] IST-WSI Project, Deliverable D10, '*Important Technological Principles and System Options for the Elements of the WSI Reference Model*,' http://www.ist-wsi.org/D10_final.pdf, December 2002.
[6] IST-WSI Project, Deliverable D11, '*Timeline and Roadmap for the Coming of the Wireless World*,' http://www.ist-wsi.org/D11_final.pdf, December 2002.

9.7 Credits

The work described in this chapter has been undertaken by partners of the IST Project WSI (Wireless Strategic Initiative, www.ist-wsi.org). The support of the European Commission and all partners are gratefully acknowledged. The original text was jointly written by: Stefan Arbanowski (Fraunhofer FOKUS, Germany); Michael Lipka (Siemens AG, Germany); Klaus Moessner (The University of Surrey, UK); Karl Ott (Ericsson Eurolab Deutschland, Germany); Ralf Pabst (Aachen University, ComNets, Germany); Petri Pulli (University of Oulu, Finland); Andreas Schieder (Ericsson Eurolab Deutschland, Germany) and Mikko A. Uusitalo (Nokia Research Centre, Finland).

Appendix 1: Glossary

AAA	Authentication, Authorization and Accounting
ACL	Access Control List
ACL	Asynchronous ConnectionLess
ADC	Analogue to Digital Conversion
AEP	Application Environment Profile
AP	Access Point
API	Application Programming Interface
AS	Access Stratum
ATM	Asynchronous Transfer Mode
BAN	Body Area Network
BB	Baseband
BGP	Border Gateway Protocol
BIOS	Basic Input Output System
BMM	Bandwidth Management Module
BS	Base Station
BS	Bearer Services
BT_AP	Bluetooth Access Point
BTS	Base Transceiver Station
CALLUM	Combined Analogue Locked Loop Universal Modulator
CAN	Community Area Network
CAST	Configurable radio with Advanced SW Technology
CDMA	Code Division Multiple Access
CF	Core Framework
CM	Configuration Management
CMM	Configuration Management Module
CN	Core Network
CODEC	CODer/DECoder
COM	Component Object Model
CoNet	Cooperative Network
CORBA	Common Object Request Broker Architecture
CPU	Central Processing Unit

Technologies for the Wireless Future. Edited by R. Tafazolli
© 2005 John Wiley & Sons, Ltd ISBN: 0-470-01235-8

CSW	Computer Software Components
DAB	Digital Audio Broadcast
DAC	Digital to Analogue Conversion
DHCP	Dynamic Host Configuration Protocol
DL	Downlink
DNS	Domain Name Server
DNS	Domain Name Service
DRM	Digital Rights Management
DSP	Digital Signal Processor
DVB	Digital Video Broadcast
EIR	Equipment Identity Register
ETSI	European Telecommunications Standards Institute
EU	European Union
EVM	Error Vector Magnitude
FDD	Frequency Division Duplex
FEC	Forward Error Correction
FLP	Flexible Linearity Profile
ForCes	Forwarding and Control Element Separation
FPGA	Field Programmable Gate Array
FTP	File Transfer Protocol
GGSN	Gateway GPRS Support Node
GIOP	General Inter-ORB Protocol
GLL	Generic Link Layer
GP	Generic Protocol
GPP	General Purpose Processor
GPRS	General Packet Radio Service
GPS	Global Positioning System
GSM	Global System for Mobile Communications
GSMP	General Switch Management Protocol
HIP	Host Identity Protocol
HiperLAN/2	High Performance LAN type 2
HLR	Home Location Register
HO	Handover
HRM	Home Reconfiguration Manager
HTTP	HyperText Transport (or Transfer) Protocol
HW	Hardware
IDL	Interface Definition Language
IETF	Internet Engineering Task Force
IF	Intermediate Frequency
IGMP	Internet Group Management Protocol
IIOP	Internet Inter-ORB Protocol
IMD	Intermodulation Distortion
IMEI	International Mobile Equipment Identity
IMSI	International Mobile Subscriber Identity
IMT-2000	International Mobile Telecommunications 2000
IP	Internet Protocol

ISA	Instruction Set Architecture
ISDN	Integrated Services Digital Network
IS–IS	Intermediate System–Intermediate System
ISO	International Standards Organisation
ISP	Internet Service Provider
ISV	Independent Software Vendor
LAN	Local Area Network
LINC	Linear Amplification with Nonlinear Components
LLC	Logical Link Control
LNA	Low Noise Amplifier
LO	Local Oscillator
LR	Location Register
MAC	Medium Access Control
MAN	Metropolitan Area Network
MDA	Model Driven Architecture
ME	Mobile Equipment
MEMS	Micro ElectroMechanical Systems
MExE	Mobile Execution Environment
MIB	Management Information Base
MIMM	Mode Identification and Monitoring Module
MIP	Mobile IP
MIPS	Million Instructions Per Second
MMS	Multimedia Message Service
MN	Mobile Node
MNSM	Mode Negotiation and Switching Module
MOBIVAS	downloadable MOBIle Value Added Services
MP3	MPEG-1 Audio layer 3
MR	Mobile Router
MS	Mobile Station
MSC	Mobile Switching Centre
MT	Mobile Terminal
MVCE	Mobile Virtual Centre of Excellence
NAT	Network Address Translator
NBS	Network Bearer Services
NE	Network Element
NEMO	Network Mobility
NO	Network Operator
OA&M	Operation, Administration and Maintenance
OEM	Original Equipment Manufacturer
OMG	Object Management Group
ORB	Object Request Broker
OS	Operating System
OSPF	Open Shortest Path First
OTA	Over The Air
PA	Power Amplifier
PAN	Personal Area Network

PBA	Parametrizable Basic Architecture
PCIM	Policy Core Information Model
PDA	Personal Digital Assistant
PHY	Physical Layer
PIM	Platform Independent Model
PLMN	Public Land Mobile Network
PN	Personal Network
PRM	Proxy Reconfiguration Manager
PSM	Platform Specific Model
QoS	Quality of Service
RAN	Radio Access Network
RAT	Radio Access Technology
RCS	Radio Control Server
RF	Radio Frequency
RM	Reconfiguration Manager
RMA	Reconfiguration Management Architecture
RMM	Reconfiguration Management Module
RNC	Radio Network Controller
ROM	Read-Only Memory
RPC	Remote Procedure Call
RRC	Radio Resource Control
RSMM	Resource System Management Module
RSSI	Receive Signal Strength Indicator
RX	Receive
SAP	Service Access Point
SAW	Surface Acoustic Waves
SCOUT	Smart user-Centric cOmmUnication environmenT
SDL	System Description Language
SDM	Software Download Module
SDP	Service Discovery Protocol
SDR	Software Defined Radio
SDRC	Software Download and Reconfiguration Controller
SDRF	SDR Forum
SDU	Service Data Unit
SEG	Security Gateway
SIM	Subscriber Identity Module
SIP	Session Initiation Protocol
SIR	Signal to Interference Ratio
SLA	Service Level Agreement
SOAP	Simple Object Access Protocol
SPRE	Software Download and Profile Repository
SRA	Software Radio Architecture
SRM	Serving Reconfiguration Manager
S-RNC	Serving RNC
SW	Software
TCP	Transport Control Protocol

TDD	Time Division Duplex
TOI	Third Order Intercept
TRSA	Terminal Reconfiguration Serving Area
TRUST	Transparently Reconfigurable UbiquitouS Terminal
TX	Transmit
UCD	User-Centred Design
UE	User Equipment
UI	User Interface
UL	Uplink
UML	Unified Modelling Language
UMTS	Universal Mobile Telecommunication System
UP	User Profile
UPS	User Plane Server
URL	Uniform Resource Locator
USIM	UMTS Subscriber Identity Module
UTRAN	UMTS Terrestrial Radio Access Network
VAN	Vehicular Area Network
VCO	Voltage Controlled Oscillator
VHE	Virtual Home Environment
VLR	Visitor Location Register
VM	Virtual Machine
VoD	Video on Demand
VoIP	Voice over IP
VPN	Virtual Private Network
WAN	Wide Area Network
WAP	Wireless Application Protocol
W-CDMA	Wideband Code Division Multiple Access
WG	Working Group
WLAN	Wireless Local Area Network
XML	eXtensible Markup Language

Appendix 2: Definitions

Accounting A management task for measuring the consumption of resources or other occurring events during resource usage. It is used as information to be related to user specific cost information like tariffs. These could include the amount of system time or the amount of data a user has sent and/or received during a session. Accounting is carried out by logging of session statistics and usage information and could be used for authorization control, billing, trend analysis, resource utilization and capacity planning activities.

Address A key to access an entity for communication purposes determining a location in space and time.

Application A program (or function implementation by software) designed to perform a specific function directly for a user.

Authentication A process by which an entity (person, software, etc.) is verified to be the entity they claim to be. The different realizations of authentication are referred to as authentication mechanisms.

Authorization A process by which it is verified that an entity (person, software, etc.) is allowed to take part in an activity. The different realizations of authorization are referred to as authorization mechanisms.

Context information Information that can be used to characterize the situation of an entity – the entity being a person, location or object – that is important to the user interaction with an application.

Domain A substructure in the environment that is related by a characterizing relationship to a controlling object. Consequently, every domain has a controlling object associated with it.

Note: For example, a security policy determines the characterizing relationship of a security domain. Another example is a technology domain, which is based on a common choice of system hardware or software.

Technologies for the Wireless Future. Edited by R. Tafazolli
© 2005 John Wiley & Sons, Ltd ISBN: 0-470-01235-8

Handover The process to change association from one access point (or network) to another access point (or network) ensuring service session mobility, irrespective of the used access network technologies and irrespective of the administrative domains the access networks belong to.

Harmonization Inventing standardized interaction elements, functions, services, configuration and personalization processes that are independent of the specific sort of device.

Identifier An unambiguous name in a given naming context.

Interaction elements Icons, sounds (*earcons*) and menus which help to organize and memorize specific functions of the device.

IP address A technical specialization of an address that identifies an interface of a technical entity. An IP address may be used for routing purposes or in the authentication of an interface.

Look and feel The UI's outer appearance and way of operating.

Mobility

> **Network mobility** The ability of a network to change physical location while maintaining service availability or a service session for the user of the network capabilities.
>
> **Service mobility** Refers to the end user ability to maintain current sessions and obtain services in a transparent manner. The service mobility includes the ability of the service provider to maintain control of the services it provides to the user.
>
> **Service session mobility (Service continuity)** Service session mobility is a specialization of service mobility where the session is not terminated due to the fact that the used terminal or network provider has changed.
>
> **Terminal mobility (Host mobility)** Refers to the capability allowing a mobile terminal to change its point of attachment to the network, without interrupting data delivery to/from that terminal.
>
> **User mobility** Refers to the ability of end users to originate and receive services and access current network services on any terminal in any location, and the ability of the network to identify end users as they move across administrative domains.

Name A term which, in a given naming context, refers to an entity.

Name space A set of terms usable as names.

Naming context A relation between a set of names and a set of entities. The set of names belongs to a single name space.

Personalization Configuring a device according to personal preferences and synchronization of Personal Information Management (PIM) data.

Policy A formal set of statements defining a choice in the behaviour of a system.

Profile, User A set of user-specific information that defines the user's working environment.

Note: It may include application settings, and network connections, which files, applications and directories the user can have access to, what his preferences are, etc.

Robustness Reliable and error-tolerant operation of a device even in difficult situations (outdoor, hands-free, background noise, maloperation).

Scenario A narrative describing a specific user interacting with a product, service, or system to do specific things.

Service The term service is context dependent. For a user, a service is the useful effect, which is made available on invocation. In implementations, a service is the full set of functionality that ensures that the expected result of the user is achieved when being invoked. Network services are effects being caused by using network devices or capabilities of networks. This includes, for example, communication support for users as well as network management events handled inside the network only.

Subscriber Role of a party that legally signs responsible for a user with respect to service usage and the consequences of the service usage for a service provider. A subscriber defines the authorization of one or more users. A subscriber can itself be a user.

Ubiquitous computing Ideally, having access to any information at any time and any place using any device.

Usability/Ease of use Intuitive, logical, fast, standardized and goal-directed operation of a device.

User (of a system) A role of a party or system in the environment of the system, which consists of interactions with the system.

User interface (UI) Hard- and software which is used to exchange information between human and machine.

Virtual Personal Assistant (VPA) Future vision of an artificial agent in the mobile device which communicates with the user over natural communication channels.

Y-Gen Abbreviation for 'Wireless Generation'. This term is used to address the rapidly growing group of mobile end users focusing on their specific needs, preferences and market potential.

Index

Technologies for the Wireless Future. Edited by R. Tafazolli
© 2005 John Wiley & Sons, Ltd ISBN: 0-470-01235-8